T0269343

CAMBRIDGE LIBRARY COLLECTION

Books of enduring scholarly value

Earth Sciences

In the nineteenth century, geology emerged as a distinct academic discipline. It pointed the way towards the theory of evolution, as scientists including Gideon Mantell, Adam Sedgwick, Charles Lyell and Roderick Murchison began to use the evidence of minerals, rock formations and fossils to demonstrate that the earth was older by millions of years than the conventional, Bible-based wisdom had supposed. They argued convincingly that the climate, flora and fauna of the distant past could be deduced from geological evidence. Volcanic activity, the formation of mountains, and the action of glaciers and rivers, tides and ocean currents also became better understood. This series includes landmark publications by pioneers of the modern earth sciences, who advanced the scientific understanding of our planet and the processes by which it is constantly re-shaped.

Outlines of the Geology of England and Wales

In the early nineteenth century, the gifted stratigrapher and amateur geologist William Phillips (1773–1828) gave several lectures to interested young people in Tottenham on the subject of geology. These lectures were later collected into a book, which Phillips expanded in later versions. This reached its peak in 1822 when the clergyman William Daniel Conybeare (1787–1857) collaborated with Phillips to produce this rigorous and improved assessment of the geological composition of England and Wales. Although no second volume was ever published, the book had a tremendous impact on geologists throughout the United Kingdom and Europe, inspiring foreign scholars to produce equivalent volumes about their own countries. Conybeare's concern for the stratigraphy of fossils is especially remarkable for the time. William Fitton, later president of the Geological Society of London, praised the book highly, remarking that 'no equal portion of the earth's surface has ever been more ably illustrated'.

Cambridge University Press has long been a pioneer in the reissuing of out-of-print titles from its own backlist, producing digital reprints of books that are still sought after by scholars and students but could not be reprinted economically using traditional technology. The Cambridge Library Collection extends this activity to a wider range of books which are still of importance to researchers and professionals, either for the source material they contain, or as landmarks in the history of their academic discipline.

Drawing from the world-renowned collections in the Cambridge University Library and other partner libraries, and guided by the advice of experts in each subject area, Cambridge University Press is using state-of-the-art scanning machines in its own Printing House to capture the content of each book selected for inclusion. The files are processed to give a consistently clear, crisp image, and the books finished to the high quality standard for which the Press is recognised around the world. The latest print-on-demand technology ensures that the books will remain available indefinitely, and that orders for single or multiple copies can quickly be supplied.

The Cambridge Library Collection brings back to life books of enduring scholarly value (including out-of-copyright works originally issued by other publishers) across a wide range of disciplines in the humanities and social sciences and in science and technology.

Outlines of the Geology of England and Wales

With an Introductory Compendium of the General Principles of that Science, and Comparative Views of the Structure of Foreign Countries

W.D. Conybeare
and William Phillips

CAMBRIDGE
UNIVERSITY PRESS

University Printing House, Cambridge, CB2 8BS, United Kingdom

Cambridge University Press is part of the University of Cambridge.
It furthers the University's mission by disseminating knowledge in the pursuit of education, learning and research at the highest international levels of excellence.

www.cambridge.org
Information on this title: www.cambridge.org/9781108075107

This edition first published 1822
This digitally printed version 2014

ISBN 978-1-108-07510-7 Paperback

The original edition of this book contains a number of colour plates, which have been reproduced in black and white. Colour versions of these images can be found online at www.cambridge.org/9781108075107

The material originally positioned here is too large for reproduction in this reissue. A PDF can be downloaded from the web address given on page iv of this book, by clicking on 'Resources Available'.

OUTLINES

OF THE

Geology of England and Wales,

WITH AN

INTRODUCTORY COMPENDIUM

OF THE

GENERAL PRINCIPLES OF THAT SCIENCE,

AND

Comparative views of the Structure

OF

Foreign Countries.

ILLUSTRATED BY

A COLOURED MAP AND SECTIONS.

&c.

OPINIONUM COMMENTA DELET DIES, NATURÆ JUDICIA CONFIRMAT.

Cicero.

BY THE

REV. W. D. CONYBEARE, F.R.S. M.G.S. &c.

AND

WILLIAM PHILLIPS, F.L.S. M.G.S. &c.

PART I.

LONDON:

PRINTED AND PUBLISHED BY WILLIAM PHILLIPS,

GEORGE YARD, LOMBARD STREET.

1822.

Preliminary Notice.

It seems requisite here to observe that the present work, which is in great measure original, was nevertheless founded on the little volume published in 1818, and entitled a 'Selection of Facts,' &c. Soon after its publication, I received a letter from the Rev. W. D. Conybeare, in which he offered me several corrections, and the contribution of much original information, on the assumption that a second edition would be called for. These offers were gratefully accepted; and not very long afterwards I received from him, among other important communications, the detailed view of the several Coal-formations of England and Wales, forming the conclusion of the first part of this work. When the proportion of his contribution had so materially encreased, I wrote to the Rev. W. D. Conybeare, requesting him to become the Editor of this work, and offering for his acceptance every thing which my industry as a compiler had enabled me to collect; his reply was, that he preferred giving me his assistance, and the repetition more than once of my request, produced only a repetition of his first reply; finally, his consent was obtained for the appearance of our names in the title page as joint Editors; his assistance however has been gratuitous, for he has no other interest in the work than as having contributed towards it.

The reader will readily perceive that a very large proportion of the whole is by the pen of the Rev. W. D. Conybeare, whose name or initial is annexed to his contributions. Not only have these contributions given to it the air of an original work; but the grand division of the whole series of our Formations, their sub-divisions, and the accounts of analagous formations in other countries, together with the Synoptical Tables prefixed to each Book, as well as the principal part of the Plate of Sections, and many material corrections and additions to the little Map, are exclusively his.

Feeling therefore how great has been the share of my coadjutor, and how comparatively little my own (for mine has chiefly been that of a compiler), I have not hesitated thus to lay before the reader, a statement of the facts so intimately connected with the production of the work.

I wish also to state the obligations I feel to the liberality of G. B. Greenough, Esq. late President of the Geological Society, for the presentation to me of a copious selection of his Notes on the Geology of our island, of which considerable use has been made.

PRELIMINARY NOTICE.

A small type and a thin paper have been preferred for the advantage of the traveller; who, it is presumed, will not fail to find the volume, an interesting companion of his travels. Some copies have been printed with the same types on a larger and more substantial paper.

W. P.

The following statement, drawn up by the Rev. W. D. Conybeare, will sufficiently explain the principles which have influenced the new arrangement he has given to the plan of the work, and the communications above alluded to.

It has been endeavoured to render the present volume useful not only as an account of the physical structure of England but also as a general manual of Geology. With this object an Introduction has been prefixed containing an elementary view of the general principles of that science, and a compendious survey of the various topics which a complete system of it ought to embrace.

The principles thus generally laid down are, in the body of the work, illustrated in the detail by their application to the geological phœnomena of our own island; the full developement of these forms the principal object of the work, but to avoid partial and incomplete views we have subjoined, wherever it was possible, concise accounts of the comparative geology of other countries: these have necessarily been brief, but it is hoped sufficiently copious to answer their subsidiary object. No single and general work which has yet appeared will indeed be found to have entered so largely upon this branch of the subject.

Every source of information with regard to the geology of England has been consulted as far as the knowledge of the editors extended; but to mere compilations from the observations of others unity of design and precision of statement must generally be wanting, without the advantage of a personal examination of the districts described; to this advantage the present work must be understood as advancing a general claim, hence a great part, perhaps the greater part, of its materials are, in the strictest sense of the term, original.

The part now submitted to the public comprises a description of the various formations as exhibited within the limits assigned to this work, from the most recent to the lowest rocks associated in the coal-districts.

The second part will embrace the series commonly called Transition and Primitive, and thus complete that branch of the subject connected with the description and distribution of the several formations; but in addition to these many other topics demand attention in an attempt to give a complete delineation of the geology of any country; the derangements which those formations have experienced, the circumstances of the vallies apparently excavated in their mass, the accumulations of gravel derived from their partial destruction, &c. are most important classes of geological phœnomena which require to be presented under proper heads in a connected view, in order to place in the full and clear light of their real evidence the inferences resulting from them; to the partial and uncombined views which have been too generally taken of these phœnomena, much of the contradictory theories which have divided geologists must be ascribed; a regular induction of them applied not to a limited district, but extending to the whole of this country, may therefore it is hoped contribute in some measure to lay a more secure and solid foundation: the manner in which it is proposed to handle these subjects will appear from the Introduction.

All other incidental matter arising from the general subject will likewise find its place in the second Part.*

The bulk and expense of the second Part will certainly not exceed and probably fall short of the present.

It is impossible to give a distinct pledge as to the period of publication; delay will be avoided as far as it can be so without injuring the character of the work.

The present part † has been presented to the public without waiting for the completion of the second, chiefly because it contains the history of those formations which have been as yet fully examined in England alone, and of which a detailed description was required to fill an important chasm in the science of Geology. The remaining formations (those of the commonly called transition and primitive districts) are of less importance under this point of view, since they are more extensively exhibited in many other countries, have been long generally known and often described, and the precision which Dr. Mac Culloch has recently introduced into this branch of the subject has left nothing further to desire.

The degree of originality belonging to the present work will appear from the following statements.

* It is proposed, as an Appendix to the second part, to give a slight sketch of the processes connected with the working the different mines, and the metallurgical operations prosecuted in the mining districts, subjects which will be found useful in a manual intended as a companion for the English geological traveller.

† The interval between the publication of the first and second part, will also be attended with an important incidental advantage in enabling the editors to subjoin, in the form of an appendix, the most material corrections and additions which during that interval may be collected.

PRELIMINARY NOTICE.

The history of the formations above the Chalk having been already fully treated in the memoirs of Messrs. Webster and Buckland, little besides the task of compilation remained to the editors with regard to the subjects embraced in the first book.

With reference to the formations comprised in the second Book, the general history of the Chalk has long been known, but no attempt to trace its details was previously in existence. The sands beneath the chalk, as far as they are exhibited on the coasts of the Isle of Wight and Dorsetshire, had been ably illustrated by Mr. Webster, but all the materials connected with their distribution in other parts of the island are entirely new, including a tolerably full description of the Wealds of Kent, Surry, and Sussex.

The general outlines of the whole Oolitic series were (as has been fully stated in the Introduction) first sketched by Mr. Smith; but, with the exception of a few extremely brief notices inserted by that gentleman in his "Strata identified by their organic remains," and the descriptions in the Rev. Mr. Townsend's Vindication of Moses, nothing had been published on the subject; almost all the details in this part of the present work are therefore strictly original.

No general accounts of the New Red Sandstone and Magnesian Limestone formations were extant, and but little assistance was to be derived from the few and partial notices which had appeared. Mr. Winch's account however of the small portion of this tract which fell under his observation, deserves favorable mention.

Though very valuable materials were extant in the descriptions of the Northumberland and Derbyshire Coal-fields by Messrs. Winch, Whitehurst, and Farey, yet a regular and connected account of the Coal-districts of this country remained a desideratum. The editors are inclined to rest the claims of the present work to public notice very principally on the information now for the first time brought together on this most important subject.

The articles pointing out the relations of the English formations to those of the continent, are also among the most original, and it is hoped most useful portions of the volume. While it was passing through the press a very able memoir on this subject by Professor Buckland appeared, which was shortly followed by a series of essays in the Annals of Philosophy by Mr. Weaver, containing an excellent outline of the researches of Friesleben, Raumer, &c. in the north of Germany. We have availed ourselves of both these sources, but the general outlines of the plan were previously laid down, and the greatest part of its details have never before been collected together.

PRELIMINARY NOTICE.

A method as systematic as the nature of the subject would admit has been adopted, and the various particulars relating to the several formations are disposed under an uniform series of general heads, which it is trusted will greatly facilitate reference. A constant order has also been pursued in tracing the local distribution of each formation, beginning with its northern extremity and proceeding regularly to the southern; thus we have avoided that frequent transition to distant geographical sites so embarrassing to the student.

The outline map which accompanies this work is chiefly compiled from Mr. Greenough's; we have not however introduced his division of the different varieties of slates in the districts of Cumberland, Wales, and Devon, regarding this point as not sufficiently ascertained; and a different division has been adopted of a part of the formations described in our third Book, we having included the limestone shale of Derbyshire under the colour of the mill-stone grit, and considering it as associated with this formation; whereas Mr. Greenough has associated it with the subjacent carboniferous limestone. This change of system produces an apparent difference through this portion of the map, where no real difference exists, for the whole question of this division is one of convenience only.

The sections will be found more comprehensive than any hitherto published of the island, and the only ones in which the relations of the older and inclined formations to the recent and more horizontal deposits are exhibited with distinctness and truth; we have purposely avoided pursuing the same lines with the various sections published by Mr. Smith, except in one limited instance, where a portion of one of the sections traverses the Weald of Surry and Sussex in a line nearly coincident with an earlier section of this observer.

The sections 5 and 6 are copied from two of Mr. Webster's. The dotted curves in the former, indicate the general lines of curvature to which (according to the very ingenious view of that writer) all the different inclinations exhibited by the strata there represented, may be reduced.

*** The two first chapters of the second book, and a part of the third chapter (altogether extending from page 59 to 185) not having been corrected by Mr. Conybeare, though principally derived from his M.S. some errors have crept into this part of the work; the most material of these are pointed out in the list of Errata, and the reader is requested to correct them with his pen.

CORRIGENDA.

Page 60, line 3, *for* Java *read* Jura
—— 25, — whole of this reposes, *read* whole of this series reposes
—— 61 —— 3, — coal transition, *read* coal and transition
—— 66 —— 13, — The great majority (perhaps eight tenths) of *read* In a great majority of instances (perhaps eight tenths)
—— 67 —— 10, — Pleacente *read* Picacente
—— 69 —— 29, 'but' to 'formation,' line 32, should follow 'bottom of the series (C).' line 26 of p. 68.
—— 74 Mr. Miller is of opinion that the specimens 2, 3, & 4, quoted line 22 *a* 26 inclusive, belong to a single species, being derived from different parts of the animal.
—— 75 line 6 *from bottom*, *for* there *read* thus
—— 76 —— 12, *for* and contracting; according to the impressing it received from *read* and contracting according to the impression received; from
—— 22, — pass direct to the inter-funnel-shaped cavity, *read* pass directly to the internal tunnel-shaped cavity
—— 79 —— 14, — upper bed *read* upper beds
—— 17, — passes that, *read* passes, and that
—— 26, — stratum *read* strata
—— 80 —— 14, — Nodder *read* Nadder
—— 8 *from bottom for* chalk, traced, *read* chalk which we have already traced
—— 81 —— 20, *for* Stowe *read* Stour
—— 26, — chalk, placed *read* chalk, though placed
—— 107 —— 8 *from bottom*, *for* advancing east, *read* advancing from east
—— last line, *for* connected *read* concealed.
—— 109 —— 2, *for* this *read* their
—— 121 —— 16, — the separate *read* these separate heads
—— 11 *from bottom for* constituent *read* imbedded
—— 125 —— 2, *for* costata *read* costatus; *for* tuberculata *read* tuberculatus
—— 10, — laris, *read* lævis.
—— 129 —— 17, — spinalosus *read* spinulosus.
—— 18, — costata *read* costatus; *for* obliqua *read* obliquus.
—— 4 *from bottom for* concurra *read* concava.
—— 3 —— —— locris *read* lævis.
—— 130 —— 13 —— —— conalus *read* conulus.
—— 132 —— 15 —— —— masked *read* marked
—— 143 —— 1, *for* Ragston hills *road* Ragstone or Greensand hills
—— 145 —— 26 27, *for* repaired, and forced, *read* filled up and thus forced
—— 10 *from bottom for* formations we, *read* formations which we
—— 149 —— 24, *for* form *read* from
—— 163 —— 18, — Newmarket, between the chalk marle and the iron-sand which occurs on the west of this county. Near Gamlingay, *read* Newmarket. Between the chalk marle and the iron sand, which occurs on the west of this county near Gamlingay
—— 167 —— 21, *after* encrinites, &c. *add* are among these remains.
—— 23, *for* in average breadth, extending *read* of average breadth, and extending
—— 169 —— 15, *for* Moroan *read* Morvan
—— 177 —— 16, — but it *read* this clay
—— 20, 21, *for* formation. The beds, *read* formation, because the beds
—— 178 —— 9, *for* Bagley wood, near Farringdon, *read* Bagley wood; and near Farringdon

INTRODUCTION.*

§ 1. *Proposal of the subject, and observations on the simplicity of the chemical and mineralogical constitution of rock formations*—§ 2. *Regular order of succession in the mineral masses.*—§ 3 *Phænomena of their disposition in strata.*—§ 4. *Methodical arrangement of the rocks into formations and general classes.*—§ 5. *Occurrence of animal and vegetable remains imbedded in the strata and phænomena of this distribution.*—§ 6. *Occurrence of beds derived from the debris of older rocks among those more recent.*—§ 7. *Considerations on the change of level indicated by preceding phænomena, and inductive enumeration of the possible hypotheses on that subject.*—§ 8. *Phænomena and origin of trap rock, with a sketch of the volcanic question.*—§ 9. *Wernerian theory on some of the above points compared with the actual phænomena.*—§ 10. *Phænomena of vallies, with an explanation of the theory of denudation.*—§ 11. *Phænomena of diluvial gravel.*—§ 12. *Changes still carried on by existing causes.*—§ 13. *Sketch of the progress of Geology and the physical allusions of the inspired writings.*

SINCE the present volume, offering itself only in the character of an elementary work, may very probably find its way into the hands of many who as yet possess but an imperfect and vague idea of the general objects of the science to which it relates, some introductory notice of the scope and design of the enquiries which it is here proposed to prosecute, appears to be requisite; and in order to supply this deficiency, the following preliminary remarks have been thrown together. In thus endeavouring to sketch the general bearings of the phœnomena which it is the business of every geological treatise to illustrate, the most simple and natural method will be to trace those phœnomena in the order in which they would present themselves to the consideration of an intelligent observer who should

* By the Rev. W. D. Conybeare.

study for himself, with the eyes of an original discoverer, this part of nature; for in following the steps of such an observer, the reader will have no difficulties arising from the assumption of a previous acquaintance with the subject which he may not possess, to surmount, but every new observation will arise from those which have preceded, in precisely that series which is best calculated to convey elementary information.

§ 1. Geology being the knowledge of the Earth's structure as far as it lies open to our observation, the fundamental point on which it rests, is, the ascertaining the order in which the materials* constituting the *surface* of our planet (for beyond this our observation cannot penetrate) are disposed. The superficial and hasty observer might suppose that these materials are scattered irregularly over the surface and thrown confusedly together, but a slight degree of attention will prove that such a conclusion would be entirely erroneous.

§ 2. If we suppose an intelligent traveller taking his departure from our metropolis, to make from that point several successive journies to various parts of the island, for instance to South Wales, or to North Wales, or to Cumberland or to Northumberland, he cannot fail to notice (if he pays any attention to the physical geography of the country through which he passes)

* A competent knowledge of Mineralogy is required to instruct the geological student in the nature of those materials as considered in themselves, and of chemistry to enable him to understand their constitution analytically; yet the number of mineral masses forming rocks of usual occurrence is so small, and the composition of those so simple, that a very limited knowledge of these sciences is sufficient for all introductory purposes, as far as the general outlines of Geology are concerned. Siliceous, argillaceous, and calcareous masses (substances with which every one is familiar under the common names of sand, clay, and limestone) constitute probably nine-tenths of these materials, and the compound rocks forming the remaining tenth consist principally of only four minerals, quartz, felspar, mica, and hornblende. These great massses contain, dispersed in various manners through them, and in comparatively small quantities, all the other substances included in the mineral kingdom; of these the various ores of the different metals are the most important; the geologist must of course, as he proceeds in his enquiries, acquire a competent knowledge of all these substances; but this knowledge which is the ultimate object of the mere mineralogist, is to the geologist only a subordinate acquisition, and forms but the alphabet by which he endeavours to decypher the part of nature which he studies. Hence the rarer varieties which will, in the estimation of the mineralogist, posses the highest interest, will, in the eyes of the geologist, attract the least regard. On this principle nearly one-half of the mineral species may be safely neglected in beginning a course of geology, nor is a knowledge of more than 100 species essential as a preliminary acquisition. The geologist therefore need not be alarmed at the extent of this *alphabet* which he must in the first place master: any common treatise on mineralogy will prove a sufficient guide in this, provided it be accompanied by the examination of a tolerable collection of specimens; but no description can possibly supply the want of actual inspection.

that before he arrives at the districts in which coal is found, he will first pass a tract of clay and sand; then another of chalk; that he will next observe numerous quarries of the calcareous freestone employed in architecture; that he will afterwards pass a broad zone of red marly sand; and beyond this will find himself in the midst of coal mines and iron furnaces. This order he will find to be invariably the same, whichever of the routes above indicated he pursues; and if he proceeds further, he will perceive that near the limits of the coal-fields he will generally observe hills of the same kind of compact limestone, affording grey and dark marbles, and abounding in mines of lead and zinc; and at a yet greater distance, mountainous tracts in which roofing slate abounds, and the mines are yet more valuable; and lastly, he will often find, surrounded by these slaty tracts, central groups of granitic rocks.

The intelligent enquirer, when he has once generalised these observations, can scarcely fail to conclude that such coincidenences cannot be casual; but that they indicate a regular succession and order in the arrangement of the mineral masses constituting the Earth's surface; and he must at once perceive that, supposing such an order to exist, it must be of the highest importance to œconomical as well as scientific objects, to trace and ascertain it.

§ 3. If with these views he is led to investigate the subject still further, he will find these mineral masses disposed for the most part in stratified beds, not exactly parallel to the horizon, but more or less inclined with reference to that plane; so that the edges of these beds, emerging in succession from beneath each other, make their appearance one after the other on the surface, thus:

This emergence is called the outcrop or basset of the strata. The other technical terms connected with this disposition, will be found in the subjoined note.*

It is obvious that by this arrangement a much greater thickness of strata is exposed to our observation than could have been had their planes preserved an horizontal direction; for in

* The angle of inclination between these planes and that of the horizon, is called their dip, or pitch; the strata are indifferently said to dip and pitch from, or to rise towards the horizontal plane—an horizontal or dead level line drawn along the planes of the strata; or in other words their intersection with an horizontal plane, is called their line of bearing or drift line.

that case one single stratum would have covered the planes of a medium elevation throughout extensive districts (if not the whole globe), and we could have been acquainted with those above it only by the structure of mountains rising above that level, and with those beneath it only by the natural excavations of the vallies, or artificial ones of wells and mines; but by the actual arrangement, the beds which in one point lie at an impenetrable depth, are in others brought up to the surface, and thus become subject to our examination, and (which is much more important) yield us those various mineral products which are often essential to the most necessary of human arts.

§ 4. When, however, the observer commences his attempt to trace more in the detail the succession of these mineral beds and masses, he will at first find himself perplexed by their almost infinite numbers; but he will soon discover that these individual strata are arranged together, in such a manner as to afford natural and easy grounds for classing them in a limited number of series, each series comprehending numerous indi-

Beds of rock are occasionally subject, from their mechanical structure, to split into smaller laminæ not parallel to the plane of stratification; thus

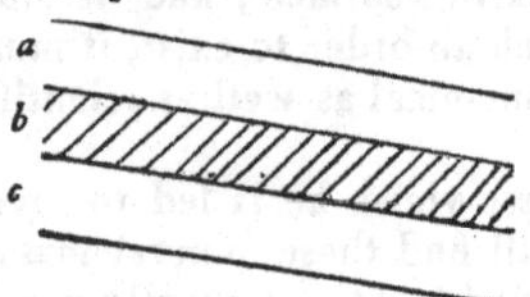

This structure is called the cleavage of the bed. Where only part of a bed is exposed, it is often difficult to distinguish the lines of cleavage from the true planes of stratification, but the doubt may be cleared by observing the upper and under surface of the bed at the line of its junction with its superstratum and substratum, especially if these be of a different substance; for instance, if the bed exhibiting the cleavage *b* be calcareous, and the beds above and below it *a* and *c* argillaceous; for then there can be no question but that these dissimilar beds are the true strata, and that the lines not parallel to them are merely lines of cleavage.

The law of stratification, as above stated, extends to all the rocks and mineral masses forming the Earth's surface, excepting perhaps the rocks of the granitic class (which generally form the lowest rocks with which we are acquainted), and those of the trap family, which are irregularly interposed throughout all the other formations; but these are points upon which it would be premature now to enlarge.

Although all the rock-masses occur forming strata, yet many of the minerals employed in the arts of life do not constitute the entire substance of such beds, but are disposed in lines irregularly traversing them; such lines are called veins, and have the appearance of having originally been open cracks or fissures ranging across the beds, subsequently filled up by the mineral substances they contain. Most metals are found in veins of this kind; if the direction of the vein approaches to a vertical plane it is called a *rake* vein, if to the horizontal a *pipe* or *flat* vein; its angle of inclination is called the *hade* of the vein.

vidual strata naturally allied and associated together To explain this by an example: if Derbyshire be the country under examination, he will find a series of twenty or more alternations of beds of coal, sandstone, and slaty clay, repeated over and over; and beneath these beds a like alternation of limestone strata with beds of the rock called toadstone. Here, then, all the individual beds at once resolve themselves into two comprehensive series—the upper containing coal, the lower limestone; each series being characterised by the repetition of its own peculiar members. Such series are called FORMATIONS; and by the aid of these general relations, the unmanageable number of the individual strata is readily reduced within convenient limits; and this division must appear perfectly natural, inasmuch as the beds composing each formation, being identical in character, must have resulted from the same order of causes.

Still further, by comparing several of those formations together, a resemblance of relations and an association in position will be observed between many of these also, which will lead to a still greater simplification by the introduction of a smaller number of yet more comprehensive classes.

Of these more comprehensive classes, five will perhaps be sufficient; the first or upper series will comprehend the beds of sand and clay which repose upon, and partially cover the great and conspicuous formation of chalk. The second class is of a less uniform character, and comprehends many formations in some respects dissimilar, which yet possess many common relations, and which the fear of constituting too large a number of general classes forbids us to separate; yet four subdivisions of it require enumeration; 1st. the chalk formation; 2nd. a series of sands and clays beneath the chalk: 3rd. a series of calcareous freestones (such as the Portland and Bath stones) and clays; 4th. beds of red marle and sandstone containing occasionally alabaster and rock salt. The third general class comprises the beds affording coal, and the limestones and sandstones on which these repose. The fifth class is characterised by the prevalence of common roofing and writing slates. The sixth, and lowest, by that of some finer varieties of slate and granite. These divisions are the same with those generally recognised by geological writers, excepting that the third is by some combined with the second, by others with the fourth; but all geological analogies and relations are grossly violated by the former of these methods; and though the latter is less open to objection, yet we shall best consult that convenience to the student which it is the great object of all such arrangements to promote, by assigning to so important a series a distinct place in the general system. Different writers have assigned different names to these classes, for the most part borrowed from theoretical views, or conveying descriptive ideas which are far from being universally applicable: in order to avoid these objections, we have taken the terms by which they are designated in the present work, from the unquestionable fact of their relative position. Regarding the third, or carboniferons series, as the middle group, we have assigned the term supermedial to the second series, as being next above it, and submedial to the fourth as being next below it. To the highest and lowest series, the terms superior and inferior, which require no commentary, have been applied. The reasons which have guided us in the details of this arrangement, will be found fully stated in the introductory chapters of the several books, and it would at present be premature to enlarge upon them; a comparative view of this arrangement, and that of other writers, will be seen in the subjoined note.*

* The most general relation under which the various formations present themselves, is that whence they have been denominated *primitive* and *secondary*; the former comprising the lowest series of rocks, which serve as the fundamental basis upon which the rest repose, never containing any traces of organised beings (i e. animals or vegetables) imbedded in them, and being entirely of a chemical composition: these therefore, it was inferred, constituted the materials of the Earth's surface at its first formation, while on the other hand the series which covered them were observed to contain, often in great abundance, the imbedded remains of the vegetable and animal kingdoms, and to be often also made up of fragments apparently torn by some convulsion from the primitive rocks and cemented again together under a new form; these therefore were necessarily considered as of subsequent and secondary origin. This distinction was first perceived by Lehman (about the year 1759), and made the basis of his system. In its principles it is philosophical and just, but does not carry the subdivision far enough for practical purposes, leaving all the secondary rocks under a single class. Werner, observing that between the primitive rocks and those which exhibited the characters of the secondary class in the most striking manner, a series of intermediate character (containing comparatively few organic remains and approaching more nearly to the chemical structure of the primitive than the mechanical of the secondary rocks) intervened, introduced the title *transition rocks*, as descriptive of this intermediate series; and a similar idea appears to have occurred perhaps yet earlier to Rouelle in France, who applies to it the designation 'travaille intermediate.' As these so-called transition rocks were of course taken from those which, strictly speaking, belonged to the secondary class, the introduction of this class made it necessary to abandon that term. Werner accordingly employed in its stead, for the rocks reposing on his transition series, the term *flœtz rocks*, derived from the belief that they generally were stratified in planes nearly horizontal, while those of the older strata were inclined to the horizon in considerable angles. But this holds good only with regard to the structure of countries comparatively low: in the Jura chain, the borders of the Alps and Pyrenees, Werner's flœtz formations are highly inclined: should we therefore persist in the use of this term, we must prepare ourselves to speak of vertical beds of flœtz (i. e. horizontal) limestone, &c. As the enquiries of geologists extended the knowledge of the various

§ 5. Thus far we suppose the observer to have been chiefly occupied in considering the character of the rocks as they are in themselves, and developing their arrangement with a view to

formations, Werner or his disciples found it necessary to subdivide the bulky class of flœtz rocks into flœtz and newest flœtz, thus completing a fourfold enumeration: other writers, adopting the transition class, have yet retained the term secondary as applied to the flœtz rocks of Werner, but this nomenclature lies open to the heavy objection already indicated, namely, that the term secondary, being opposed to primitive only, ought to include all rocks not of that class, and of course the transition order among the rest. These writers have bestowed the name tertiary on the newest flœtz class of the Wernerians. A synoptical and comparative view of the arrangement proposed in the present work and those of former writers is subjoined.

Character.	Proposed names.	Wernerian names.	Other writers.
1. Formations (chiefly of *sand & clay*) *above the chalk.*	*Superior order.*	Newest flœtz class	Tertiary class.
2. Comprising *a. Chalk.* *b.* sands & clays *beneath the chalk* *c.* calcareous freestones (*oolites*), & argillaceous beds. *d. New red sandstone, conglomerate & magnesian limestone.*	*Supermedial order.*	Flœtz class.	Secondary class.
3. Carboniferous rocks, comprising *a. Coal-measures.* *b. Carboniferous limestone.* *c. Old red sandstone*	*Medial order.*	Sometimes referred to the preceding, sometimes to the succeeding class by writers of these schools; very often the coal-measures are referred to the former—the subjacent limestone and sandstone to the latter.	
4. *Roofing slate, &c. &c.*	*Submedial order.*	Transition class.	Intermediate class
5. *Mica slate.* *Gneiss.* *Granite, &c.*	*Inferior order.*	Primitive class.	Primitive class.

In all these formations, from the lowest to the highest, we find a repetition of rocks and beds of similar chemical composition, i. e. siliceous, argillaceous, and calcareous, but with a considerable difference in texture, those in the lowest formations being compact and often crystalline, while those in the highest and most recent are loose and earthy. These repeti-

ascertain the exact disposition of the mineral materials they afford; but a circumstance cannot fail to have struck him during the course of his researches which opens to his view a far more extensive and interesting field of enquiry with regard to the relations of these rocks to the general revolutions of nature; for he will have found in many of these beds spoils of the vegetable and animal kingdom imbedded, particularly the remains of marine zoophytes and shells, and often in such abundance as to constitute nearly the entire mass of particular strata. If he is led by the interest thus excited to examine more closely the phœnomena attending the distribution of these remains, he will find them as remarkable in the detail as they are striking in a general point of view. In some countries he will perceive that none of these remains occur (for instance in Cornwall and the Scotch highlands), in others (as in the south-eastern counties of England) not a well can be sunk, or pit opened, without presenting them in abundance; and pursuing the enquiry, he will arrive at the conclusion that the lowest series of rocks, which have therefore been considered as primitive, are *entirely destitute of those remains.** That the next contains them sparingly, while they abound in the three succeeding series, although not without the occasional interposition of beds in

tions, form what the Wernerians call formation suites. We may mention 1st, the limestone suite; this exhibits, in the inferior or primitive order, crystalline marbles; in the two next, or transition and carboniferous orders, compact and subcrystalline limestones (Derbyshire limestone); in the supermedial or flœtz order, less compact limestone (lias), calcareous freestone (Portland and Bath stone), and chalk; in the superior or newest flœtz order, loose earthy limestones

2d. The argillaceous suite presents the following gradations; clay-slate, shale of the coal-measures, shale of the lias, clays alternating in the oolite series, and that of the sand beneath the chalk; and lastly, clays above the chalk.

3. The siliceous suite may (since many of the sandstones of which it consists present evident traces of felspar and abundance of mica, as well as grains of quartz, and since mica is more or less present in every bed of sand) perhaps deserve to have granite placed at its head, as its several members may possibly have been derived from the detritus of that rock; it may be continued thus; quartz rock and transition sandstone, old red sandstone, millstone-grit and coal-grits, new red sandstone, sand and sandstone beneath the chalk, sand above the chalk. In all these instances a regular diminution in the degree of consolidation may be perceived in ascending the series.

* Some appearances of organic remains have indeed been said to have been observed among primitive rocks, but they may very possibly have been deceptive; the only observation of this kind which requires notice is one of Dr. Mac Culloch's; that most accurate geologist describes a bed of gryphite limestone as underlying gneiss in one of the Hebrides; but when the extreme contortions of the strata of gneiss, as figured by himself, are

which they are still rare, if not altogether wanting. In the examination of these interesting remains, he must call the science of the zoologist and botanist to his aid, and thus he will discover that a great part of the genera, and a vast majority of the species, are entirely different from the animals and plants with which we are at present acquainted, as covering the face of the earth or occupying its waters.* Hence geology presents

taken into account, it will be obvious that in consequence of the flexures by which they are often bent backwards, a bed really superior in its general position, may appear to be inferior in partial observations: thus let *a a a* be

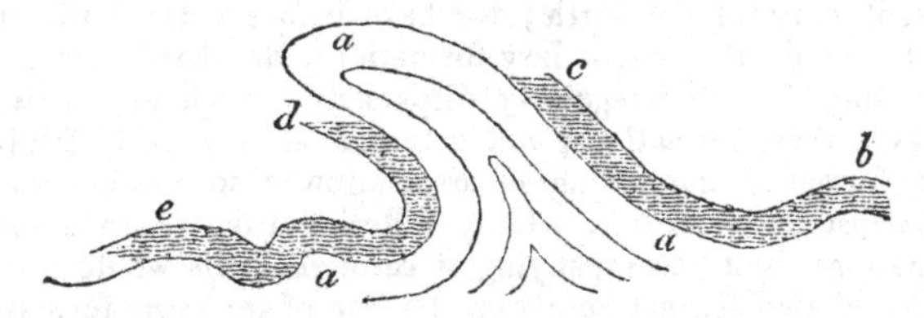

the contorted substratum of gneiss, and *b c d e* an incumbent bed of gryphite limestone following its flexures; it is clear that if this bed be visible only at the point *d*, it will appear to underlie the gneiss. A comparison with Dr. Mac Culloch's figures referred to, will shew that these contortions are not exaggerated.

* In speaking of the difference between recent and fossil species, it becomes us to be cautious in pronouncing that the latter do not at present exist because we are not acquainted with them in a recent state, and this caution is still more necessary with regard to those genera which the "dark unfathom'd caves of ocean" may possibly conceal in their recesses: we must remember that we were long acquainted with the encrinites, terebratulæ, and trigoniæ in a fossil state, before the analogous beings in a recent state had found their way to our collections; yet the general facts seem too strong to be entirely thus accounted for. With the exception of those contained in the most recent beds (the crag) only, nine out of ten fossil shells belong to species decidedly different from any known to exist. The family of ammonites, for instance, contains more than two hundred fossil species according to many authors, and it does not seem possible to reduce this estimate above one half; yet of all these not one is known recent, and the only recent species of the whole genus is a very minute shell; yet the fossil species sometimes measure three feet in diameter. Is it probable that a genus so numerous, and having species of such large size, can have been overlooked, especially as they are furnished with an apparatus whose use was evidently to give them buoyancy, like their allied family the nautilus? so that it is not likely they can remain concealed from inhabiting deep waters only. The same remarks will apply to the belemnites, of which no recent species is known. It would not be possible to point out the main features of the case in a more striking manner, than by referring an observer, well acquainted with recent conchology, to the shells contained in the carboniferous limestone of Derbyshire; he would at once recognise the total want of general resemblance: the difference also which we shall shortly notice between the shells in the different formations, affords a strong cor-

to the comparative anatomist and botanist, but particularly to the former, a rich fund of new materials, and adds to the several departments of natural history supplements, the knowledge of which is indispensable to complete our views of them: indeed, in many instances, important peculiarities of organisation, and remarkable links in the chain of animated beings are presented in these fossil remains, and many chasms which must otherwise have existed, are filled up in a satisfactory manner.

But the principal concern of the geologist is with the manner in which these remains are distributed in the strata forming the present crust of the earth; we have before noticed that they are confined to the secondary formations, and have now to add that they are not irregularly dispersed throughout the whole series of these formations, but disposed as it were in families, each formation containing an association of species peculiar in many instances to itself, widely differing from those of other formations, and accompanying it throughout its whole course; so that at two distinct points on the line of the same formation, we are sure of meeting the same general assemblage of fossil remains. It will serve to exemplify the laws which have been stated, if the observer's attention is directed to two of the most prominent formations of this island; namely, the chalk, and the limestone which underlies the coal in Northumberland, Derbyshire, South Wales, and Somerset. Now, if he examines a collection of fossils from the chalk of Flamborough head or from that of Dover cliffs, or, it may be added, from Poland or Paris, he will find eight or nine species out of ten the same;

roborative presumption that they are, *a fortiori*, different from those of the present ocean. The nearest approach to recent species appears to exist in some of the coralline and madreporal remains; but these classes have not as yet undergone an examination sufficiently rigorous, either in a recent or fossil state, to enable us to pronounce with certainty.

The remains of marine oviparous quadrupeds (Ichthyosaurus, Plesiosaurus, Maestricht animal, &c.) are referable to new genera widely different from any thing with which we are acquainted, and the fossil species of crocodile are strongly distinguished from the recent. These enormous and singular animals (sometimes almost rivalling the whale in size) which must often come to the surface to breathe, cannot surely have eluded the observation of all our voyagers. The land quadrupeds found in some of the most recent strata, and many of those even mingled in the diluvial detritus with the bones of animals still existing in the same countries, are often of genera widely distinct from any with which we are acquainted (e. g. Palœotherium, Megatherium, Mastodon, &c.) or of distinct species, as the fossil bear, rhinoceros, and elephant; and M. Cuvier has shewn at large the little probability there is that any of them exist in an unknown condition. It must be carefully remembered that an accurate and rigorous knowledge of Zoology is requisite in any one who ventures to discuss this subject; a superficial acquaintance with it can only lead into confusion and error.

he will observe the same echinites associated with the same shells; nearly half these echinites he will perceive belong to divisions of that family unknown in a recent state, and indeed in any other fossil bed except the chalk. If he next proceeds to inspect parcels of fossils from the carboniferous limestone, from whichever of the above localities they may have been brought, he will find them to agree in the same manner with each other; that is, he will find the same corals, the same encrinites,* the same productæ, terebratulæ, spiriferæ, &c.; but if he lastly compares the collection from the chalk with that from the mountain lime, he will not find one single instance of specific agreement, and in very few instances any thing that would even deceive an unpractised eye by the superficial resemblance of such an agreement.

The difference between these distant formations, in this respect, is indeed much greater than that between those which are more nearly contiguous; but still even between these, there are generally considerable, though less striking distinctions, as to the species of organic remains contained.

If we cast a rapid view over the phœnomena of this distribution, the subject must appear to present some of the most singular problems which can engage the attention of the enquirer into nature; first, we have a foundation of primitive rocks destitute of these remains; in the next succeeding series (that of transition) corals, encrinites, and testacea, different however from those now known, appear at first sparingly; the fossil remains of the carboniferous limestone are nearly of the same nature with those in the transition rocks, but more abundant; the coal-measures, however, themselves, which repose on this limestone, scarcely present a single shell or coral; but on the contrary abound with vegetable remains, ferns, flags, reeds of unknown species, and large trunks of succulent plants, strangers to the present globe. Upon the coal rest beds again containing marine remains (the magnesian limestone); then a long interval (of new red sandstone) intervenes, destitute almost, if not entirely, of organic remains, preparing as it were the way for a new order of things. This order commences in the lias, and is continued in the oolites, green and iron sands, and chalk. All these beds contain corals, encrinites, echinites, testacea, crustacea, vertebral fishes, and marine oviparous quadrupeds, yet widely distinguished from the families contained in the lower beds of the transition and carboniferous class, and par-

* Of that important division of the encrinital family, Crinoidea inarticulata of Miller, which appears confined to the older rocks, i. e. carboniferous and transition limestone, as are the Crinoidea articulata to the more recent lias, oolite, chalk, &c.

tially distinguished among themselves according to the bed which they occupy. Hitherto the remains are always petrified* (i.e. impregnated with the mineral substance in which they are imbedded; but lastly, in the strata which cover the chalk we find the shells merely preserved, and in such a state, that when the clay or sand in which they lie is washed off, they might appear to be recent, had they not lost their colour, and become more brittle. Here we find beds of marine shells alternating with others peculiar to fresh water; so that they seem to have been deposited by reciprocating inundations of fresh and salt water. In the highest of the regular strata, the crag, we at length find an identity with the shells at

* It would afford an interesting subject of enquiry to trace the various changes which organised substances have undergone in consequence of this inhumation. Bones have generally lost their phosphoric acid and gelatine if in regular strata, and have their spongy texture impregnated with the matrix in which they lie, limestone, clay, and iron pyrites; one instance of a bone penetrated by silex has occurred to the author, on the beach at Reculver. The calcareous substance of shells, echinites, encrinites, corals, &c. in its slightest change seems only to have lost its colouring matter and gelatine; next they become impregnated with the mineral matrix in which they lie, especially if that matrix be calcareous; hence they become much more compact; often at the same time their original calcareous matter undergoes a change of internal structure, assuming a crystalline form, and in some cases, viz. asteriæ, encrinites, and echinites, a calcareous spar of very peculiar character results, of an opaque cream colour: it would be desirable to ascertain the circumstances in the original texture of these three families, whence this uniformity in the spathose structure of their remains arises; often the original matter of the shell has entirely disappeared, leaving a vacant cavity. It is a curious question what menstruum can have dissolved the shell when buried in a calcareous matrix which must have been equally liable to be attacked by any agent which could have attacked the shell, and no less so to account for the hollow casts in solid nodules of flint or blocks of chert, completely environing these casts on all sides. In this case how did the testaceous matter which has disappeared escape from its apparently close prison? The space left by these hollow casts has often been filled up by an infiltration of some new substance, e. g. chalcedony which thus forms a model of the original shell; the chalcedony is generally disposed in those concentric rings which mark the stalactitic variety: this is generally the state of the fossils in the green sand of Blackdown, which are chalcedonic substitutions in place of the original shell, and are exquisitely beautiful. A similar substitution of chalcedony for the original matter of the shells imbedded in the lias of St. Donats, Glamorganshire, is much less easily accounted for as the matrix itself is there not siliceous but calcareous. In many of these instances some singular play of affinities, and the removal of the original substance in a state of solution through the pores of the surrounding rocks, must have taken place in the laboratories of nature in a manner which our own imperfect chemistry is scarcely competent to explain. It is much to be desired that Dr. Mac Culloch, or some writer possessing his accurate chemical knowledge and precision of thought, would undertake a full investigation of the phœnomena here alluded to.

Some remarks on the changes undergone by vegetable remains will be found in the first chapter of the book on the carboniferous strata.

present existing on the same coast; and lastly, over all these strata, indiscriminately, there is spread a covering of gravel (seemingly formed by the action of a deluge which has detached and rounded by attrition fragments of the rocks over which it swept) containing the remains of numerous land quadrupeds, many of them of unknown genera or species (the mastodon and the fossil species of elephant or mammoth, bear, rhinoceros, and elk) mingled with others equally strangers to the climates where they are now found (hyænas, &c.), yet associated with many at present occupying the same countries.*

The lists of organic remains given in the present work, may, it is hoped, promote this important branch of geological enquiry; which, notwithstanding the rapid advances lately made in it, can as yet only be considered in a state of progressive improvement; indeed, when it is remembered that it requires all the resources of a perfect acquaintance with many departments of Zoology, and those especially which are as yet least understood (namely, the history of invertebral animals), we have rather reason to be surprised that so much has been accomplished, rather than that much still remains to be done; and enough has been said to demonstrate that it is a subject which can be treated with advantage only by those who bring to it a matured and precise knowledge of the branches of natural history with which it is connected, a remark extorted by the flippant manner in which some writers have treated con-

* The general laws of the distribution of organic remains which have been above stated, are chiefly derived from the structure of England, the only country which has been accurately examined in this respect. Von Schlotheim's materials as to Germany are greatly deficient in the precision which is so essentially requisite both as to zoological and geological details; indeed that ingenious author can only consider them as an hasty sketch. Smith has published a useful stratigraphical arrangement of English fossils, in the preface to which he pointedly observes, that the various species of fossil shells may be found with nearly as much readiness and certainty in the natural strata, as in the drawers of a well arranged cabinet.

It is to be regretted that we have as yet no means of ascertaining whether a similar succession of secondary beds takes place in very distant countries (America for instance), and whether these are characterised by similar families of organic remains As the recent animals of these countries are widely different, one would naturally suppose that the fossils would be different also; yet in some instances we have reason to believe that this, in the earlier of the secondary strata at least, is not the case, for the most extensively diffused assemblage of organic remains with which the present author is acquainted, is that which characterises the transition limestone; namely, the chain coral, the alveolaria, some peculiar encrinites, several species of terebratula and spirifer, the orthoceratite, and the trilobite; all of these, except the latter, he has seen from the Canadian lakes, many from Melville island; and all, including the trilobite, from Sweden and the islands of Gothland and Oeeland; the identity of remains in the chalk in very distant points has been already noticed.

clusions, the premises of which they were incompetent to comprehend.*

§ 6. Another class of substances imbedded in the secondary strata, and throwing light on the convulsions amidst which they have been formed, are the pebbles or rolled fragments of rocks, older than themselves, which they are often found to contain; thus the lower beds of the supermedial order (namely, the conglomerate rocks of the new red sandstone) contain in great abundance rolled fragments of the carboniferous limestone belonging to the class next below it (the medial order), as well as of many still older rocks; being in fact only a consolidated mass of gravel, composed of debris of these rocks.†

The necessary inferences from this fact are, first, the rock whence the fragments were derived must have been consolidated, and subsequently to that consolidation have been exposed to the mechanical violence (probably the action of agitated waters), which tore from it these masses and rounded them by attrition, before the rock in which these fragments are now imbedded was formed; and secondly, since loose gravel beds (and such must have been the original form of these, though now consolidated into conglomerate rocks) cannot be accumulated to any extent (from the action of gravity) on an highly inclined plane. We are sure when we find such beds, as we often do, in nearly vertical strata, that this cannot have been their original position, but is one into which they have been forced by convulsions which have dislocated them subsequently to their consolidation.

* The present writer regrets that he is obliged to mention the hasty speculations contained in a Geological Survey of the Yorkshire coast as in some degree liable to this censure. The descriptive part of that work merits considerable praise as a valuable and interesting contribution to the geological branch of local history; but it will at once be evident to those who read the catalogue of fossil shells there given, that a knowledge of conchology is not the author's strongest point, and equally so that his ideas of geology are formed only from the inspection of a very limited district. The imperfect acquaintance with his subject displayed in his concluding part may readily be accounted for; its flippancy (see particularly page 302) is less easily to be excused: had the whole of the third part of this work been suppressed, the remainder, like the books of the Sibyl, would have acquired a much higher value.

† This example has been selected because it is not open to any doubt, for it has been sometimes said that the supposed derivative pebbles were in fact original concretions mistaken for such; but no suspicion of this kind can be brought against the alleged instance, since the pebbles of the carboniferous limestone are marked by the characteristic organic remains of that rock, and the angles of these remains, where they approach the surface of these pebbles, are broken down and rounded off, so that the proof of their origin is complete.

These consolidated gravel beds are called conglomerates, breccias, or pudding-stones; we find them among the transition rocks, in the old red sandstone, in the millstone-grit and coal-grits, in the lower members of the new red sandstone, in the sand strata beneath the chalk, and in the gravel beds associated with the plastic clay, and interposed between the chalk and great London clay.

§ 7. From the occurrence of the marine remains lately noticed, occupying, as they do, rocks spread over two-thirds of the surface of every part of our continents which have been explored, and rising to the highest situations, even to the loftiest summits of the Pyrenees and still more elevated points on the Andes, it is an inevitable inference that the greater part of those continents have not only been covered by, but have been formed of materials collected beneath the bosom of the ocean; that we inhabit countries which we may truly call factas ex æquore terras. The great and fundamental problem, therefore, of theoretical geology is obviously to assign adequate causes for the change of level in this ocean which has permitted these masses which once formed the bottom of its channel to rise in hills and mountains above its waves. The causes which it is possible to imagine are reducible to two general classes; first, the decrease of the absolute quantity of water; this must have resulted from causes entirely chemical, namely, the decomposition of some portion of the water, its constituents entering into new forms of combination, and its fixation in the rocks formed beneath it; it is probable that these causes have operated to some degree, but it seems impossible to ascribe to them the very great difference of level for which we have to account. The second class of possible causes is entirely mechanical; those, namely, which may have produced a change of relative level without any diminution of absolute quantity in the waters. The causes of this kind which have been proposed, are, first, the absorption of the waters into a supposed central cavity, but the now ascertained density of the earth (being greater than that which would result from an entirely solid sphere of equal magnitude of the most compact known rock) renders the existence of any such cavity very doubtful; secondly, a writer in the Journal of the Royal Institution, vol. 2. has proposed the very ingenious hypothesis that a change of temperature of a few degrees will, from the unequal expansibility of the materials of land and water sufficiently account for this change of level; thirdly, it has been ascribed to violent convulsions which have either heaved up the present continents, or, which amounts to the same thing (as the same relative change must have taken place in either view), depressed the present channel of the ocean. It is not the business of the present work to propose theories, but to record facts; these facts are thus connected with the above discussion; if the violent elevation of the continents (or depression of the channel of the ocean) supposed in the last mentioned hypothesis really took place, it must have left traces in the disturbed, contorted, and highly inclined position of the strata, and these disturbances must be the greatest where the change of level has been the greatest, i. e. in the neighbourhood of the loftiest mountains.

The enquirer with this view will be led to examine what is the actual position of the strata; how far that position can be considered as having resulted from original formation, and how far it must have resulted from subsequent convulsions and derangement.

When beds recomposed from the fragments and detritus of older rocks (such as are called conglomerates and pudding-stones) which must previous to their consolidation have existed as loose gravel, occur among vertical or highly inclined strata, we may conclude with absolute certainty that this inclined position cannot have been original, but must have resulted from subsequent disturbance; for it is obviously physically impossible to support an aggregation of loose gravel in vertical or nearly vertical planes. A similar argument will apply where, among the inclined strata, thin beds distinguished by peculiar organic remains, are interposed; for we cannot imagine any combination of circumstances under which (previously to the consolidation of the matrix containing them) the detached joints of encrinites, or the loose shells of testacea, or the scattered pinnulæ of ferns, should have disposed themselves in thin vertical layers. It is manifestly absurd in these cases to attribute the vertical direction to the action of any crystallizing force, or any cause of the kind; no such causes could have placed a vertical bed of limestone containing encrinites, in contact with a vertical bed of coal-shale containing canes and fern leaves. Now such arguments will be found to apply strictly to a very large class of highly inclined strata, and it will therefore deserve consideration whether we can in any case (for the phœnomena are always similar) ascribe the occurrence of vertical beds to these supposed causes.

Those remarkable dislocations of the strata called faults, are connected with the same question; these are breaks or fissures cutting across a mass of strata, accompanied by a sinking or depression of the portion of that mass on one side of the break, often amounting to many hundred feet. These phœnomena have been from the nature of the workings most fully explored in our coal mines.

It will be therefore one object of the present work to collect all the phœnomena of this nature which have been observed in this country, under a general point of view; they are partly treated of under the heads ‘ inclination’ and ‘ stratification,’ in the account of each formation; the combined results will be presented in a distinct chapter in the last book.

§ 8. Another class of facts, implicated with the questions arising from the convulsions to which the great change of the ocean's level has been ascribed, are those which relate to the rocks known by the generic designation of the Trap formation; for these rocks, being in the opinion of a large class of geologists decidedly of volcanic origin, the partisans of these views will undoubtedly attribute to the explosions which produced them, the principal agency in forcing up the strata, and heaving from the depths of the waves the ponderous masses of the continents. It is not however the object of this work to dwell upon theoretical views, further than to point out the manner in which the facts collected from observation may be brought to bear upon them. The phœnomena of the English trap rocks will be fully discussed in treating of those associated with the coal-formation; they will not be found perhaps very decisive, nor so illustrative of the great points at issue, as the appearances presented by similar rocks in many other countries. Without pronouncing any judgment on the controversy, we may observe that the weight of geological authorities decidedly preponderates, at present, in favour of the igneous origin of these rocks. In the subjoined note we have shortly stated the general question, as to the extent and manner in which volcanic agency may possibly be supposed to have operated in the convulsions which appear to have affected the Earth's surface, but we wish to keep these conjectural speculations entirely distinct from that positive knowledge acquired from observation, which is as yet the only certain portion of geological science.*

* In support of the hypothesis which ascribes an important part to volcanic agency in modifying the surface of our planet, the following at least plausible arguments might be adduced; we submit them, however, without venturing to determine what real weight they possess.

1. It must first be kept in view that the question is, to assign an adequate cause for the undoubted fact of the emergence of the loftiest mountains of the present continents; and that when so mighty an effect is to be accounted for, the mind must be prepared to admit, without being startled, causes of a force and energy greatly exceeding those with which we are acquainted from actual observation.

2. The broken and disturbed state and inclined position of the strata composing those continents, many of which must have been at the time of their original formation horizontal, indicate (as we have seen) that one, at least, of the causes operating to effect this great change of relative level

§ 9. As connected with the great problem of the change in relative level between the continents and the ocean, we have to

between the land and waters, was the elevation of the former by mechanical force.

3. The only agent with which we are acquainted, whose operation bears any analogy to the effects above specified, is the volcanic energy which still occasionally forms new islands and elevates new mountains.

4. Although these effects are indeed now partial and limited, yet there is certain proof that volcanic agency has formerly been much more active; the extinct volcanoes of the Rhine, Hungary, and Auvergne, as well as those which occupy so large a portion of Italy, where one only now remains in activity, concur in proving that we now experience only the expiring efforts, as it were, of those gigantic powers which have once ravaged the face of nature.

5. If to this certain proof of the greater prevalence of volcanic convulsions in earlier, but still comparatively recent, periods of the history of our planet, we add the presumption that the trap rocks (so singularly intruded among the regular strata, and producing where they traverse those strata so precisely the effects of heat acting under compression, and so different in all their phœnomena from formations decidedly aqueous) were of volcanic origin, we shall find that scarcely a country exists which has not been a prey to the ravages of this powerful principle. If with many of the best geological observers (Dr. Mac Culloch, Von Buch, Necker, &c.) we incline to extend the same conclusions to granitic rocks, a mass of volcanic power, clearly adequate to all the required effects, is provided.

6. The question will undoubtedly present itself, what is the source of volcanic action; and sufficient proof exists that this source is deeply seated beneath the lowest rocks with which our examination of the Earth's surface makes us acquainted; for in Auvergne the lavas have evidently been erupted from beneath the primitive rocks.

7. The very important recent discoveries with regard to the increased temperature noticed in descending deep mines, &c. by Messrs. Fourrier and Fox, will, if confirmed by further examination, prove that some great source of heat exists beneath the Earth's crust. Mr. Fox's observations have been disputed by Mr. Moyle, who considers him to have been led into error, by the higher temperature of the portions of the mine where it had been raised by the animal heat of the workmen employed; but it is obvious that this can never account for the regular gradation of increased temperature said to have been noticed in every successive level examined: the subject, however, cannot fully be cleared without reiterated and continued observations. While this paragraph was passing through the press, Mr. Fox has returned what appears to be a satisfactory answer to the objections of Mr. Moyle, in the Ann. Phil. for May 1822.

8. A degree of presumption may be thought to arise from these considerations, that the crust of the Earth rests on an heated nucleus, the true source of volcanic energy. If this nucleus be in a fluid or viscous state, its undulations would readily account for the convulsions which have affected that crust both in originally dislocating and elevating portions of its strata, and in the actual phœnomena of earthquakes, (of many of which phœnomena no other hypothesis appears to offer a sufficient explanation), while at the same time it would assign an adequate reason for the figure of the globe as a spheroid of rotation.

9. On this supposition, we should at once perceive a reason why the effects of the volcanic force may have been much more violent in earlier periods, while that mass of deposits which now covers the supposed vol-

notice an important, but too hasty generalisation of an opposite school of geologists (the Wernerians), which supposes the basset edges of the strata to occupy levels successively lower and lower in proportion as they are of less ancient formation and recede from the primitive chains, forming the edges of the basins in which they have been deposited. The accompanying diagram will assist us in understanding these views, and comparing them with the real fact as it exists in nature.

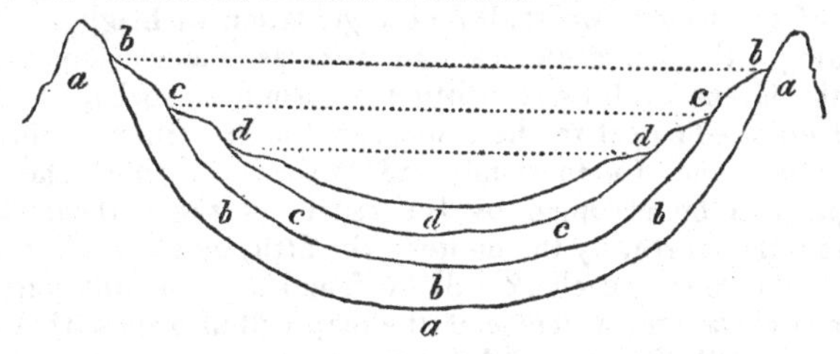

Here *a a a* is the supposed fundamental basin of primitive rocks once occupied by the original ocean; when that ocean had sunk to the level of the dotted line *b b* it deposited the bed *b b b*, which conformed itself to the form of the bottom of the basin, and therefore rose in a steep angle against the ridges forming its side; when the ocean had further sunk to *c c* it deposited the bed *c c c*, and in like manner at the level *d d*, formed the bed *d d d*; but as the new deposits recede further from the steep primitive ridges, they would continue more horizontal even to their very edges.

From this hypothesis two corrollaries necessarily arise; first, the basset edges of each formation must every where be of the same level; secondly, the basset edges of the strata, when traced horizontally across the surface of a country, will form parallel zones, in such a manner that the central area will be occupied by the most recent formation, encircled in regular order by successive zones of the older formations, the edge of each of which will take a wider and wider sweep.

canic nucleus was only gradually forming over it, than at present; and we shall also find a reason for the higher temperature which many of the remains both of the animal and vegetable kingdoms, found in the strata of countries, now too cold for the existence of their recent analogues, appear to indicate as having formerly prevailed.

10. It must be remembered that one of the essential conditions of the theory above sketched is, the operation of volcanic agency beneath the pressure of an incumbent ocean; and that it does not, therefore, in any degree question the Neptunian origin of the majority of the rocks which have evidently been formed in the bosom of that ocean. With regard to the trap rocks, and perhaps the granitic, alone, does it venture even to insinuate an opposite mode of formation.

Let us examine, then, how far the facts ascertained by observation accord with this view. It will at once be seen on inspecting the map and sections of this island appended to the present work, that there is an approach towards the structure required by this theory, but yet attended with irregularities, which must prevent our adopting it without great modification. In the first place, we will consider the favorable side of the question. If we examine the mountain ridges and chains of hills of the island, we shall find a gradation in height corresponding with their geological age, such as is above supposed; for the Grampian hills, consisting of primitive rocks, are the most elevated; next to these succeed the transition chains of South Scotland, Cumberland, and Wales; the third class in height will be occupied by the ranges of the carboniferous series; the fourth, by the oolites; the fifth, by the chalk; and the strata above the chalk will be found to form hills only of inferior elevation; a glance at the map will also shew that the bassets of all the strata above the new red sandstone form successive zones, fulfilling in a general manner (though not without irregularities) the conditions of the hypothesis.

But on the other hand, if we compare the basset edges of the same strata on the opposite sides of the great European basin (assuming the primitive ranges of our own island as one of its borders, and those of the Alpine chains as the other), we shall find their level totally different. The oolite, for instance, whose highest point with us is less than 1200 feet, attains an height of more than 4000 in the Jura chain, and in the mountains of the Tyrol has been observed by Mr. Buckland crowning some of the loftiest and most rugged summits of the Alps themselves. Again, if we compare the inclination of the strata at the edges of the basin, we shall find any thing but the supposed regular gradation from an highly elevated to an horizontal position; on the contrary, we shall see the horizontal beds generally reposing at once upon the truncated edges of those which lie at very considerable angles; and in place of the general conformity or parallelism which ought to prevail between the several formations, we shall observe in many instances appearances of the greatest irregularity in this respect; and these irregularities will be found to increase in approaching those chains which are the most elevated. In order to enable the reader to compare the real structure of a mineral basin on the great scale, with that resulting from the Wernerian hypothesis, we refer to the sections of England which accompany this work as exhibiting a portion of the western border of the principal basin of Europe, and subjoin in this place a rough sketch of the Alpine border of that basin, taken from a section of Ebel's.

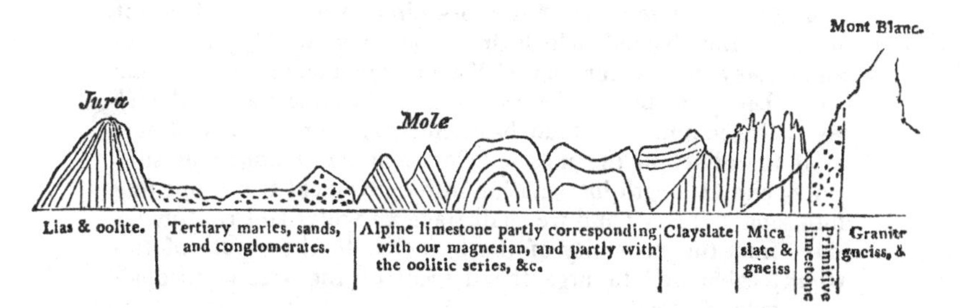

It is impossible to conceive a more striking picture of derangement, apparently resulting from the operation of violent convulsions, than is here presented; and however we may doubt whether the details of the section are perfectly ascertained, it is impossible that such a representation could have been formed, did not the greatest disturbance actually exist.*

§ 10. The appearances exhibited by the numerous vallies which furrow the earth's surface, and certain allied phœnomena presently to be stated, will properly form some of the concluding objects of enquiry to the geological observer, as being connected with the most recent of those causes which have modified that surface, previously to its passing into the state in which we now behold it, and becoming subject to the order of causes which still prevails.

The first thing that will strike him will be the regular con-

* It seems impossible to deny that, if the only change which took place during the formation of the strata, were a regular and gradual subsidence of the level of the ocean, the phœnomena ought to be exactly in accordance with the above Wernerian exposition; but if we suppose that, during their formation, the continents were elevated by mechanical forces acting in a series of great convulsions, we shall perhaps obtain a nearer approximation to agreement with the actual phœnomena, as deduced from observation. If these convulsions resulted from volcanic agency, we have before seen that there is every reason to believe this cause to have acted with most violence in the earliest periods; and this will sufficiently account for the greater derangement of the older rocks. In many instances, as might be anticipated, in countries least raised above the level of the sea, the subsequent formations seem, from their nearly horizontal position and regular conformity to each other, to have experienced only a gradual and gentle action of the elevating forces, and here and here only we may expect to find and do find an accordance with the Wernerian view; but in the more lofty and mountainous countries, where we should conclude (if we suppose these chains to have been raised by convulsions of the nature described) that those convulsions must have raged with violence during much longer periods, we shall accordingly find that the derangements do actually extend through a far more extensive range of geological formations; and observe comparatively recent strata exhibiting the same phœnomena, of high inclination, contortion, and irregularity.

figuration of these vallies * in those circumstances which adapt them for the channels which drain the countries they traverse, and convey their waters to their final receptacle, and at the same time their principal source, the ocean. In almost all of them† we shall see numerous branches ramifying over extensive tracts as if to collect in the most efficient manner the aqueous supplies, and at length inosculating into a principal trunk opening into some æstuary; and we shall trace a regular and continuous slope from the extremity of the branches to the mouth of the trunk, calculated to urge the descent of the waters through the whole system.

Now this configuration is exactly that which would necessarily be produced by the action of waters scooping out channels for their passage in draining themselves off from the face of a country. We may daily see the same operation repeated in miniature by the drainage of the retiring tide on muddy shores, especially in confined æstuaries where the fall is considerable and rapid.

That such has been in fact the agency which has in every instance greatly modified, and in many entirely produced, the inequalities which now mark the earth's surface, giving rise to all its beautiful variety of hill and valley, phœnomena of the most decisive character, constituting a body of evidence as nearly approximating to demonstration as the nature of the case can admit, leave no reasonable ground to doubt. Of these we shall proceed to give a brief exposition, reserving for the body of the work their fuller consideration, and their illustration by a copious induction of examples from every part and every formation of the island.

We must begin, however, by remarking that many of the theories advanced on this subject appear defective in two points. First, because, ascribing every thing in the formation of vallies to the agency of running waters, they entirely overlook the effect which must have been produced by the violent convulsions which appear in so many instances to have broken and elevated the strata, and must in so doing have necessarily formed a surface diversified by many and great inequalities; we should therefore perhaps take a more just as well as more com-

* There are some excellent remarks on this subject in one of those earlier memoirs which anticipated the regular and full developement of geological science, Packe's Memoir for a Map of East Kent, published 1737: it is need ess to refer to the later materials accumulated in Playfair's illustrations.

† The only exceptions are, first, the vallies which terminate in inland lakes unconnected with the ocean, and secondly, with regard to the regular and uniform slope of the districts in which chains of lakes abound; these will be hereafter considered.

prehensive view of the subject in combining the agency of these two orders of geological causes. The vallies of mountainous countries (where every sign of disorder and disturbance prevails in the strata) owed in all probability at least their first outline to the disruptive forces which acted around them; and here accordingly we find that regular and systematic conformation, which has been already noticed, far less clearly marked in the structure of the vallies; for instead of the uniformly descending slope of their channels, this line is broken by deep hollows, the receptacles of large lakes. But although on these grounds we may refer the original formation of the vallies of such districts, in part at least, to the convulsions alluded to, yet there are the strongest proofs that even here also the vallies have subsequently been greatly modified by the rush of mighty currents of water through them; and in lower countries, where the horizontal and undisturbed position of the strata shew that other convulsions cannot very sensibly have affected the figure of the surface, we must refer its present inequalities almost exclusively to the excavating action of such currents.

The second defect which calls for animadversion in some of the theories which in other respects have given the clearest views of the phœnomena under discussion, is, that while they correctly ascribe the excavation of vallies to the agency of aqueous currents, they look to no other supply of that agency than the streams (often inconsiderable rills) which now flow through them, borrowing liberally from time what they confessedly want in force. The advocates of this view imagine, that in a long lapse of ages the incessant action of this minute cause would be sufficient to account for the mighty effects observed; but not to dwell on the difficulties which the truly immense periods required, must present to any one who imagines he has less than an eternity of past time to calculate upon; yet even conceding that eternity, it is easy to shew that the phœnomena attendant on vallies are very commonly of such a nature, that to believe them to have been formed by their actual rivers, however long their action may have endured, involves the most direct physical impossibilities. In fact (as we shall presently see) this hypothesis must be abandoned at once by any one who will take the trouble of subjecting it to a rigorous application to the vallies of any extensive district, or to any map of those vallies in which the configuration of the surface is accurately represented; and it must principally be ascribed to the imperfection of all but the most recent maps in this respect, and to the circumstance that the eye seldom takes in enough of the surface of a country to judge correctly of the totality (if we may so speak) of its configuration, that it could ever have been seriously defended. It is indeed the more extraordinary that a cause so manifestly inadequate, should ever have been embraced, since the fundamental fact of geology, namely, that the continents, now dry land, were once covered with the ocean, which is of necessity (however differently explained) common to every geological theory, involves in itself the admission of a cause fully adequate: for, however that ocean may have been brought to its present level, it could never (on any view of the matter) have drained off the surface of the lands it has deserted, without experiencing violent currents in its retreat; and in those currents (the existence of which no one can on any hypothesis dispute) might have been found a force far more commensurate to the effects to be accounted for.*

It is further necessary, as a preliminary to the discussion of this subject, to state a distinction to which we shall often have occasion to recur, arising from the direction of vallies relatively to that of the chains of hills among which they range: considered under this relation they may be divided into two classes, commonly distinguished as *longitudinal* and *transverse* vallies. The longitudinal valleys are those which pursue a course parallel to the direction of the chains which bound them; the transverse, those which cut transversely across the hill chains. Since the direction of the chains of hills is very generally the same with the direction of their constituent strata, it is unnecessary to add that the longitudinal vallies will in such cases be parallel to, and the transverse cut across, the strata among which they range.

To proceed to the phœnomena on which the proofs of this agency of great aqueous currents in the formation of our vallies depend, we should begin with those which shew that they have been excavated in the strata subsequently to their original consolidation.

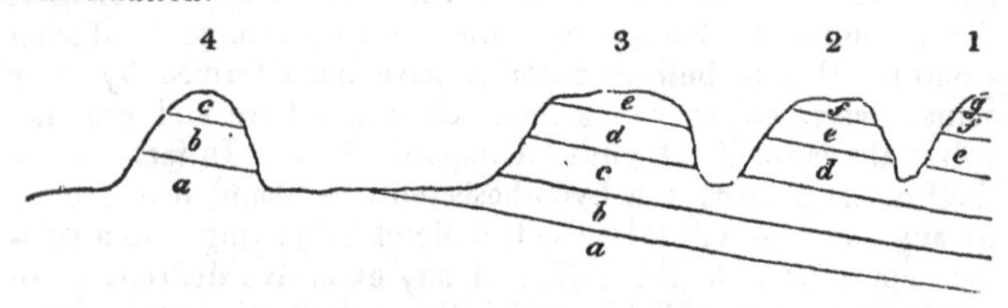

The nature of these proofs will be most clearly explained by the inspection of the accompanying diagram, which represents

* We do not at present enter into the question whether the vallies may not have been modified by other currents subsequent to the first retreat of the ocean from our continents.

the sectional profile of a country composed of stratified rocks and traversed by several vallies. One of these vallies separates the hills No. 1 & 2, a second separates 2 & 3; at the further base of the hill No. 3, a plain succeeds, in the middle of which rises the insulated hill No. 4, crowned by the same strata which appear in the neighbouring chain. Such insulated hills are termed *outliers*, and constitute a phœnomenon of much importance as connected with the present question.

If then we examine the structure of the vallies as here represented, we find precisely the same series of strata repeated on both their sides, in exactly the same order, and under circumstances which indicate them to have been once continuous, and to have been subsequently removed from the intervals occupied by the vallies, by some cause which has here excavated or scooped away the materials which once filled those intervals. The phœnomena are exactly similar, on the large scale, to those which would be exhibited in the small by a block of laminated marble in which the tool of the sculptor had chisselled out deep furrows; and as we should not doubt in the latter instance that the laminæ now interrupted by these furrows had been once continuous, and the interruptions effected by subsequent violence, so we have the same or stronger evidence in the case before us; for the strata broken through by the vallies are, in a majority of instances, evidently the result of aqueous deposition; now we cannot possibly suppose that such a cause could have deposited exactly the same beds, in the same order, and the same planes, throughout the mass of the hills, and yet have abruptly ceased to deposit them in the narrow intervals now possessed by the vallies; undoubtedly then those intervals were once filled by the same deposits whose truncated edges now appear on their sides; and the intervals themselves (i. e. the vallies) have been formed by the subsequent excavation or erosion of the strata in these points.

In the treatise on the Deluge by Mr. Catcot (a physico-theological writer of the last century belonging to the Hutchinsonian school) are the following forcible remarks on this subject.

" If a person were to see the broken walls of a palace or castle that had been in part demolished, he would trace the lines in which the walls had been carried, and in thought, fill up the breaches, and reunite the whole. In the same manner, when we view the naked ends or broken edges of strata on one side of a valley, and compare them with their correspondent ends on the other, we cannot but perceive that the intermediate space was once filled up, and the strata continued from mountain to mountain."

The whole subject of the excavation of vallies by diluvial currents is discussed with great ability by this writer; but it is unnecessary to add, that much of erroneous observation, and more of unfounded inference, will be found in geological speculations of that age and school.

This passage has been already cited by Mr. Greenough in his Geological Essay, in a section devoted to the consideration of the present subject, which condenses much valuable information on the several points connected with it.

The proof, as above stated, is still further strengthened by the occurrence of broken fragments of the materials which once filled up these intervals, scattered over their surface. Not only do we observe these natural breaches bearing every mark of the violence which has produced them, but we find the ruins themselves strewn around; immense accumulations of debris torn from the adjacent rocks, and generally more or less rounded (as if by attrition against one another while rolled along by the action of strong currents), very generally cover the bottom of the vallies which traverse, and the plains which stretch beyond the base of the elevated chains. To the consideration of the phœnomena presented by these accumulations we shall presently return.

On these grounds, then, the proof that the vallies have been in many instances entirely excavated by the agency of powerful aqueous currents, and in all greatly modified by the same cause, seems as strong and complete as the nature of the case can possibly admit.*

* Mr. De Luc, in his travels through England, endeavouring to meet these arguments, adduces many instances of vallies in which, according to his account, the opposite sides consist of different strata, as a contradiction of the above theory of their formation; but since we can only expect to find any given stratum in the continuation of its plane, it by no means follows that all vallies excavated by water must necessarily present the same strata on both sides; this in fact is only a necessary consequence when the vallies are transverse, or cut across the direction of the strata; because in this case the plane of those strata will necessarily range along both the opposite sides. But in longitudinal vallies, ranging parallel to the direction of the strata, especially when the beds are very sensibly inclined to the horizon, we ought not to expect the same strata on both sides; for in this case the truncated edge of a superior bed may form the escarpment on one side the valley, and the opposite slope may be formed by the ascending plane of the stratum which emerges from beneath it; the figure illustrating the nature of stratification, page iii, may serve to shew the sectional profile of a series of such vallies. M. De Luc's instances are either of this kind or founded entirely in mistake (the fact being exactly the reverse of his representation) as we shall shew at length in the part of the work dedicated to this subject. In a great majority of instances (ninety-nine out of the hundred at least) the strata are regularly found in the continuation of their planes, whenever and however these planes are cut by the vallies. The

But although we may safely attribute these effects to running water, yet we shall find on examination that the agency of the streams which at present flow through our vallies (to whose long continued action they have sometimes been ascribed) is quite inadequate to afford a satisfactory solution of the phœnomena. The proofs of this assertion are not merely the apparent disproportion between this cause and the effect to be accounted for, and the entire absence of any streams in many vallies (those of the chalky districts especially), but we find a still more decisive refutation in a phœnomenon of common occurrence,—the intersection of two series of vallies, the one extending longitudinally along the base of a chain of hills, and the other cutting transversely across that chain, under such circumstances that no stream could have risen to a sufficient height to form the transverse vallies by excavating a passage through the crest of the chain, but must have discharged its waters at a level far inferior to that required for this effect, through the longitudinal valley at its base. The details of this configuration of surface, and the arguments arising from it, will be treated with the detail they require in the body of the work under the proper head; here they can only be alluded to; and it will thus be seen that large sheets of water sweeping over the face of an extensive tract at once, can alone account for the phœnomena. Almost all the vallies of the Weald of Kent, Surrey, and Sussex, present this combination of circumstances, as do many others of those which traverse the chalk range in various parts of the island; and a circuit of a few miles round Bristol, alone affords no less than ten instances of the same kind.

The same agency that has excavated the vallies, appears also to have swept off the superior strata from extensive tracts which they once covered; the proofs of this are to be found in insulated hills, or *outliers* of those strata placed at considerable distances from their continuous range, with which they have every appearance of having been once connected; in the abrupt and truncated escarpments which form the usual terminations of the strata; and in the very great quantity of their debris scattered frequently over tracts far distant from those where they still exist in situ. This stripping off the superstrata is appropriately termed *denudation*.

---

only excepted cases are, when the direction of the valley coincides with that of a fault, or dislocation and subsidence of the strata, and these are of the very rarest occurrence A glance at Mr. Smith's or Mr. Greenough's map will at once shew that all the principal transverse vallies throughout the island, do actually exhibit the same strata on both sides.

The surfaces of the strata appear to have been exposed partially, at least, more than once to the action of these denuding causes, and even at very early epochs, while many of the more recent beds were as yet only in the process of being deposited; for among those beds we find many, as we have already observed, made up of water-worn debris which must have resulted from causes of this kind. Indeed whenever, and in whatever manner, the waters first receded from the emerging continents, currents which could hardly have failed to produce such effects must have taken place; but the most important agency of this kind appears to have been exerted at a more recent period, and subsequently to the consolidation of all the strata, by an inundation which must have swept over them universally, and covered the whole surface with their debris indiscriminately thrown together, forming the last great geological change to which the surface of our planet appears to have been exposed.

§ 11. To this general covering of water-worn debris derived from all the strata, the name of *Diluvium* has been given from the consideration of that great and universal catastrophe to which it seems most properly assignable. By this name it is intended to distinguish it from the partial debris occasioned by causes still in operation; such as the slight wear produced by the present rivers, the more violent action of torrents, &c. &c.; to the latter the name *Alluvium* has lately been appropriated; but many authors confound the two classes of phœnomena together, describing them generally as alluvial. The phœnomena of the diluvial debris, or gravel, are highly important and interesting. Its existence, as we have already seen, demonstrates the nature of the causes which have modified the present surface of our planet; its quantity may serve in some degree as a measure of the force with which they have acted; and its distribution may indicate the direction in which the currents swept it onwards. For instance, when we find rounded pebbles derived from rocks which exist in situ only in the mountains of the north and west, scattered over the plains of the midland counties, we may be sure that the currents drifted from the former point to the latter; and it often affords a curious and interesting problem to the geologist to trace these travelled fragments to their native masses, often hundreds of miles distant. The accumulations of this gravel above referred to, in the midland counties, especially along the plains at the foot of the escarpment of the chain of the inferior oolite on the borders of Gloucestershire, Northamptonshire, and Warwickshire, are of surprising extent, and the materials brought together are from so many quarters, that it would not be difficult to form a nearly complete suite of the geological formations of England from their fragments here deposited. Portions of the same gravel have been swept onwards through transverse vallies affording openings across the chains of oolite and chalk hills, as far as the plains surrounding the metropolis; but the principal mass of the diluvial gravel in this latter quarter, is derived from the partial destruction of the neighbouring chalk hills, consisting of flints washed out from thence, and subsequently rounded by attrition.*

On a general investigation of these and similar phœnomena, it does not seem possible to assign any single and uniform direction to the currents which have driven this debris before them; but they appear in every instance to have flowed (which indeed must of necessity be the case with the currents of subsiding waters) as they were determined by the configuration of the adjoining country; from the mountains, that is, towards the lower hills and plains. As far as England is concerned, this principle will produce a general tendency to a direction from north and west towards south and east, greatly modified however by obvious local circumstances.

Another circumstance connected with the distribution of these travelled fragments is, that we often find them in masses of considerable size, accumulated in situations now separated by the intervention of deep vallies from the parent hills (if we may so speak) whence we know them to have been torn. This appears to be a demonstrative proof that these intervening vallies must have been excavated subsequently to the transportation of these blocks; for though we can readily conceive how the agency of violent currents may have driven these blocks down an inclined plane, or, if the vis a tergo were sufficient, along a level surface, or even up a very slight and gradual acclivity, it is impossible to ascribe to them the Sisyphean labour of rolling rocky masses, sometimes of many tons in weight, up the face of abrupt and high escarpments. The attention of geologists was first directed to this phœnomenon by the discoveries of Saussure, who noticed one of its most striking cases—the occurrence of massive fragments torn from the primitive chains of the Alps, scattered at high levels on the escarpment of the opposite calcareous and secondary chains of the Jura, although between the two points the deep valley containing the lake of Geneva is interposed.

The occurrence of colossal blocks of granite scattered over the plains of northern Germany, which may be traced up to the

* See some excellent observations on diluvial gravel appended by Professor Buckland to his memoir on the Lickey quartz rock, whence much of the gravel of the midland counties seems to have been derived. (G. T. vol. 5.)

Scandinavian chains on the opposite side of the great gulf occupied by the Baltic, is a well known instance of the same kind.

It is true indeed that in these two cases partial solutions of the problem have been attempted, by supposing that these blocks may have been drifted across on ice-bergs; yet even with regard to the case of the shores of the Baltic we see nothing similar occurring in the actual order of things, and with regard to that presented by the Jura the hypothesis seems to destroy itself; for if the surface of the ocean were there raised to the level at which these blocks are found (although under present circumstances ice-bergs might exist in a river flowing at that level), yet with the general rise of the ocean's surface, the body of the atmosphere, and the lines of temperature would of necessity rise also, and ice-bergs could not exist in an ocean under that latitude.

But in truth this phœnomenon is one of very common occurrence, and is exhibited, though on a less striking scale, on almost every chain of hills throughout England, under circumstances which admit of no escape from the obvious inferences first stated. The chain of the middle oolites in Oxfordshire, for instance, is extensively covered through Bagley wood with debris from almost every class of rocks from the transition series to chalk, and among them many blocks of great size and weight occur. Farey's list of the insulated hills in Derbyshire, having gravel scattered over their surface, affords many similar instances. The downs surrounding Bath, (Hampton Down for example), though abruptly scarped and surrounded by vallies more than 600 feet deep, have yet on their very summits flints transported from the distant chalk hills; but it is unnecessary to multiply examples farther. The simplest explanation of the fact will be, that these fragments were transported by the first action of the currents, before they had effected the excavation of the vallies now cutting off all communication with the native rocks whence they were derived. In some instances also it may appear that these transported masses have not accomplished their whole journey at once, but may have been detached by some of those earlier convulsions to which the conglomerate beds, associated with so many of the formations, must be ascribed; and having reposed in these strata for a time, have been washed out afresh and moved forwards to a new destination.

The organic remains of land animals dispersed through this diluvial gravel have been already mentioned: they must with the highest probability be referred to the races extinguished by the great convulsion which formed that gravel; many of them are of species still inhabiting the countries where they are thus found; some of the species now only inhabiting other climates, and some few of species and genera now entirely unknown.

To the same period we may ascribe the bones *of the same species with the above*, found in many caverns; but in many of those instances it is probable that some of the animals now found there, previously inhabited them as their dens. Professor Buckland appears satisfactorily to have proved that this must have been the case in the remarkable instance of the cavern lately discovered near Kirby Moor Side, Yorkshire. Here the remains found in the greatest abundance are those of hyænas; with these are mingled fragments of various animals, from the mammoth to the water rat, all the bones present evident traces of having been mangled and gnawed, and the whole are buried in a sediment of mud subsequently incrusted over by stalactitical depositions. Mr. Buckland's explanation is, that this cavern was occupied by the hyænas; who, according to the known habits of those animals, partially devoured even the bones of their prey, and dragged them for that purpose to their dens: around their retreats a similar congeries of mangled bones has been noticed by recent travellers. The proofs of these points, deduced from the circumstances of the cavern, the state of the bones, and the ascertained habits of the animals in question, appear to be decisive; the sediment in which the bones are imbedded, and the occurrence of the remains of the mammoth, and other species only known (in these climates at least) in a fossil state in the diluvial gravel, clearly refer their remains to the same æra.

Caverns containing bones of a similar class (the mammoth, the fossil species of rhinoceros, &c.) have been found near Swansea, at Hutton hill (on the Mendip chain in Somersetshire), and near Plymouth.

---

§ 12. We have finally to examine the local changes which have taken place subsequently to this last great and general convulsion, and which still continue to take place under the influence of the order of causes at present in actual operation. In these we may often observe a balanced and compensated effect of destruction and renovation; for instance, in the most powerful of these causes—the agency of the sea upon the coasts,—we find the headlands and projecting promontories undermined and washed away by the waves, towards which their sections present in consequence, scars of mural cliffs: but the materials thus absorbed are usually thrown up again and constitute extensive tracts of newly formed marsh-land along the less exposed points of the coast; in part also forming new banks and shoals in the adjoining channel of the sea. The agency of rivers is of a similar description; these, when their higher branches assume the force of wintry torrents, carry away sometimes in considerable quantity the looser materials of the soil through which they rush, but they deposit these materials again in the formation of alluvial flats and deltas near their mouths. The action of atmospheric causes, the frost by rifting and detaching portions of the outer surface of rocks, and the rain by washing the finer parts away, either contribute to the agency of the torrents, or accumulate the fragments detached in a slope or talus of debris, at the foot of the hills whence they are derived.*

These actions, however, appear to be circumscribed within very narrow limits; over a great part of the earth's surface the influence of these wasting causes is absolutely null, the mantle of green-sward that invests it being an effectual protection. The barrows of the aboriginal Britons, after a lapse of certainly little less, and in many instances probably more, than two decads of centuries, retain very generally all the pristine sharpness of their outline; nor is the slight fosse that sometimes surrounds them in any degree filled up. Causes, then, which in two thousand years have not affected in any perceptible manner these small tumuli, so often scattered in very exposed situations over the crests of our hills, can have exerted no very great influence on the mass of those hills themselves in any assignable portion of time, which even the imagination of a theorist can allow itself to conceive; and where circumstances are favorable to a greater degree of waste, still there is often a tendency to approach a maximum at which

* The results of these causes, that is to say the alluvial deltas formed at the mouth of rivers, and the talus of debris accumulated at the foot of precipices, may, from observations of their known increase within certain periods, be submitted to a species of retrograde calculation, by which we may reason as to the length of the total period during which these causes have operated under their present conditions; or (in other words) the period which has elapsed since the last great convulsion which has given to our continents their present general form. For instance, the alluvial tract at the mouth of the Po has been ascertained by observation to have a regular rate of increase in a century, and the line where this tract begins, against what must have been the original coast, is capable of being determined: these data afford, it is obvious, sufficient grounds for calculating the length of time requisite to produce the whole of this alluvial tract, and it is satisfactory to observe that the period thus deduced, agrees with that assignable to the deluge recorded by the inspired historian. These and similar phœnomena have been designated by De Luc as geological chronometers. M. Cuvier has a very interesting chapter on this subject.

farther waste will be checked; the abrupt cliff will at last become a slope, and that slope become defended by its grassy coat of proof. It should appear that even the action of the sea, certainly the most powerful and important of all those we have surveyed, has a similar tendency to impose a limit to its own ravages. It has obviously in many instances formed an effectual barrier against itself, by throwing up shingle banks and marsh lands in the face of cliffs against which it once beat; and after the destruction has been carried to a certain point, it appears necessary from the mode of action (excepting where very powerful currents interfere) that the very materials resulting from the ruin should check its farther encrease: even where these currents exist, these also have a tendency to throw up barriers of shingle in their eddy. Historical records, and the very nature and physical possibilities of the case, alike compel us to dissent entirely from those crude and hasty speculations which would assign to the causes now in action, the power of producing any very material change in the face of things; and which would refer to these alone, acting under their present conditions, and with only their present forces, the mighty operations which have formed and modified our continents.

It is a curious object of enquiry arising from this subject, whether the materials thus carried into the sea have produced any effect on its level. The materials derived from the destruction of lofty cliffs by the waves, having been deposited in the formation of low marshes, must of course occupy a greater surface in the latter than in the former condition; and if they have formed banks in the shoal waters, the same consequence must have resulted; so that the sea must in either case have lost more room by the diffusion of the materials, than it can have gained by encroaching on the cliff; to this must be added the materials brought down to the sea coast, and there deposited by rivers; so that it cannot be doubted but that the basin of the ocean must, by the combined action of these causes, been in some degree (however trifling that degree may be) narrowed. If therefore, as seems probable or rather indeed certain, the quantity of water in the ocean is permanent, its level cannot but have been slightly raised by this reduction of the superficial space allotted to it. It is probable, however, that this effect can only be spoken of theoretically, being so small in proportion to the total mass of waters in the ocean, and the area they occupy, as to be absolutely imperceptible. But although it is little probable that any perceptible change can thus have been produced in the general level of the ocean, it is nevertheless very possible that the level of the tides along the coast may have been sensibly affected, since the effect of these depositions, which are chiefly accumulated along the line of coast, must be to create shoal water along the same line; and it is very possible that the mass of waters flowing forwards with the impulse of the flood, may be forced, by the resistance it thus encounters, to rise higher than would be the case were the bottom more deep. It seems certain that the bottom of all narrow æstuaries into which extensive rivers discharge themselves, have been sensibly raised by these depositions; and in such situations it must follow as a necessary corollary, that the level of high water has been raised also.

The occurrence of submarine forests, i. e. the remains of forests traceable along the line of coast considerably beneath the high water level, affords a phœnomenon of great interest, which does not seem readily explicable except on the views just stated. The remains of such a forest were traced by Sir Joseph Banks, (see Philosophical Transactions for 1799), along the Lincolnshire coast; the same thing may be observed beneath the marshes of the Thames from the Isle of Dogs to Purfleet, at several points along the southern coast, in the Bristol channel, at Blue Anchor near Dunster, and at Shurton bars and Stolford, where they have been accurately described by Mr. Horner (Geol. Trans. vol. 3); at Newgill Sands, Pembrokeshire; on the Lancashire coast, &c. &c. It is evident in many of these instances that the trees have not been drifted to their present place, but have grown where they now are found, since not only are the remains of their stems in an upright direction, but the roots may often be traced spreading in an undisturbed position through the substratum.

In many instances we find vallies opening towards the sea which have evidently once been æstuaries, completely filled up by these depositions, and the phœnomena they present are often interesting. Usually the bottom is coarse gravel, upon which rests a finer silt; often one or more beds of vegetable matter alternate with the silt; it appears that these have been derived from drift wood, which has floated into the æstuary, become saturated with water, or as it is called water-logged, and perhaps covered by alluvial debris, drifted on its surface by the wintry torrents, and thus sunk. Bones of animals, and sometimes of men, and rude implements of art, have been found among these accumulations, having drifted probably upon the wood. The stream-works, as they are called, of Cornwall, are evidently æstuaries thus filled; in these, rounded fragments of tin ore are sufficiently abundant among the gravel which forms the bottom bed, to render the laying open these deposits a profitable speculation; some interesting sections of these ancient æstuaries will be found in the third volume of the Geological Transactions.

The tract around Bovey Tracey in Devon, which exhibits several beds of wood-coal, alternating with the alluvial debris of granitic rocks, is another instance of such an æstuary.

The formation of peat bogs is another of the geological changes in actual progress. Dr. Mac Culloch in a most able paper published in the Edinburgh Philosophical Journal for 1820, has treated this subject in the most precise and satisfactory manner: he considers peat bogs as divisible (from the consideration of the vegetables which compose them, and the causes which concur in their formation) into the following varieties; mountain, marsh, forest, lake, and marine peat: in the progress of the work we shall have occasion to explain more fully, and to apply his principles. The formation of peat bogs often occasion, or at least materially contribute to, the partial filling up of lakes, the extension of marsh lands along the coast, &c. Much valuable information concerning these processes may be found in De Luc's Travels in the north of Europe.

The materials accumulated in alluvial districts are usually in a loose form, as sand, marle, and clay. The formation of compact rocky masses is of more rare occurrence; under favorable circumstances, however, and especially along the coast, sandstones are formed from the consolidation of the drift sand, and where the oxide of iron is present to act as a cement, the process goes on rapidly: the northern coast of Cornwall affords extensive examples of this process (see vol. 1 of the Cornish Geological Transactions). Captain Beaufort describes a line of petrified beach as extending along various points of the coast of Caramania; the same thing is common in the Bahama Isands, and the human skeleton brought from the beach at Guadaloupe, and preserved in the British Museum, was imbedded in a mass of this description*: it is unnecessary to add that marine remains are commonly dispersed through rocks of this description; occasionally comminuted shells form almost the entire mass, in which case the rock is of course calcareous: a near resemblance to some of the oolites may also be observed in these recent formations. Calcareous rocks are often also deposited from the waters of existing rivers or lakes in large masses. The celebrated Travertino of Rome is thus formed, and Germany affords some similar instances. The deposits of petrifying waters, such as the Baths of San Filippo, the stalac-

* "The occurrence of the Bulimus trifasciatus, a very common West Indian shell, in this mass, strongly indicates its modern formation."

titical incrustations of caverns, &c. may be referred to the same general head. In many instances these recent calcareous deposits include land or freshwater shells, leaves, and other substances which have become accidentally entangled in them.

But of all the changes resulting from causes in actual operation, those depending on the agency of volcanos are the most remarkable. By this cause new mountains are still occasionally elevated, and new islands formed; thus the Monte Nuovo, elevated above 600 feet, rose in the space once occupied by the Lucrine lake in the year 1538. See Hamilton's letters on Volcanos, p. 127; and more recently (1759) in Mexico, from similar causes, a tract of ground of four square miles in extent, swelled to the height of 524 feet, and in the centre of a thousand burning cones, six large masses elevated themselves more than 1200 feet; the loftiest summit, known as the Volcano of Jorullo, rising no less than 1695 feet. What renders the case more remarkable is, that the point where this extraordinary convulsion happened, is more than 42 leagues distant from every other volcano. See Humboldt's New Spain, vol. 2. p. 165. The islands of the Greek Archipelago, especially those of Hiera and Santorini, have received accessions by the extension of those previously existing, and the formation of new islets in the adjoining sea by volcanic eruptions, in 726, 1427, 1573, and 1650. New islands are recorded as having been thus formed among the Azores in 1628, and very recently in 1811.

By the same cause extensive districts are covered with thick beds of volcanic ashes; those resulting from successive eruptions being regularly disposed, stratum super stratum, and overflowed by long currents of lava, or sometimes by streams of ejected mud. Mackenzie's and Henderson's travels in Iceland, Fortis, Spallanzani, Dolomieu, and Ferrara in Italy and Sicily, and Humboldt's in South America, should be referred to on these heads.

Closely connected with volcanic phœnomena are those of earthquakes. These often produce subversions of the strata; which represent, though on a smaller scale, and may perhaps account for, those grand dislocations and derangements which we generally observe in the beds composing mountain chains. See particularly the description of the effects produced by the great earthquake in Calabria.

The agency of volcanic powers appears to have been much more extensive at an earlier period, even under the existing order of things, and subsequently to the last great convulsion of the Earth's surface, than they are at the present moment; for we observe craters now extinct of the most indisputable character, which have covered with showers of scoriæ many countries of Europe, at present remote from any active volcanos. Such craters are found in Auvergne, on the Rhine between Bonn and Andernach, in Hungary, and along the west side of Italy, where Vesuvius alone still retains its energy. In all these instances the fact that the currents of lava and showers of ashes have descended the present vallies, proves the activity of these volcanos subsequently to the complete excavation of those vallies.

Another actual cause of change operating on the surface of the planet, is one which we can scarcely include among geological phœnomena, and which it yet seems necessary to mention in order to present a complete view of those changes: we allude to the coral reefs and islets formed in the midst of the Pacific ocean and some other seas, by the minute but combined labours of millions of marine zoophytes. The following extract from Kotzebue's voyages (as cited in the Quarterly Review) gives the latest and perhaps the best view of this interesting subject.

"As soon as it [the ridge or reef] has reached such a height, that it remains almost dry at low water, at the time of ebb, the corals leave off building higher; sea-shells, fragments of corals, sea hedge-hog shells, and their broken off prickles are united by the burning sun, through the medium of the cementing calcareous sand, which has arisen from the pulverisation of the above-mentioned shells, into one whole or solid stone, which, strengthened by the continual throwing up of new materials, gradually increases in thickness, till it at last becomes so high, that it is covered only during some seasons of the year by the high tides. The heat of the sun so penetrates the mass of stone when it is dry, that it splits in many places, and breaks off in flakes. These flakes, so separated, are raised one upon another by the waves at the time of high water. The always active surf throws blocks of coral (frequently of a fathom in length, and three or four feet thick) and shells of marine animals between and upon the foundation stones; after this the calcareous sand lies undisturbed, and offers to the seeds of trees and plants cast upon it by the waves, a soil upon which they rapidly grow to overshadow its dazzling white surface. Entire trunks of trees, which are carried by the rivers from other countries and islands, find here, at length, a resting place, after their long wanderings: with these come some small animals, such as lizards and insects, as the first inhabitants. Even before the trees form a wood, the real sea-birds nestle here; strayed land-birds take refuge in the bushes; and at a much later period, when the work has been long since completed, man also appears, builds his hut on the fruitful soil formed by the corruption of the leaves of the trees, and calls himself lord and proprietor of this new creation." vol. iii. pp. 331–3.

§ 13. Having thus brought to a conclusion our survey of the phœnomena which it forms the object of Geology to investigate, we subjoin a rapid sketch of the progress of that science especially in our own country.

In the limited number of physical subjects which attracted the attention of classical antiquity, we can only venture to include a few insulated phœnomena from among those which fall within the province of Geology. Such striking natural appearances as earthquakes and volcanos could not indeed entirely escape notice, and we occasionally find crude theories proposed to account for their causes :* the formation of new lands by the alluvium of rivers, † the birth of new islands, ‡ and the encroachment of the sea on the land, || are also topics often handled by them which at least border on Geology. With the great and leading fact of geological speculation, the occurrence of the spoils of marine animals imbedded in the solid strata of the continents, they were acquainted,§ but it seems not to have excited much interest or curiosity ; and it is singular that when adducing other arguments to prove the favorite tenet of some of their schools, that the face of nature was undergoing a perpetual change, so that what was then land had once been sea, this, which amounted to occular proof, is very often overlooked.‡

* Pliny. Hist. Nat. lib. 2. 81 ad 86. Aristotle, Meteorologica. lib. 2. 14, 15, 16. Lucretius, lib. 6. Seneca, Nat. quæst. a cap. 4. Plutarch de Placitis, Philos. lib. 3.

† Pliny. lib. 2. 87. Aristotle. Meteor. lib. I. 14. Herodotus, lib. 2. in initio. Polybius gives an interesting account of the increase of alluvial deposits in the Palus Mœotis (lib. 4). Strabo mentions the volcanic formation of a new mountain of enormous height in the neighbourhood of Methone. (Ed. Ab. tom. 1. p. 102.)

‡ Pliny. lib. 2. 88. 89.

|| Pliny. lib. 2. 90. 92. 94.

§ Pausanias (Attica) describes a quarry of shell limestone, ΛΙΘΟΣ ΚΟΓΧΙΤΗΣ, at Megara as a solitary example of this kind of stone in Greece; he observes that it was soft, white, and quite full of shells.—Xenophon, in the Anabasis, more than once mentions temples and other buildings constructed of a similar stone. The professed writers on minerals, however, scarcely allude to the subject. Theophrastus mentions an ostracites, but in such a manner that it is not clear whether he meant an organic remain or not. Pliny only says of a fossil of the same name that it had the figure of a shell; and though he describes under various fanciful names, what appear to have been Belemnites, Ammonites, Echinites, Encrinites, Corallites, and casts of Cardia, &c. does not betray any suspicion of their true origin. The same writers mention the occurrence of fossil bones and ivory equally without remark.

‡ The arguments which Aristotle brings in support of this are deduced from a fanciful notion that the Earth was partially subject to a change like

Ovid alone, perhaps, alleges it with this view among the illustrations of the above doctrine, which he has put into the mouth of Pythagoras; his words have been so frequently cited, that it is almost unnecessary to repeat them.

> "Vidi factas ex æquore terras,
> Et procul a pelago conchæ jacuere marinæ."

In some of their general physical notions we may almost fancy we see the germ of more modern theories; thus that of the deplacement of the sea, afterwards adopted and adorned by Buffon, may be traced in Aristotle, who appears to have considered it as a periodical revolution of nature; ‡ and the wild but splendid conception embraced by many of their schools, but particularly by the Stoics, that the Earth had experienced frequent destructions and renovations from the agency of igneous devastations (ecpyroses) and inundations (cataclysmi) recurring after intervals of distant ages, reminds us in so many respects of the Huttonian theory * that we might almost suppose it to have been adopted from the consideration of the same natural appearances: but it is more consistent with the general genius of their philosophical speculations to believe that it was deduced from the principles they assumed on the high priori road, than introduced by any train of inductive reasoning grounded on observation.

that from youth to age, by which its moisture dried up; from the earlier habitation of the cities in the upper than lower regions of Egypt and Greece; the formation of the former in great part from the Nile; the gradual filling up the Palus Mœotis, &c. and the necessity that such changes, though unobservable to the eye of man which contemplates but a moment, must yet take place in the lapse of time, which he believed to be of infinite duration. Meteorologica. l. 1. 14.

‡ ου αει τα μεν γη τα δε θαλασσα διατελει παντα τον χρονον αλλα γιγνεται θαλασσα μεν οπου χερσος ενθα δε νυν θαλασσα παλιν ενταυθα γη. κατα μεντοι τινα ταξιν νομιζειν χρη ταυτα και περιοδον. Meteor. l. 1. 14.

* See particularly Lipsius de Physiologia Stoicorum. This writer quotes the following passage from Censorinus, a philosopher of that school:—Est preterea annus quem Aristoteles maximum potius quam magnum appellat, quem solis & lunæ vagarumque stellarum orbes conficiunt, cum ad idem signum ubi quondam simul fuerant unà revertuntur. Cujus anni hiems summa est Cataclysmus quem nostri diluvionem vocant; æstas autem Ecpyrosis quod est mundi incendium. Nam his alternis temporibus mundus tum exignesere tum exaquescere videtur. The reader who wishes for a fuller account of this doctrine, may be referred to a chapter dedicated to its investigation in Dr. Prichard's Egyptian Mythology, which ably condenses all the collections of Lipsius, and to which we are indebted for the above quotation. The whole of Dr. Prichard's work is a model of judicious, sober, and philosophical criticism, applied to subjects where we are accustomed to meet only with extravagant conjectures, and still more extravagant etymologies.

On the practical operations of mining among the ancients much curious information may be found in an interesting paper by Mr. Hawkins, published in the first volume of the Cornish Geological Transactions.

The Arabian writers in the middle ages appear to have cultivated mineralogy with some success; the first foundation of a rational arrangement of minerals was laid by Avicenna at the close of the tenth century.

Several Italian writers are cited as having noticed the occurrence of fossil shells in the hills of that country at an early period. The celebrated Boccacio is among these; and in the fifteenth century Alessandro degli Alessandri proposed the hypothesis that the axis of the earth's rotation might originally have had a different situation from the present,* as a means of accounting for the change in the place of land and sea indicated by this circumstance. Fracastoro, in 1517, enters largely into this subject, and observes that the phœnomena are such that they cannot be satisfactorily explained by a transient convulsion, such as the deluge alone. Palissy, a French writer in 1580, has been cited with high praise by Fontenelle as an original discoverer, on the ground of similar observations; but the priority of Fracastoro is evident.

George Agricola, a native of Misnia, who flourished during the first half of the sixteenth century, published on several branches of Mineralogy; in particular he has illustrated in a full, precise, and clear manner, the various phœnomena of metallic veins.

Before the close of that century, an Englishman, George Owen of Pembrokeshire, left behind him a very valuable manuscript work on the topography of his native county. In this he has traced with much accuracy the direction and extent of the strata of coal and the limestone which accompanies them through the whole of South Wales, and pointed out the connections of this tract with similar districts in Gloucestershire and Somersetshire. This appears undoubtedly the earliest attempt to establish the important and fundamental geological fact, that the same series of rocks succeed each other in a regular order

* A similar notion found an advocate in Voltaire, who even believed in the wild tradition of the Egyptians, that the Sun had *twice risen in the west* within the memory of that nation, ascribing this to a supposed revolution of the Earth's axis round one of the equatorial diameters, which he imagined was completed in four millions of years. It is needless to add that astronomical observation does not afford the slightest ground for these extravagant speculations; the real change of the obliquity of the ecliptic is a phœnomenon arising from causes of a very different nature: it is a secular variation confined within very narrow limits, and inadequate to account for any geological appearances.

throughout extensive tracts of country, and to elucidate the geological structure thus indicated; but his work having continued in manuscript till recently published in the second volume of the Cambrian Register, remains a striking instance of those anticipations of subsequent discoveries which may often be noticed in the history of science, but can in no degree have contributed to forward them.

During the seventeenth century we find little but theoretical writers, like Burnet, without observation, or collectors without general views. Among the collectors, however, Woodward deserves very honorable mention. While he enriched one of our Universities, Cambridge, Llwydd labored to rival him at the other. Llwydd appears to have been acquainted partially, at least, with the occurrence of particular shells in particular strata; having observed that the same varieties of Echini are peculiar to the chalk of England and the north-east of Ireland.

But Lister chiefly demands our notice in this century, having been undoubtedly the first proposer of regular geological maps. (See Philosophical Transactions for 1684.) The very idea of this proposal indicates an acquaintance with the regularity of geological structure prevailing over extensive districts; it does not appear that he ever carried his design into execution, but he illustrates it by mentioning the divisions he would have adopted for Yorkshire, and a map coloured according to these divisions would afford a fair delineation of its true structure. He also shews that he was well acquainted with the extent of the chalk formation in this island and France; and from some of his notices, it farther appears that he had recognised, at least in one particular instance, the distinction of strata by their organic remains.

Early in the following century we find the occurrence of the chalk and sandy hills in parallel zones in Bedfordshire, observed by Holloway (Philosoph. Trans. for 1723); and the same fact still more ably illustrated by a masterly description of the triple range of hills of chalk, of Kentish rag stone, and of clay traversing the county of Kent, by Mr. Packe, author of a chorographical chart of East Kent, published 1730. About the same period, Mr. Strachey, in a series of communications to the Royal Society, had well described the coal-district of Somersetshire. He notices the inclined position of the carboniferous strata, and the horizontal direction of the beds of red ground and lias which cover them; his sections also demonstrate that he was acquainted with the regular succession of beds in this district, namely, chalk, freestone (Bath oolite), red ground, coal-measures, metalliferous limestone of Mendip,

&c.: he however explains it on an absurd principle, which must have prevented his applying it to any general purpose.

As we approach the middle of the eighteenth century, we find the scattered rays of information, which alone can be discerned previously, converging into a more condensed and steady light; the disjoined atoms falling, as it were, into a regular system. The splendid genius of Buffon, though on this subject it wasted its strength in the unprofitable pursuit of theoretical speculations, and added little or nothing to the solid and accumulating mass of inductive observations, yet undoubtedly by the very brilliancy of those speculations, and perhaps by their extravagance also, strongly tended to kindle a more general attention to this branch of philosophical enquiry. Guettard, in 1746, first carried into execution the idea proposed by Lister years before, of geological maps: he divided the surface of the earth into three grand zones; the schistose, which nearly coincided with the primitive and transition districts of later geologists; that of marles, which included generally the secondary limestones; and that of sand, which in like manner comprised what have been since termed tertiary formations: the localities of individual minerals were expressed by signs, analagous to those employed in chemistry. He appears to have endeavoured to extend these principles to the structure, not only of a considerable part of Europe, but of Canada also, and Asia Minor. Such extensive generalisations at that period of the science could not be otherwise than extremely hasty, and incorrect; and accordingly the attempt to accomplish too much, appears to have brought his method into much discredit: indeed in his later publication, the Atlas Mineralogique de France, conducted in conjunction with Monnet, he nearly confines himself to indicating the localities of individual minerals.

Lehman, in a work published in 1756, was the first to introduce generally, and to establish firmly, the great distinction between primitive and secondary rocks. It had however previously been obscurely indicated in the writings of the Tuscan Steno, and more clearly by his successor at Florence, Targioni; Desmarest, in the Encyclopedie Methodique, also claims in favour of his countryman Rouelle, in a course of lectures delivered about the same period, the honour of having established a still more complete division into primitive, intermediate, and secondary rocks, and of having distinctly announced that the organic remains contained in the two latter classes, were distributed in regular assemblages (termed by him *amas*), each containing peculiar genera and species; but these lectures appear never to have been printed, nor to have exercised any influence beyond the sphere in which they were delivered. Besides establishing the great distinction of primitive and secondary rocks, Lehman illustrated his doctrines by many details concerning the disposition of the carboniferous rocks, and those associated with cupriferous marle-slate (the same since fully described by Friesleben) as exhibited around the Hartz and Erzegebirge: he fell into the common error (which Werner himself afterwards imitated) of supposing that the few rocks which had fallen under his observation in a limited district, were all that the world afforded; and that he had sufficient data, from the inspection of the former, to reason as to the universal structure of the latter.

In 1760 the Rev. J. Michell, in a paper on the Cause and Phœnomena of Earthquakes published in the Phil. Transactions, delivered the whole doctrine of the regular succession of the stratified masses constituting the crust of the earth, in a manner still more satisfactory and compleat: he observes that this structure is such that we always meet with successive zones of the various mineral masses lying parallel to, and rising towards, the crest of the principal mountain-ridge: he illustrates this position by instances derived from the mountains of North and South America, and from England, in which latter country, he observes, the general direction of the strata, and the chains formed by them, runs from east-north-east to south-west; he also particularly remarks (as indeed Lister had done previously) the length and continuity of the range of chalk extending more than 300 miles on the opposite sides of the Channel in England and France; he adds that it would be easy to have carried the proof of the regular succession of the strata of England much farther into detail, and among his papers was found the following list of English strata, bearing date about 1788 or 1789.

	Yards of thickness.
Chalk	120
Golt	50
Sand of Bedfordshire	10 or 20
Northamptonshire lime and Portland limes, lying in several strata	100
Lyas strata	70 or 100
Sand of Newark	about 30
Red clay of Tuxford, and several red marles	100
Sherewood Forest, pebbles and gravel	50 *unequal*
Very fine white sand	*uncertain*
Roche Abbey and Brotherton limes	100
Coal strata of Yorkshire	—

Whitehurst, in his enquiry into the original formation of the Earth (1778), insists on the regular succession of the strata, and confirms this doctrine by a full and exact account of the geological structure of Derbyshire, which he illustrates by several good sections. He has left little for succeeding enquirers to glean concerning the general history of the carboniferous limestone and coal-formations of that district; but, as has been well observed, a great part of his work is infected with that taste for cosmogony which had misled so many of his predecessors: his friend Mr. Kier shortly afterwards imitated successfully the descriptive and valuable part of his writings, in an able memoir on the district of limestone, coal, and basalt, in the south of Staffordshire, published in Shaw's history of that county.

In the year 1788 Hutton published his *Theory of the Earth*, a work which has exercised a lasting influence over the writings of a large class of English geologists. It is unnecessary here to recapitulate the heads of a system so generally known through the elegant "Illustrations" of its ablest advocate. Hutton had the merit of first directing the attention of geologists to the important phœnomena of the veins issuing from granitic rocks, and traversing the incumbent strata, and of bringing forward in a striking point of view the circumstances which seem to corroborate the igneous origin of trap rocks: the wildness of many of his theoretical views, however, went far to counterbalance the utility of the additional facts which he collected from observation. He who could perceive in the phœnomena of geology nothing but the *ordinary* operation of actual causes, carried on in the same manner through infinite ages, without the trace of a beginning or the prospect of an end, must have surveyed them through the medium of a preconceived hypothesis alone.

We have now arrived at the period at which Werner first published his researches: his 'Kurze Klassification' appeared in 1787, but his system seems to have received various accessions between that time and 1796. It is difficult to estimate his real and independent merits, since he himself published but little, and we are acquainted with his system only by later publications of his pupils, which are at the same time compilations from other sources: those merits appear to have consisted chiefly in a superior acquaintance with the mineralogical characters of rocks, in having traced with more minuteness the succession of primitive and transition rocks, together with the few flœtz rocks which he had opportunities of studying in that part of Germany with which he was alone acquainted, and which he fondly imagined to form the type of the whole globe; and (which was of more importance) in having reduced the hitherto irregular elements of geological science into a stricter method and more systematic form. His attempts at theorizing must now appear to all but his most devoted adherents, among the most unsuccessful and unphilosophical ever made, and even these are gradually abandoning one by one all his most characteristic opinions. There was, however, in his character an energetic determination of all his powers to the advancement of his favorite pursuit, which communicated itself to his whole class, and doubtless he has done more than any other individual to promote its career.

The travels of Saussure in the Alps, and of Pallas in various parts of the Russian territory, but especially the former, afforded however contributions to the advancement of true geological science, more important perhaps in themselves than the methodical arrangements of the Freyberg school, and certainly much more so than all its theoretical accompaniments.

In 1790 Mr. William Smith, (a name which can never, in tracing the history of English geology, be mentioned without the respect due to a great original discoverer) appears to have commenced his researches in the neighbourhood of Bath, having in that year drawn up a tabular view of the strata exhibited in that district, which in fact contained the rudiments of his subsequent discoveries. Ten years afterwards he circulated proposals for publishing a treatise on the Geology of England to be accompanied by a coloured map and sections, and in the interval had freely communicated the information he possessed in many quarters, till in fact it became by oral diffusion the common property of a large body of English geologists, and thus contributed to the progress of the science in many quarters where the author was little known. In this same interval, between 1790 and 1800, several volumes of reports were published by the Board of Agriculture, many of them containing much local geological information; and *to this board must undoubtedly be ascribed the honour of having produced the earliest geological maps of any part of England*; for its first series of reports published in 1794 contains very adequate geological maps of the North riding of Yorkshire, of Derbyshire, and of Nottinghamshire, and a less perfect one of Devonshire; that of Kent, published in 1796, has a regular geological map of that county (which indeed after the treatise of Packe in the beginning of the century it was easy to construct). Between this date and 1813, the same board has also given useful maps of Sussex, Surrey, Berks, Bedford, Gloucester, Wilts, Lincoln, Durham, and Cheshire, besides publishing a second report of Derbyshire dedicated exclusively to its mineralogy by Mr. Farey. Maton's tour through the western counties, published in 1796, has also a regular though of course imperfect geological map of the west of England.

These are certainly the earliest published geological maps of any part of this island; but it is probable that Mr. Smith had already commenced the manuscript of his own, which after many delays at length made its appearance in 1815, and was succeeded by various county maps on a larger scale, sections, &c.

D'Aubuisson has liberally said of this great performance of an unassisted individual—

"Ce que les minéralogistes les plus distingués ont fait dans une petite partie de l'Allemagne, en un demi-siècle, un seul homme (M. William Smith, ingénieur des mines) l'a entrepris et effectué pour toute l'Angleterre; et son travail, aussi beau par son résultat, qu'il est étonnant par son étendue, a fait conclure que l'*Angleterre est régulièrement divisée en couches, que l'ordre de leur superposition n'est jamais interverti*; *et que ce sont exactement des fossiles semblables qu'on trouve dans toutes les parties de la même couche et à de grandes distances.*——

"Tout en payant au travail de M. Smith le tribut d'admiration qui lui est dû, il me sera permis de désirer que des observations ultérieures en confirment l'exactitude, et déjà, sur plusieurs points, les travaux des minéralogistes anglais l'ont confirmée."

Nor is this praise in any respect too high; to say indeed that the first geological map of any country is likely to be free from material imperfections, is to maintain a position which every one acquainted with the ordinary progress of science must feel to be untenable. This is an object only to be gained by a series of gradual approximations, and it is by no means a small tribute of commendation to say that Mr. Smith has commenced that series with a performance, in which the trifling errors of detail which it may exhibit, bear no proportion in importance to the great general views which it correctly lays down. If we cast a rapid glance over this and his other publications, beginning with his representations of the more recent strata, and descending the geological series, we shall at once see what he has achieved, and added to our previous information, and what he has left for others. The tertiary beds above the chalk he has represented only generally, their more accurate division having been reserved for the researches of Mr. Webster, &c.: the chalk formation he has laid down with great precision, but its limits had, as we have seen, been before generally stated by many authorities: hence through the series of sands and oolites, down to the new red sandstone or red ground, the whole field is, with few and immaterial exceptions, entirely his own. Before his researches it would have been known only under the vague designation of a district of secondary shell limestones and sandstones, and to him we owe the attempt, in most instances successfully made, to ascertain by precise determinations the various and important members of this series, and to trace them from one extremity of the island to the other. In this enterprize (sufficiently arduous to try the powers and establish the reputation of any individual entering upon ground hitherto untrodden) he may perhaps in some instances have suffered a few omissions to escape without detection, and more rarely have identified too hastily beds in distant parts of the country really belonging to different formations; but still the great mass of his divisions remains unquestioned and unquestionable, and has been adopted, though with an occasional change of nomenclature, and a few requisite corrections, by all the geologists who have followed his steps, as well as in the present work. The carboniferous districts are also on the whole represented with a near approach to correctness, but are far inferior in this point to those occupied by the series last mentioned, and here there was also extant a greater quantity of previous materials; the districts of old red sandstone, and those occupied by the transition and primitive rocks, are very inadequately represented.

Subsequently to the publication of Mr. Smith's map in 1819, another on nearly the same scale was published by Mr. Greenough; the execution of this is more minute and delicate, and the details more exactly laboured; the general configuration of the surface of the country, its hills and vallies, are represented with far more precision than had previously been attempted in any general map of the island,—points which did not enter into the construction of Mr. Smith's map; and many of the imperfections of the former are removed: to this therefore we have referred as a general standard throughout the work, and have therefore studiously noted every remaining incorrectness which a careful collation of it in the course of our enquiries with the materials derived from subsequent observation has enabled us to detect: from this also we have copied the slight outline map prefixed to this volume, with a trifling change in the system of colouring which a different view of the division of a part of the carboniferous series of rocks has obliged us to introduce, and some other deviations which will be accounted for in the explanation of the map and plates.

The able investigation of the formations above the chalk by Mr. Webster, and his comparative views of their relation to those similarly situated in the basin of Paris, have supplied an important desideratum in the history of our strata; which has been also still more largely indebted to the general and extensive researches of Professor Buckland; who has materially improved the arrangement of Smith; has been the first to effect a combination between the progress of Geology in England and on the continent, by a distinct and masterly survey of their comparative structure; and has first treated the action of the diluvial currents with fulness and precision.

To Dr. Mac Culloch we owe our best information on the structure and phœnomena of the interesting class of trap rocks and of granite, and the substitution of a rigorous determination of their mineralogical character for the vague nomenclature and descriptions so commonly applied to many of the older formations. While he has been elucidating the geological history of Scotland, that of Ireland has received similar contributions of the highest merit from Messrs. Weaver and Griffith.

The Edinburgh school has to boast of several distinguished geological names; it is impossible to mention that of Playfair without the admiration demanded by a genius of a very high philosophical order, or that of Jameson without the respect due to a long and meritorious career of labours devoted to the advancement of this science; but we cannot but feel the injurious effects which have in this instance been produced by that excessive addiction to theoretical speculations, which has converted the members of that school into the zealous partisans of rival hypotheses, and led them to contribute far less than they otherwise must have done, to the real progress of inductive geology. To this cause we must ascribe it, that it has fallen so far behind the schools of London and Oxford;* in the latter case a striking and satisfactory proof has been afforded in opposition to the misrepresentations of shallow sciolists, that the institutions of academical education are far from unfavorable to the cultivation of the physical sciences, and that an ignorance of the rules of classical composition, and of the languages, and philosophy of polished antiquity, are by no means essential advantages in researches of this nature: it has been rather seen

> quid mens rite, quid indoles
> nutrita faustis sub penetralibus
> possit.

Cambridge has yielded a similar evidence in the valuable memoirs of Messrs. Sedgewick and Henslow.

* It will be readily permitted to the present writer at once to discharge a debt of justice and enter a gratifying memorial of private friendship, by here recording the fact that we owe the introduction of these pursuits into the University above mentioned, to lectures delivered between 1805 and 1810 by the present Professor of Chemistry Dr. Kidd. His more private exertions in encouraging the rising talents of others, and promoting their co-operation, were as successful in effect as liberal in design. The Oxford school may claim the important observations of Mr. Strangeways on the Russian Empire as among its fruits.

In the above slight and hasty sketch of the progress of this study, we have been obliged to pass over in silence many less important sources of information; such are the class of writers who have given local natural histories; for instance Plott, Moreton, Borlase, Price, &c. and writers of general topography, many of whom have contributed in some degree to the accumulation of materials for English Geology. Among these Stukeley the antiquary, the father of our home tourists, seems to have been intent on pursuing Lister's project of a mineral map, since he refers in his index to the particulars he had collected as intended to facilitate such a design. We have also omitted Huchinson and his school of physico-theological writers, who have however collected many important observations; we have already quoted a striking passage from Catcott, one of this school, on the subject of denudation. The continental writers on this subject of the present day, we have not endeavoured to include; for it would be impossible in so limited a compass to give an adequate idea of the labours and merits of Cuvier, Brongniart, Daubuisson, Humboldt, Von Buch, Brocchi, and a long train of other observers who have illustrated this science in France, Germany, and Italy: the Anglo American States have also produced a very able geological observer, Mr. Maclure.

The necessity of these and other omissions we the less regret, as it is known that the whole subject is at present employing the pen of a writer who has shewn his competency for the task by an able article connected with it in the Edinburgh Review for 1818, to which (as will have been perceived) the present sketch is largely indebted.

And here we cannot conclude this rapid sketch of the general bearings of geological science, without some allusion (imperfect as from our limits it must necessarily be) to those highest interests which the eager attacks of an half-informed scepticism, and sometimes also the injudicious defences of those, whose sincerity of intention ill supplied the want of a precise acquaintance with the phœnomena under consideration, have seemed to involve in the discussions of this branch of physics. With respect to the former class, the characteristic to which we have just alluded, their impatience, namely, to avail themselves of the immature results of an imperfect knowledge, opposed as it is in every respect to that persevering and reflective spirit of enquiry which marks genuine philosophy, and can alone lead to the ultimate discovery of truth, must create a reasonable suspicion of their opinions; for no sooner has any new discovery, whatever might have been its subject, occurred, (whether it was a fragment of Indian chronology, or an Egyptian zodiac, or the mechanism of the universe, or that of living bodies, or lastly some new fact relating to the structure of the earth) than the first aspect under which some minds have seemed anxious to view it has been, whether it would not furnish some new weapon against Revelation. Whether such a mode of proceeding was more likely to arise from a genuine desire to remove prejudice and bigotry, or rather was itself the fruit of a prejudiced and bigotted eagerness to propagate peculiar opinions, we do not feel called upon to decide. Its result is matter of history, and it would perhaps be instructive to form a collection of the attacks which have been made on all these subjects, and the theories which have been broached concerning them, under such views: it would form a curious monument of the aberrations of reason, and be quite as humiliating to its pride as the records of the most unenlightened enthusiasm; no other single principle could perhaps be pointed out, by which so much crude and absurd speculation would be brought together; in whatever degree the physico-theological reveries of the Hutchinsonian school deserved this character, they have been infinitely surpassed in every respect.* All this, however, it may be justly said, leaves the real merits of the question untouched; but it is necessary, nevertheless, to dissipate the illusions which often hover over it, and to assist us in forming a fair estimate how far such attacks are to be attributed to pure reason, and how far to prejudice and passion.

But let us proceed. Before we examine the bearings of physical science on Revelation, our ideas should first be settled as to what may be reasonably expected from Revelation in this respect. Both its opponents and some of its defendants often argue as if it should have included the discovery of a system of physical truth; which it would not be difficult to shew, gives an entirely erroneous view of its professed object; to treat, namely, of the history of man only, and that even but as far as affects his relations to his Creator, and the dealings of Divine Providence in regard to him. To have made physical truth generally the subject of Revelation, would have been to destroy its great use, namely, that its investigation might form at

* See for the proofs of this not merely the earlier theories on many of these subjects of Buffon and Bailly, and the notions of Voltaire, but more especially in the present day the Hydrogeologie of Lamarck (before which old Burnet and Whiston fade into sober reason) and all the metaphysical speculations which load the first volume of the same author's otherwise valuable 'Système des Animaux sans Vertebres.' De credulitate infidelium would be the proper title for such a collection; on the whole the cavils of many an objector might be effectually silenced, by moving the previous question of his real acquaintance with that science from whence he professed to draw them.

once the most delightful resource and the most invigorating exercise of those powers of reason bestowed on us as our distinguishing prerogative. The remark of the Poet becomes still more strikingly just when applied to intellectual cultivation.

> ——— ——— Pater ipse colendi
> haud facilem esse viam voluit ————
> ———————— curis *acuens* mortalia corda
> nec torpere gravi passus sua regna veterno.

And who would willingly exchange the play of mind which the task of invention affords, for a system which should leave no province to reason beyond learning that which was already plainly recorded? did it not carry us beyond our limits, it would be easy to shew how much of the interest of science arises from the former cause, and that, not only in the minds of the few who are gifted to achieve original discoveries, but of the many who in following their steps in some degree, 'pursue the triumph and partake the gale'. How little comparative curiosity should we feel concerning the course of the Niger, or the northern coast of America, could they be as easily examined as the Thames and the Channel.

The general connection of physical science will therefore be rather with natural than revealed religion; for in the former the great problem is, to trace the Author of Nature in his works, and our interest in the evidences thus furnished is materially (as we have seen) kept alive by their being made the matter of gradual and successive discovery; so that the mind is continually presented with *fresh* proofs, extending as its general knowledge extends. Thus this connection is essential; but that with Revelation is incidental only, and confined to such single facts as happen to be mentioned in relation to the providential history of man, its great object: difficulties arising from these, its advocates are of course bound fairly to meet, and this must be required in geology as in other cases; but before we attempt this, we will, for the reasons above stated, cast a hasty glance at the relations of that science with Natural Religion.

This important subject has been very fully and ably handled in the inaugural lecture published by Professor Buckland, and we feel persuaded that we cannot pursue a more satisfactory course than by presenting some short extracts from that work to the reader, desiring at the same time to refer him to the original for a fuller view of the whole argument.

"In being introduced then to a new kingdom of nature, we can scarce fail to inquire, whether we shall here also find the same proofs of subserviency to final causes, which are so strikingly exhibited in the animal and vegetable creation. And the answer will be found in the affirmative. Such proofs, though, from the nature of the subject, less obvious than in the two former instances, are nevertheless plainly discernible and capable of demonstration. To enter at large into those proofs would require more ample space than can now be devoted to it, and presupposes a knowledge of the subject of which we are but beginning to treat; but some few may be briefly alluded to.

"A great majority of the strata having been formed under water, and from materials evidently in such a state as to subject their arrangement to the operation of the laws of gravitation; had no disturbing forces interposed, they must have formed layers almost regularly horizontal, and therefore investing in concentric coats the nucleus of the earth. But the actual position of these beds is generally more or less inclined to the horizontal plane, though often under an angle almost imperceptible. By this arrangement many strata affording numerous varieties of mineral productions are made to emerge in succession on the surface of the earth; whereas the inferior must have been buried for ever beneath the highest, had their position been strictly horizontal; and in such case we should have wanted that variety of useful minerals almost indispensable to the existence of man in a state of civil society, which this succession of different strata now presents to us.

"In the whole machinery also of springs and rivers, and the apparatus that is kept in action for their duration, through the instrumentality of a system of curiously constructed hills and valleys, receiving their supply *occasionally* from the rains of heaven, and treasuring up in their everlasting storehouses to be dispensed *perpetually* by thousands of never-failing fountains; we see a provision not less striking or less important. So also in the adjustment of the relative quantities of sea and land in such due proportions as to supply the earth by constant evaporation, without diminishing the waters of the ocean; and in the appointment of the atmosphere to be the vehicle of this wonderful and unceasing circulation; in thus separating these waters from their native salt, (which, though of the highest utility to preserve the purity of the sea, renders them unfit for the support of terrestrial animals or vegetables,) and transmitting them in genial showers to scatter fertility over the earth, and maintain the never-failing reservoirs of those springs and rivers, by which it is again returned to mix with its parent ocean: in all these we find such undeniable proofs of a nicely balanced adaptation of means to ends, of wise foresight and benevolent intention and infinite power, that he must be blind indeed, who refuses to recognize in them proofs of the most exalted attributes of the Creator.

"Another valuable contrivance in the structure of the globe is, that nearly all its materials are such as to afford by their decomposition a soil fit for the support of vegetable life; and that they are calculated to undergo and have undergone a superficial decomposition. Here is an instance of relation between the vegetable and mineral kingdoms, and of the adaptation of one to the other, which always implies design in the surest manner: for had not the surface of the earth been thus prepared for their reception, where would have been the use of all that admirable system of organization bestowed upon vegetables? And it is no small proof of design in the arrangement of the materials that compose the surface of our earth, that whereas the primitive and granitic rocks are least calculated to afford a fertile soil, they are for the most part made to constitute the mountain districts of the world, which, from their elevation and irregularities, would otherwise be but ill adapted for human habitation; whilst the lower and more temperate regions are usually composed of derivative or secondary strata, in which the compound nature of their ingredients qualifies them to be of the greatest utility to mankind by their subserviency to the purposes of luxuriant vegetation.

"Thus Geology contributes proofs to Natural Theology strictly in harmony with those derived from other branches of natural history; and if it be allowed, on the one hand, that these proofs are in this science less numerous and obvious, it may be contended, on the other, that they are calculated to lead us a step farther in our inferences. The evidences afforded by the sister sciences exhibit indeed the most admirable proofs of design and intelligence originally exerted at the Creation: but many who admit these proofs still doubt the continued superintendance of that intelligence, maintaining that the system of the Universe is carried on by the force of the laws originally impressed on matter, without the necessity of fresh interference or continued supervision on the part of the Creator. Such an opinion is indeed founded only on a verbal fallacy; for "laws impressed on matter" is an expression, which can only denote the continued exertion of the will of the Lawgiver, the prime Agent, the first Mover: still however the opinion has been entertained, and perhaps it nowhere meets with a more direct and palpable refutation, than is afforded by the subserviency of the present structure of the earth's surface to final causes; for that structure is evidently the result of many and violent convulsions subsequent to its original formation. When therefore we perceive that the secondary causes producing these convulsions have operated at successive periods, not blindly and at random, but with a direction to beneficial ends, we see at once the proofs of an overruling Intelligence continuing to superintend, direct, modify, and control the operations of the agents, which he originally ordained.*

* "Examples of this kind are perhaps nowhere more strikingly afforded than in the instance of those fractures or disturbances called *faults*, which occur in the alternating beds of coal, slaty clay, and sandstone, which are usually associated under the name of Coal-measures.

"The occurrence of such *faults*, and the *inclined position* in which the strata composing the coal-measures are usually laid out, are facts of the highest importance as connected with the accessibility of their mineral contents. From their *inclined position* the thin strata of coal are worked with greater facility than if they had been horizontal; but as this inclination has a tendency to plunge their lower extremities to a depth that would be inaccessible, a series of faults, or traps, is interposed, by which the component portions of the same formation are arranged in a series of successive tables, or stages, rising one behind another, and elevated continually upwards towards the surface from their lowest points of depression. A similar effect is often produced by undulations of the strata, which give the united advantage of inclined posision and of keeping them near the surface. The basin-shaped structure, which so frequently occurs in coal-fields, has a similar tendency to produce the same beneficial effect.

"But a still more important benefit results from the occurrence of *faults* or *fractures*, without which the contents of no deep coal-mine would be accessible. Had the strata of shale and grit-stone that alternate with the beds of coal been continuously united without fracture, the quantity of water that would have penetrated from the surrounding country into any considerable excavations that might have been made in the porous grit beds, would have been insuperable by the powers of the most improved machinery: whereas by the simple arrangement of a system of faults, the water is admitted only in such quantities as are within control. Thus the component strata of a coal-field are divided into numberless insulated masses, or sheets of rock of irregular form and area, not one of which is continuous in the same plane over any very large district, but each is separated from its next adjacent mass, or sheet, by a dam of clay impenetrable to water, and filling the narrow cavity produced by the fracture which caused the fault.

"If we suppose a thick sheet of ice to be broken into fragments of irregular area, and these fragments again united after receiving a slight degree of irregular inclination to the plane of the original sheet, the united fragments of ice will represent the appearance of the component portions of the broken masses, or sheets, of coal-measures we are describing, whilst those intervening portions of more recent ice by which they are held together, represent the clay and rubbish that fill the faults, and form the partition walls that insulate these adjacent portions of strata, which were originally formed like the sheet of ice in one continuous plane. Thus each sheet or inclined table of coal-measures is inclosed by a system of more or less vertical walls of broken clay, derivative from its argillaceous shale beds at the moment in which the fracture and dislocation took place: and hence have resulted those joints and separations, which, though they occasionally interrupt at inconvenient positions, and cut off suddenly the progress of the collier, and often shatter those portions of the strata that are

" The consideration also of the evidences afforded by Geological phœnomena may enable us to lay more securely the very foundations of Natural Theology, inasmuch as they clearly point out to us a period antecedent to the habitable state of the earth, and consequently antecedent to the existence of its in-

in immediate contact with them, yet are in the main his greatest safeguard, and indeed essential to his operations.

" The same faults also, whilst they prevent the water from flowing in excessive quantities in situations where it would be letrimental, are at the same time of the greatest service in converting it to purposes of utility, by creating on the surface a series of springs along the line of fault, which often give notice of the fracture that has taken place beneath.

" A similar interruption of continuity in the masses of the primitive rocks, and rocks of intermediate age between these and the coal-formation, is found to occur extensively in the working of metallic veins. The vein is often cut off suddenly by a fault or fracture crossing it transversely, and its once continuous portions are thrown to a considerable distance from each other. The line of fracture is usually marked by a wall of clay consisting of the abraded fragments of the rock, whose adjacent portions have been thus dislocated. Such faults are universally known in the mines of Cornwall by the term *flukan*, and they produce a similar advantage to those that traverse the coal-measures in guarding the miner from inundation, by a series of natural dams traversing the rocks in various directions, and intercepting all communication between that mass in which he is conducting his operations, and the adjacent masses on the other side of the flukan or dam.

" It is probable that the greater number of springs, that issue from those rocks which are unstratified, are kept in action through the instrumentality of the faults by which they are intersected.

" It may be added also, that the faults of a coal-field, by interrupting the continuity of the respective beds of coal, and causing their truncated edges to abut against those of uninflammable strata of shale or grit, afford a preservative which prevents the ravages of accidental fire from extending beyond the area of that sheet in which it may take its beginning, but which, without the intervention of such a provision, might lead to the destruction of entire coal-fields.

" It is impossible to contemplate a disposition of things so well accommodated, and indeed so essential to the various uses which the materials of the earth are calculated to afford to the industry of its inhabitants, and even to the supply of some of their first wants, and entirely to attribute such a syetem to the blind operation of fortuitous causes Although it be indeed dangerous hastily to introduce final causes, yet since it is evident that in many branches of physical knowledge, more especially those which relate to all organized matter, the final causes of the subjects with which they are conversant form perhaps that part of them which lies most obviously open to our cognizance, it would surely be as unphilosophical to scruple at the admission of these causes when the general tenor and evidence of the phœnomena naturally suggest them, as it would be to introduce them gratuitously unsupported by such evidence. We may surely therefore feel ourselves authorized to view, in the Geological arrangement above described, a system of wise and benevolent contrivances prospectively subsidiary to the wants and comforts of the future inhabitants of the globe, and extending itself onwards, from its first formation through all the subsequent revolutions and convulsions that have affected the surface of our planet."

habitants. When our minds become thus familiarised with the idea of a beginning and first creation of the beings we see around us, the proofs of design, which the structure of those beings affords, carry with them a more forcible conviction of an intelligent Creator, and the hypothesis of an eternal succession of causes is thus at once removed. We argue thus—it is demonstrable from Geology that there was a period when no organic beings had existence: these organic beings must therefore have had a beginning subsequently to this period; and where is that beginning to be found, but in the will and *fiat* of an intelligent and all-wise Creator?

"With what acuteness of argument, and what obstinacy of perseverance, the extraordinary notion of an eternal succession was maintained in ancient times, even by some of the greatest philosophers, it is quite unnecessary here to state: and if some writers on Geology in later times have professed to see in the earth nothing but the marks of an infinite series of revolutions, without the traces of a beginning; it will be quite sufficient to answer, that such views are confined to those writers who have presumed to compose theories of the earth, in the infancy of the science, before a sufficient number of facts had been collected; and that, if possible, they are still more at variance with the conclusions of Geology, (as a science founded on observation,) than they are with those of Theology."

We have seen then that the evidences of Natural Religion are still further confirmed by the discoveries of Geology, as indeed could not fail to be the case; for every effort that has carried forward the land-marks of human knowledge, has at the same time disclosed to our view a widening range of this proof, and such is its cumulative nature that it regularly grows with the growth and strengthens with the strength of true science. Let us next enquire in what manner the observations of Geology bear upon the few physical facts recorded in the writings which we receive as inspired.

Two only points can be in any manner implicated in the discussions of Geology.

I. The Noachian Deluge.
II. The Antiquity of the Earth.

With regard to the first of these points, Geology, far from affording the slightest ground to question the truth of the Mosaic record, brings to its support (if that which rests securely on its more appropriate ground—a solid and immoveable foundation of moral evidence—can be said to require or receive much additional support from physical arguments) a strong collateral testimony.

Without this auxiliary evidence it might have been, and indeed often has been, objected to the fact of an universal deluge, that such a convulsion involved supposed physical impossibilities; but no one can have attentively considered the monuments of the great changes with which Geology makes us acquainted,* without at once perceiving that they prove the existence of an order of things, in which such convulsions not only might, but actually did, take place. Let us again quote the words of an author who has himself examined with the fullest precision the important phœnomena to which we allude.

* The geological appearances in which we are entitled to look for the traces of a catastrophe violent and transitory, are obviously not those presented by the original formation of the strata constituting the Earth's surface, but those connected with the accidents they have subsequently experienced, their partial destruction, the erosion and excavation of their surface, and the dispersion of fragments torn from them, under the form of water-worn pebbles, over the general face of the continents. In these phœnomena, and the remains of terrestrial animals buried beneath these debris, the genuine geological evidence of this great convulsion resides, and not in occurrence of those marine remains which form constituent parts of all the vast series of secondary strata; for the agency of the deluge could not have been to form these immense deposits of which the greater part of the Earth's crust, as far as it is known, consists. On the contrary, it was evidently as far as it went a destroying agency, although limited in its effects. It was natural indeed that the earlier observers, while the phœnomena of the distribution of these marine remains, and the depth of the masses formed by them, remained unknown, should refer them to this cause; but these points being ascertained, it is obvious that their hypothesis became untenable, for

1st. Had these remains been brought to their present situation by diluvial currents, they ought to be mingled confusedly together; we ought to have found the same genera and species in the lowest limestones and the highest beds above the chalk; and those remains of land animals which appear undoubtedly to be diluvial, should have been mixed amongst them; but the fact is notoriously otherwise, the organic remains being distributed in distinct assemblages, in such a manner that each formation is characterized by its peculiar assemblage, without confusion or intermixture. No transitory inundation can account for the circumstances of this distribution; they are such as indicate, beyond the possibility of reasonable doubt, that the animals imbedded in the strata lived and died in the spots where they are now found, while these continued for a long period under the waters of the ocean; and that they were there buried under successive deposits formed beneath those waters during the progress of many ages. The perfect state of many of the most fragile shells also proves that they could not have been drifted from a distance by any violent convulsion.

2dly. There is every reason, as we have seen, to ascribe the gravel debris derived from the partial destruction of the strata to the action of the deluge; but the strata must evidently not only have been formed, but also consolidated, before solid fragments, such as could have assumed the present form of the gravel pebbles, could have been torn off them. Now it does not seem within the limits of physical possibility to ascribe the formation of

"The grand fact of *an universal deluge* at no very remote period is proved on grounds so decisive and incontrovertible, that, had we never heard of such an event from Scripture, or any other authority, Geology of itself must have called in the assistance of some such catastrophe, to explain the phœnomena of diluvian action which are universally presented to us, and which are unintelligible without recourse to a deluge exerting its ravages at a period not more ancient than that announced in the Book of Genesis.

"It is highly satisfactory to find the following strong statement on this subject, published by one who deservedly ranks in the very first class of natural observers, and in the very centre of continental philosophy. 'It may be seen,' says Cuvier, 'that nature every where distinctly informs us that the commencement of *the present order of things cannot be dated at a very remote period;* and it is remarkable that mankind every where speak the same language with nature.' And in another place he adds, 'I am of opinion with M. Deluc and M. Dolomieu, that if there is any circumstance thoroughly established in Geology, it is that the crust of our globe has been subjected to a great and sudden revolution, the epoch of which cannot be dated much farther back than five or six thousand years ago; and that this revolution had buried all the countries *which were before inhabited by men and by the other animals that are now best known.*' Theory of the Earth, § 34."

these strata and their consolidation (a process which must have evidently required time) to one and the same *transient* convulsion with their subsequent partial destruction: this argument becomes stronger when we remember that there are interposed among the strata themselves many beds of similar gravel (for instance beds consisting of rounded fragments of carboniferous limestone associated with the more recent deposits of the second red sandstone), the unavoidable inference is, that the rock whence these pebbles were formed must in every instance have been consolidated before the rock containing them was deposited; yet in the instance before us the deposition of the conglomerate rock must have preceded that of the highest strata, by the whole interval necessary to account for the formation of all the constituent beds of the oolite, sand, and chalk series; and all these again must have been consolidated before they were exposed to the action of the deluge. It matters not whether the time assigned to these effects be comparatively long or short; it seems manifest that a single year must have been totally inadequate.

Deeply convinced how injurious to the real evidence of the deluge it must prove to mingle with it phœnomena which cannot, without violating every rule of physical reasoning, be ascribed to that convulsion; we have been the more particular in urging these considerations, and again most earnestly deprecate the injudicious interference of advocates, the sincerity of whose intentions cannot compensate for the want of full information concerning the real state and essential conditions of the problem whose solution they undertake.

The second point in which the facts delivered in the sacred record are brought into contact with the deductions of Geology is, the Autiquity of the Earth. It has been objected to the authority of that record that it does not allow a sufficient period for the successive deposition of the secondary strata, containing as they do the remains of successive races of animals, which appear to have lived and died where they are now found, while the deposits in which they are buried were gradually accumulating.

Before proceeding to consider the force of this objection, we are first desirous again to recal to our readers, that the great subject to which revelation relates, is the Providential history of man: the antiquity of the human race is therefore an essential feature of that revelation; but the questions whether any other state of our planet preceded that in which it became the habitation of intellectual and moral agents, and if so, what convulsions may have happened to it during that state, are points with which it has no direct connection; a perfect knowledge of these could have furnished no topics calculated either to awaken the slumbering, or to reassure the penitent, conscience.

Now with regard to the antiquity of the human race, the conclusions deducible from geological reasoning appear strictly in accordance with the declarations of Revelation, no human remains having yet been found excepting in beds of undoubtedly very little antiquity.*

With regard to the time requisite for the formation of the secondary strata, we have the choice of the following hypothesis.

1st. If we adhere to the common interpretation of the periods of creation as having been literally days of twenty-four hours, and refuse to admit the existence of another order of things previous to that recorded by the inspired writer, we might still perhaps find a sufficient space of time for the purposes required in the interval between the creation as thus limited, and the deluge. Upon this hypothesis we must suppose the present continents (in the greater part of their extent) to have been included in the channel of the primitive ocean, and to have gradually emerged thence during this period, becoming occupied, as they appeared, by the land animals whose remains we find among the diluvial gravel; the primitive continents may upon this supposition either have been limited por-

* It may be enquired why we have not met with human remains in the diluvial gravel: to this it may be answered that there appears little reason to conclude from the sacred narrative that the antidiluvian population had become numerous, and that it appears to have been concentrated in countries which have not as yet received examination.

tions of the present (such as present no secondary rocks), for at first it seems evident that a limited space only would be requisite; or if more extensive, they may have been submerged in whole or in part,* during those great convulsions which accompanied the deluge.

Or secondly, We may perhaps without real violence to the inspired writer, regard the periods of the creation recorded by Moses and expressed under the term of days, not to have designated ordinary days of twenty-four hours, but periods of definite but considerable length; such a mode of extending the signification of this term being not unexampled in other parts of the sacred writings. Those who embrace this opinion will of course assign the formation of the secondary strata, in great part at least to these *Days of creation*; and we have the authority of several divines in favor of such an interpretation.

Or thirdly, It does not seem inconsistent with the authority of the sacred historian to suppose that after recording in the first sentence of Genesis the fundamental fact of the original formation of all things by the will of an intelligent Creator, he may pass, sub silentio, some intermediate state whose ruins formed the chaotic mass he proceeds to describe, and out of which, according to his farther narrative, the present order of our portion of the universe was educed; upon this supposition the former world whose remains we explore may have belonged to this intermediate æra.

It does not become us to propose hypotheses of such a nature with any feeling of confidence. It is amply sufficient for our purpose to shew that there exists more than one mode by which the appearances presented by the structure of the globe may be satisfactorily reconciled to the facts recorded by Moses, in order to remove the objection which has been drawn from them: other hypotheses tending to the same effect may perhaps present themselves to other minds. They who have experienced the limits which so soon present themselves to all the researches of human philosophy (limits which will ever be the most distinctly recognised by those who are able to cast the most comprehensive survey over its whole field, and who have approached the utmost boundary most nearly) will ever be content to acquiesce in what has well been termed "*a learned ignorance*" on many subjects; while at the same time attributing a due weight to moral evidence on the one hand, and physical evi-

* The notice with regard to the rivers flowing from Eden appear to indicate at least a partial identity between the antidiluvian and postdiluvian continents; but this argument is perhaps not decisive, since the names in question may perhaps (as is common in the appellatives of countries) have been generic terms; the whole context, however, certainly favors the former idea.

dence on the other, they will be far indeed removed from a gloomy and cheerless scepticism.*

* 'The true state of the question respecting the difficulties that arise from the periods of time in which the creation is said to have taken place, has been set forth with much ability and fairness by Mr. Sumner, a divine whose rational and sober piety no person will venture to dispute, and whose admirable work on the Records of Creation, from its originality of sentiment, accuracy of argument, and elegance of writing, ranks amongst the most able productions of the present day.'

"'Any curious information as to the structure of the Earth ought not," he says, "to be expected by any one acquainted with the general character of the Mosaic records. There is nothing in them to gratify the curiosity or repress the researches of mankind, when brought in the progress of cultivation to calculate the motions of the heavenly bodies, or speculate on the formation of the globe. The expressions of Moses are evidently accommodated to the first and familiar notions derived from the sensible appearances of the earth and heavens; and the absurdity of supposing that the literal interpetation of terms in scripture ought to interfere with philosophical inquiry, would have been as generally forgotten as renounced, if the oppressors of Galileo had not found a place in history. The concessions, if they may be so called, of believers in Revelation on this point have been amply remunerated by the sublime discoveries as to the prospective wisdom of the Creator, which have been gradually unfolded by the progressive improvements in astronomical knowledge. We may trust with the same confidence as to any future results from Geology, if this science should ever find its Newton, and break through the various obstacles peculiar to that study, which have hitherto precluded any general solution of its numerous and opposite phenomena.'"

'After following up these general remarks with a more detailed exposition of the harmony which subsists between the facts observable in the structure of the earth, and a fair and liberal interpretation of the Mosaic account of the creation, Mr. Sumner concludes his statement with the following satisfactory result of his investigations.

"'All that I am concerned to establish is the unreasonableness of supposing that Geological discoveries, so far as they have hitherto proceeded, are hostile to the Mosaic account of the creation. No rational naturalist would attempt to describe, either from the brief narration in Genesis or otherwise, the process by which our system was brought from confusion into a regular and habitable state. No rational theologian will direct his hostility against any theory, which, acknowledging the agency of the Creator, only attempts to point out the secondary instruments he has employed It may be safely affirmed, that no Geological theory has yet been proposed, which is not less reconcileable to ascertained facts and conflicting phenomena, than to the Mosaic history.

"'According to that history, we are bound to admit, that only one general destruction or revolution of the globe has taken place since the period of that creation which Moses records, and of which Adam and Eve were the first inhabitants. The certainty of one event of that kind would appear from the discoveries of geologers, even if it were not declared by the sacred historian. *But we are not called upon to deny the possible existence of previous worlds, from the wreck of which our globe was organized, and the ruins of which are now furnishing matter to our curiosity.* The belief of their existence is indeed consistent with rational probability, and somewhat confirmed by the discoveries of Astronomy, as to the plurality of worlds.'" Records of Creation, vol. 2. p. 356.

OUTLINES

OF THE

Geology of England and Wales,

&c.

SYNOPTICAL TABLE.

BOOK I.

SUPERIOR ORDER.

Synonymes. Newest Flœtz or Tertiary Rocks,

Comprising the Formations above the Chalk.

Chapter I. *Preliminary.*

Section I. General view of the highest and most recent de-deposits: *a. alluvial; b. diluvial; c.* regular strata and their division.

Section II. *Strata above the Chalk* generally considered; *a.* nature and extent: *b.* subdivisions; *c.* analagous formations in other countries.

Chapter II. *View of the Upper Marine formation.*

Section I. *Crag; a.* chemical and external characters; *b.* mineral contents; *c.* organic remains; *d.* range and extent; *e.* elevation; *f.* thickness; *g.* inclination; *h.* agricultural character; *i.* phœnomena of water and springs; *k.* miscellaneous remarks.

Section II. *Bagshot Sand, a* to *k.* as above.

Section III. *Isle of Wight, a* to *k.* as above.

Chapter III. *Freshwater formations.*

Section I. General view, with a note on Freshwater shells in other formations.

Section II. *Upper Freshwater formation*; Isle of Wight *a* to *k* as above.

Section III. *Lower Freshwater formation*; Isle of Wight, *a* to *k.* as above.

Chapter IV. *London Clay.*

Section I. Preliminary view.

Section II. *a* to *k* as above.

Chapter V. *Plastic Clay.*

Section I. General view, *a* to *k.* as before.

Section II. Local details; *a.* neighbourhood of Reading; *b.* of London; *c.* Newhaven; *d.* Dorsetshire; *e.* Isle of Wight.

Appendix.. On the formations above the Chalk in Lincolnshire and Yorkshire.

Organic remains of the beds above the Chalk.

This list is inserted on a separate sheet instead of being incorporated in the text, because at the time this part of the work passed through the press it was not intended to enter so much into detail upon this branch of the subject as was afterwards judged expedient; by this addition the work will now be found to contain full catalogues of the Organic remains contained in the various formations, so far as it was possible to refer to figures: unfigured species have seldom been noticed, because a bare name could give little information.

These lists include all the species figured by Mr. Sowerby in the three first volumes of his Mineral Conchology: in the second volume it is intended to give additional lists embracing the figures which may be published during the interval. The present list was chiefly drawn up by Mr. Miller.

CRAG.

ZOOPHYTES.

Numerous *Retepores*, *Flustræ*, and *Escharæ*. Aldborough.

UNIVALVES.

Tubular Univalves; *Annelidæ* of Lamarck.

Dentalium costatum. S. lxx. f. 8. Holywell.

Clypeaceous Univalves.

Patella equalis. S. cxxxix. f. 2.
unguis. cxxxix. f. 7.
Emarginula crassa. S. xxxiii. Ipswich.
reticulata. S. xxxiii. ibid.
Calyptræa Sinensis. Park. Org. Rem. P. v. f. 3.

Spiral Univalves.

Buccinum elongatum. cx. f. 1. Essex.
rugosum. cx. f. 3. Holywell.
reticosum. cx. f. 2. ibid.
granulatum. cx. f. 4. Ipswich.
Murex striatus. S. t. xxii. & cix. Essex, &c.
contrarius. S. t. xxiii. ditto
corneus. S. t. xxxv. Aldborough, &c.
trilineatus. S. t. xxxv. f. 4 & 5.
S. cxcix. f. 1 & 2.
rugosus, Park. Org. Rem. t. v. f. 16.
costellifer. S. cxcix. f. 3.
echinatus. S. cxcix. f. 4.
Cassis bicatenata. cli. Bawdsey.
Natica depressa. v. Woodbridge.
Turbo rudis. S. t. lxxi. f. 2. Aldborough.
littoreus. S. lxxi. f. 1. Bramerton.
Scalaria similis. S. xvi. Bramerton.
Turritella incrassata. S. li. f. 6. Holywell.
conoidea. li. f. 5. ibid.
Vivipara suboperta. S. xxxi. ibid.

CRAG.

Voluta Lamberti. S. cxxix. Park. P. v. f. 10.
Trochus similis. clxxxi. f. 2. Holywell.
lævigatus. clxxxi. f. 1. ibid.
Infundibulum rectum. S. xcvii. f. 3. Holywell.

BIVALVES.

Mya lata. S. lxxxi. Suffolk.
Unio crasciusculus. clxxxv.
Listeri? cliv. f. 1. Suffolk.
Lingula ovalis. xix. f. 4. Pakefield.
Mytilus antiquorum. cclxxv. f. 1 to 3.
alæformis. cclxxv. f. 4.
Cardium augustatum. cclxxxiii. f. 2.
edulinum. cclxxxiii. f. 3.
Parkinsoni. S. xxxxix. Holywell.
Venericardia senilis. cclviii.
Venus rustica. cxcvi. Suffolk.
lentiformis. cciii. Suffolk.
turgida. cclvi.
gibbosa. clv. f. 3. Suffolk.
æqualis. xxi. Holywell.
Astarte obliquata. clxxix. f. 3.
Mactra arcuata. clx. f. 1 & 6. Holywell.
dubia. clx. f. 2 to 4. Ipswich.
ovalis. clx. f. 5. Suffolk.
cuneata. clx. f. 7. Bram.
Tellina obliqua. clxi. f. 1.
ovata. clxi. f. 2.
obtusa. clxxix.
Nucula Cobboldica. clxxx. f. 2.
lanceolata. clxxx. f. 1.
similis. cxcii. f. 1 & 2.

MULTIVALVES.

Balanus crassus. S. lxxxiv. f. 2. Holywell.
tesselatus. lxxxiv. f. 1.
Pholas cylindricus. cxcviii.

BAGSHOT HEATH SAND.

We have not as yet any figures or regular description of the shells of this formation.

FRESHWATER FORMATIONS.

UNIVALVES.

Lymnæus fusiformis. clxix. f. 2. 3. Cowes.
. . . . clxix. f. 1. ibid.
Planorbis cylindricus. cxl. f. 2. ibid.
euomphalus. cxl. f. 7. ibid.
lens. cxl. f. 4. ibid.
Cyclostoma obtusus. cxl. f. 3. ibid.
Helix globosus. clxx. Shalcomb?
Potamides.
Paludina.
Melanopsis.
Phasianella? angulosa. clxxv. f. 2. Shalcomb.
minuta. clxxv. f. 3. ibid.
orbicularis. clxxv. f. 1. ibid.

BIVALVES.

Unio.

We have not felt ourselves able in this list to distinguish the first and second freshwater formations, not being certain how far the species are common or peculiar; we are happy to understand that Mr. Webster is engaged in a work on this subject which will supply the requisite information.

LONDON CLAY.

ZOOPHYTES.

Several *Escharæ* and *Flustræ* as figured in Brander.
Turbinolia sulcata. Lamoroux. Barton.

LONDON CLAY.

SHELLS UNIVALVES.

Tubular Univalves. *Annelidæ* of Lamarck.

Serpula crassa. S. xxx. Highgate.
Dentalium incrassatum. S. lxxix. f. 3 & 4. Richmond.
striatum. S. lxx. f. 4. Hordwell.
entale. S. lxx. f. 3. ibid.
nitens. S. lxx. f. 1 & 2. Highgate.

Spiral Univalves Chambered.

Nummulites. Stubbington.
Nautilus imperialis. 1. u. Richmond.
centralis. 1. b. Highgate.
ziczac. 1. b. ibid.
Ammonites acutus. xvii. f. 1. Minstercliff.

Not Chambered.

Seraphs convolutus, cclxxxvi. *Bulla* sopita. Brand.
Cypræa oviformis. S. iv. f. 1—4. Highgate.
pediculus. Geol. Trans. vol. ii. p. 204.
Conus dormitor. ccci. Bart.
concinnus. cccii. ibid.
scabriculus. ccciii. ibid.
Terebellum fusiforme. cclxxxvii. ibid.
Auricula simulata. clxiii. f. 5—8.
turgida. clxiii. f. 4.
Voluta luctator. S. cxv. f. 1. } Strombus luctal.
spinosa. cxv. f. 3. } Brand.
ambigua. cxv. f. 5. *V.* ambigua, Brand.
costata. ccxc. Bart.
magorum. ccxc. Bart.
Oliva Branderi. cclxxxviii. *V.* Ispidula.
Salisburiana. cclxxxviii.
Ancilla aveniformis. S. xcix. *Voluta* anglica?
Lin. Trans. vol. vii. f. 1. Barton
turritella xcix. ibid.

LONDON CLAY.

Buccinum defossum. Lin. Trans. vol. vii. f. 2.
Melania costata. ccxli. f. 2. Bart.
sulcata. xxxix. m. Stubbington.
Cassis carinata. S. t. vi. Highgate. Buc. Nodosum? Brand.
striata. St. t. vi. ibid.
Murex interruptus. S. ccciv. Lin. Trans. v. 7. f. 5. Barton.
scopulorum. Lin. Trans. f. 6. Barton.
porrectus. .... f. 7. ibid.
punctatus. .... f. 3. ibid.
nitidus. .... f. 8. ibid.
asper. .... f. 9. ibid.
Bartonensis. S. xxxiv.
coniferus. S. clxxxvii. 1. Highgate.
regularis. clxxxvii. 2. Bart.
carinella. clxxxvii. 3. 4. d°
fistulosus. clxxxix. 1 &2. d° } M. pungens.
tubifer. clxxxix. 3 to 8. d° } Brand.
curtus. cxcix. f. 5. Highgate.
gradatus. cxcix. f. 6. Plump.
tuberosus. ccxxix. f. 1. Highgate.
minax. ccxxix. f. 2. ibid.
cristatus ccxxx. f. 1 & 2. ibid.
coronatus. ccxxx. f. 3. ibid.
trilineatus. xxxv. f. 45. ibid.
Pleurotoma attenuata. S. cxlvi. f. 1. Stubbington.
exorta. cxlvi. 2. Bart. *Murex* exorta. Brand.
rostrata. cxlvi. f. 3. d° *M.* rostratus. d°
acuminata. cxlvi. f. 4. Highgate.
comma. cxlvi. f. 5. Stubbington.
semicolon. cxlvi. f. 6. ibid.
colon. cxlvi. f. 7. Bart.
Fusus acuminatus. cclxxiv. f. 1 2 & 3. *Murex* porrectus. Bart.
asper. cclxxiv. f. 4 to 7. ibid.
rugosus. cclxxiv. f. 8 to 9. ibid.
bulbiformis. ccxci. { *Murex* Bulbus. Brand. / Pyrus. ibid. }
bifasciatus. ccxxviii. Highgate.
longævus. S. t. lxiii. *Murex* longævus. Brand.
Rostellaria lucida. S. xci. f. 1 to 3. High.
rimosa. S. xci. f. 4 to 6. Bart.

LONDON CLAY.

Murex rimosus. Brand.
Rostellaria macroptera. S. ccxcviii. Bart.
Strombus amplus. xxx. Barton.
Cerithium cornucopiæ. clxxxviii. f. 1 3 & 4. Stubb.
giganteum. clxxxviii f. 2. ibid.
pyramidale. cxxvii. f. 1. Bart.
geminatum. cxxvii. f. 2. ibid.
funatum. cxxviii. ibid.
dubium. cxlvii. f. 5. Stubb.
Infundibulum obliquum. S. xcvii. f. 1. Bart.
tuberculatum. xcvii. f. 4 & 5. *Trochus* apertus. Brand.
spinulosum. xcvii. f. 6.
Trochus agglutinans. xcviii. *T.* umbelicatus. Brand.
Bennettiæ. xcviii. Bart.
extensus. cclxxviii. f. 2 & 3. ibid.
Solarium patulum. S. xxxv. Highgate.
discoideum. S. xi. Barton.
cuniculatum. Geo. Trans. vol. ii. p. 204.
Turritella elongata. S. t. li. f. 2. Barton. *T.* vagans? Brand.
edita. S. t. l. f. 7. Barton. *T.* editus. Brand.
brevis. S. t. li. f. 3. Barton.
conoidea. S. li. f. 1. 4. & 5. Barton.
Turbo sulcatus. Lin. Trans. vol. vii. f. 9. Barton.
Scalaria semicostata. S. t. xvi. Barton.
acuta. S. t. xvi. ditto.
Ampullaria acuta. cclxxxiv. } *Helix* mutabilis.
patula. ccixxxiv. } Brand.
sigaretina. cclxxxiv.
Vivipara lenta. S. t. xxxi. f. 3. Barton. Helix lentus. Brand.
concinna. S. xxxi. f. 4 & 5. Barton.
Natica glaucinoides. S. v. Highgate.
similis. S. v. ibid.
hantoniensis. Lin. Trans. vol. vii. f. 10.

Bivalves.

Ostrea gigantea. S. t. lxiv. Bart.
flabellula. S. ccliii. *Chama* plicata. Brand.
.... decussata. S. xxvii f. 1. Highgate.
costata. S. xxvii. f. 2. Bart.
.... nitens. S. xiv. Highgate.
.... semigranulata. S. cxliv. Barton.

LONDON CLAY.

Pecten carneus. cciv. Stubbington.
Pectunculus costatus. xxvii. f. 2. Hordwell.
decussatus. xxvii. f. 1. Highgate.
Avicula media. S. t. ii. Highgate.
Modiola elegans. S. t. ix. Bognor and Highgate.
depressa. S. viii. Highgate.
subcarinata. ccx. f. 1. ibid.
Nucula similis. cxcii. f. 3 4 & 10. Barton.
trigona. cxcii. f. 5. ibid.
minima. cxcii. f. 8 & 9. ibid.
Arca Branderi. cclxxvi. f. 1 & 2. Barton.
appendiculata. cclxxvi. f. 3. ibid.
Chama lamellosa. Geol. Trans. vol. ii. p. 204.
Mya subangulata. S. t. lxxvi. f. 3. Barton.
intermedia. S. lxxvi. Bognor.
Lingula tenuis. xxix. f. 3. ibid.
Solen affinis. S. t. iii. Highgate.
Cardium semigranulatum. cxliv. Barton.
nitens. xiv. Highgate.
Cardita margaritacea. ccxcvii. f. 1. 2. 3. Richmond.
Isocardia sulcata. ccxcv. f. 4. Islington.
Venericardia planicostata. Sowerby. t. l. Brukelsambay.
carinata. f. 2. Stubbington.
globosa. cclxxxix. *Chama* sulcata. Brand.
oblonga. ccxxxix. Bart.
deltoidea.
Venus incrassata. clv. f. 1. 2. Brockenhurst.
Corbula globosa. ccix. f. 3. Highgate.
pisum. ccix. f. u. Hordwell.
revoluta. ccix. f. 8 to 13. ibid.

Multivalves.

Teredo antenautæ. cii. Highgate.

A slight notice of the vegetable remains of this formation will be found in the text, to which we also refer for the Crustaceous and Vertebrated animals. To the latter we have to add that the remains of a Crocodile very nearly approaching to the characters of existing species, and especially to the Crocodile *a museau aigu*, has recently been discovered in the London clay at Islington.

PLASTIC CLAY.

Univalves.

Infundibulum echinatum. S. xcvii. fig. 2. Plumstead.
Murex latus. S. xxxi. ibid.
gradatus. cxcix. fig. 6. ibid.
rugosus. cxcix. fig. 1. 2. ibid.
Cerithium funiculatum. cxlvii. f. 1. 2. ibid.
intermedium. cxlvii. fig. 2. 4. Charlton.
melanioides. cxlvii. fig. 6. 7. ibid, Isle of Wight, &c.
Turritella.
Planorbis hemistoma. cxl. fig. 6. Plumstead.

Bivalves.

Ostrea pulcra. cclxxix. Bromley.
tener. cclii. fig. 2. 3.
Pectunculus Plumstediensis. xxvii. f. 3.
Cardium Plumstedianum. xiv.
Mya plana. lxxvi. fig. 2.
Cytherea.
Cyclas cuneiformis. clxii. fig. 2. 3. Plumstead.
deperdita. clxii. fig. 1. ibid.
obovata. clxii. fig. 4 to 6. ibid.

The sand at the bottom of this formation rarely contains shells, but at Feversham in Kent, silicified casts of *Cucullæa* decussata, S. ccvi. and *Natica* canrena have been observed.

The material originally positioned here is too large for reproduction in this reissue. A PDF can be downloaded from the web address given on page iv of this book, by clicking on 'Resources Available'.

Book I. SUPERIOR ORDER.

Formations above the Chalk.

Chap. I. Preliminary.*

Section I. A. *General view of the highest and most recent deposits.*

(*a*) *Alluvial.* In delivering, according to the plan proposed in the introduction, an account of the Geological situation of England, and tracing the disposition of the materials which constitute its mineral masses, following the order of that disposition in descending from the formations which occupy the highest place in the series and are therefore of the most recent origin, to those which serve as the basis to all the rest and must therefore have been formed at the earliest period; the first distinction which claims our attention, is that between those formations which result from causes still in operation and which actually proceed under our own observation, and those which have proceeded from a former and different order of things. The former, which of course where they occur must occupy a position superior to all other formations, are however of very limited extent; they consist in the accumulation of sand and shingle along the sea coast and particularly in estuaries; in the formation of new lands on the banks of rivers and lakes by the alluvial depositions they carry down, assisted by the growth of aquatic plants, in the growth and increase of tracts of marshland, from the cause last specified; in the accretion of calcareous tufa from the deposits of springs running through limestone rocks and the like. These formations appear to have proceeded uninterruptedly, as at present, from the period when our continents assumed their present form, and the actual system of what may be called geological causes began to ope-

* By the Rev. W. D. Conybeare, F.R.S. &c.

rate; they will of course constitute the first term of our geological division, and may be conveniently designated by the term *Alluvial.*

The products of active volcanos, though of so different an origin, will of course be referable to the same æra, but of these in treating of this country it is unnecessary to speak.

(*b*) *Diluvial.* Next in order to these, we find a mantle as it were of sand and gravel indifferently covering all the solid strata, and evidently derived from some convulsion which has lacerated and partially broken up those strata, inasmuch as its materials are demonstratively fragments of the subjacent rocks, rounded by attrition. The fragmented rocks constituting these gravel deposits are heaped confusedly together, but still in such a manner that the fragments of any particular rock will be found most abundantly in the gravel of those districts, where the parent rock itself appears *in situ* among the strata. In these deposits, and almost in these alone, the remains of numerous land animals are found, many of them belonging to extinct species, and many others no longer indigenous to the countries where their skeletons are thus discovered.

Between these accumulations of fragmented rocks and the vallies traversing the present surface of the earth, there clearly exists a close relation; that, namely, between the breaches which have been opened in the ruined strata, and the materials which have been removed from those breaches. The same causes that have excavated the one, have heaped up the other; and these causes have evidently (as appears from a general examination of the phenomena) acted at once on all the strata, and at a period subsequent to their original formation and consolidation: hence they must be assigned to the last violent and general catastrophe which the earth's surface has undergone, whatever has occurred since being either the quiet action of causes still continuing to operate, or convulsions, violent indeed, but of very limited and local extent.

It has therefore, from the most probable views concerning the nature of this great catastrophe, been proposed to designate these formations, which naturally constitute the second term of our geological series, *Diluvial.*

A strict adherence to the method of treating the formations regularly according to a descending series, would naturally lead us to detail the several facts, I. connected with the alluvial formations, II. connected with the diluvial formations; but many circumstances concur to render it advisable to separate the history of these posterior formations, from that of the regular series of strata which they cover. The history of the diluvial fragments of the pre-existing strata, could scarcely indeed be rendered intelligible until some acquaintance with the parent strata themselves had been acquired; we shall therefore refer the fuller consideration of these subjects to an Appendix; and contenting ourselves for the present with the above remarks, proceed at once to the history of the regular strata.

(*c*) *Regular Strata and their division.* These regular strata consist, as has been observed in the introduction, of various beds of sand, clay, limestone, and other mineral substances; deposited, as is evident from the exuviæ of marine animals contained in nearly all of them (with the exception of the rocks constituting the lowest and earliest group), at the bottom of the ocean; superimposed on one another in regular order, and making their appearance on the surface of the earth, by emerging in succession from beneath one another as the line of that surface cuts the planes of their stratification, which are seldom strictly horizontal, although in the more recent rocks they approach very nearly to such a position; but in the older an approach to a vertical position is very frequent.

In treating of these strata, the immense number of the individual beds might seem at first to defy the powers of enumeration or examination, but we shall soon find that these individual beds naturally form themselves into assemblages of similar strata; e. g. we shall find 50 or 100 beds of chalk alternating with the same number of flint, and thus constituting a single though compound whole. Several even of these compounded assemblages which occupy a neighbouring position will also be found to possess so many points of common analogy with one another, that it will be convenient, by a further composition, to constitute still more general classes for their reception. By this process, the almost infinite extent of the subject becomes reduced within manageable limits; the grand divisions thus obtained form the landmarks that guide us in the enquiry, and simplify and generalise without confounding our conceptions. Assuming these divisions then, it will be the object of the following pages, inverting the above process, to analyse each of them into its most essential elements, that the accuracy of particular, may be added to the simplicity of general, knowledge.

The great range of Chalk hills extending through the island from Yorkshire to Dorsetshire, is so prominent a feature, naturally, as well as geologically, of its surface, that it forms a most convenient line of demarcation; and the relations of the deposits which occur reposing upon the chalk, and occupying the areas circumscribed by it, are such that they naturally form themselves into one of the grand divisions above alluded to; this division will comprise the highest in the English series of regular strata,* and therefore occupy the first place in our arrangement.

Section II. *Strata above the Chalk.*†

(*a*) *Nature and extent.* These consist of various beds of sand, clay, marle, and imperfectly consolidated limestone. They occupy two extensive tracts, each circumscribed by the hills of the chalk formation, excepting where the line of sea coast traverses their areas and conceals their continuation.

Since the chalk formation, dipping on all sides though generally at a scarcely perceptible angle beneath these formations, forms, when viewed on the large scale, concave areas in which they have been deposited, those areas have been denominated basins; but it must not be inferred from this term that the chalky edge of the basin can be traced completely round, since the interference of the line of coast prevents this. The most northerly of these basins has been denominated that of London, from its including the metropolis; the southern that of the Isle of Wight, because the northern portion of that island falls within it.

* No superior or more recent regular formations are known to exist in any part of the earth yet examined, with the exception of some trap rocks probably of volcanic origin.

† The earliest published account which conveys any distinct information of these strata, is a series of papers by Mr. Middleton on the Mineral Strata of Great Britain, inserted in the Monthly Magazine for October 1812 and following months; which, although not written in scientific language, and containing some inaccuracies, yet possesses very considerable merit. He rightly enumerates the beds above the chalk in the following order: 1. Vegetable mould; 2 and 3. Brick earth and shells, sand, and gravel, (the upper marine formation of this volume); 4. London clay; 5. Shells, pebbles, sand, and pipe clay containing wood coal occasionally, and resting on the white sand which covers the chalk (our plastic clay formation). The account given of each of those beds, though short, exhibits fairly all the most striking features and localities, and requires correction in very few instances.

Mr. Webster's interesting discovery however of the alternation of fresh-water formations with those of marine origin, thus establishing the perfect analogy of the French and English series, and the more scientific character of his memoir, have undoubtedly given to his name a just precedence amongst the observers of these formations; and his materials were completed long before the publication of Mr. Middleton's paper. Each indeed appears to have been perfectly unacquainted with the researches of the other.

Subsequently, Mr. Buckland has completed the history of the lower members of these formations; but we still remain without any particular memoir on the upper members as they appear in Suffolk, and round Bagshot; Mr. Warburton however is understood to be engaged in supplying this deficiency.

The boundary of the first of these basins may be stated generally as a line running from the inner edge of the chalk south of Flamborough Head in Yorkshire nearly south till it crosses the Wash, then south-west to the upper part of the valley of the river Kennet near Hungerford in Wiltshire, and thence tending south-east to the north of the Thames and the north-west angle of the Isle of Thanet: in all these directions the bounding line is formed by the chalk hills; on the east side the boundary is the coast of the German ocean.

The boundaries of the Isle of Wight basin may be generally assigned by the following four points: on the north, a few miles south of Winchester; on the south, a little north of Carisbrook in the Isle of Wight; on the east, Brighton; and on the west, Dorchester. It is every where circumscribed by chalk hills, excepting where broken into by the channel between the Isle of Wight and the main land.

(*b*) *Subdivisions.* The nature of the beds occupying these areas, has before been generally stated; we have now to consider them more attentively in the detail; in doing this we shall find the occurrence of a very thick bed, or rather formation of clay, marked by peculiar fossils, near the middle of the series, to afford a very convenient division. This bed has been called the London clay, as forming the substratum of the metropolis. The beds above this are remarkably distinguished by the agreement of the greater part of the shells they contain, with those still existing in the adjacent seas; three-fourths of the fossils exhibiting this agreement, while scarcely one-fourth of those in the London clay, and still fewer in the lower strata, can be referred to such originals. Moreover it is in this upper part of the series in the Isle of Wight, that that most important phenomenon, the alternation of beds containing the shells peculiar to fresh water with those of marine origin, occurs: hence the following subdivisions of the strata above the chalk will naturally result.

A. above the London clay	1. Upper marine formations. 2. Freshwater formations.
B.	3. The London Clay. 4. Plastic clay and sand between the London clay and the chalk.

As a farther proof of the propriety of this division, it deserves attention that in the Isle of Wight, where a great convulsion has elevated the chalk into a vertical position, the London clay has been similarly affected; while the upper strata are placed horizontally over these, and have been undisturbed, which clearly indicates that an interval elapsed between the two formations.

(*c*) *Analagous formations in other countries.* Before proceeding to consider these subdivisions, a few words may be added on the general extent of formations of this æra in Europe.

They may generally be recognised at once from the loose and unconsolidated state of the beds composing them, from the greater abundance and variety of their fossil shells, especially univalves, and from the high state of preservation of those remains; which, being seldom in any degree mineralized, have undergone no further change than the loss of their colouring matter, and perhaps of part of their animal matter and phosphoric acid, and might often be mistaken for recent shells.

The extensive tract called the basin of Paris, exhibits a close conformity to that of the Isle of Wight, as will be hereafter pointed out, both in the fresh-water and marine formations; an exact coincidence being observable in the shells of both districts.

Another similar tract of fresh-water and marine formations occurs in the south-west of France surrounding Agen (department Lot et Garonne); and hence these modern beds stretch to the very foot of the Pyrenees to Pau on the south-west, and Carcassone on the south-east. There is yet another of these recent districts near Aix (Bouches du Rhone), where fish resembling those of Monte Bolca are found.

On the north-east, the beds above the chalk range from near Calais through the Netherlands to Aix la Chapelle; at Cassel, Brussels, Maestricht, and Aix, great deposits of their shells may be seen, but these differ considerably from those of the English series; and the Maestricht beds appear to be older, and approach more nearly to the age of the chalk.

The north of Germany, to some distance from the shores of the Baltic, seems occupied by these formations; they are however generally concealed by an immense accumulation of diluvial debris; but near the lake Swerin in Mecklenburg, deposits of shells are found which must be referred to this class; and the gypsum of Luneburg, close to which a chalk pit is opened, must belong to the same formation with that of the basin of Paris.

Along the course of the Rhine, near its junction with the Maine, fresh-water and marine beds of the same class are found.

In the great valley of Switzerland, and especially near the lake of Constance, similar formations occur, and Professor Buckland has discovered that the Nagleflue of that district, which often forms considerable mountains, is a conglomerate of the same age. These recent formations stretch through the whole subalpine valley to Vienna, and may be traced in Hungary and Moravia in detached localities.

In Italy, formations of the same age constitute the subappenine hills occupying the wide expanse of the valley of the Po, and thence forming a zone skirting the shores of the Adriatic even to the point of Otranto; on the west side of the Appenines they occur less extensively, being very generally covered by the products of the volcanoes which have ravaged that side of Italy; they may however be traced in the Val d'Arno, in Rome itself mingled with volcanic products, and in other places. The description of the Italian formations agrees very closely with that of the basin of London, but there is often considerable difference in the species of shells they contain; perhaps not more than one-third can be considered as common to the two districts.

Sicily, Dalmatia, and parts of Greece appear to exhibit similar beds, and Malta to be composed entirely of them.

In Spain, several tracts of fresh-water formation have been observed, but they seem to be of very local and limited occurrence; the deposit of shells at the foot of the hill of St. Julians near Alicant, described by Mr. Townshend, seems to be clearly of this age.

In the north of Europe, Count Von Buch mentions that at the head of several of the Norwegian Firths surrounded by primitive hills, he found small deposits of clay full of shells resembling the recent, but at the height of more than 100 feet above the actual level of the sea; these seem to be the traces, faint indeed, of the action of the causes which have produced more extensive similar deposits in others quarters, and to be referable to the most recent of these deposits.

In Iceland, a deposit of this nature appears to constitute the solitary exception to the general prevalence of igneous products in the composition of that country; for we find a hill of clay, full of shells of the Venus Islandica, near Husavik, mentioned as the only instance of the occurrence of marine remains in that island: the shells are of a smaller species than those now found in the neighbouring sea.

In North America, the tract extending between the Atlantic and the Alleghany mountains, appears to be composed principally of formations of this character: organic remains from this quarter are preserved in the Woodwardian collection at Cambridge.

A series of specimens has lately been presented to the Geological Society from the plain at the foot of the Himmaleh mountains, which exhibit a close agreement in character with those of the London clay.

In Lower Egypt, analagous formations probably occur, since a coarse marly limestone, full of nummulites, is found near the pyramids.

Although in England these deposits contain no important minerals, a few specimens of selenite, of sulphate of barytes and of resinous substances making up the list, on the continent they are more productive; the beds of gypsum frequently become very important, sulphur is also often found in considerable quantity, amber and other resins probably of vegetable origin, also occur.

CHAPTER II.

General View of the Upper Marine Formation.

In the divisions above assigned, the reasons have been given for making the beds which repose upon the London clay, the first object of our detailed enquiry; the propriety of separating the consideration of those of marine and fresh-water formation is sufficiently obvious, to which it may be added that the latter occur only in a single district.

Strata, distinct in character from the London clay, and reposing upon it, may be traced in two separate tracts within the London basin; first, occupying a considerable district on the east coast of Suffolk, of which it forms the low cliffs, and secondly, forming the substratum of Bagshot heath, and other adjoining tracts of similar character. In the Isle of Wight basin, it has already been observed that the marine strata alternating with those containing fluviatile shells, occupy a similar, i.e. superior, position with regard to the London clay; and sandy tracts, seemingly holding the like place, may be found in other parts of that basin. That the beds occurring in the several districts above mentioned, agree in the relation of their posterior formation to the London clay, is manifest; but that they are strictly identical or contemporaneous, does not hence follow as a certain inference, since they may all have been local and unconnected deposits: it will therefore in the present state of our information be most advisable to dedicate a separate article to the exposition of the geological facts connected with each of the above districts.

1. That of Suffolk, where the deposit resting on the London clay appears as a sand or gravel enclosing shells of peculiar characters; the whole mass having obtained the local appellation of Crag.
2. The sandy beds of Bagshot and the neighbouring heaths.

3. The marine stratum, alternating with those of fresh water in the Isle of Wight, which consists of a shelly marle; the other tracts within the Isle of Wight basin have not been examined, but appear to agree nearly with the sand of Bagshot heath.

These beds, but more especially the Bagshot sand, seem to present near analogies to the sand of the upper marine formation in the basin of Paris. (C.)

Section I. *Crag of Suffolk.*

(*a*) *External characters.* At Walton Naze, a point of land about 16 miles south-east of Colchester in Essex, it constitutes about 30 feet of the upper part of the cliff, the lower 15 feet being of the London clay. It there consists of sand and gravel enclosing shells, and the same characters prevail also beyond the Nase in the projecting cliff of Harwich; but it also includes friable masses of ferruginous sand, somewhat cemented together, and also enclosing shells. The same occurs again on the Suffolk side of the river Stour (G. T. vol. i. p. 327). Crag is a local name for gravel.

(*b*) *Mineral contents.* The only remark under this head appears to be, that the sand and gravel and organic remains enclosed in them, often exhibit tints which bespeak a considerable impregnation of iron.

(*c*) *Organic remains.* Among the fossils which have been enumerated as belonging to these beds, many agree with those in the upper marine formation in the Paris basin. (G. T. vol. ii. p. 218). The shells are found in an excellent state of preservation, and though generally in a confused mixture, are sometimes so disposed that patches of particular genera and species appear, as is the case with the small *pecten*, the *mactræ*, and the *left-turned whelk.* Like fossils of most other strata, this assemblage of shells manifests a peculiar distinctive character. *A few shells only, which may be placed among those which are supposed to be lost, or among those which are the inhabitants of the distant seas, are here discoverable; the greater number not appearing to differ specifically (as far as their altered state will allow of determining,) from the recent shells of neighbouring seas.* Among those of which *no recent analogue* is known, appears to be the *terebratula* figured in Dale's history and antiquities of Harwich, &c. tab. 11, fig. 9, p. 294, and described Phil. Trans. No. 291, p. 1578. This shell is in general about an inch and a half long, thick, nearly oval, roughly striated transversely, and has its large foramen defined by a distinct border. It appears to differ from every known recent or foreign terebratula. Another of the probably lost shells of this stratum is the fossil *oyster*, figured Organ. rem. &c. vol. iii. pl. 14, fig. 3, and which is there considered to be the same oyster as that which is described by Lamarck as *ostrea deformis*: also the *volute*, Org. rem. vol. iii. pl. 5, fig. 13, and the Essex *reversed whelk*, as it has been termed, *murex contrarius* Linn. Hist. Conch. of Lister, tab. 950, fig. 44, b c, is here very abundant; but the fossil shell, with the whirls in the ordinary direction, is sometimes also found in this stratum. Among the *recent shells*, the resemblance of which to the fossil ones of this stratum is such as appears to render a comparison by an experienced conchologist necessary, may be enumerated, *Patella angarica*, *Patella militaris*, *Patella sinensis* (*Calyptrea* Lam.), *Patella fissura* (*Emarginula* Lam.), one or two species of patellæ with a perforation in the apex, *Fissurella* (Lam.), *Nerita glaucina*, *Nerita canrena* (*Natica* Lam.), *Turbo terebra* (*Turritella* Lam.), *Murex corneus*, *Murex erinaceus*, *Strombus pes pelicani*, *Cypræa pediculus* with no sulcus along the back, *Pholas cuspatus*, in fragments, *Solen ensis* and *Solen siliqua* in fragments, *Cardium edule*, *Cardium aculeatum?* bearing the size and form of this shell, but having from 34 to 36 ribs with no depressed line down the middle nor vestiges of spines; *Mactra solida*, *Venus exoleta*, *Venus scotica*, *Venericardia senilis* Lam., *Arca glycemeris*, *Arca nucleus*. Besides these remains of *marine animals*, the fossil hollow tubercles, having lost the spines, of the *thornback* are here found; also fragments of the *fossil palate* (*Scopola littoralis* of Lhwydd) and fossil remains of sponge and alcyonia, particularly a very fair specimen of the *reticulated alcyonium* (Org. rem. vol. ii. pl. 9. fig. 9.) In this bed, and among the gravel and the shells, are frequently found fragments of *fossil bones*, which possess some striking peculiarities. They are seldom more than half an inch in thickness, two inches in width, and twelve in length: always having this flat form, and generally marked with small dents or depressions. Their colour, which is brown, light or dark, and sometimes inclining to a greenish tint, is evidently derived from an impregnation with iron. From this impregnation they have also received a great increase of weight and solidity; from having been rolled they have acquired considerable polish; and on being struck by any hard body, they give a shrill ringing sound. These fragments, washed out of the stratum in which they have been imbedded, are found on the beach at Walton, but occur in much greater quantities at Harwich. To what animal these bones belonged is not known; but a large *fossil tooth*, probably of the mammoth, was found within the last few years on the beach at Harwich. (G. T. vol. i. p. 327—336.).

(*d*) *Range and extent.* The extent of this bed has not yet been completely ascertained. The nearest point to the metropolis at which it is seen, is Walton Nase in Essex, where it is exposed on the cliff for 300 paces in length; it caps the cliffs on both sides of Harwich. Quarries of it are worked on the southern bank of the river Orwell in Suffolk, and near Southwold, which is about two miles south of Lowestoff in the same county, it appears in the cliff, together with sand and red loam covering the London clay. But it may be concluded that the extent of this stratum is very considerable, since the same bed of shells is found on digging, through Suffolk and a great part of Norfolk; thus appearing to extend over a tract of at least 40 miles in length. (G. T. vol. 1. p. 377 & 529.)

(*e*) *Elevation.* The country formed by this bed is extremely flat; its surface may be considered as rarely exceeding 50 or 60 feet above the level of the sea.

(*f*) *Thickness.* The thickness of this bed is visible only perhaps on the coast of Essex, near Walton Nase, where it amounts to about 30 feet, and on the cliffs on both sides of Harwich, where its thickness is nearly the same.

(*g*) *Inclination.* It is nearly horizontal.

(*h*) *Agricultural character.* The tract occupied by this formation in Suffolk and Norfolk, affords a sandy soil blended with rich loams, and constitutes one of the most fertile and highly cultivated districts in England. The shelly mass of the crag itself also forms a useful improvement to the poorer sands. (C.)

(*i*) *Phænomena of water and springs.* This formation is generally porous, and affords no quantity of water except when pierced through; the water is then thrown up by the retentive substratum of the London clay. (C.)

(*k*) *Miscellaneous Observations.* The crag seems to claim priority of description, as being probably the newest bed of the upper marine formation, to which it is generally considered to belong by the best observers. The striking peculiarities belonging to the organic remains it encloses, are worthy of particular note. It has not hitherto been observed in immediate connection with either of the beds which are here associated with it as belonging to the same formation.

Section II. *Bagshot Sand.*

(*a*) *External Characters.* This bed consists chiefly of siliceous sand and sandstone without any cement, but occasionally includes, or is associated with, brick earth (G. Map).

The *Grey-weathers* or *Druid-stone*, of which the principal masses of Stonehenge, and those near Amesbury in Wiltshire, and the sandstones overlying the chalk in many other places, consist, do not appear to have been seen in a continuous bed, but were probably once imbedded in sand. At Stonehenge, the largest exceed seven feet in breadth, three in thickness, and fifteen in height. The sandstone is so hard as to require blasting by gunpowder, and is used for paving and building. Windsor Castle and Lord Carrington's house are constructed of it. It is also sometimes termed *Sarsden-stone* (G Notes).*

Chalk flints are sometimes imbedded in the Grey-weathers of Wiltshire, (G. T. vol. ii. p. 224). And small flint pebbles often occur in the sandstone of High Wycombe, and St. Anne's Hill. At Purbright, and many parts of the surrounding country, loose blocks of a stone are found, resembling those termed the Grey-weathers.

(*b*) *Mineral contents*; none.

(*c*) *Organic remains.* Casts of various shells are found in this sand on Bagshot heath, but are frequently so imperfeet as to preclude the possibility of ascertaining their species. Mr. Warburton, who has presented an interesting collection from that place to the Geological Society, remarks that they exhibit circumstances of agreement with those of the Upper marine formation of Paris. (C.)

(*d*) *Range and extent.* The term Bagshot sand is here used rather in a generic than specific sense, for it is intended to include not only the extensive covering of sand which overlies the London clay, and which is known by the name of Bagshot Heath, and that of Frimby and Purbright Heaths and other patches on the south of them in Surrey which are scarcely disjoined from it, but also the sand of Hampstead in Middlesex. No details of these considerable deposits of sand have hitherto been made public, except in regard to their extent, which is shewn on Mr. Greenough's map; whence, we may observe that in its usual form of sand, it covers the London clay from near Finchley on the north, to the south of Hampstead, and on the east of that line, forms a part of the eminences on which Highgate and Hornsey are built. The same sand is again visible

* It is not however certain that these sandstones may not rather be referable to the sand of the plastic clay. (C.)

near Egham on the north, extending south, though not uninterruptedly, to within a few miles of Guildford in Surrey, and from near Kingston to about seven miles west of Bagshot, occupies a part of Windsor forest.

(*e*) *Height of hills, &c.* This sand is found forming or covering a few eminences of no very considerable height, such as Hampstead and Highgate hills in Middlesex, St. Anne's and Shrubs hill west of Chertsey, Chobham ridges and Romping downs north of the Hog's back, and St. George's hill on the south of Weybridge in Surrey: several other elevations are visible in the same county, as on the south of Esher and several parts of Bagshot and Frimby heaths. Its greatest elevation (C.) amounts to 463 feet.

(*f*) *Thickness.* The thickness of these sands does not appear to have been determined.

(*g*) *Inclination of the beds.* Appears to be conformable to that of the subjacent London clay and nearly horizontal.

(*h*) *Agricultural character.* The agricultural character of this district differs materially from that of the crag, presenting a poor hungry sand and producing only heath and furze: hence the numerous tracts of unimproved and unimprovable wastes which are allowed to remain even in the immediate neighbourhood of the metropolis. (C.)

(*i*) *Phænomena of water and springs.* These beds, being loose and porous, yield water only when thrown up by the substratum of London clay. (C.)

Section III. *Isle of Wight.**

(*a*) *Chemical and external characters.* All that is known of this bed has been published in the interesting paper on the strata overlying the chalk, in the second volume of the Geological Society's Transactions, by Mr. Webster, Secretary to the Geological Society, who describes this bed as consisting chiefly of marle of a light greenish colour.

(*b*) *Mineral contents.* It does not appear to contain any.

(*c*) *Organic remains.* The marine shells found in this bed in Headon hill consist of cerithia, many species, and of cyclas, cytherea, ancilla, &c. The shells are so numerous that they

* In the continental portion of the basin denominated from the Isle of Wight, there are some tracts of sand and gravel occupying Beaulieu, and a part of the New Forest in Hampshire, and reposing apparently on the London clay; these therefore are probably contemporaneous with the beds described in this article. Mr. Greenough considers them as agreeing with the sand of the former section: the soil of these tracts is peculiarly adapted to the growth of oak. (C.)

may be gathered by handfulls, and are in general extremely perfect; some of them can scarcely be distinguished from recent shells, and few of them are found in the London clay; they are accompanied by roundish nodules of green indurated marle. In Bramble Chine, the banks of which belong to this bed, there is a large bed of fossil oyster shells, the greater part of which are locked into each other in the natural way; agreeing in this, and in other respects with the Upper Marine formation of the Paris basin. On comparing the fossils of this bed with those of the Crag, it will be observed that the natica canrena occurs in both, and that this bed contains two species of ampullaria, the Crag one other.

(*d*) *Extent.* This bed is most readily observed and studied on the cliff of Headon hill on the north-west coast of the Isle. It appears about half way up the cliff, and is separated from the Upper Fresh-water bed, which covers it, by a coat of sand a few inches thick. It passes from thence round Totland and Colwell bays, and may be traced all round the north side of the Isle, and is visible at Cowes, Ride, and Bembridge. The shells of this bed are numerous on the shore near Cowes. But the whole north shore of the island has been for ages in a state of constant ruin by the action of the sea and the sliding down of the soil. It is difficult therefore to find any part of the strata in their original situation; on this account fresh-water and marine shells are frequently found together in confusion.

(*e*) *Height of hills, &c.* In the Ise of Wight it appears only to form a bed, not distinctly alone constituting any considerable tract; its actual boundaries from the forementioned causes, being unknown.

(*f*) *Thickness.* The thickness of this bed in Headon hill is 36 feet; it does not appear to have been determined in any other part of the Isle.

(*g*) *Inclination.* In Headon hill, this bed dips a few degrees to the north.

CHAPTER III.

FRESH-WATER FORMATIONS.

Section I. *General view of these formations.*

For a description of these interesting formations, and also, we may add, for a knowledge of their existence in England, we are indebted to a communication to the Geological Society, 'on the Freshwater formations of the Isle of Wight,* with some observations on the strata over the chalk in the south-east part of England,' by Thomas Webster, Sec. G.S. inserted in the second volume of its Transactions; from which the following information is chiefly extracted.

Headon hill (see Section, pl. 2, fig. 6), as has already been mentioned in treating of the Upper Marine formation of the Isle of Wight, is situated on the north-western coast of that Isle, and a natural section of this hill, effected at some remote period by the action of the sea, has laid open to us the extraordinary circumstances of its formation, shewing that it consists of alternate deposits by salt and fresh water, the evidence of which rests upon the decided characters of the shells they

* *Note by the Rev. W. D. Conybeare.* The occurrence of distinctly marked fresh-water formations in England, may, generally speaking, be considered as confined to the part of the Isle of Wight described in this article; there are however some apparent indications in two or three other places in England of a Fresh-water formation above the chalk: but it may be doubted whether the circumstances are such as to warrant such a conclusion, the phœnomena observed in these instances being generally rather an occasional intermixture of a few fresh-water shells in the midst of marine formation, than the occurrence of strata decidedly distinct, and containing remains of the former class only: such an intermixture has occasionally been traced in the following places, and may be accounted for on the supposition of dead shells carried down the æstuaries of rivers, and there mingled with those of the sea.

The cliffs on the south of Harwich, in Essex, consist of red loam, fresh-water shells and crag, resting on the London clay (G. Map.) On the same authority, it may be quoted that fresh-water shells occur on or near the cliff of London clay, east of Southend on the coast of the same county.

Among the pyritous casts of shells at Sheppey, Mr. Webster found one resembling the lymneus, and another the planorbis, but too imperfect to decide the species. Mr. Brande mentions also three fresh-water shells from the same place, the lymneus, melania and nerita. These shells however, observes Mr. Webster, which are very few in number, do not prove the existence of a fresh-water formation in this place similar to those of the basins of Paris and of the Isle of Wight; for, being found among the remains of vegetable and marine animals, we may suppose these fresh-water shells were carried down together with the branches of trees and fruits by the numerous streams and rivers that must have flowed into this gulph.

It might be most naturally expected that we should find on the coast of Hampshire opposite to the Isle of Wight, traces of the same formations which assume characters of so much importance in the Isle of Wight itself; and we certainly do discover many fresh-water shells on this side in the Harwich cliffs, but still they appear only to be casually intermixed: here however, if any where, the research for distinct fresh-water strata might be undertaken with some prospect of success, and the whole line of cliff on this side of Hampshire certainly deserves a more careful examination than it has yet received.

We should be careful not to confound with these fresh-water strata, which from their alternation with the regular marine formations clearly belong to a period anterior to that in which our continents finally passed into their enclose. Many of the shells found imbedded in one of these strata are quite entire; whence it is argued that, considering their thinness and extreme liability to injury, they must have lived in the very place now occupied by the strata which enclose them. These strata it will be recollected do not lie in a hollow, but a hill; it is nevertheless concluded that this hill must have been the bottom of a fresh-water lake.

But as in this hill there are two strata containing fresh-water shells, separated by that which has already been described as belonging to the Upper marine formation, it may be concluded that this hill must twice have been within the bosom of a fresh-water lake, and between the periods been covered by salt water. Hence there are two formations by fresh water, which by their discoverer have been designated, in reference to their

present condition, with the occasional occurrence of fresh-water shells in alluvial tracts—belonging possibly to a very recent date, and certainly to one less ancient than the above. Near Kew in Surrey, land and river shells were found in sand and gravel overlying the remains of elephants, &c. and therefore certainly posterior to the catastrophe which inhumed these animals.

Traces of fresh-water shells have been supposed to be observable even in some of the strata of an antiquity greater, and in one instance considerably so, than that of the chalk itself; but the observations are of a doubtful character; they may be stated as follows:

1. The clay stratum separating the green and iron sand in the weald of Kent, and containing concretions of a compact argillaceous limestone known under the name of Petworth marble, abounds in a univalve supposed to be Helix vivipara, and a fluviatile shell; but the identity does not seem to be fully ascertained.

2. The strata immediately below the iron sand, known by the name of Purbeck marble, present a smaller species also referred to the Helix vivipara. Mr. Webster has given the following observations on these beds.

In the Isle of Purbeck a series of strata of shelly limestone, known by the name of the Purbeck stone, alternates with shale and marle. Some of the fossils of these strata strongly resemble fresh-water shells; they appear to be cyclostoma, planorbis, &c. (G. T. vol. ii. p. 166.) It was long ago observed by Woodward, in his history of fossils, that the shells in the Purbeck marble consist chiefly of the Helix vivipara; and it is rather surprising that this very ancient fresh-water formation should not have excited more attention. Beautiful impressions of fish are frequently met with by the quarry-men between the laminæ of the limestone; and I saw abundance of fragments of bones, some of which belonged to the turtle. Complete fossil turtles have been found, and lately one extremely perfect. (W. p. 192.)

3. Even among some almost of the oldest strata exhibiting organic remains, a similar occurrence of fresh-water shells has been supposed to be observable; for the only species of shells yet discovered in association with the coal measures is a kind of unio, much approaching that of fresh water, and the absence of marine remains seems to favour the supposition, nothing else but vegetables (certainly those of the land) being discoverable in the coal strata. It seemed necessary, although involving an anticipation of points to be examined hereafter, to make some allusion to these facts at present, in order to bring the whole subject of fresh-water formations under the reader's eye at once.

position, the *Upper* and *Lower*.* It will be most advantageous to treat of these separately.

Section II. *Upper Fresh-water formation.*

(*a*) *Chemical and external characters.* This formation lies above the bed already described as belonging to the Upper marine formation, from which it is separated by a bed of sand six inches thick; it is covered by alluvium of which the summit of Headon hill consists, and is in great measure composed of a yellowish white marle, that does not endure the frost, enclosing masses which are more indurated, and appear to contain a greater proportion of calcareous matter. It is the most remarkable and best characterized of all the beds of Headon hill.

(*b*) *Mineral contents.* Through all these strata run veins, frequently several inches in thickness, of a very pure carbonated lime, which is crystallized, frequently in a radiated form.

(*c*) *Organic remains.* This stratum every where contains fresh-water shells in abundance, without any admixture whatever of marine exuviæ. They consist, like those of the lower fresh-water beds, of several kinds of lymnei, helices and planorbes, and other fresh-water shells which are extremely thin and friable; together with seeds of a flattened oval form, and parts of coleopterous insects.

(*d*) *Range and extent.* This stratum is not confined to the upper part of Headon hill, where it appears as a bed of very considerable thickness, but also may be seen in many parts north of the middle range of chalk hills. On the western coast it does not extend further than Totland bay, but occurs again at Warden point, forming the summit of the cliff. Numerous blocks of it lie loose in the soil in many parts about Cowes,

* The English beds belonging to these formations may in general terms be considered as consisting of marle, argillaceous limestone, and sand, crossed by veins of calcareous spar. The external and chemical characters of the English and French beds, are considered to be sufficiently different from every other known rock, to render them distinguishable even without the shells they enclose. The beds of France are described as white or yellowish, sometimes as tender and friable as chalk or marle, and sometimes very hard, compact, and solid, with a fine grain and conchoidal fracture. In the latter case the stone breaks into sharp fragments like flint, and cannot be worked, but sometimes it will even admit of being polished as a marble. It is also frequently filled with infiltrations of calcareous spar. Both the English and French beds contain the cyclostoma, planorbis, and limneus. The Fresh-water formation of England is altogether without the beds of gypsum, which are even numerous in that of France, and on the uppermost of which were found the bones of unknown birds and quadrupeds, together with a few fresh-water shells.

Bembridge and Binstead: but in the neighbourhood of Calbourne, and between that place and Thorley, several quarries are opened in it, and afford excellent stone. The rocks at Bembridge ledge and Whitecliff bay on the east side of the island, are referred to this formation. The beds of this formation, or more correctly of these formations, are now considered as generally occupying the northern half of the Isle.

(*e*) *Height, &c.* This bed occupies the upper part of Headon hill, which is 400 feet above the level of the sea, and the summit of which consists of alluvium. In other parts of the Isle, the elevation is very inconsiderable.

(*f*) *Thickness.* This bed is described as being 55 feet thick in Headon hill; but on it lies a stratum of clay, 11 feet thick, containing an unknown bivalve shell; above which is another bed of yellow clay without shells, and upon it a stratum of friable calcareous sandstone, also without shells. Above the latter lies another calcareous stratum enclosing a few fresh-water shells. These beds are occasionally very compact and of a porcellanous character. Other parts contain masses of loose chalky matter, most of which are of a roundish form, and among these also are many beds of a calcareous matter, extremely dense, and much resembling those incrustations that have been formed by depositions from water on the walls of ancient buildings in Italy.

(*g*) *Inclination.* This bed, in common with all the others of which Headon hill consists, dips gently to the north.

(*h*) *Agricultural character.* (*i*) *Phœnomena of water and springs.* On these heads we have nothing to observe.

(*k*) *Miscellaneous remarks.* The more compact parts of this stratum are durable, and have been long employed as a building material in the Isle of Wight, and in many places on the opposite coast, as at Portsmouth, Southampton, Lymington, &c.

Section III. *Lower Fresh-water formation.*

(*a*) *Chemical and external characters.* In Headen hill it consists of a series of beds of sandy, calcareous, and argillaceous marles, sometimes with more or less of a brownish coaly matter. Some of them appear to consist almost wholly of the fragments of fresh-water shells, many of which however are sufficiently entire to ascertain their species. In the quarries of Binstead near Ryde, the characters of the beds differ considerably from the above, and from each other, as the following section will evince. The strata of East Cowes quarries almost exactly resemble those of Binstead.

Section of Binstead quarries.

On the summit lies	feet	in.
1. Blue clay, in which are many large and loose masses which appear to belong to the Upper Fresh-water formation		
2. Limestone composed of coarse fragments of fresh-water shells	2	—
3. Ditto, the fragments being less coarse	4	—
4. Ditto............still less coarse	2	—
5. Ditto	1	6
6. Siliceous limestone	1	6
7. White shell-marle	—	10
8. Siliceous limestone	—	6
9. Sand	—	8
10. Siliceous limestone rag	—	6
11. Sand	50	—
12. Blue clay	depth unknown.	

The fragment beds of the above quarries consist of hollow moulds of fresh-water shells held together by calcareous spar.

This stratum is separated in Headon hill from the Upper fresh-water, by the Upper marine already described, and is sometimes separated from the bed of black clay on which it rests, by a black coaly matter. This bed of clay is thirty-five feet thick, nearly horizontal, contains fossil shells, and sometimes selenite, and it reposes on a bed of white sand 30 to 50 feet thick, which forms the base of Headon hill, and which is largely employed in glass-making. But neither the clay nor sand has hitherto been distinctly referred to any particular formation.

(*b*) *Mineral contents.* Under this head it may be observed that at Headon these beds contain more or less of a brownish coaly matter, and that in common with the other strata of which the hill consists, they are traversed by veins of calcareous spar.

(*c*) *Organic remains.* These are lymneus, planorbis, and cyclostoma, and perhaps the helix, and a bivalve resembling the fresh-water mytilus. The casts of the shells in the East Cowes beds are frequently entire, and belong to the genus cyclostoma, Lam. No marine exuviæ have been observed in these beds.

(*d*) *Range and extent.* The beds of this formation in Headon hill are extremely irregular, and are not to be traced distinctly from each other for more than a few hundred yards, the remaining part being so hid by the mouldering slope, that the formation can only be observed in mass. It may be seen, however, extending round the north side of Headen into Totland bay, where it forms the upper part of the cliff; and at the point called Warden-ledge, it is found in a more uniform and indurated state. In Colwell bay, it dips to the north, is soon lost, and is not to be seen any more on that side of Yarmouth.

The whole of the northern part of the Isle of Wight is considered to belong to the fresh-water formation, but the ruin is so considerable that it is difficult to find any of the beds in situ, and hence fresh-water and marine shells are often found intermixed on the west and north coasts, though in places they occur in alternate layers: and although the precise boundary of these beds in the Isle of Wight is unknown, they may be traced a considerable distance east of Ryde, perhaps as far as Nettlestone.

(*e*) *Height.* The greatest height of this formation above high water mark, may be taken at about 90 feet at Headon hill; it is scarcely found at so great an elevation in other parts of the Isle.

(*f*) *Thickness.* In Headon it is 63 feet thick: in the Binstead quarries it appears to be only 14 feet by the foregoing section, without reckoning the sand on which it reposes.

(*g*) *Inclination, &c.* It dips gently towards the north.

(*h*) *Agricultural characters.* (*i*) *Phænomena of water and springs.* On these heads we have no information.

(*k*) *Miscellaneous remarks.* The quarries of Binstead near Ryde, of which a section has been given, were formerly of great celebrity, and furnished materials for many ancient edifices, both civil and religious, in the Isle of Wight and the counties contiguous to it. They are now very little worked, but their extent may be traced by the broken ground where they have been filled in. The mansion of Lord Henry Seymour near East Cowes, and a wall raised for the purpose of preventing the encroachments of the sea, were built of the rag limestone from a quarry near the mansion.

CHAPTER IV.

THE LONDON CLAY.

Section I. *Preliminary view of this Stratum.*

The great argillaceous formation designated by this name next claims our attention. It is rendered highly interesting by the variety of its organic remains, both animal and vegetable, and by the inferences deducible from them—the smaller number of species which can be completely identified with recent analogues, point out the greater antiquity of this, than of the preceding formation; and the circumstance of its having been thrown into a vertical position in the Isle of Wight, by convulsions which must have taken place previously to the deposition of the upper beds which repose horizontally against its truncated edges, proves that a sufficient interval at least to allow of its assuming a considerable degree of consistency must have elapsed, before these newer strata were accumulated upon it. (C.)

Its name is derived from its forming the general substratum of London and its vicinity, occurring immediately beneath the vegetable soil, excepting when occasional deposits of alluvial on diluvial, gravel, sand, &c. intervene.*

Section II.

(*a*) *Chemical and external characters.* This formation consists chiefly and sometimes wholly of bluish or blackish clay, which is mostly very tough. Its chemical and external characters are however subject to some partial and local variations, though these never appear to be very considerable. Some of its strata, for instance, occasionally partake of the nature of marle, since they efferyesce when exposed to the action of an acid,

* Perhaps another exception (C), but it is a solitary one, may be thought to be indicated by the following curious section preserved in Sir C. Wren's Parentalia (p. 285), and obtained in preparing the foundation of the present Cathedral of St. Paul's in London.

The surveyor observed that the foundations of the old church stood upon a layer of very close hard pot earth, which he therefore judged firm enough to support the new building; and on digging wells in several parts he found this pot earth to be about six feet thick or more on the north side of the church-yard, but thinner and thinner towards the south, till it was scarce four feet upon the declivity of the hill. Below this he found nothing but dry sand, mixed sometimes unequally, but loose, so that it would run through the fingers. He went on till he came to water and sand mixed with periwinkles and other sea-shells: these were about the level of low water mark. He continued boring till he came to natural hard clay.

The upper stratum of pot earth had been used as a Roman pottery near the north-east angle of the present church, where they found urns, sacrificing vessels, and other pottery in great abundance, and were interrupted in digging the foundation of the north-east angle of the church, by the quarry from which the pot earth had been extracted: the subjacent sand and gravel beds being considered too loose to support the weight of the intended building, it was thought necessary to secure this part of the foundation by erecting it upon an arch. The outer, or north-east pier of this arch stands in the old clay pit, in a shaft sunk to receive it more than 40 feet below the stratum of pot earth that had been removed, and descending through the beds of sand and gravel above mentioned, to the subjacent stratum of hard clay. (G. T. vol. iv. p. 288).

and sometimes strongly. In the Isle of Wight it contains much green earth. Occasionally it includes beds of sandstone. At Bognor, on the coast of Sussex (G. T. vol. ii. p. 190,) are several rocks now appearing as detached masses in the sea, though evidently forming portions of a stratum once continuous. The lowest part of these rocks is a dark grey limestone, or perhaps rather a sandstone, containing much calcareous matter, enclosing many fossils belonging to the blue clay. The upper part is a siliceous sandstone. Bognor rocks resemble much the nodules and beds of limestone that are found in the blue clay of Alum bay, and are no doubt owing to the great abundance of calcareous matter in this part of the bed. The Barns rocks between Selsea and Bognor, the Roundgate and Street rocks on the west, and Mixen rocks to the south of Selsea, are portions of the same bed: similar masses also appear at Stubbington. The cliff near Harwich in Essex (G. Notes) contains beds of stratified limestone.*

* The fullest idea of the nature and composition of this formation will be conveyed by the following actual sections, generally obtained from the sinking of various wells in the neighbourhood of the metropolis.

Section afforded by a well sunk at Josiah Forster's, at Tottenham in Middlesex. (P.)

	feet.
1. Immediately below the surface was found brick-earth and coarse yellow sand (from which washed malm bricks are made) stiff clay, and gravel	20
2. Blue clay of various intensity of colour and degrees of stiffness, adapted for tile-making. It effervesced slightly, and enclosed hard and irregular masses of a lighter colour, full of minute appearances of charred vegetable matter, and septaria, which also effervesced	60
3. Blue clay of a greasy aspect and somewhat greasy to the touch; it did not effervesce	20
4. Purple, blue, red and brown clay mixed, having greatly the appearance of some varieties of lithomarge: it did not effervesce	10
5. Blue, white, and brown clay mixed, much heavier than the preceding: it effervesced strongly and contained very compact, hard, and nearly cylindrical masses, six to 12 inches long and of a yellowish white colour, which effervesced strongly	10
6. Yellowish white clay, frequently compact and hard, and equally heavy with the preceding: it effervesced strongly..	3
Rock bored through	2
	125

Water rose to within a short distance of the surface in a few hours.

No rock was found in sinking several wells, not half a mile distant; in these wells only blue clay, and the alluvium covering it, were perforated; the water rose immediately from beneath the former, and has in some instances thrown up so great a quantity of white sand as to obstruct the

The limestone bed underlying Paris, and affording the materials of which many of its buildings are composed, known by the name of Calcaire grossier, corresponds most nearly in its relations and fossils, with the London clay of our series; yet no

rising of the water: but after taking up several tons of the sand, the water rose to its first level. No specimen of the rock forming the lowest stratum of the above section was obtained; it may perhaps be considered analogous to the calcareous sandstone of the Bognor rocks, and the beds of limestone in the cliffs near Harwich in Essex, as may also the rock bored through for 10 feet in depth, after sinking through 122 feet of blue clay, at the foot of a small eminence, near Sewardstone in Essex. In both instances the water rose rapidly from beneath the rock; but at Bromley near Stratford le Bow in Essex, a bed of rock, one foot thick, occurred 24 feet above the main spring.

Section of a well at Twyford near Acton in Middlesex, from papers in possession of the Geological Society. (C)

	feet.
1. Yellow clay	38
2. Lead coloured clay, containing some fossil wood at 188, and shells at 200 feet from the surface	170
3. Rock } 4. Sand and clay with pebbles }	2
5. Variegated clay, red, blue and black	32
6. Sand and water........thin	
7. Clay.	

The Section obtained at the Highgate Tunnel presented the following strata. (C.)

	feet.
1. Flint gravel, sand and loam, occasionally concreted by iron (diluvial)	10
2. Loam with a few marly concretions, but no septaria, pebbles, or shells	30
3. Blue clay with septaria and shells, hardest in the lowest part	65
	105

The first attempt at Highgate was to drive a tunnel (like that of Pausillipo) through the hill; but after a tunnel of small dimensions had been driven, this plan was abandoned in consequence of finding the substratum sandy and loose, and incapable of supporting an arch of the dimensions required, and resisting the superincumbent and lateral pressure. Mr. Middleton is of opinion that the London clay had been cut entirely through; and that this stratum was the sand of the inferior plastic clay formation: but it is highly improbable (unless a great undulation in the strata be supposed) that that sand should have occurred in this point at such a level; and indeed the nature of the contents of this stratum, consisting of septaria, fossil fish, and lobsters, &c. identify it with the London clay formation, and prove that it was merely one of those subordinate sandy beds which often occur in a similar situation.

The well at White's Club-house, St. James's-street. (Ralph Walker, Esq. P.)

142 feet blue clay
93 feet of red clay

235

Water rose to about 45 feet from the top.

beds have been discovered in England which bear a nearer approach to them in chemical and external characters than the faint resemblance of the above imperfect calcareous beds. (C.)

Wherever this clay is visible in the form of a cliff, or has been perforated in sinking wells, it has uniformly been found to contain nearly horizontal layers * of ovate or flattish masses of argillaceous limestone; which, as they mostly exhibit, though not always, the appearance of having been traversed in various directions by cracks since partially or wholly filled by calcareous spar or sulphate of barytes, have obtained the name of Septaria. These masses abound so greatly, that they have been

That at Mr. Fulk's in the Regent's Park. (Ibid.)

140 feet of blue clay
42 feet of red clay

182

That of Mr. Cooke's, in Swallow-street, Westminster. (Ibid.)

18 feet gravel and loam
12 feet yellow clay
130 feet blue clay
50 feet red clay

210

Water to within 60 feet of the surface.

At Epping in Essex, on the north-east of the metropolis, the clay seems to possess the greatest depth, viz. 392 feet; the particulars of the sinking of a well there are given in a note under the head *(f)* thickness of this stratum.

In the Agricultural Survey of Middlesex, several sections may be found: near Kensington the borings present, 1. vegetable mould; 2. flint gravel, 5 to 10 feet; 3. lead coloured clay, 300 feet; 4. oyster shells, &c. sometimes concreted by a calcareous cement, 3 to 5 feet; watery sand. On the east and south-east of London, the clay formation grows rapidly thinner, the subjacent beds approaching more nearly to the surface. In some sections (the particulars of which will be given in treating of the inferior plastic clay), the thickness of the London clay was found as follows. 1. At Liptrap's and Smiths' Distillery, one mile on the east of London, 77 feet; 2. at Bromley near Stratford le Bow, 44 feet; 3. on the north shaft of the unfinished tunnel of Rotherhithe, about one mile east of Bromley, 49 feet; in the southern shaft, in which direction the strata rise, only 9 feet. (C.)

Hordwell cliff, on the coast of Hampshire, is about 150 feet high and a mile and a half in length; for nearly 60 feet from the surface it consists of red gravel. The remainder of the cliff is clay, through which land-springs are continually trickling, so that it is constantly falling. Towards the bottom, are large nodules of a hard reddish iron-stone, being no other than an entire mass of shells, with which the church, &c. are built. (G. Notes from Gents. Mag. 1757.)

* An exception to this rule occurs at Alum Bay in the Isle of Wight, where the layers of septaria, like the beds containing them, are in a vertical position.

considered as being characteristic of the London Clay; but it is not the only one of the English beds which contains them.‡

By the preceding remarks and sections, it will appear that this formation is uniformly marked as consisting of a vast argillaceous deposite, containing subordinate beds of calcareous concretions, sometimes passing into solid rocks, or exhibiting some local variations from the occasional mixture of sand or calcareous matter in the mass of the clay: these local changes however never prevail to such a degree as to interfere materially with the general identity of character.

(*b*) *Mineral contents.* It contains interspersed through it, sulphuret of iron,‖ selenite, and occasionally phosphate of iron; hence, on account of its containing these salts, the water issuing from this stratum is unfit for domestic purposes. (G. T. vol. ii. p. 188). It is also suspected to contain sulphate of magnesia.*

The fossil copal or resin discovered in excavating Highgate archway, though evidently derived from the vegetable kingdom, still in its present state may be classed among the mineral contents of this formation. Amber also is said occasionally to be found in the gravel about London, and is also found in the cliffs of brown clay, probably belonging to this stratum, which occur on the coast of Holderness in the south of Yorkshire. (Pennant's Arctic Zoology). It probably exists in this stratum, since it is so found in the contemporaneous formations of other parts of Europe, as Italy, France, and especially Prussia, where it

‡ The septaria on the south of Walton on the coast of Essex, are very imperfect; they are collected into heaps on the Nore, and shipped to Harwich, where they are manufactured by government into a cement. (G.Notes.) Dale, in his history of Harwich, (p. 101), speaking of the septaria so abundantly found in the cliffs of the neighbourhood, says, 'with these the walls of the town were for the most part built, and the streets generally pitched, they by ancient custom belonging to the town as their right.' The long cliff of the London Clay extending along the northern side of Sheppey Isle furnishes abundance of septaria, from which that excellent material for building under water and for stucco, is made, and which is known by the name of Parker's Cement. Being separated from the clay by the action of the sea, they are collected on the beach, and exported to various places where they are calcined and ground. (G. T. vol. ii. p. 193.)

In Hampstead and Highgate hills in Middlesex, and in Boughton hill in Kent, the layêrs of septaria occur about 50 feet below the summit (G.Notes.) In Shooter's hill they are very near the surface. At Epping, they were not found nearer than 100 feet beneath the grass, and continued to the depth of 300 feet. (P.)

‖ The manufactory of sulphuret of iron from the pyrites contained in this stratum, is carried on in many places; as near Walton on the Suffolk coast, the Isle of Sheppey, &c. &c.

* Said to have been noticed by the late Mr. Tennant. See (*i*) *Phænomena of water*, &c.

occurs in conjunction with a large accumulation of vegetable remains. (C.)

(*c*) *Organic remains.* † Few formations claim a greater interest from the organic remains preserved in them than this. In the higher order of animals, it presents us with the crocodile and turtle among the amphibious class; a proof that the shores of some dry land where these animals might have deposited their eggs, must have existed at the period of its formation within a distance easily accessible. Of Vertebral fish, several species are found beautifully preserved, to which it is much to be regretted that the attention of the comparative anatomist has not yet been sufficiently directed; of crustaceous fish, many species of the lobster and crab occur.

The testaceous molluscæ are also very numerous and beautifully preserved, often retaining nearly the appearance of recent shells. There are very few genera of recent shells which have not some representation imbedded in this formation, but the specific character is usually different, that difference being often however so minute as to escape an unpractised eye; on the other hand but few of the extinct genera, so common in the older formations, occur in this, so that it seems to hold a middle character in this respect between the earlier and more recent beds. Thus though nautilites resembling those of the Indian seas are common, specimens of the cornu ammonis and the belemnite are so rare, that it is in a very high degree doubtful whether they ever have really been found.* The nummulite, so common in the contemporaneous strata of France, is in England found only in a few places in this formation, but it is by no means ascertained that this should be considered as an extinct genus. Echinites, so common in the chalk, are very rare in this formation.

Zoophytes are likewise extremely rare; perhaps none have yet been found in England (with the exception of a few minute corallines investing the surface of the shells); in France, however, a few species of madrepore have been discovered: few if any of the numerous fossil family of the Encrinites have yet

† This article is by the Rev. W. D. Conybeare.

* The Cornua ammonis figured in Mr. Sowerby's work from Norfolk, are beyond a doubt alluvial or diluvial, having been drifted from the inferior oolite. The specimens called belemnites in Mr. Jacob's catalogue are likewise doubtful: no belemnites are mentioned among the fossils of the contemporaneous formations in France or Italy; a few genera of minute multilocular shells, supposed to be extinct, discorbites, rotalites, lenticulinites, as well as the nummulites, are enumerated, but such small shells, though still existing, may well have escaped observation. Ammonites are however found in the beds immediately reposing on the chalk at Maestricht, but these are probably older than the London clay.

been enumerated as occurring any where in this formation.* astroitæ are however mentioned as found at Sheppey, but it does not appear whether madrepores, or joints of the pentacrinus, are meant.

A catalogue arranged in the Linnæan method, with accurate engravings of the shells of this formation, was published by Brander under the title Fossilia Hantonensia. Mr. Jacobs has added a short list of those found in Sheppey Island to his Hortus Favershamiensis; very interesting catalogues will also be found in Mr. Webster's excellent memoir, G. T. vol. ii; and the figures and descriptions published in the French Annales de Museum, by Lamarck, of the Parisian fossils, may be consulted also with advantage, since these generally agree with the English.

It frequently contains small portions and even masses of wood, more or less retaining the woody fibre, but more often having the appearance of being charred, and of a black colour. They sometimes exhibit the perforations, and even contain the casts, of an animal which is considered to be analagous to the teredo navalis or borer, still infesting the seas surrounding the West India Islands. The wood occasionally appears to have formed a nucleus, around which have been deposited those masses of argillaceous limestone, mostly in the form of septaria, and which have already been mentioned as being characteristic of the London clay: they often contain shells still exhibiting the pearly lustre.

But the most interesting facts connected with the vegetable remains of this formation, are those which have been observed in the Isle of Sheppy, of which some account was published by Dr. Parsons in the 50th volume of the Philosophical Transactions: the quantity of fruit or ligneous seed vessels is prodigious. Mr. Crowe of Faversham has procured from this fertile spot a very large collection, and by carefully comparing each individual specimen by its internal as well as external appearance, he has been enabled to select 700 specimens, none of which are duplicates, and very few of which agree with any known seed vessels. These vegetable remains have also been

* It will give some idea of the variety and number of the testaceous fossils of this formation to state, that at Grignon (near Paris) in a single spot, a single individual, Mr. Defrance, collected 500 different species of shells, besides serpulæ, siliquaria, dentalia, and a few echinites and madrepores. The shells of Hordwell cliff in Hampshire, which closely agree with those of Grignon, are probably equally numerous.

Mr. Webster observed an exact coincidence in all the fossils discovered in the lower beds of the French calcaire grossier at Liancourt, and the corresponding beds of the London clay at Stubbington.

found on the opposite Essex shore, but in very small numbers, and in that part of the stratum which has been examined at Kew. Among Mr. Crowe's specimens are many which appear to belong to tropical climates, some which seem to be a species of cocoa-nut, and other varieties of spices.

The existence of a neighbouring region of dry land seems attested by these vegetable remains) which, from the state in which they are found can scarcely have been supposed to have drifted from any great distance), as well as by the occurrence of the amphibia before mentioned. We can scarcely resist the temptation of asking "What was that ancient land? had any part of England then raised its head above the waves? does it not sound extravagantly, even to enquire whether its oldest and highest mountain tracts then formed a groupe of spice islands frequented by the turtle and crocodile?" * Speculations like these, though unavoidably suggested, almost give the features of romance to the sober walks of science.

The fossils of this formation are abundantly found wherever the sea has laid open natural sections, or artificial excavations have been carried on to any great extent; the following localities may be mentioned.

The cliffs on the south of Harwich in Essex; those which skirt the north of the Isle of Sheppy; those of Stubbington near the eastern entrance of the Southampton water, and more especially those between Hordwell and Christchurch in the south-west of Hampshire; the rocks of Bognor in Sussex (though scarcely deserving the name, as they are mere blocks just peeping above the tide) are likewise very productive. The great excavation for the Highgate archway near London afforded numerous and beautiful specimens; some excavations at Kew are recorded in the same account in P. T. vol. 103, p. 134. The Pits at Richmond are often cited in Woodward's catalogue. Shells have been also found in the wells on Shooter's hill. The Canal lately undertaken, but since abandoned, for the drainage

* The figure of the Earth as a spheroid of rotation precludes the idea that the line of its equator can have shifted, and many physical reasons concur to render it in the highest degree improbable that the obliquity of the ecliptic has undergone any material change. We cannot therefore refer the indication of a change of climate, which geology seems to present in the higher latitudes, to astronomical causes: but it seems by no means improbable, when we take into consideration the proofs of the much greater extent and energy of the volcanic fires which have acted on our planet at an earlier period, that its general temperature may then have been higher; the number of the points universally allowed to be extinct volcanos will authorise this remark, without having recourse to the controverted, but certainly probable, theory, of the igneous origin of basaltic rocks.

of the low grounds of East Berkshire, near Bray, presented a rock and fossils exactly agreeing with those of Bognor.

(*d*) *Range and extent.* This clay forms the superior stratum of the chalk basin of London, except where it is partially covered by the sands of the upper marine formation, already described as appearing on the surface of Highgate hill, Bagshot, Frimby, and Purbright heaths, &c. or by alluvial sands, gravel, and loam.*

It extends uninterruptedly and in a south-westerly direction from Orford on the coast of Suffolk, about 20 miles north-east of Harwich, and a little to the north of Ipswich, in that county, to the South of Coggeshall, and thence to Roydon, in Essex; whence it turns nearly south, extending to a little on the west of Edmonton in Middlesex, and thence in a north-westerly direction by Chipping Barnet and South Mims to the north of Ridge hill; here it suddenly turns southward, and afterwards south-west by Harefield and Uxbridge to a little on the east of Colnbrook: it then turns nearly west, crosses the Thames by Windsor to Twyford, and thence passes to its extreme point on the west, which is about three miles south-west of Reading in Berkshire. It then turns to the south-east in an irregular line to within a very short distance of Farnham and Guildford in Surry, and by Epsom and a little north of Croydon to Deptford in Kent; in which direction, at the distance of about five miles beyond its general boundary, an outlying and insulated mass of this formation constitutes the upper regions of the conspicuous height called Shooter's hill, forming its rounded summit and reposing on the platform of Blackheath where the inferior beds of the plastic clay, &c. appear.

The London clay therefore constitutes a very large part of the soil of Suffolk, nearly the whole of Essex, including Hainault and Epping forests, quite to the sea, the whole of Middlesex, and portions of Berkshire, Surrey and Kent; in the latter county it appears on the northern side of the Medway, constitutes the Isle of Sheppy, the cliff from Whitstable to the Reculver north of Canterbury, and extends in a south-westerly direction nearly to that city, and thence westward about six miles to Boughton hill; a small patch of it is visible on the south-west of Ramsgate in Pegwell Bay. (G. Map).

In the chalk basin of the Isle of Wight it is also very ex-

* The beds of the Plastic clay are often so similar to those of the London clay, that it is by no means easy to trace in every instance their demarcation. The boundaries indicated above, are therefore only given as the nearest approximation to accuracy which the present state of our information admits, and may very probably allow of many local corrections; the position of the springs, to which the sands underneath this formation serve as an immense, reservoir affords the best guide in tracing its limits. (C.)

tensive, forming the whole coast from Worthing in Sussex to Christchurch in Hampshire, and extending from the latter place, inland by Ringwood, Romsey, Fareham, and passing a mile or two south of Chichester to Worthing. It forms the whole of this tract from the above named places to the coast, except where it is covered by alluvium, or by a sandy tract, analagous to the Bagshot sands, constituting Beaulieu forest, and a part of the New forest. It is also found in the Isle of Wight, and extends, nearly *in a vertical position*, along the island from Alum Bay on the west to Whitecliff Bay on the east, between the nearly horizontal freshwater and upper marine beds on the north and the nearly vertical beds of the Plastic clay formation on the south of it. See Plate 2, fig. 6. (G. Map). To this it may be added, that this clay forms the bottom of the channel termed the Solent, which is between the Isle of Wight and coast of Hampshire. Bognor rocks consist of it.

The occupation of the very large tracts just recited, by the London clay, has been proved over a great portion of it by the sinking of wells, as already adverted to in treating of the chemical and external characters of this stratum, and by the cutting of roads, as over Shooter's hill and Boughton hill in Kent; and the cliffs which it forms, and which are numerous and extensive, have been mentioned as being particularly favorable to the collection of the organic remains it encloses.

(*e*) *Height of hills, &c.* The country formed by this stratum is generally low, and may for the most part be considered as flat, or at the most as consisting of very gentle undulations. Here and there however it rises; the highest point it attains is the summit of High Beech in Essex, being 759 feet above the level of the sea; Langdon hill on the coast of the same county is 620 feet high: Danbury hill is not greatly inferior; Shooter's hill in Kent is 446 feet high; Richmond hill and St. Ann's hill in Surrey, which consist of this clay, are less elevated.

The northern half of the isle of Sheppy consists of a range of hills of above 200 feet in height, presenting to the sea cliffs of the London clay about 90 feet in height aud four miles in length, and declining gradually towards the east and west. These cliffs have deen formed by the action of the sea, and of which whole acres sometimes fall at once. (G. T. v. ii. p. 192.) The cliffs at the Reculver on the north-west coast of the Isle of Thanet (C.) consist of this clay and are about 70 feet high.* Hordwell cliff in Hampshire has already been mentioned as about 150 feet high.

* These are remarkable for the very rapid encroachment of the sea; it is said that in the reign of Henry the Eighth the church stood nearly a

(*f*) *Thickness, &c.* The London clay has been pierced in various places to different depths. One mile east of London it was found to be only 77 feet thick: at Tottenham, in the sinking of several wells it was found to be 110 to 120 feet: at White's Club-house, St. James's, 235 feet: at the Dock-yard at Portsmouth 102 feet, while in Portsmouth 266 feet were sunk down, without getting through the clay. It is said that at Lord Spencer's at Wimbledon in Surrey, it was pierced to the depth of 530 feet without passing through it. Most of the wells in Essex are very deep: at Colchester barracks 108 feet, Chelmsford barracks 300 feet (P. T.), East Hanningfield 474 feet (G. Notes), Epping 392 feet.

The actual thickness of this clay in Sheppey is estimated, by adding the height of the cliff to the depth of the wells, at 530 feet, but it may be supposed to be much thicker in Essex; for if to the depth of this clay at Epping (392 feet), we add 300 feet, the superior height of High Beech, (which is about five miles from that place and scarcely one mile nearer the termination of the clay), we may assume it at High Beech to be about 700 feet thick. (P.)

(*g*) *Inclination.* The beds of this clay are so nearly horizontal, that no perceptible difference from that position has been observed that we are aware of in the chalk basins of London and the Isle of Wight; except in the Isle itself (see Pl. 2. fig. 6.), where this bed is nearly vertical. This extraordinary deviation in regard to position will be treated of hereafter in speaking of the numerous accompanying beds of sand and clay belonging to the plastic clay formation, and of the chalk, of which the position is nearly the same.

(*h*) *Agricultural character.* This clay chokes the plough and rolls before it in a broken and muddy state; after rain it is not slippery, but adheres to the shoes; after drought it presents cracks nearly a yard in depth and several inches in breadth. On the Nore, south of Walton it forms a sort of pavement in many places, and divides by desiccation into small columns resembling, on a small scale, the Giant's causeway. (G. Notes.)

mile distant from the coast; in the Gentleman's Magazine is a view of it about the middle of the last century, which still represents a considerable space as intervening between the north wall of the church-yard and the cliff; that wall and half the church-yard have since been washed away and the church is yearly threatened. The accumulation of ruin at the foot of the cliff is striking, the whole area being included by the massive walls of the Roman station Regaltium; large fragments of that wall are mingled with the wreck of modern cottages and the boulders of marlstone washed from the clay, and the beach is strewn with fragments of Roman pottery and bones from the modern cemetery. (C.)

According to Townsend it is sometimes called *Wood-sowerland*, because, although it is productive of the finest elm, oak, and ash timber, it requires chalking before it can produce good corn. (G. Notes.)

Barren as this clay naturally is, it is rendered by prodigality of manure in the neighbourhood of the metropolis, excellent garden ground. (G. Notes.)

On Epping Forest, most of Windsor Forest, and much of the New Forest, the oaks are finest where the clay is intermixed with the sand lying above and below it. (G. Notes.)

The surface of the vegetable mould does not however in most cases rest immediately upon the London clay, but upon beds of rich marles and loams, which often alternate with gravel and sand, and sometimes to the depth of 30 or 40 feet.

(*i*) *Phænomena of water and springs*. This clay is so extremely dense, as to be almost impervious to water. Hence but few springs issue immediately from the stratum itself, and whenever these are found, the water is impure owing to its containing salts, which render it hard, and unfit for domestic purposes: the pyrites which in some places abounds in this clay, undoubtedly contributes, by its decomposition, to render the water impure.

The late Mr. Tennant is said to have noticed the existence of sulphate of magnesia in the London clay. It has not however been ascertained whether the springs of Epsom in Surrey, which arise from some of the strata above the chalk, arise from this stratum. (G. T. vol. ii. p. 138.) Many other localities are mentioned by the writers on mineral waters, in which this formation is said to yield springs impregnated with the same salt, viz. Bagnigge-wells, St. George's-fields, Kilburn, Kensington, Pancras, and Richmond; these springs must in all probability rise from the London clay, being situated far within its boundary. The position of the following, being near the border of the formation, render their source more doubtful; Acton (Middlesex), Barnet (Herts), Brentwood, Upminster, and Colchester (Essex), Dulwich, Streatham, (Surrey). Muriate of soda is said also to be contained in some of these springs. It is probable however that they have never been correctly analysed; but these indications imperfect as they are, deserve notice. (C.)

The dense nature of this stratum is of vast importance to the metropolis and its vicinity. The alluvium covering the surface of this clay is full of water, and the quantity daily drawn from it alone in the metropolis itself is almost incredibly great. Many of the wells furnishing to the inhabitants a plentiful supply of remarkably limpid but somewhat hard water, drawn by the public pumps, are not deeper than the alluvium; which also furnishes, by means of very shallow wells, many of the large distilleries, sugar houses, and some of the breweries with astonishing quantities, of which some notice will hereafter be taken in treating of Alluvium generally. The soft pump water afforded by some of the wells in London, and very many on its north and north-east, (and of which several sections have already been given under the heads *a* and *f* rises from beneath the London clay); and many places, now abundantly supplied by these perforations, were without water well adapted for domestic purposes, until within the last 30 years. Wherever a well is sunk, the immediate rise of the water has some effect in depressing for a time that of the neighbouring wells; the sinking of one on the south of the Thames above London Bridge, has even lowered the water in one on the northern border of the river, proving that the currents of the river flows over this stratum. (P.)

The water afforded by these wells, and which rises from the sands of the Plastic clay formation underlying it, is very limpid and free from salts; it is therefore what is termed *soft* in a remarkable degree, is adapted to every domestic purpose, and never fails. It frequently rises so instantaneously on passing through the clay, as not to suffer the well-digger to escape without rising above his head. It appears to rise in different places to different heights: at Liptrap's distillery at Mile End near London, it stands in the well precisely at the level of high water mark in the Thames; but at Tottenham, four miles north of London, it rises 60 feet above that level, for the water has stood for twenty years in my own well within ten feet of the summit, which is 70 feet by barometrical measurement above high-water mark at London bridge; while in a well at Epping, about fifteen miles north-east of London, the water rises to within 26 feet of the summit of a well, 340 feet above high water mark in the Thames, and therefore 314 feet above that level.* (P.)

* The history of this well, which was sunk by my friend Isaac Payne at Epping, and of another at Hunter's Hall, two miles from Epping, furnish some facts, not readily explained on the supposition that the water of both is derived from the common source of the wells sunk through the London clay, viz. the sands of the plastic clay formation.

The summit of the well at Epping, as above stated, is 340 feet above high water mark. The first 27 feet from the surface consisted of gravel, loam, and yellow clay, then blue clay 380 feet, then alternating beds of sandy beds of blue clay, and of blue clay unmixed with sand, and three or four feet thick, continued for 13 feet more; in the whole 420 feet, of which 200 feet were sunk through, and 220 bored, four inches diameter. As no water was found, it was considered as a hopeless labour: the boring was abandoned and the well covered over: at the end of five months it was found that the water had risen to within 26 feet of the surface, and it has so continued. The sinking was therefore 340 feet above the level of the Thames, and 80 feet below it. The water is limpid and soft.

The summit of the well at Hunter's Hall was found by levelling to be about 70 feet above that of the well at Epping, and therefore 410 feet above the high water mark of the Thames: but the depth of the well was only 350 feet; it therefore did not reach the level of the Thames by 60 feet: and the water stands in it 130 feet above the bottom of the well.

The facts relating to these wells are expressed by the following diagram. (P.)

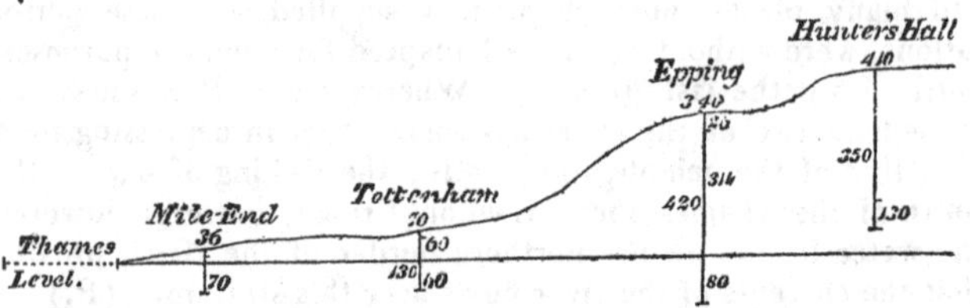

The depth to which it is necessary to sink in order to procure water beneath any retentive stratum, will of course vary, as the inclination or undulation of the inferior water-bearing stratum brings it nearer to the surface. The only difficulty then is, to explain why water issuing from the same stratum should stand at different levels, since it would appear on the laws of hydrostatics, that the water in every well so sunk, ought to have a common surface, that surface being determined by the lowest level at which the water-bearing stratum appeared exposed to the day along its basset edge; for since such a stratum would necessarily shed its waters from that point, it may seem difficult to understand how it can act as a reservoir to retain them at any higher point in the interior of its area. But a little consideration will shew, that this conclusion supposes a full and instantaneous communication to exist between the various portions of the water-bearing stratum, whereas in nature no such case exists, the porosity of every stratum being imperfect, and greater in some parts than others; hence, when any adjacent valley is excavated so deep as to expose a portion of the water-bearing stratum, it will drain it in that part, and reduce the level of the water in the wells immediately contiguous to that of such drainage. Thus the valley of the Thames, denudating the plastic clay near Deptford, brings down the water of the wells on the east of London to its own level; but the communication between the points so drained, and the more distant parts of the stratum, not being perfect, these will not be drained to the same degree. The water therefore procured from them will rise to a higher level; thus the waters which have been collected on the basset of the sand accompanying the plastic clay against the chalk hills of Essex, &c. percolating gradually through and saturating their mass, will rise to different heights in proportion as the low points of drainage are near or distant. The only general rule that can be deduced is,—that the water of wells can in no case rise to an higher level, than the highest point of the strata collecting them; but the local circumstances of the drainage, effected in the vallies traversing those strata, may compel them to assume various inferior levels in the proximity of such vallies; faults which may cut off portions of the water-bearing strata from their general mass, or dam up their waters in particular directions, will also afford other causes of local variation. (C.)

CHAPTER V.

PLASTIC CLAY FORMATION.

Section I. *General view of the Plastic Clay.*

(*a*) *Chemical and external characters.* Viewing it on the great scale, we may consider this formation as composed of an *indefinite number of sand, clay, and pebble beds, irregularly alternating.* Of these, the sands form in England the most extensive deposition; *in which* the clay and pebbles are interposed subordinately, and at irregular intervals. An attentive examination of the general points of resemblance in the substance of the clays, sands, and pebbles, forming these irregular alternations above the chalk, leaves no doubt as to their being members of *one great series of nearly contemporaneous depositions,* intermediate between the chalk and the London clay. (G. T. vol. iv. p. 209.)

This formation was named in conformity with the term by which an analagous deposite has been designated by M. M. Cuvier and Brongniart, who discovered it overlying the chalk in the basin of Paris. For some remarks on the analogy between the Plastic clay formations of England and France, the reader is referred to a communication by Mr. Webster on the strata lying over the chalk in the second volume of the Transactions of the Geological Society (p. 200 & seq.), from which however, it may be advantageous here to insert the following quotation. 'The Plastic clay of the Paris basin is described as sometimes consisting of two beds separated by a bed of sand. The lower bed is properly the plastic clay. It is unctuous, tenacious, contains some siliceous but no calcareous matter, and is absolutely refractory in the fire when it has not too great a portion of iron. It varies much in colour, being very white, grey, yellow, grey mixed with red, and almost pure red. This clay is employed, according to its quality, in making coarse and fine pottery and porcelain. The French sands are of great variety of colours. A species of imperfect coal occurs in the lower strata of the Paris basin.'

It is noticed above that this formation consists in England primarily of beds of sands, clays and pebbles. The sands are of an almost infinite variety of colours, at Alum Bay on the coast of the Isle of Wight; but the sand appears to pass into sandstone at Studland point on the opposite coast of Dorsetshire. The clays also are of various colours and degrees of purity, and are sometimes laminated: thus we have fire clay, brick clay, pipe clay and potters' clay,* the two former at Cheam in Surrey and at Poole in Dorsetshire, the second also at Reading in Berkshire; the two latter near Poole and in the Isle of Wight: an argillaceous rock appears as one of the beds at Newhaven on the coast of Sussex: layers of chalk flint pebbles are found in Alum bay, at Newhaven, and under the clay beneath London. Wood coal occurs in beds in the Isle of Wight, in Dorsetshire, and at Newhaven. Fullers' earth in the beds at Catsgrove near Reading and on the Edgeware road. These particulars will serve to shew the close analogy subsisting between the beds of this formation in England and France.

(*b*) *Mineral contents.* The mineral contents of these beds do not appear as veins, but are generally intermixed with the clays and sands of which the formation consists. If indeed the coal found in this stratum is to be ranked among its mineral contents, it forms an exception; being uniformly found in layers or beds parallel with the sands and clays above and below it. This coal however, is a very imperfect one, and still bears decided marks of its vegetable origin: nevertheless it seemed to demand at least some notice under the head mineral contents. It will be spoken of more at large at treating of the Organic remains common to this formation.

Pyrites occurs both in the pure white clays of the trough of Poole and the impure beds of the Isle of Wight; in the latter it abounds, and owing to its decomposition alum works formerly existed in Alum bay. (G. Notes.) A thick dark blue clay, very near the vertical chalk of Alum bay, on the north, contains green earth, and nodules of a dark coloured limestone. Selenite and fibrous gypsum occur both in Alum bay and near Newhaven on the coast of Sussex, and mica in the sand of the latter, as well as tubular ironstone.

(*c*) *Organic remains.* The organic remains of this formation

* Patches of *Plastic clay* are frequently found over the chalk in different places in the south of England; some of these are yellow, and are employed for the common sorts of pottery; but others are white, or greyish white, and are used for finer purposes. The coarser clay is very frequently met with, nor are the finer kinds of very rare occurrence. In the Isle of Wight two species of plastic white clay are worked for the purpose of making tobacco pipes, for which purpose also there is an extensive quarry, situated in several beds of white clay, on the banks of Poole harbour in Dorsetshire, and about two miles west of that place. A similar clay, which is used for making gallipots, is dug from the banks of the Medway A fine light ash-coloured, nearly white clay, which is employed in the pottery works, is also dug at Cheam, near Epsom in Surrey. (G. T. vol. i. p. 344.)

consist of ostreæ, cerithiæ, turritellæ, cythereæ, cyclades, &c. together with the teeth of fish, imperfect coal partaking more or less of the woody fibre, and sometimes exhibiting even the branches and leaves of plants:* and on the authority of Wilson Lowry, Esq. we are enabled to state, that in some beds belonging to the plastic clay formation in the immediate vicinity of Margate, fossil bones have lately been discovered.

The occurrence of organic remains in the different beds of this formation, is, like the alternation of the strata composing it, exceedingly irregular; sometimes they occupy the clay, at other times the sand or pebbles, and very frequently are wanting in them all. (G. T. vol. iv. p. 299.) There is not the smallest trace of animal or vegetable remains in any of the strata of this formation at Reading in Berkshire, except in the green sand. The same barrenness of organic remains is noticed in the purest beds of the French plastic clay, and by Mr. Webster (G. T. vol. ii. p. 200) in the plastic clay of the Isle of Wight and Corfe Castle.

(*d*) *Range and extent.** The sands and clays of this formation, but chiefly the former, are visible overlying the chalk of the London basin, and for the most part skirt the whole district occupied by the London clay, beneath which they also lie. The highest northern point at which this formation is seen, is two or three miles south-west of Hadleigh in Essex, whence it borders the clay to about five miles south-west of Braintree, but the line between these deposits is not well ascertained; Halstead and Coggeshall, and the intermediate tract are both on the plastic clay. Again, it extends from Ware on the north to near Edmonton on the south, over Enfield Chase, and passing close to St. Albans, it skirts the London clay to Uxbridge, on the north of which it takes a westerly direction to a little on the north of Beaconsfield in Buckinghamshire, and thence about south to the banks of the Thames. It is seen again at Reading in Berkshire, and extends thence, though not in a straight line, to a few miles beyond Hungerford, which

* Mr. Webster ascertained that the vegetables in this formation at Newhaven, agreed with those found in the Paris basin; one of these was the fruit of the Palm tree, another instance of the occurrence of the exotics of a warmer climate. (C.)

The north of France, from Beauvais to Reims presents a marly and sandy tract containing five beds of wood coal resembling those of the Isle of Wight, and like them supporting alternations of marine and fluviatile shells: similar facts are common among the contemporaneous formations of the continent. (C.)

* This indication of the exact limits of this formation on the side of its junction with the London clay, must, for the reasons assigned in treating of that formation, be regarded only as an approximation. (C.)

may be said to be its extreme point on the west, except a few outlying masses, south of a line from the latter place to Marlborough in Wiltshire. Turning south from a little on the west of Hungerford, to the foot of the chalk hills, it passes east by Kingsclere, Basingstoke, and Odiham in Hants, and Guildford in Surrey; thence rather in a north-easterly direction a little to the south of Croydon, it continues to skirt the foot of the chalk hills by Farnborough and Chatham in Kent, and thence by Milton and Ospringe to the foot of Boughton hill, where it divides, passing on the one hand in a north-easterly direction, it skirts the London clay to Whitstable on the coast, and on the other nearly east to Canterbury (which stands on the beds of this formation,) to the coast of the Reculver; whence again it passes to the south-west, except where marshy lands intervene, by Sandwich, which is built upon it, a little to the south of Deal.

In all this tract, the breadth of the surface occupied by the beds of this formation is not very considerable, scarcely exceeding four or five miles, except near its extreme point on the west; where it may be said to occupy a surface of 20 miles in length, on the north of the chalk range extending from Basingstoke to beyond Inkpen beacon, and 10 miles broad on the average. On the south of Woolwich, the tract also is considerable.

This formation likewise overlies a part of the chalk in Dorsetshire, Hampshire, and Sussex, included in the Isle of Wight basin. It there also skirts the London or Blue clay, and lies beneath the considerable tract of that clay on which Lymington and Portsmouth are built, and which occupies a large area surrounding them. The northernmost point in which the beds of this formation appear in this basin is Houghton hill, about 10 miles on the west of Salisbury in Wiltshire, except that it constitutes Chidbury hill and a few other outlying patches on the north-west. From Houghton hill it passes eastward, south of a line by Bishopstoke and Bishops Waltham over Bere forest, which consists of it, to Arundel in Sussex, which it touches on the south, quite to the western extremity of Brighthelmstone; at which place it terminates, except an outlying portion of it occupying the summit of the chalk cliff on each side of Newhaven harbour. From Houghton hill it passes in a direct line westward towards Salisbury, within about a mile of which place it turns south to Fordingbridge in Hampshire, and thence in a south-easterly direction by Cranbourne and Bere Regis to within two or three miles of the north-east of Dorchester in Dorsetshire, which is its extreme point on the west, except that it covers the summits of two or three chalk hills on the south-west of it. From near Dorchester it turns eastward skirting the foot of the chalk range on which stands Corfe castle, to the shore of the Isle of Purbeck, in Studland Bay, and again forms the shore of Poole Bay on the west of Poole to Christchurch head.

The surface occupied by these beds in the tract between Houghton hill in Wilts and Brighton in Sussex, does not exceed five miles for the first half of the way, but in the latter half it tapers off to a very narrow point at the latter place South of the line between Houghton hill and Dorchester its breadth is more considerable, averaging perhaps eight miles, and occupying a considerable tract on the west of Poole harbour, termed the Trough of Poole.

It is again visible in the Isle of Wight, where the extraordinary appearance of this formation in the vertical and many coloured sands and clays of Alum Bay, will require further notice. In a word these sands and clays being seen nearly in the same position in that bay on the west, and in White Cliff Bay on the east, leads to the conclusion, that they range along the island between the London clay on the north and the Chalk on the south. (G. Map).

(*e*) *Surface of country.* The surface of this formation may in the general be said to be flat: on the north-east of London it is remarkably so, but on the north and north-west, as on the south-west of Hertford, it attains considerable elevation: between Deptford in Kent and its extreme point in the west, it assumes the form of gentle eminences. Between Houghton hill in Wilts and Brighton in Sussex, the country it occupies is very flat and even low; but between that hill and Dorchester its undulations are considerable: it rises into a range of considerable eminence some miles on the south-east of Salisbury, of which Dean hill is the westermost extremity, and again a few miles on the west of Ringwood in Hants, as well as in the Trough of Poole.

(*f*) *Thickness.* The thickness of this formation does not appear to have been ascertained in many places. At Wormley End near Wormley Bury in Hertfordshire, and just beyond the boundary of the London clay in that neighbourhood, Sir Abraham Hume found chalk at 100 feet below the surface. (G. Notes.) At the sinking of a well at Liptrap and Smith's distillery, one mile east of London (of which the sinkings are given under the head of local details), the Beds of this formation appeared to be 104 feet thick, between the London clay and the chalk; and from 100 to 200 feet seems the average thickness round Woolwich. But the thickness of these beds appearing in the vertical cliffs of Alum bay in the Isle of Wight (see Pl. 2. fig. 6) is no less than 1100 feet. It was however very possible (C.)

that their vertical position has, by pressure, induced a very considerable expansion; indeed, considering the loose state of the sandy beds, this can hardly have failed to be the case.

(*g*) *Inclination.* Sections of the beds of this formation appear in the cliffs between Lymington and Poole in Dorsetshire, where they lie in a horizontal direction. (G. T. v. 2. p. 189.) But on the opposite shore of the Isle of Wight in Alum bay, they are in a vertical position, the probable cause of this remarkable circumstance will be adverted to in treating of the nearly vertical chalk of that island. It must however here be observed, that although at the east and west shores of the Isle, viz. at White cliff and Alum bay, all the beds of this formation appear to be vertical, yet in the more central parts they differ, for Newport in the Isle of Wight is on horizontal strata. The organic remains which have been found in perforating these strata to the depth of 200 feet, are the same as those found in the vertical marles and clays of Alum bay. (W. p. 178.)

The dip of the beds at Chimting near Newhaven on the coast of Sussex, is conformable with that of the chalk on which they lie, viz. about 20° to the west. (G. T. vol. iv. p. 295.) That of the beds traversed by the Redriffe Tunnel is 1. in 30. south. (C.)

SECTION II.

Local details.

Although the forementioned facts may be considered as offering a general view of the principal phœnomena of the Plastic clay formation of England, it will be impossible to do that justice which the importance of the subject demands, without entering considerably at large into the particulars which we have been furnished by various observers; especially since they will serve to direct the attention of the traveller to the places best adapted for observation, which are, near Reading in Berkshire, near London, at Newhaven on the coast of Sussex, near Poole in Dorsetshire, and in the Isle of Wight.

(*a*) *Near Reading.*

The Catsgrove brick kilns, distant about half a mile from the town of Reading, on the south-west, where the works have been carried on for more than a century, at this time present the following section, beginning from the chalk on which the beds constituting the plastic clay formation rest. (G. T. vol. iv. p. 278.)

Section of Catsgrove Hill.

No.		*thickness in feet.*
1.	Chalk containing the usual extraneous fossils and black flint	unknown
2.	Siliceous sand mixed with granular particles of green earth, and containing both rolled and angular chalk-flints, oysters, and many small and nearly cylindrical teeth of fish from a line to an inch in length	3
3.	Quartzose sand of a yellowish colour with a few small green particles, and containing no pebbles or organic remains	5
4.	Fullers' earth	3
5.	White sand used for bricks	4
6.	Lowest brick clay of a light grey colour mixed with fine sand, and a little iron-shot	5
7.	Dark red clay,* mottled with blue, and occasionally a little iron-shot. It is used for tiles.	6
8.	Bed called the White vein. A fine ash coloured sand mixed with a small portion of clay and in some parts passing into loose white sand. It is used for bricks	5
9.	Fine micaceous sand laminated and partially mixed with clay, and occasionally iron-shot. It is used to make tiles	4
10.	Light ash-coloured clay, mixed with very fine sand of the same colour. It is used for bricks.	7
11.	Dark red clay partially mottled and mixed with grey clay	4
12.	Soft loam, composed in its upper region of fine yellow micaceous sand, mixed with flakes of a delicate ash coloured clay, which become more abundant in the deeper portions of the stratum, and having its lower regions much iron-shot, and occasionally charged with ochreous concretions, and decomposing nodules of iron pyrites. It is used to make soft bricks for arches	11
	Total	57

* The red clay of Reading, on the north of the Hog's back, and at East Horsley, is perfectly identical with that of Meudon in France, nor have I found this colour equally intense in any other clay. The bricks made of this clay are of a bright Roman ochre colour. (G. Notes.)

13. Alluvium composed of clay, sand and gravel, the gravel chiefly consisting of chalk flints, both rolled and angular, with a few pebbles of quartz, and of brown compact sandstone. This alluvium is covered with vegetable mould.............

The oysters of No. 2 are remarkably perfect when first laid open, and seem to have undergone no process of mineralisation; they soon fall to pieces by exposure to air and moisture. The chalk flints contained in it are many of them in the state of small rounded pebbles; in others the angles are unbroken. Both varieties are covered with a crust of greenish earth of the same nature with the green particles in the sand. The angular flints appear to have been derived from the partial destruction of the bed of chalk immediately subjacent, of which the upper surface in contact with the sand is considerably decomposed to the depth of about a foot, and its fissures and numerous small tubular cavities (the latter derived apparently from the decay of organic substances) are filled with granular particles of the green earth and siliceous sand, the incumbent stratum.

In other large quarries of brick earth on St. David's hill, west of Reading, and only a quarter of a mile distant from the Cats-grove, many of the subdivisions noted in the latter do not appear.

The prevailing organic remains found near Reading are oyster shells, and these are found only in the lower bed consisting of sand mixed with green particles. Teeth apparently of sharks are mingled in the mass.

(*b*) *Near London.*

Very little attention has yet been given to the beds of this formation immediately on the north or north-east of the metropolis, but Woodward mentions oysters as being found on the north side of the Thames, in a stratum of sand that covers the chalk near Hertford: this probably is one of the oyster beds of this formation. (G. T. vol. iv. p. 293).

A section of the beds of this formation, underlying the Blue clay near London, is given below, in an account of the beds passed through in sinking a well at Messrs. Liptrap & Smith's distillery.*

* Section afforded by the sinking of a well at Liptrap & Smith's distillery, one mile on the east of London. (P.)

Surface of ground 36 feet above high-water mark.

Alluvium....	7 feet of earth and other substances
	22 feet of gravel and sand—containing *land springs*
London Clay	77 feet of Blue clay

On the south of London, these beds occur with well defined characters, at Blackheath, Lewisham, Charlton, Woolwich, and on the west of Plumstead. In all these places, *the thin bed next*

Plastic Clay Formation....	5 feet of sand—*a spring*
	6 feet of soft Blue clay
	2 feet of a kind of marl—yielding *some water*
	9 feet of yellow clay
	12 feet of Blue clay intermixed with gravel
	9 feet of hard brown clay mixed with gravel
	12 feet of green-coloured sand and gravel
	12 feet of small gravel and sharp sand—*a good spring*
	$1\frac{1}{2}$ feet of green-coloured clay
	$35\frac{1}{2}$ feet of light-coloured sand
Chalk.......	2 feet of flint & chalk; mostly flint; very hard; *a spring*
	36 feet of chalk & flint—hard
	23 feet of ditto .... somewhat softer
	62 feet of ditto .... *a spring*
	37 feet of ditto .... no water
	370

The first 29 feet a circular well 12 feet diameter,
The next 77 feet (i. e. 106 feet from surface) 10 feet diameter,
The remaining 264 feet bored 8 inches in diameter.

The water from each of the springs in the Plastic clay formation and the Chalk, rose precisely to the same height, viz. exactly to 36 feet from the surface, and therefore to the high water-mark of the Thames. The springs of the Chalk yielded the greatest quantity of water.

Section afforded by the sinking of a well at Joseph Foster's, at Bromley, near Stratford-le-Bow, in Middlesex. (P.)

Alluvium, &c.	18 feet of loam, clay, gravel and sand
London Clay	44 feet of Blue clay—*water from beneath it*
Plastic Clay	2 feet of Blue clay } *soft water*
	1 foot clay, sand and shells, mixed } *soft water*
	4 feet gravel, sand and shells } *soft water*
	4 feet fine sand
	9 feet blue & yellow clay
	4 feet sand & shells, with large lumps of pyrites—*a spring* of soft water, sufficient to fill a $2\frac{1}{2}$ inch pipe
	9 feet Blue clay with abundance of broken shells, some resembling oysters, and pyrites
	1 foot solid limestone
	22 feet { Black sand passing into small round pebbles like the Blackheath pebbles; Black sand veined; Some small pebbles in the sand, which is still hard and compact
	2 feet of Blue clay very hard and firm—*copious spring* of water, which threw much fine white sand into the pipe
	120

above the chalk, which at Reading contains fishes teeth and oysters, is composed of a similar substance of loose sand mixed with chalk flints, both rolled and angular, and generally coated with a dark green crust; but *here they contain no organic remains*, and seldom exceed two feet in thickness. Above this thin bed, is a stratum of fine-grained ash-coloured sand, destitute of shells or pebbles, and varying in thickness generally from thirty to forty feet. The stratum is seen to the greatest advantage in the Woolwich sand-pits, which present an enormous artificial section. (G. T. vol. ii.)

The clay beds of the Woolwich pits, and of Loam Pit hill near Lewisham, contain cerithia, turritellæ, cythereæ, ostrea, &c. which occur in patches in the stratum of rolled flints, and sand covering the clay beds near Woolwich. This shelly stratum may be traced hence, along the slope of the hills overhanging the marshes of the Thames through the parish of Charlton; a part of the Woolwich excavation is called New Charlton hill, and there are

The following account of the tunnel at Rotherhithe, is by the Rev. W. D. Conybeare.

The remarkable attempt to carry a tunnel beneath the Thames at Rotherhithe, in order to facilitate the communication between the two sides of the river, was carried on in the beds of this formation about a mile on the south of the last section, with which therefore it furnishes a good point of comparison.

Two shafts were sunk, one on each bank, between 50 and 60 feet deep, then a horizontal tunnel was carried from the bottom of the southern towards the northern shaft, running about 30 feet below the bottom of the river; more than two-thirds of the distance had been excavated when a quicksand supposed to communicate with a hole in the bottom of the channel was penetrated, the works were in consequence immediately flooded and abandoned.

The strata rose from north to south about one foot in 30, so that the upper bed of London clay which exhibited a thickness of 49 feet in the north shaft, was reduced to nine feet in the southern; and the strata penetrated in sinking the latter shaft, were again cut in driving the tunnel, being brought down successively to its level by their inclination.

1. The London clay had its usual characters, the upper part being blue, the lower variegated, the strata occurring beneath it were as follows:

	feet	in.
2. Loose watery sand and gravel	26	8
3. Blue clay	3	0
4. Loam	5	1
5. Blue clay with shells, chiefly cytherea	3	9
6. Hard conglomerate rock, consisting of flint gravel with a calcareous cement	7	6
7. Light blue laminated clay with pyrites	4	6

It is obvious that no ground could be worse chosen for the purpose than this, consisting of loose strata of sand, &c.: had the attempt been made where the channel of the river is entirely in the firm London clay, it would probably have succeeded, and added another to the wonders of the metropolis.

others almost equally extensive immediately below Charlton church: on the east the same shells may be traced as far as Plumstead; they are well exhibited in a ravine just beyond the east side of Woolwich common in that direction; the detail of these beds is given in the following sections. (C.)

1. *Near Plumstead in Kent.*

In a pit about a mile south-east of Plumstead were found Cerithia, Turritellæ, Cythereæ and Calyptræa trochiformis Lam. Trochus apertus Brander. Arcæ glycemeres, Arcæ naticæ, and many minute shells in good preservation but extremely brittle, and having their surfaces every where indented with impressions of the minute sand, which is especially obvious in the Cyclades. On the heath near Crayford, about four miles east of Charlton, long vaulted oysters are found; and about two miles further in the parish of Stone, is *Cockleshell bank*, so called from the great number of shells observable in it, which are of the three kinds first mentioned above. Fragments of these shells are frequently turned up by the plough in the neighbourhood, and the same shells have been found at Dartford, Bexley and Bromley to the southward. Near Bromley stone is found near the surface, formed of oyster shells still adhering to the pebbles to which they were attached, and which are similar to those occurring in the beds of Charlton and Plumstead, the whole being formed by a calcareous cement into a coarse shelly limestone containing numerous pebbles: a quarry of this stone is worked in the grounds of Claude Scott, Esq. The Strombus pes pelicani, and a species of cucullæa, in a siliceous state, have been found at Faversham in Kent, in a bed of dark brown sand slightly agglutinated by a siliceous cement, and lying over the chalk. (G. T. v. i. p. 343—4.)

2. *Woolwich Pits.* (See G. T. v. iv. pl. 13. No. 1.)

No.		feet.
1.	Chalk with beds and nodules of black flint.....	—
2.	Green sand of the Reading oyster bed, containing green coated chalk flints, but no organic remains	1
3.	Light ash-coloured sand without shells or pebbles	35
4.	Greenish sand with flint pebbles	1
5.	Greenish sand without shells or pebbles	8
6.	Iron-shot coarse sand without shells or pebbles, and containing ochreous concretions disposed in concentric laminæ..........................	9
7.	Blue and brown clay, striped, full of shells, chiefly cerithia and cythereæ..............	9

8. Clay striped with brown and red, and containing a few shells of the above species 6
9. Rolled flints mixed with a little sand, occasionally containing shells like those of Bromley; e.g. ostrea, cerithium and cytherea, disseminated in irregular patches 12
10. Alluvium. ——

Total thickness 81

No. 1 and 2 are not laid open in the great sand pits, but are seen in a chalk pit adjoining to the eastern extremity of the sand-pit.

The Woolwich shell beds may be again seen on the west of Blackheath at a lock of the canal about a mile above New Cross towards Croydon, in the plain that lies under the east side of the Sydenham hills. At this lock have been observed ancilla buccinoides, cerithium denticulatum, cyclas deperdita, a small buccinum, and a small nerite. (G. T. vol. iv. p. 292.) The same shells (C.) are found in the wells between Camberwell and Deptford.

It is mentioned by Woodward that the Woolwich shells are found at Camberwell and Beckenham, on the north-west and south-east sides of the Sydenham hills. They also occur at the following places in the same line; namely, Redriffe tunnel, Lewisham, Blackheath, Woolwich, Beckenham, Chislehurst, Bexley, Cockleshell Bank, two miles south of South Fleet, Windmill hill near Gravesend, and Higham on the Thames and Medway canal. They are also found at Rungewell hill near Epsom, and at Headley between Epsom and Dorking. (G. T. vol. iv. p. 293.) All these localities (C.) are precisely indicated by figures in the map accompanying the memoir on this formation in the volume referred to. Further in Kent, near Faversham, the ash-coloured sand of this formation, though usually destitute of fossils, contains silicified arcæ, glycemeres and naticæ canrenæ.

3. *Loam Pit hill.* (G. T. vol. iv. pl. 13. No. 2.)

The following Section at Loam Pit hill, near Lewisham, about three miles south-west of Woolwich, presents analogies that identify many strata in that and the preceding section, as from the chalk upwards to No. 8, in each inclusively; the principal difference consists in the presence of fewer or no pebbles, in beds of sand evidently contemporaneous. The beds presented by the following sections cannot all be observed at one place, but may be traced along the sloping surface of the hill, at three successive apertures near each other, in which the upper stratum of each lower pit is dug into, and forms the floor of the next above it.

No.	*Lower pit.*	feet.
1.	Chalk with beds and nodules of flint..........	—
2.	Green sand identical with the Reading oyster bed, and in every respect resembling No. 2 at Woolwich..............................	1
3.	Ash-coloured sand, slightly micaceous, without pebbles or shells..........................	35
	This bed, though below the general floor of the middle pit, is sunk into it by deep shafts.	
	Middle pit.	
4.	Coarse green sand containing pebbles	5
5.	Thick bed of ferruginous sand, containing flint pebbles	12
6.	Loam and sand, in its upper part cream-coloured, and containing nodules of friable marle, in its lower part sandy and iron-shot	4
7.	Three thin beds of clay, of which the upper and lower contain cythereæ, and the middle oysters	3
	Upper pit.	
8.	Brownish clay containing cythereæ	—
	This is the lowest bed sunk into the upper pit, and is not there penetrated to a depth exceeding one foot. The interval between this and No. 7, which occupies the summit of the middle pit, is not exactly ascertained, but cannot be considerable, probably	6
9.	Lead-coloured clay containing impressions of leaves..................................	2
10.	Yellow sand............................	3
11.	Striped loam and plastic clay, containing a few pyritical casts of shells, and some thin leaves of coaly matter	10
12.	Striped sand, yellow, fine and iron-shot	10
13.	Alluvium.	——
	Total	91

At a point still higher in the hill lies a mass of the London clay.

Similar beds may be traced round the sloping terrace that bounds the north-west and south sides of the plain of Blackheath, the surface of which consists of a bed of rounded pebbles,* about 20 feet thick; beneath this is a bed of sand, identical with No. 12 of Loam Pit hill, resting on another of plastic clay, which supports the water of all the wells on Blackheath, and possesses the same peculiar dark red colour with the plastic clay of Reading, Corfe Castle, and Paris, and has been used for pottery.† Beneath this clay, the Woolwich shell beds, and subjacent thick ash-coloured sand, are to be seen in several parts of the sloping terrace that surrounds the Blackheath plain. Under these, on the north side, appears the chalk, separated from the ash-coloured sand by the same thin pebble bed as at Reading.

* The bed of pebbles covering Blackheath consists almost wholly of rolled chalk flints, such as the neighbouring strata of the plastic clay formation contain abundantly, and from which they were probably derived. (G. T. vol. iv. p. 291.)

The *Hertfordshire puddingstone* is composed of ovate siliceous pebbles of various colours imbedded in a siliceous cement. These pebbles appear to be no other than altered chalk flints of the same æra with those found at Blackheath, and differing only in their being united by the cement. Many of the purest varieties of the Blackheath pebbles, if polished, are exactly similar to those of the Hertfordshire puddingstone. Large blocks of a coarse variety of the same siliceous puddingstone are not uncommon on the surface of the chalk in the south of England; as at Bradenham near High Wycombe in Bucks, at Nettlebed in Oxfordshire, at Portesham near Abbotsbury in Dorsetshire, and in Devonshire. They have not yet been found imbedded in their native stratum, which seems to have been destroyed extensively above the English chalk, and to have been a member of that series of irregular alternations of beds of clay, sand, and gravel, either separate or mixed together, which has been designated the Plastic clay formation. (G. T. vol. iv. p. 301.)

† It is probable that the plastic clay contains at Blackheath, as at Corfe Castle, Alum bay, and Loam Pit hill, the remains of vegetable matter approaching to coal; and that this circumstance has given origin to the erroneous opinion so prevalent, that there is good coal at Blackheath, if government would allow it to be worked.

The very high improbability of finding good coal above the chalk, is acknowledged by all who have even the smallest acquaintance with the English coal mines. The presence of black vegetable matter in a state approaching charcoal, in almost all our secondary argillaceous strata, has caused endless vain attempts to search for useful coal, in formations where the discovery of that substance would be contrary to all experience in this country. No good coal has I believe been yet found in England in any stratum more recent than the new red sandstone, or red rock marle. That of the Cleveland moors in Yorkshire being above the lias and in the oolite formation, is of so bad a quality as scarcely to form an exception to this position. (G. T. vol. iv. p. 289.)

4. *Near Bromley in Kent.*

At Sundridge park near Bromley in Kent, is an immense deposit of shells peculiar to the Plastic clay formation, accumulated confusedly in a bed of loose sand and pebbles. Of these shells some are broken, and others entire and delicately preserved. They are also sometimes fixed together by a calcareous cement (derived apparently from the substance of the shells themselves), forming a hard breccia with the siliceous pebbles and sand in which they are imbedded. A similar breccia was sunk into in the workings of the Redriffe tunnel. I have from the bed at Bromley, adds the Rev. W. Buckland, a specimen in which five oyster shells are so affixed to the opposite sides of a large kidney-shaped pebble, that they seem to have commenced their first growth on it, and to have been attached to it through life, without injury by friction from the neighbouring pebbles: we cannot but infer then that these pebbles received their form during a long period of agitation, which was succeeded by a period of repose, in which latter they were in a state of sufficient tranquillity for the shells in question to live and die undisturbed in the midst of them. (G. T. vol. iv. p. 300.)

5. *Near Ewell in Surrey.*

At Ewell in Surrey, the clay of this formation is worked as a fire clay; it occurs in two or three layers of different qualities, rising from under the edge of the London clay. The uppermost of these beds is of a reddish colour with blue veins; the next is a bed of clay about three feet thick, not unlike fuller's earth; and this rests upon sand of a similar brown colour; beneath which may be seen the lower bed of white sand, and under that the chalk. These beds of clay and sand, mixed in various proportions, are manufactured into tiles and bricks for ovens, furnaces, &c. where great heat is to be withstood. The lower sand may be seen to rest on the chalk on the south side of Addington hills, at Croomhurst, and in the neighbouring part of Surrey. (Middleton. C.)

(c) *At Newhaven in Sussex.*

The beds of sand at Newhaven on the coast of Sussex, enclose shells belonging to the genera cerithium, cytherea and ostrea, together with pyritous casts of them; fruit of one of the palm tribe with the fibres distinct, and impressions of leaves. There is also a thick bed of blue clay containing marine fossils, which are different from those usually found in the London clay, overlying beds of marle and clay containing coal, which are proved by the organic remains they contain, to belong to the Plastic clay formation. (G. T. vol. ii. p. 191.)

A similar deposition of sand to that of Reading, containing a breccia of chalk-flints as its lowest stratum (about three feet thick), lies between Newhaven and Beachy Head on the coast of Sussex, in the cliff at Chimting castle, half a mile on the east side of Seaford. The sand is here fawn-coloured, with small flakes of mica, and occasionally contains irregular veins and masses of tubular concretions of iron-stone. Its greatest thickness is under 50 feet. Under this sand, the breccia of the lowest bed forms an ochreous pudding-stone, composed of sand and chalk flints (the latter both rolled and angular), the whole being strongly united by a ferruginous cement, and the flints being covered externally by a green coating like those in the oyster bed at Reading in Berkshire. The chalk rises suddenly to a lofty cliff on the east side of the flat ground that lies between Newhaven and Seaford, dividing the beds of the Plastic clay formation at Newhaven from their outlying fragments at Chimting, with which they were probably connected before the excavation of the valley of the Ouse. At Chimting castle there is but a small insulated portion of the sand and breccia incumbent on the chalk; the sand soon ceases in ascending the hill eastward from the castle, and afterwards the breccia, having formed a thin cap on the chalk for a short distance, disappears a little below the signal-house, about one mile east of Seaford; between which and Beachy Head, it is believed that nothing appears on the chalk but occasional patches of alluvial sand and gravel. But on the west side of the Ouse at Newhaven, the breccia appears covered by the sand as at Chimting castle; differing only in being less firmly cemented, but appearing equally identical with the oyster bed at Reading. (G. T. vol. iv. p. 294—5.)

Section * of the Castle hill at Newhaven, commencing at the lowest bed. (G. T. vol. iv. p. 296.)

No.		feet
1.	Chalk, containing alumine, in hollows on its surface	50
2.	Breccia of green sand and chalk-flints, the latter covered with a ferruginous crust	1
3.	Sand varying from yellow to green and ash-coloured	20

* These beds appear again on the opposite shores of France, near Dieppe, in the same relative position.

4. Series of clay beds containing coaly matter, selenites and fibrous gypsum, also leaves of plants, and sulphur-coloured clay 20
5. Foliated blue clay containing cerithia and cyclades and a few oysters. In this clay is a seam of iron pyrites about an inch thick, with pyritical casts of cyclades and cerithia....... 10
6. Consolidated argillaceous rock full of oysters, with a few cyclades and cerithia 5
7. Alluvium full of broken chalk flints mixed with sand 10

(*d*) *Dorsetshire.*

Potters' clay is described by Dr. Berger as alternating with the loose sand of this formation in the trough of Poole, in beds of various thickness at different depths, and he says that from a cursory chemical examination by Dr. Marcet, the existence of alumine, lime, magnesia, oxide of iron and silica were discovered in it. It feels greasy and smooth, its colour varies from ash-grey to blue, its fracture is a little shining and uneven. It contains cylindrical blue nodules, (called *pins* by the workmen) of a more close texture, in which there is probably a greater proportion of the oxide of iron. This clay is sent to Staffordshire where it is mixed with ground flints, and employed in the finer kinds of pottery. Its specific gravity is 1.723. Beneath this potters' clay lies a seam of some thickness, of an extremely friable earthy brown coal, which crumbles to pieces when put into water. It burns with a weak flame, emitting a particular and rather fragrant smell of bitumen, somewhat analogous to that of Bovey coal, but is of less specific gravity. *Pins* of clay frequently traverse it. (G. T. vol. i. p. 254.)

All along the north side of the range of chalk hills which extend from Handfast point to beyond Corfe Castle, there is an extensive stratum of pipe clay in a horizontal position. It contains a bed of coal so exactly resembling that of Alum bay, that this circumstance, added to the quality of the clay, and its position above the chalk, is sufficient to identify it. The same stratum of clay, though not of equal quality, may be traced in the hills near Poole, and is found in many parts of that extensive tract called the trough of Poole. (W.)

I have been favored by S. L. Kent, Esq. M.G.S. with the following section afforded by a Quarry of Pipe clay, situated on the borders of Poole harbour, and about two miles west of Poole.

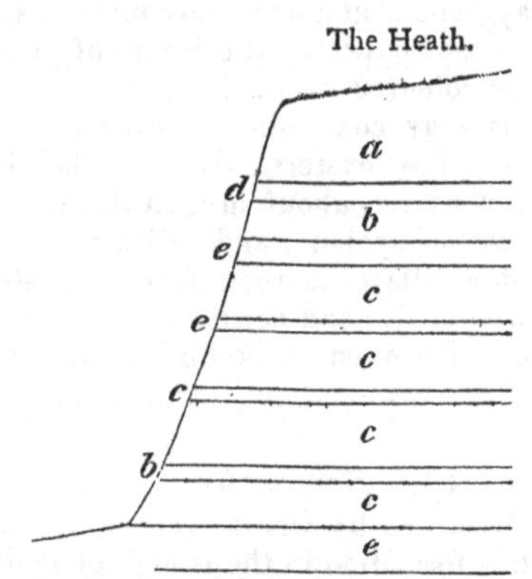

a. White sand and frit about 10 feet. *b b*. Brown clay 4 f. and 1 f. each. *c c c c*. Beds of white pipe clay, 3 to 5 feet each. *d*. Red sand 2 feet. *e e e e*. Black sand, 1 to 3 feet thick.

The *frit* (as it is termed by the quarrymen) appears to consist of white and brown sand, agglutinated by an iron-shot cement. The upper stratum of brown clay is used for fire bricks; of the other and thinner beds of brown, and also of the black clay, no use is made. In the white pipe clay are found nodules of indurated clay with pyrites. The sand on which these beds lie is nearly level with high water mark in Poole harbour. (P.)

Studland is near a romantic range of cliffs which end in a narrow neck of land called the Southhaven point, and forming the southern entrance to Poole harbour: the cliffs are of a compact yellow sandstone in which are several grottos and cavities. (G. Notes.)

(e) *Isle of Wight.*

The stratum next on the north to the nearly vertical chalk of Alum bay, is chalk marle (see Pl. 2. fig. 6.), then succeed green, red, and yellow sand, together about 60 feet in thickness, and in the same position,* and afterwards a bed of dark blue clay, about 200 feet thick, containing much green earth and nodules of a dark coloured limestone enclosing a few fossil shells mentioned below: next on the north follows a vast suc-

* The probable cause of the vertical position of these beds in Alum bay will be mentioned in treating of the chalk of the Isle of Wight.

cession of beds of sand of various colours, 321 feet in thickness; next to which, and in the middle of the bay, is a very numerous succession of beds, which contain a large proportion of pipe clay of various colours, white, yellow, grey and blackish: these alternate with beautifully coloured sands. The clay is sometimes in beds several feet in thickness, without any admixture, and sometimes in laminæ not a quarter of an inch thick, with sand between them: the whole being about 543 feet in thickness. Near the middle of these latter beds are three beds of a sort of wood coal, the vegetable origin of which is distinctly pointed out by the fruits and branches still to be observed on it. It sometimes splits into irregular layers in the direction of the bed, and the cross fracture is dull and earthy. It burns with difficulty and with very little flame, giving out a sulphureous smell. About 150 feet in the north are five other beds of coal similar to the preceding, each a foot thick. On the north of the whole of these beds of sand, clay and coal, of 543 feet in thickness, are several layers of large water-worn black flint pebbles, imbedded in deep yellow sand; to which succeeds a stratum of blackish clay with much green earth and septaria, 250 feet in thickness, and analagous to the London clay. (G. T. vol. ii. p. 184.)

These strata extend quite across the isle in a vertical position, keeping parallel to the chalk, and appearing again at White Cliff bay on the east end; where, however, they are much concealed by grassy slopes. The sands, marles, and clays are of every possible variety and shade of colour, giving to the cliff in Alum bay, which is about 200 feet high, when viewed at a little distance, a very beautiful appearance. The number and variety of these vertical layers is endless: they may be compared to the stripes on the leaves of a tulip, and are almost equally bright. On cutting down pieces of the cliff, it is astonishing to see the brightness of the colours, and the delicacy and thinness of the several layers of white and red sand (some of which do not exceed the eighth of an inch), shale and white sand, yellow clay, and white or red sand, and indeed almost every imaginable combination of these materials. The forms of particular parts of this cliff, when viewed near, and from the beach, are often of the most picturesque and even sublime kind, resembling the weather-worn peaks of Alpine heights. The fact is, that the rain and weather has worn away the softer parts, leaving the more solid, sharp and pointed. (W. p. 160.)

Among other fossil shells, too imperfect for the discovery of their genus, were found cythereæ and turritellæ in masses of dark coloured limestone in the thick stratum of blue clay, a little on the north of the nearly vertical chalk of Alum bay. (G. T. vol. ii. p. 191.) The fossil coal of the several beds already mentioned, and of which the vegetable origin is discoverable by the fruits and branches still observable, burns with difficulty and with very little flame, giving out a sulphureous odour. Eight beds occur, each about a foot thick. (W.)

APPENDIX.*

On the Formations above the Chalk in Yorkshire and Lincolnshire.

The northern extremity of the formations above the chalk in Lincolnshire and Yorkshire, separated from the main tract of the London basin by the æstuary of the Wash, has not yet been satisfactorily explored, and has been too hastily considered as entirely concealed by alluvial or diluvial deposits; but, from the following sources of information it should appear that this is far from being universally the case, and that the regular substrata may be observed through considerable tracts; this district therefore seems to merit a particular notice: since however the intelligence to be gleaned is too scanty to identify the exact place of the strata mentioned in the regular series, it has been determined to throw them together in this article.

The south-east angle of Yorkshire from Bridlington bay to Spurn head, known under the name of Holderness, is peninsulated by the Hull river, which is skirted on either side by a broad tract of marsh land, protected from the incursions of the high tides by its embankments only; a breach in which would reduce it to the state of an island, separated by a strait about six miles broad from the chalk hills of the Wolds. These marshes are remarkable for the occurrence of the remains of an extensive forest at the level of the present low-water-mark, which may be traced at the same level round the coast of Holderness and through Lincoln; this will, on a future occasion, form an interesting subject of inquiry (see Book VI); at present our concern is with the regular strata.

The district of Holderness rises from these marshes into low hills, which face the sea with a line of cliffs sometimes exceeding 100 feet in height, remarkable for the rapid encroachments of the sea, which in these quarters seems more than once to have gained on the land, and again relinquished its contests.

* By the Rev. W D. Conybeare.

Pennant, in the introduction to his Arctic Zoology, describes these cliffs as being very lofty; as extending from near the village of Hornsey on the south of Bridlington bay to Spurn Head, as consisting of brown clay, and containing amber; these particulars seem to indicate the formations of either London clay, or Plastic clay.

Dr. Alderson, in a paper on the geology of the neighbourhood of Hull and Beverley, published in the third volume of Nicholson's Philosophical Journal, considers the hills of Holderness as alluvial accumulations heaped upon the sub-marine forest before mentioned; which, from its level on both sides, he thinks must extend beneath them, and mentions a spot on the coast (at Seathorm and Withernsea) where that forest may now be traced within 100 yards of the cliff; which, from its rapid wasting away by the sea, must he thinks formerly have covered it; but this observation seems scarcely strong enough to warrant his conclusion: he describes the district as composed partly of gravel, partly of clay, variously mixed with *shells*, with here and there particles of *culm or powdered coal*, a description much resembling that of the Plastic clay strata. The superficial gravel, however, is most probably an alluvium resting on these strata, since it is said to contain fragments of older rocks, apparently brought by a current from the north.

In boring beneath the marshes, they are found to rest on a stratum of sand; beneath this occurs a bed of clay, which finally reposes on the chalk. There we undoubtedly have the lower members of the Plastic clay formation—these beds are horizontal, the chalk beneath dips slightly to the east, five yards in the mile—the springs which percolate through the chalk rising, on penetrating the clay, to the level of the marshes. There can be no doubt that the substrata concealed beneath the marshes along the adjoining coasts of Lincolnshire, are similar to the above.

SYNOPTICAL TABLE.

Book II.

SUPERMEDIAL ORDER.

Synonymes.—This class includes generally all the secondary formations more recent than the great Coal-deposit, and between it and the Tertiary or Newest Flœtz class. As the first Flœtz Limestone of Werner corresponds with the lowest calcareous formation of this order, it may be said to be co-extensive with the Flœtz class of his school, as distinguished from the Newest Flœtz; but some Wernerians include the Coal-deposits among the Flœtz, while others refer them to the Transition order.

Introduction, *comprising a General View of this Order.*

Chapter I. *Chalk.*

Section I. General and introductory observations: (*a*) on the connection of the Chalk and more recent beds; (*b*) on the Foreign localities of this formation.

Section II. *Particular account of the Chalk formation; a.* chemical and external characters; *b.* mineral contents; *c.* organic remains; *d.* range and extent; *e.* elevation; *f.* thickness; *g.* inclination; *h.* agricultural character; *i.* phœnomena of water and springs; *k.* miscellaneous remarks.

Section III. Particular view of the Sections of this formation in the *cliffs of the Southern coast;* (*a*) *Isle of Thanet,* (*b*) near *Dover,* (*c*) *Sussex,* (*d*) *Isle of Wight,* (*e*) *Isle of Purbeck,* (*f*) between *Lyme and Sidmouth;* (*g*) Comparative view of the opposite French coast.

Chapter II. *Beds between the Chalk and Oolitic series.*

Section I. General and Introductory; (*a*) general enumeration of the series as ascertained in England, (*b*) Foreign localities.

Section II. *Chalk Marle, a* to *k.* as before.

Section III. *Green Sand, a* to *k.* as before.

Section IV. *Weald Clay, a* to *k.* as before.

Section V. *Iron Sand, a* to *k.* as before.

Section VI. VII. VIII. & IX. Particular account of the distribution of these formations in the several districts occupied by them. Sect. 6. *in the Weald*; Sect. 7. in the *Isle of Wight;* Sect. 8. *Isle of Purbeck;* Sect. 9. *Midland counties.*

Chapter III. *Oolitic series, including all the Strata between the Iron Sand and Red Marle, or New Red Sandstone.*

Section I. General view; (*a*) of these formations in England; (*b*) Foreign localities.

Section II. *Upper division.* 1. *Purbeck beds, a* to *k.* as before.
2. *Portland Oolite, a* to k.
3. *Kimmeridge Clay, a* to *k.*

Section III. *Middle division.* 1. *Coral Rag, a* to *k.*
2. *Oxford Clay, a* to *k.*

Section V. and VI.* *Lower division.* Sect. V. Upper members of the series associated with the *Great Oolite*, including *Cornbrash, Stonesfield slate, Forest Marble*, and *Great Oolite, a* to *k.* as before.
Sect. VI. Its lower members: *Fullers' Earth, Inferior Oolite, Sand* and *Marlestone.*

Section VII. *Lias* (forming the base of the Oolitic series) *a* to *k.* as before.

Chapter IV. *Formations between the Lias and Coal Strata.*

Section I. *New Red Sandstone, a* to *k.* as before. (An account of the *Conglomerates and Amygdaloid of Devonshire* is given among the local particulars under the head range and extent.)

Section II. *Magnesian Limestone, a* to *k.*

Section III. Comparative view of analogous Formations in other countries.

* Printed V. & VI. by mistake, being IV. & V.; the Lias consequently should have been Section VI.

Book II. SUPERMEDIAL ORDER.

Introduction. *

IT is the object of this division of our work, to comprise the several formations which intervene in descending, from those which have been described in the former book, to the coal measures.

This series of strata comprises several distinct groupes well marked from each other, and therefore entitled to the name of separate formations; but since many general relations and analogies pervade the whole,—since they are naturally linked together and appear to have been the successive products of the same order of causes, acting gradually indeed and accumulating deposits of various kinds in distinct periods, but yet without the intervention of any violent and abrupt change,—it has been judged expedient to include them all under one title in our more general classes, in order to avoid the needless multiplication of these greater divisions, the principal use of which consists in the assistance they afford to the memory of the student. To this class we have assigned the name Supermedial, from the position it occupies immediately above the Coal series, which forms the middle order in our arrangement. The formations which compose it, have been included by other authors under the term Secondary or Flœtz rocks; but since both these names have been made to comprise also the series of rocks constituting the coal districts, which (as forming a leading and strongly marked natural division, and belonging evidently to a very different order of things) clearly demand a separate classification, it became necessary to adopt a new and distinctive denomination. And it may be further observed, that of the old appellations, that of secondary is objectionable, because it stands properly distinguished from primitive rocks only, and never ought to have been retained after the Wernerian interpolation of the transition class; since, strictly speaking, the rocks of that class are themselves universally of secondary formation. The appellation flœtz, or horizontal, is still worse, because it conveys a descriptive idea which is only partially in accordance with the facts of the case, being founded on circumstances falsely assumed to be universal or even general;

* By the Rev. W. D. Conybeare.

for although in England and other low countries these formations are usually found with an horizontal stratification, yet in the borders of the Alps, in the Java chain, and generally in the vicinity of all very lofty mountain tracts, they are as usually found in elevated and contorted strata,— an important fact, which, as we shall hereafter have occasion to see, throws great light on the questions which have been agitated concerning the causes of such phœnomena and the elevation of mountain chains.

The general class thus formed, admits of four principal divisions; and all of these, excepting the first, may be further subdivided into smaller aggregates of similar strata, each of which aggregates has usually been considered as entitled to the rank of a distinct formation. The following list gives a general view of these four divisions and their subdivisions, beginning with the highest, and will sufficiently explain these remarks.

A. The Chalk formation. This requires no observation.

B. The series of ferruginous sands; the upper containing interspersed green particles, the lower of a rusty brown colour, divided by an intermediate bed of clay.

C. The series of oolites, consisting of three aggregates of coarse shelly limestones, often oolitic in their texture, alternating with argillaceous deposits often containing beds of argillaceous limestone; the lias clay and lias, on which the whole of this reposes, may be considered as one of these deposits.

Before proceeding to the fourth division, we would pause to introduce a few remarks illustrative of the connexion and general relations of the three preceding.

They all contain marine organic remains, which have a general similarity of character, and are in the same state: they are not simply preserved, as were those in the superior formations, but, strictly speaking, lapidified; being always (unless where the shell has perished and left a cast only,) penetrated with the stony matter of their envelope; often, however, in a finer form, so as to admit a crystalline arrangement of its particles. In these beds, not only the great majority of species are different from those now known to exist, but a very large proportion of the genera are in the same predicament. Although each of these series, and indeed almost every member of each series, contains many species of organic remains peculiar to itself, which constantly accompany its course throughout this island at least, and form (to use the lowest term) a good *empirical* character by which it may be distinguished; yet several of the species, and a majority of the genera, are common to the whole suite: but if we compare this suite, either with the more recent shelly beds above the chalk, or the older associated with the coal transition series, a marked and striking difference is instantly perceived, and the common genera will be found to bear a comparatively small proportion.

The natural connexions of these series may be further illustrated by the gradual transition of the one into the other, often observable near the point of junction; thus the lower beds of the chalk often become charged with sand and interspersed with green particles, and thus pass into the green sand of the second series; this sand is often associated with calcareous beds nearly approximating in character to some of those in the oolitic or third series; and the clays which subdivide this series, are generally only to be distinguished by the specific differences of their organic remains.

Having thus stated the analogies which connect the above members of our supermedial class, it remains to notice the fourth and lowest series of beds which we have comprised in it; although this is distinguished from the preceding by many peculiar circumstances, it will yet be seen that there are sufficient grounds for including it in the same general division.

D. This consists of the series of the new red sandstone and magnesian limestone. The sandstone is characterised by the deposits of gypsum and rock salt contained in its upper marly beds, and by the conglomerates abounding in its lower beds. It is entirely destitute of organic remains, and is thus strongly distinguished from the preceding formations; but the magnesian limestone beds which form the lowest strata of this series, again present these remains, and the genera are more nearly allied to those contained in the oolites than those of the older formations. This circumstance, taken together with the general conformity of stratification in this and the preceding series, and the appearances of a gradual transition which may be observed between the lowest marles of the lias formation, and the upper marles of this series at their junction, render it more advisable to class them together, than to constitute a separate class for the reception of this series only, which would otherwise be necessary.

A comparative view of the distribution of the several members of this class in England and other countries, will be found in the introductory sections of the chapters in which each of them is separately considered.

CHAPTER I.

THE CHALK FORMATION.

Section I.

(*a*) *General and introductory observations on the connexion of the Chalk with the more recent beds.* This rock, which forms one of the most remarkable features among the deposits of England, both on account of its extent and its perfectly distinctive characters, follows immediately beneath the strata described in the preceding chapter. In general, an interval seems to have taken place between the completion of this formation, and the deposition of those which repose upon it; and the surface of the chalk, at the line of junction, usually bears marks of having undergone, during that period, a partial destruction subsequently to its consolidation; a bed of debris being spread over it, consisting chiefly of flints washed out of its mass, and the surface being irregularly worn into frequent cavities, many of them of considerable depth, filled with similar debris.* On this debris rests the Plastic clay, or the

* The chalk of the numerous quarries, and where the roads are cut through it, along the south side of the Thames, as at Rochester, Gravesend, North Fleet, Greenhithe, &c. is remarkable at its junction with the sand and gravel [of the Plastic clay formation] for the deep indentations on its surface, which upon examination are ascertained to be sections of long furrows and of cavities, that were apparently occasioned by the powerful action of water on the surface of the chalk, prior to the deposition of the superior strata. (G. T. vol. ii. p. 176.)

The enormous quantity of completely rolled and rounded chalk-flint pebbles that occur in the Plastic clay formation on the south of London, corroborate the arguments adduced by M. M. Cuvier and Brongniart, from the irregular projections and furrowed surface of the French chalk, and from the fragments of chalk forming a brecchia with the Plastic clay at Meudon, to prove the consolidation of the chalk to have been completed before that partial destruction of its upper strata by the force of water, to which they justly attribute these furrows and the Meudon brecchia. These English beds of chalk-flint pebbles (the wreck of strata thus destroyed) afford additional evidence of the immense scale on which this aqueous destruction was carried on, and confirm also the conjecture (which by them is chiefly grounded on the total difference of the organic remains in the two formations), that a long period of time has probably intervened between the deposition of the chalk and the plastic clay. (G. T. vol. iv. p. 301.)

ash-coloured sand formerly described: here, therefore, the transition from the chalk to the more recent formations appears to have been abrupt, not gradual; in a few instances however a bed of intermediate character, a cretaceous marle * is interposed at the junction, which may seem to countenance this idea,—that where the series of deposits was permitted, from the circumstances under which they were formed, to proceed quietly, such a gradation may have taken place. † (C.)

The result of these repeated destructions of the chalk, has probably reduced the extent actually occupied by this formation to much narrower limits than those which it formerly possessed. On the north of Northamptonshire, and borders of Rutland and Leicestershire, and in the vale of Shipston in Warwickshire, extensive accumulations of chalk-flints, mixed with rounded fragments of hard chalk, occur in such quantities as almost to warrant the inference that this formation once existed in situ on these spots, now nearly 50 miles from its nearest point. In the fields on the south of Sywell in Northamptonshire, the fragments of chalk are so abundant as to give the appearance of a regular substratum of that substance, turned up by the plough. In the Philosophical Transactions for 1791, is an account of a chalk pit found at Redlington in Rutland; which, if correct, must be considered as a relic of this destroyed tract. The account is very precise, indicates a sufficient knowledge of the general nature and localities of the formation, and is such as to render the testimony very respectable; but the point is so important that further inquiries are desirable. The chalk is described as regularly interstratified with flints; and the surrounding district being entirely occupied by the ferruginous sands of the inferior oolite, it is not easy to conceive that it could have afforded any rock which could have been mistaken for chalk. Another detached patch of chalk is said, in the same place, to exist near Stukely in Huntingdonshire on the banks of the Turnpike road, but no particulars are given, and here soft varieties of other calcareous beds might be confounded with this substance. (C.)

* A chalk-marle without flints, is the stratum which in Alum Bay in the Isle of Wight, lies immediately next on the flinty chalk. It pulverizes with the frost, and as the rains have washed it down, its situation is marked by a deep hollow. (G. T. vol. ii. p. 178.) There appears also to be indications of its existence in the same position in other parts of the chalk basin of the Isle of Wight (hereafter to be described); for in many parts of Sussex, south of the South Downs, as at Emsworth, Lavant, Siddlesham, South and North Bersted, Middleton, &c. there are pits of a marle without flints, which is evidently over the chalk: the same marle has also been found in Dorsetshire on the west of Corfe castle; but it has not been discovered upon it in the London basin. (G. T. vol. ii. p. 178.)

† On the Continent, the calcareous beds which repose on the chalk at Maestricht, though they cannot properly be classed as a portion of the chalk formation (an attempt which has been made by some foreign writers, and can only tend to confusion), yet they certainly approach much more nearly to it than any other of the superior formations, and their organic remains seem to indicate a greater antiquity than belongs to any of the upper beds in England. (C).

(*b*) *On the foreign localities of this formation* * The Zone of chalk which sweeps across our eastern and southern counties from Yorkshire to Dorsetshire (as will be more particularly described under the proper head), must be considered only as a part of the western edge of a most extensive tract of this formation; which, stretching from the Thames to the Don, occupies the interior of what may be termed the great central basin of Europe; understanding by that denomination, the lower countries bounded by the following primitive and transition mountains and districts. On the north, the primitive districts of Russian Finland, Sweden, Norway, and Scotland; on the west, the transition and primitive chains of Cumberland, Wales, Devonshire, and Brittany; on the south, the primitive mountains branching from the Cevennes in the centre of France, the Alps, with the various insulated ancient groupes of Germany, &c. as the Black forest, the Rhingau and the Vosges, the Bohemian, Thuringian, Saxon, Silesian, and Carpathian mountains; on the east, the Ural and its branches. It is not to be understood that the chalk immediately reposes on these mountains, (for an interval of greater or less breadth, in which the intermediate formations make their appearance, always occurs), but that within the area so circumscribed, an interior area may be traced, throughout which there is every reason to believe the substratum of chalk extends. We proceed to trace it.

It may be observed through the northern coasts of France,† occupying an extent exactly corresponding to its line on the southern coast of England (as will be more particularly shewn in the course of this chapter). At the north of the Seine, its outer edge (which reposes on green sand, having oolite and lias in the neighbourhood) turns south, and so continues to Blois, where the formations above the chalk overlie and conceal its southern extremity: it reappears at Montargis, and turning again north (for the whole chalk district of France forms a sort of Cape protruding to the south of its general line), runs east of Troyes, Rheims, and Valenciennes, having the green sand, oolites, and lias on its east, till it approaches the latter town, where most of these formations are wanting (an instance of want of conformity in their direction), and the chalk, with a few beds of green sand, there called *Turtia*, rest horizontally on the truncated edges of the coal-formation, which extends thence along the banks of the Meuse to Liege and

* By the Rev. W. D. Conybeare, F R.S. &c.

† A general account of these formations has been given by Omalius d'Halloy: a translation of his memoir is printed in Thomson's Annals: some particulars are added from private sources.

Aix: the coal is here even worked beneath the chalk. North of Valenciennes, the edge of the chalk appears to trend to the east, but it is generally overlaid by the sandy superstrata through the Netherlands; it may however be seen on the south of Maestricht, and at Henri Chapelle near Aix. From hence, it may be considered as ranging beneath the sandy and diluvial tracts of North Germany, towards Berlin; the whole of this district is well known to present the appearance of an uniform and vast sandy heath, covered with a deep accumulation of diluvial gravel, in the midst of which occur enormous rounded blocks of granite, for which a source cannot be found nearer than the opposite shores of the Baltic—thus exhibiting one of the most striking problems submitted to the investigation of geology. The great mass of this gravel, however, consists of chalk flints, well marked and bearing traces of all the characteristic fossils: at least nine-tenths of the whole consist of these; a sign that the parent formation can be at no great distance. In such a tract, a rock *in situ* is like an oasis in the desert; at Luneburg however the fortifications are partly constructed on a rock of gypsum, and about a quarter of a mile hence, on the left of the road to Hamburg, the writer of this article was gratified by detecting a chalk-pit which had escaped the attention of former observers: it contains the usual alternation of flints, and affords good specimens of the inoceramus, echinites, and most of the characteristic fossils. On entering Poland, the chalk throws off the mantle which has for a time concealed it, and reappears in a line of hills running nearly parallel to the Carpathians: it is finely exhibited at Cracow: it contains abundant flints, affords the usual organic remains, and rests on green sand: it was here examined by Professor Buckland. Hence, passing by Lemberg, it appears to extend into Russia. Hills of chalk were noticed by Dr. Clarke at Kasankaiya on the Don, and the town of Bielogorod, signifying the white city, is said to take its name from white hills of the same substance in its neighbourhood. Engelhardt observed chalk, containing its usual flints and fossils, even in the Crimea.

No particulars can be gathered of the eastern or north-eastern boundaries of this formation. We may conjecture however that they pass by the Valday hills to the mouths of the Vistula; thence, the northern border must run eastward through the Baltic to the island of Rugen, where chalky cliffs present themselves on its northern coast, being found also on the neighbouring continent, in Pomerania and Mecklenburg. Hence the line appears to pass to the south of Sweden, where a small chalk tract occurs near Malmo, crossing to the opposite coast of Zealand, and including the small isle of Mona on the south. Some account of these localities may be found in De Luc's travels.

From Mona, the line of the chalk has not been traced: it probably traverses Holstein, (where it is said to occur, probably near the gypsum of Kiel) to the mouth of the Elbe, and thence crosses the German ocean to Flamborough head in Yorkshire; thus completing the circuit in which we have endeavoured to follow it.

Throughout this extensive tract, the chalk appears to preserve a remarkable uniformity of character, affording a satisfactory instance of the vast areas over which geological causes have operated in the formation of continuous deposits of a similar, or rather identical, nature. The great majority, (perhaps eight-tenths) of the organic remains also, which occur in this rock, are common* to all the localities in which it has been particularly observed; viz. England, France, the Netherlands, Germany and Poland: indeed we have never seen a fossil from any foreign chalk-pit, to which an analogue might not be produced from those of this island. These facts are interesting, as shewing that the comparison of formations in very distant tracts, rests on firm and satisfactory grounds, and as illustrating the importance of organic remains in establishing that comparison. Still it should be remembered, that the tract to which the above observations refer, extensive as it is, is yet but a small portion of the whole surface of the globe, is limited to a single basin, and lies under nearly a similar climate throughout. The progress of the science may be alike impeded by too hastily extending our generalisations beyond the boundaries strictly warranted by our induction, and by too sceptically rejecting the conclusions so warranted, merely because they are general.

Beyond the limits of the great cretaceous area already described, local tracts of chalk occur in the following places.

In Ireland, a remarkable deposit of chalk forms the basis of the great basaltic area in the north-east angle of that island; it contains flints; the organic remains agree with those of England; the thickness of the whole deposit does not exceed between 200 and 300 feet; it rests on green sand.

In Italy, the *Scaglia*, which covers the extreme secondary chains of the Alps in the Veronese, may perhaps be a variety

* Mr. Schlottheim, a German writer on the geological distribution of organic remains, has a remark which appears to contradict this opinion; but this arises from his considering the limestone beds above the chalk, in St. Peter's hill, Maestricht, which contain many peculier fossils, as a part of the chalk formation.

of chalk: it is described* as a calcareous bed, containing nodules and beds of variously coloured flints, resting on the oolites and white limestones, and dipping under the tertiary hills (i. e. those consisting of the formations more recent than the chalk); it re-appears against the volcanic groupe of the Euganean hills near the mouth of the Po, which appear to have forced it upwards.

In Spain, chalk is said to occur near Cervera, on the road from Barcelona to Lerida; gypsum abounds in the same neighbourhood, and at Pleacente two miles from Valencia, but the descriptions are too vague to be relied on; the gypsum mentioned seems to be rather that of the red sandstone, than of the formation above the chalk, and possibly a cretaceous marle may have been mistaken for the latter rock.

We have no precise information of the existence of chalk beyond the limits of Europe. It may be mentioned, however, that the maps accompanying Rennell's memoir on the retreat of the ten thousand, mark a chalky tract as existing on the south bank of the Euphrates, somewhat above the well known bitumen spring of Hit, and opposite the whole range of hills in that quarter which are denominated 'White'; in corroboration of this locality, Professor Buckland saw at Sir Joseph Bankes's an amulet, found among the ruins of Babylon, evidently cut from a chalk-flint.

This rock is also said to have been observed in China, 20 miles north of Pekin, by Sir George Staunton (vol. ii. p. 172), but the description is too vague to be relied on.

Chalk has not been found in any part either of North or South America yet explored, and Mr. Macluren positively asserts that it does not exist on that continent.

Section II.

Particular account of this formation.

(*a*) *Chemical and external characters.* The nature and qualities of Chalk, generally speaking, are too well known to need description, but some varieties of it are found which may render it requisite. The purest and best chalk, commonly selected for economical purposes, is nearly of a pure white; it has an earthy fracture, is meagre to the touch, and adheres to

* See Ferber's Travels in Italy, Letter V.

the tongue and even to the fingers: it is dull, opake, soft, light (its specific gravity being only 2.3), and it always occurs massive. A specimen analysed by Bucholz, yielded lime 56.5, carbonic acid 43, water 0.5. Magnesia has been detected in some of the French chalk, and may perhaps occasionally occur in that of England. Such are its mineralogical characters, and in its purest state it may be considered only as a carbonate of lime, liable, from the slight cohesion of its particles, to absorb a small quantity of moisture. Sometimes however it has been found to contain a very small proportion of alumine, but it frequently involves a considerable portion of sand, of which it may be freed by pounding and washing. The less pure varieties are yellowish and even yellow, and are sometimes so hard as to return a ringing sound to a blow from the hammer, as will be perceived in the subsequent account of some of the beds near Dover. Occasionally some of the lower beds of this formation are of a reddish or red colour, probably derived from the presence of a small quantitiy of iron, as in Lincolnshire, and at its northern termination in Yorkshire. (P.)

The specks of green earth more commonly found in the lower than in the upper beds, are probably derived, like the red variety, from the presence of iron in different states of oxidation. The lower beds of the chalk occasionally increase in hardness * until they afford a tolerably compact limestone; sometimes, however, these harder beds alternate, even near the bottom of the series. C.)

Pliny describes this substance under the title 'Creta argentaria,' and adds, 'petitur ex alto in centenas pedes actis plerumque puteis, ore angustatis intus, ut in metallis spatiante vena. Hac maxime Britannia utitur.' This very method is used in Hertfordshire, and other parts of the kingdom at present. The farmer sinks a pit, and, in the terms of the miner, drives out on all sides, leaving a sufficient roof, and draws up the chalk in buckets through a narrow mouth. Pliny informs us in his remarks on the British marles, that they will last 80 years, and that there is not an example of any person being obliged to marle the same land twice in his life. An experienced farmer, whom Mr. Pennant met with in Hertfordshire, assured him that he had about 30 years before made use of this manure on a field of his, and that were he to live the period mentioned by the Roman naturalist, he should have no occasion to repeat it. Pennant's Chester, p. 303—(G. Notes.)

According to Smeaton, chalk, when well burnt, will make as good lime as the hardest marble. The harder kinds are used in building, as in the Isle of Wight. (G. Notes.) That of the cliff on the east of Dover is blasted, squared, and used for the purposes of the harbour, and in the construction of a dock, especially such parts of it as are mostly under water.

The abbey of Hurley in Berkshire, and its parish church, anciently a chapel, are said to be constructed of chalk, and the remains of these are still as fresh and sound as if they had been the work of the last century. The mullions and arches of St. Catharine's chapel near Guildford, are of chalk that contain flints. (G. Notes.) The abbey at St. Omar's, ruined during the revolution, is entirely constructed with chalk, and retains all its beautiful gothic ornaments in great perfection. (C.)

Chalk occasionally contains subordinate beds of other substances; as of soft and indurated chalk-marle, which occur near Dover, and of fuller's earth, (G. Notes), which is found in the north of England, at Claxby in Lincolnshire, and in Sussex. Mr. Webster has noticed thin parting seams of clay between the strata of chalk in some of the pits in the Isle of Wight.

The upper chalk appears in most places in England, wherever occasional openings have been made in the very large tract occupied by it, notwithstanding the extensive ruin which its beds must have suffered at remote periods, of which we find manifest proofs in the valleys by which the chalk is intersected, and especially in the immense deposits of rounded flint gravel in the eastern and south-eastern parts of England.

The inferior beds* may be traced along the lower regions of the escarpment presented by the hills of this formation, in the cliffs on the coast near the junction with the older formations, and in a few other places where exposed by denudation; but such beds must be considered rather as anomalies of local and partial occurrence, than as belonging to the general character of the formation.

This formation, being composed throughout of a series of homogeneous beds of a tender earthy limestone, does not admit of those subdivisions which many others require.

The occurrence, however, of those numerous layers of nodular flints which alternate through the greater part of its mass, constitutes one of its most remarkable and essential features; being constantly present in all the upper beds of the formation, which may thus be in most instances readily distinguished from the lower in which the flints are rare, and most usually entirely wanting; hence the well known division into the *upper* and *lower chalk*.

* The ensuing part of this article is by the Rev. W. D. Conybeare, F.R.S.

The chalk even yet often contains a mixture of silex; at the period of its formation, a considerable quantity appears to have been precipitated with it, in a state of such minute division as to allow the chemical attraction of its molecules to have effect; these (from the same causes which produced the formation of layers of calcareous concretions in beds of clay) separating from the cretaceous pulp, and uniting together, particularly where the presence of any imbedded organic remains, (e. g. alcyonium, sponge, or shell) offered a nucleus for them to form upon, constituted the layers of nodular siliceous concretions in question.

Such appears the most probable origin of these appearances; it may be noticed in confirmation, that the extraordinary figures so often assumed by these nodules, will on examination be found to arise from their having been formed on some variety of sponge or alcyonium as a nucleus: the manner in which the siliceous matter has penetrated the most minute pores of the organic substances enveloped in it, is truly admirable.

These flinty nodules occur in strata alternating conformably with those of the chalk.

Frequently also, but less commonly, the flinty strata consist of tabular masses of that substance.

Veins of flint, traversing the regular strata at various angles, though more rare, may still be observed almost wherever any extensive range of chalk cliffs is explored; e. g. in the Isle of Thanet, the neighbourhood of Brighton, and the Isle of Wight; the flint filling such veins is usually tabular.

It is unnecessary to give a formal description of the external characters of a substance in every one's hand. The specific gravity of flint is stated at 2.594: it consists according to Klaproth of 98 parts of silex, 0.5 of lime, 0.25 alumine, 0.25 oxide of iron, and 1 of water. It is infusible, but whitens, and becomes opake; besides the water chemically combined, it appears to contain some, mechanically distributed through its pores, as its recent fracture, when freshly taken from its native bed, always exhibits an appearance of moisture.

The flints which have been washed out from the chalk at a remote period, occur in the various deposits of gravel; and thus, from their superficial position, and the loose texture of the mass in which they occur, exposed through ages to the influence of atmospherical action, and the percolation of water, have often become much altered, and undergone changes somewhat analagous to those which may be produced in them by fire: the black colour is rendered less intense, or changed to brown, yellow, or red, probably by a change induced in the oxidation of the iron contained in it; these different colours are sometimes arranged in zones parallel to the outward surface of the pebble, the altering action having operated with different intensities at different depths; a singular fissured appearance may sometimes be observed in such pebbles.

The interior of the flinty nodules, often contains drusy cavities lined with tubercular chalcedony or quartz crystals; these generally appear to have formed the nidus of some organic remain, which has permitted only the finer particles of the siliceous matter to percolate, and thus favored their assuming a more delicate texture, or even a crystallised arrangement. The presence of a sponge or alcyonium seems to have been particularly favorable to the formation of chalcedony; almost whenever it occurs in flint, a careful examination will detect the traces of these zoophytes: very beautiful appearances often result from their radiated ramifications through the chalcedony, which should be dipped in water to bring them perfectly out, being partially hydrophanous.

(*b*) *Mineral contents.* The beds of flint, so numerous in the upper chalk, have already been mentioned; and these, together with the occasional beds of hard and soft chalk marle, occurring near Dover, and of Fullers' earth in the north of England, at Claxby in Lincolnshire, and still more lately discovered in the same situation in Bepton hill near Midhurst* in Sussex, are interstratified with the beds of chalk in parallel layers: these substances may therefore be considered as of contemporaneous formation with the chalk itself, and therefore not strictly as mineral contents, under which head however they seem to require this notice. The chalcedonised flints are often found in cabinets. The only mineral substance common to the chalk is iron pyrites, which is found in most if not all the beds, in masses varying from the size of a pea to several inches in diameter: they are mostly crystalline, and generally exhibit, on being broken, a fibrous and diverging structure, arising from the position of the crystals of which they are composed, and of which the summits commonly appear on the surface of the

* The following account of the existence of beds of Fuller's earth in the chalk of Sussex was presented to me by Frederick Sargent, Esq: they were first noticed by the Rev. C. P. N. Wilton. The Fullers' earth is found in two beds situated in the upper chalk, and so near the summit of the South Downs, near the village of Bepton, that only one foot of chalk lies above the upper bed; the beds are nearly horizontal, and are from three to four inches thick, and the substance itself bears every characteristic of Fullers' earth; it does not effervesce with acids, is unctuous to the touch, somewhat translucent on the edges, and falls to pieces when thrown into water. Below these beds, and near the middle of the chalk escarpment, lies a bed of soft chalk marle three or four inches thick. (P.)

mass: they often occur of a cylindrical form, and were heretofore picked up and preserved under the name of *thunderbolts.* Often, however, these masses have undergone decomposition while in situ, so that nothing is left but an ochreous mass, which has occasionally been observed to penetrate the surrounding chalk, colouring it in concentric circles of various shades of brown. The pyrites is sometimes deposited around shells. (P.)

In a chalk quarry at Bishopston down near Warminster in Wiltshire, enormous blocks of crystalline carbonate of lime, one of which weighed 50 cwt., and measured between 30 and 40 cubic feet, was cut into slabs for chimney pieces at Mr. Noel's, a stone-mason at Warminster. They lay in the upper chalk almost close to the surface, and the flinty chalk is much thinner in this than the adjoining quarries. Calcareous spar also occurs in a chalk quarry at Nook near Heytsbury, in blocks less both in number and size. (G. Notes.)

A brown or blackish brown substance has been observed coating the chalk in several places in Suffolk. It has sometimes the appearance of a sooty powder, but is occasionally fibrous. It has been noticed by the Rev. J. Holme in a pit near Budlingham bridge in the parish of Frecklingham; in another near West Row Ferry, and also at West Row near Mildenhall, and is considered by that gentleman to be most probably the black oxide of manganese; this however requires confirmation. (P.)

The chalk of Claxby in Lincolnshire is very fibrous in the cracks, and when exposed to air, powdered with black specks, but in the interior it is beautifully white: grey flints are disseminated through it. (G. Notes)

A septarium found in chalk at Steyning in Sussex, is now in the collection of Mr. Parkinson. (G. Notes.)

(c) *Organic remains.** Under this head, the chalk presents us with phœnomena very different from those of the more recent formations.

Although numerous individual specimens are every where to be found, yet the number of genera to which they belong are restricted within comparatively narrow limits.

If variety be wanting, it is however compensated by novelty. In the abundance of the newer beds, although the species were usually different, yet the genera agreed with those still found in the ocean: but here we are presented with many new genera, and probably not a single species will be found, identical in all its characters with any now known to exist.

* The whole of this article is by the Rev. W. D. Conybeare, F.R.S. &c.

To begin with the remains of the more perfect animals, we find the remains of several species of vetebral fish; among which may be specified, teeth of a species of shark perhaps near the Squalus galeus, two varieties of the grinding palatal bones belonging to unknown genera, vertebræ, and scales.

Among the testaceous molluscæ, the order of multilocular univalves has left the following genera imbedded in this formation; Ammonites*, both ovate and circular; Scaphites*; Belemnites†; all these genera are extinct; the first occurs rarely in the upper chalk, the second only in the lower, and the varieties are peculiar and characteristic of these beds.

Of the common spiral univalves, the genera and species found in the chalk are very few and rare; a striking contrast to the abundance in which they occur in the newer beds: the genera, Trochus, Cirrus, and Turbo, are mentioned.

Serpulæ and spirorbes are common.

Among the bivalves may be numbered; Ostrea, four species, 1. resembling edulis; 2. variety of o. crista galli; 3. (not half an inch long, crenulated on each side the hinge), canaliculata*; Pecten,* two species, or more; Terebratulæ, five species, three smooth, two plicated; Magas*; Plagiostoma* spinosa; Dianchora* lata; Inoceramus (the fibrous shells) several species, of which one is figured G T. vol. 5. pl. 1. The four last genera are extinct.

Of Multivalves, a species of balanus has been found.

The important family of Echinites may be considered as characteristic of this formation, and at least as equalling in number all the other shells found in it; many of the species, and one entire genus is indeed peculiar to it. In enumerating the genera, we shall give both the names of Leske and Lamarck, distinguishing them where they differ by the respective initials, and add references to the figures in the 3d vol. of Parkinson's Organic remains. 1. Helmet-shaped; Echinocorys *Leske*; Ananchytes, several species (P. pl. 2, fig. 4) 2. Conical; Conulus, *Leske*, Galerites, Lamarck (Pl. pl. 2, fig. 10) many species. 3. Heart-shaped; Spatangus (P. pl. 3, fig. 11). 4. Spheroidal, with the mouth and vent on the opposite poles, and studded with mamillated tubercles; these constitute the Cidares of Leske, but Lamarck has divided them into two genera; in one, the tubercles are perforated to admit the passage of muscular filaments which assist in the motion of the

* The shells marked by an asterisk are figured in Sowerby's Mineral Conchology.

† When the belemnites are mentioned by Cuvier as characteristic of the French chalk, this is only to be understood in contradistinction to the more recent beds, in which they rarely or never occur.

spines, always large in this variety; to this he restricts the name Cidaris; one beautiful species, Cidaris papillata (P. pl. 1, fig. 11) is found in the chalk: in the other division the tubercles are imperforate, the spines, which are smaller, being moved by the contractions of the outer skin only; Lamarck distinguishes this genus as the Echinus properly so called: the Cidares variolatæ (P. pl. 1, fig. 5, 7, 10, and pl. 3, fig. 1) belong to it. Of these genera, all the species 1 and 2 appear to be extinct, and the former, confined to the chalk formation only: those of 3 are distinguished from the recent, only by strongly marked specific differences; those of 4 exhibit a very near approach to recent specimens, but are still to be identified.

Of the Star-fish, Asterias *Lamarck*, four species are described by Mr. Parkinson (Org. rem. vol. 3) as belonging to the English chalk. 1. (Pl. 1, fig. 1) nearly resembling the Pentagonastes semilunatus of *Linck*. 2. (Pl. 1, fig. 3) approaching Pentagonastes semilunatus of *Linck*. 3. Pentaceros lentiginosus, *Linck*. 4. Stella lumbricalis lacertosa corpore spherico, *Linck*; the species nearly resemble the recent.

Among the Zoophytes, the family Encrinus has several genera in the chalk. 1. Pentacrinus, rare. 2. Straight Encrinus (Park. O. R. vol. ii. pl. 13, fig. 34, 70—73.) 3. Bottle Encrinus, (same plate, fig. 75, 76.) 4. Stags-horn Encrinus (same plate, fig. 31, 38, 39): all these species are extinct; one species only of the genus Pentacrinus is known to exist, which differs materially from the fossil.

The fossil long known under the name of the Tortoise Encrinus, which resembles some species of the Encrinites in having a pelvis composed of pentagonal plates supporting articulated tentacula, but differs from them in wanting the articulated stem or column, and must therefore float freely, is now, on account of the essential distinction, formed into a new genus to which the name Marsupite is given; one species only has been described (same plate fig. 24); it is only known fossil, and is peculiar to the chalk.

Of the family Madrepore (Polyparia lamellifera *Lamarck*) only one species occurs, nearly approaching the madrepora cyathus of Ellis, and Caryophyllia cyathus of Lamarck (Parkinson's organic remains, vol. ii. fig. 15 & 16); a more elongated variety (perhaps only a different stage of growth) is found with this; the principal difference between the recent and fossil specim n is, that the exterior of the former is smooth, that of the latter *striated*.

The families of Alcyonium and Spongia present numerous remains: much obscurity prevails in the distribution even of the recent species of these families. Ellis makes their distinction to consist in the presence of polypi, as inhabitants of the cellules of the alcyonia, and believes the sponges to possess none of these animacules, but to be simply investited with a living gelatinous flesh. Lamarck, with greater probability, supposes the sponges to have polypi like the alcyonia, differing only in the greater solidity of the fleshy parts of the latter, which permit them to be observed when removed from the water; while those of the former dry up instantly on being taken out of their natural element, and thus escape observation. It is manifestly impossible that a distinction of such a nature should be ascertainable in petrifactions preserving only the solid parts of the animal; from general character, however, many of them appear rather to belong to the sponge than alcyonium.

The substance of these bodies consists, as to its interior texture, of a mass of interwoven fibres, penetrated by larger pores, regularly or irregularly disposed.

These fibrous reticulations sometimes run confusedly together, so that their meshes present no regular or determinate figure; sometimes they are regularly disposed, so as to give to the whole mass a plicated character; we shall therefore assume this distinction as the foundation of our division.

Of that division which is irregularly reticulated, four genera at least may be traced in the chalk formation.

1. Of a ramifying form; two varieties are figured by Parkinson, vol. ii. pl. 8. fig. 6 & 12, and pl. 11. fig. 4.

2. Palmated; the larger pores disposed in a quincuncial order; rare.

3. Irregularly turbinated and funnel-shaped masses; the varieties of this form are almost infinite, arising probably from the different contractions of the mass.

4. Fig-shaped; agreeing exactly with that figured in Solander and Ellis, pl. 59, fig. 4, as a sponge, but pronounced by Lamarck to be an alcyonium; it is pediculated at the bottom, and flattened at the top, which presents a funnel-shaped cavity penetrating in the direction of its axis; its larger pores radiate from the axis. Parkinson figures varieties found in chalk or its derivative gravel, pl. 9. fig. 11, 12, 4, and pl. 11. fig. 8. The appearance of the ramifying filaments round the funnel-shaped cavity of this genus, is seen in the specimen, pl. 9. fig. 1.; fig. 7 & 5, pl. 12, also belong to this genus.

Another genus is there characterized; it occurs "forming large irregular sessile masses, upper surface tuberculated, traversed by large irregular ramifying pores."

The most remarkable of the regularly plicated division, and that perhaps of which all the seemingly different species are only accidental varieties of form, has been ably described by

Mr. Mantell in the eleventh volume of the Linnæan Transactions, p. 401, under the name of Alcyonium chonoides (from Χωνη, a funnel); and as, from the limited circulation of the original work, the description (which is highly interesting) is probably known to very few of our readers, we shall subjoin an abridgement.

This alcyonium is, as to its general form, funnel-shaped and fixed by the root; the external coat is composed of fasciculi of muscular fibres, which, arising from the pedicle, proceed in a radiating manner towards the circumference, and by frequently anastomosing, constitute a retiform plexus capable of dilating, lengthening, and contracting; according to the impressing it received from this contractile power, arises a great variety in the general form of the specimens: when quiescent, it appears to have been funnel-shaped; when partly expanded, cyathiform; when completely so, discoidal; occasionally even throwing the surface into deeply undulating folds, so that a transverse section of it exhibits an indented outline, something like that of the heraldic nebule; these fasciculi are further connected by lateral processes, which increase the firmness of the integument formed by them; from the inner part of this envelopement, arise tubuli which pass direct to the inter-funnel-shaped cavity, and terminate on its surface in small circular pores or openings, often disposed in a quincunical order; in some specimens, a substance of a sponge-like appearance fills up the interstices between these pores, probably the remains of a spongy membrane which in the recent specimen served to connect the tubes, and give consistence to the whole mass. Each of these pores was perhaps the cell of a minute polypus.

Fig. 1. Exterior view. Fig. 2. View looking into the inner cavity.

Specimens are sometimes found invested with chalk only, but more usually enveloped in flinty nodules; Parkinson (pl. 10, fig. 14, 15, 16,) has given a representation of the root and part of the stem in this state, but the specimens were too imperfect to enable him to ascertain their true character; an impression of the inner root marked by quincuncial pores is figured by him pl. 11, fig. 12.

A similar specimen from the French chalk is figured in Ramond's travels to Mont Perdu, where it is inserted as an object of comparison with an analogous fossil found in the Pyrenean limestone.

In considering this class of the organic remains occurring in the chalk, we have exceeded the brief notice to which we usually confine our remarks on this head, since the confusion and obscurity which prevail concerning its genera and species, render every attempt to methodise and illustrate them useful.*

(c) *Range and extent.* This extensive deposit stretches, with little interruption, from Flamborough head on the coast of Yorkshire, to near Sidmouth on the coast of Devonshire, forming a range of hills often of considerable elevation, and of which the most precipitous escarpment is generally on the north-western side. Another range of hills branches from it, in the south of England. (G. Map.)

The cliff between Bridlington quay, which is about five miles south of the point called Flamborough head,† consists of chalk, as well as Speeton cliff about six miles on the north-west of it. The chalk then proceeds inland, rising into hills which for about 15 miles from the shore have nearly a westerly

* The form of those singularly shaped masses of flint known by the name of Paramoudra in Ireland, where they are most common, and also found at Whittlingham and other places near Norwich, bears a strong resemblance, but on an enormously larger scale, to the flinty nodules investing some species of tubercular alcyonia. (C.)

† These cliffs exhibit at the top the superior soft chalk containing horizontal layers of flint, at the bottom the hard variety with few flints. We may be permitted to relieve the dryness of geological detail, by quoting the striking portrait of this place given by Pennant (Arctic Zoology, Introd. p. xi.): the vast height of the precipices and the amazing grandeur of the scenes which open on the north side, giving wide and solemn admission through most exalted arches into the body of the mountain, together with the gradual decline of light, the deep silence of the place, unless interrupted by the striking of the oar, the collision of a swelling wave against the sides, or the loud flutter of the pigeons affrighted from their nests in the distant roof, afford pleasures of scenery which such formations as this alone can yield. These also are wonderfully diversified; in some parts the caverns penetrate far, and end in darkness; in others are pervious, and give a romantic passage by another opening equally superb. Many of the rocks are insulated, of a pyramidal form, and soar to a great height; the bases of most are solid, but in some pierced through and arched; all are covered with the dung of the innumerable flocks of migrating birds which resort here annually to breed, and fill every little projection, every hole, which will give them leave to rest; multitudes were swimming about; others swarmed in the air, and stunned us with the variety of their croaks and screams. (C.)

direction, and of which the escarpment is towards the north. The line of hills then sweeps suddenly round, a little to the east of south, forming the Wolds of Yorkshire, and are terminated by the river Humber. On the opposite coast of that river, at Burton in Lincolnshire, the chalk rises again from beneath the alluvial matter forming its banks, into a range of hills, the Wolds of Lincolnshire, having a south-easterly direction, and terminating at Burgh, a town about five miles north of Wainfleet, and near the easternmost point of the Lincolnshire coast, where it dips beneath alluvium. The chalk tract of the Yorkshire and Lincolnshire wolds has an average breadth of rather more than six miles. (G. Map.)

Near the shore of the north-western point of Norfolk, and bordering the Wash which separates it from Lincolnshire, the chalk re-appears† as a ridge of hills running nearly south for about 15 miles, and occupying the surface for about 10 miles on the east of the hills, to Burnham Market, but its breadth on the southern half scarcely exceeds three to five miles. This little range terminates at Castle Acre about five miles north of Swaffham, sinking beneath the diluvial sands of Norfolk. The chalk appears again on the north of Thetford, and forming a low tract, passes, except where it is covered by marshy or alluvial soil, by Newmarket in Suffolk, to a little on the east of Cambridge,‡ on the west and south-west of which appear some outlying masses, as the Coach and Horses hill, and the summit of Madingley hill which rest (G. T. vol. v. p. 115) on galt. From Cambridge, the western limit of the chalk

* In the Wolds of *Lincolnshire*, the chalk consists of two colours, red and white, each lying in regular strata, the red being generally undermost; in the white, seams of flint are frequently met with from two to six inches thick. The chalk rests upon a coarse brown pebbly sand without organic remains, consisting of quartz and oxide of iron. (G. T. vol. iii. p. 394.)

† Hunstanton cliff, though only about 80 feet high, forms, from the flatness of the adjacent country, a conspicuous object; immediately beneath the vegetable soil and chalk marle, beds of regular chalk about 30 feet in thickness occur; these rest on a hard red stone four feet deep, which is often ground and made into a red paint (evidently the same with the lower bed in the Lincolnshire wolds); seven feet of loose friable dirty yellow stone succeed, placed on a base of iron-coloured pudding-stone projecting into the sea, with vast fragments scattered over the beach. (C.)

‡ The chalk of *Cambridgeshire* is described as consisting of two varieties; the *upper* containing the common black flint in abundance, and the *lower* or *grey chalk*, which contains little or none. If a line be drawn from Royston by Balsham to Newmarket, it will pretty exactly define the limits of both varieties; the hills to the eastward of it being composed of the upper beds, while those to the west, consist of the lower or grey chalk which composes by far the greater part of the hills of Cambridgeshire, and which will be again noticed in treating of the chalk marle. (G. T.)

passes to the south-west, and at Royston downs assumes the character of a range of hills, the escarpment of which continues with little interruption, and in the same direction, by Baldock and Hitchin in Herts, Dunstable in Bedfordshire, and Tring and Wendover in Buckinghamshire to Goring in Oxfordshire, a distance of about 75 miles, where it is broken through by the valley of the Thames. (G. Map.) This range,* in passing through Buckinghamshire and Oxfordshire, is well known under the name of the Chiltern hills, in Bedfordshire the names of Royston and Luton Downs are applied to different parts of it: its average breadth is from 15 to 20 miles: from the summit of its north-western escarpment it declines gradually to the south-east, where its strata dip beneath the upper bed of the London basin; it is furrowed by many transverse vallies of which several break entirely through it; such are that through which the proposed line of the London and Cambridge Canal passes that of the Grand Junction Canal; but the principal opening through the chain, is that through which the Thames flows from Goring to Maidenhead, and which constitutes the most striking and picturesque scenery on the banks of the principal river of the island. On the west of the Thames, the chain is continued, bearing the name of Ilsey Downs and the Whitehorse hills. to Marlborough Downs, where it is broken through by the valley of the Kennet; which, like most of the rivers flowing through this formation, rises in the subjacent and older stratum, and breaks entirely through the elevated chains of the chalky downs. This valley separates the northern Marlborough Downs from the long ridge on the south of Marlborough, which stretches westwards to Bagdon hill about three miles north-west of Devizes in Wilts. On the south-east, this ridge, skirting round the vale of Pewsey (a broad opening between the chalk ranges, exhibiting the inferior strata), connects itself with the northern point of that extensive cretaceous district which occupies all the north of Hampshire and most of the south of Wiltshire; its longest diameter from east to west, being more than 50 miles; its shortest from north to south, about twenty.

This vast area has been appropriately denominated by Pennant, the great central Patria of the Chalk; the centre and source whence all the ranges of this rock traversing the island diverge.

The following points will assign its limits with sufficient precision for our present purpose. On the east, the hills above

* The remainder of the description of this range and extent of the chalk is by the Rev. W. D. Conybeare.

Selborne, (a spot familiar, from the classical pages of Mr. White, to all who can be interested either in natural history or elegant literature); on the west, those skirting the vale of Warminster; on the north, Inkpen hill, the loftiest summit of this formation, attaining the height of 1011 feet above the level of the sea on the south of Salisbury. A great part of the area so included, is well known by the name of Salisbury plain. The whole of this district consists of an elevated platform, distinguished only by gentle elevations of surface, and covered by a scanty herbage. It is intersected by fewer vallies than the chains formerly described; of these, the principal is that of the Salisbury Avon, which, rising in the substrata of the vale of Pewsey, breaks through the chalky tract, as do its tributary streams on the west, the Willy and the Nodder; in the center the river Anton, rising within the chalk, descends towards Southampton water; and on the east the river Barge flows by Winchester to the same point.

In order to illustrate the relations of this central mass, and the branches diverging from it, we may compare the whole line of the chalk with its ramifications, to the small letter k in the common printed characters, placed obliquely, thus, k: the interior angles between the stem and the two branches, will represent the spaces occupied by the superior strata in the basins of London and the Isle of Wight; the exterior angle between these branches, the denuded and protruding area of inferior strata in Kent and Sussex; while the outer side of the stem will correspond with the general escarpment of the chalk towards the inferior strata on the north-west: the inosculation of the stem and branches, indicates the position of the great central mass, the breadth of which will be easily accounted for, when it is considered that this tract exhibits the total breadth of the chalk between its opposite escarpments towards the older formations; whereas in every other point, the interior area of the chalk is concealed by its superstrata, and its edges alone exposed to view.

It will at once be seen that the range of chalk, traced up to its junction with this central mass, is that represented by the upper part of the stem, and that it forms the north-west border of the London basin.

Another similar chain is detached from the north-east angle of the great central mass near Farnham, extending to the straits of Dover near Folkestone, (this corresponds to the upper branch of the k); it is well known under the name of the

North Downs of Surrey and Kent. This chain bounds the London basin on the south, as the former did on the north; the hills composing it descend with a gentle slope along the back of the strata, towards the basin; but on the opposite or southern side, are broken down into a steep escarpment facing the older ranges of the Weald. This range is broadest at its western termination, where its strata have the least inclination, and consequently occupy by their basset the greatest horizontal space. It is here not less than ten miles across. On the other hand, at its western extremity near Guildford, where the beds are considerably inclined, it forms a narrow and steep ridge known by the name of the Hog's-back, which cannot exceed half a mile in breadth. The escarpment of this chain ranges from west to east by the following towns; Guildford, Dorking, Seven Oaks, Maidstone, Folkstone; the cliffs in the neighbourhood of Dover, which will hereafter be particularly described, are formed by the section of this chain against the coast.

This chain is broken through by all the rivers which run northwards from the Weald, viz. the Wey, the Mole, the Darent, the Medway, and the Stowe; which thus form a system of transverse valleys, crossing at right angles the great longitudinal valley which extends along the foot of its escarpment. These circumstances will be more fully described in treating in the Sixth Book of the Phænomena connected with Vallies.

The chain attains its greatest height at Botley hill.

The Isle of Thanet, which consists entirely of chalk, placed near the north-east chain, is not in fact connected with it; being separated by a trough occupied by the upper strata, from beneath which its strata rise towards the north-east.

Another corresponding chain (forming the lower branch of the k), is thrown off from the south-east angle of the central mass near Petersfield. The escarpment of this chain looks northwards, skirting the district of the Weald (occupied by the inferior strata) on the south, as did the preceding chain on the north. The line of junction at the foot of the hills, passes south of Midhurst, north of Arundel and Lewes, to the sea at Beechy Head; between which point and Brighton, the section of the chain against the coast exhibits a magnificent range of chalky cliffs, resembling those presented under similar circumstances by the section of the North Downs. On the south, this chain gently descends beneath the superstrata, occupying the basin of the Isle of Wight, which it bounds on the north-east.

This chain is known by the name of the South Downs, as contradistinguished from the North Downs before described. Its average breadth is about five miles: it is broken through, like the North Downs, by a series of transverse vallies, which convey across it to the channel, several rivers rising in the interior ridges of the Weald, viz. the Arun, the Adar, the Ouse, and the Cuckmere. Its highest point is Butser hill in Hampshire, which is 917 feet above the level of the sea.

Nearly connected with this chain is a low and insulated ridge of chalk called Portsdown hill, lying entirely within the basin of the Isle of Wight, and protruding as it were through the superstrata which occupy that basin, and on all sides surround it. Its situation is near the commencement of the chain we have just described, and it seems to be the result of a slight undulation in the strata. It lies on the north of Portsmouth, and extends in length about ten miles east and west between Havant and Fareham, while its breadth scarcely exceeds a mile.

Lastly, from the south of the great central mass near Salisbury, a chain is detached (answering to the lower part of the stem of our k); which, proceeding through Dorsetshire, bounds the Isle of Wight basin on the north-west. Its escarpment first runs west-south-west to Shaftesbury (the vale of Tisbury lying in the angle between itself and the central hills); thence it trends south towards Blandford (where the valley of the Stour breaks through the chain), but about two miles north of that town turns again westward, in which direction it extends (being however broken through by the valley of the Frome), for about 20 miles, to the north of Beaminster, forming Horn hill, which together with the hill on its north, may be said to form its extreme point of connexion in the west of England. On the west of these hills however there are some outlying masses, the probable consequences of denudation, lying upon and surrounded by the beds of the green sand. One of these masses is immediately on the south of Crewkerne, and the road from that place to Chard passes for five or six miles over the summit of another. Chard itself is situated on the edge of another outlying mass, about one mile wide and five miles long from east to west. Two or three other small patches occur on the south-west, and between Chard and Sidmouth; and on the east of the latter place are three outliers of more considerable dimension, and very near to the coast. Returning to Horn hill, we thence trace the escarpment to within a mile of the coast near Abbotsbury. The chain however does not yet actually reach the sea-shore; for suddenly turning round to the east,* so as to form an acute angle with its former course, it proceeds in that direc-

* To make our letter k represent this, we must suppose the extremity of the lower part of its stem to be bent suddenly inwards, thus

tion through the middle of the isle, or more properly, peninsula of Purbeck, forming cliffs by its section against the coast, at either extremity of that isle, viz. at Whitenose on the west and Handfast point on the east. Remarkable circumstances attend this latter part of its course; the chalk, usually nearly horizontal, becomes, throughout the Isle of Purbeck, vertical; and exhibits at Handfast point some of the most singular and interesting phœnomena in stratification which geology has yet recorded.

The chain terminates at Handfast point directly towards the Isle of Wight, which is distant about 15 miles, and appears to be there resumed by a similar chain of Chalky Downs; which, corresponding with the former in direction, and like it, having its strata forced (probably by the same convulsion) into a vertical position, traverses the island longitudinally; presenting, by its sections on the coast, magnificent cliffs at the Needles point on the western extremity, and Culver cliffs on the eastern.

An insulated mass or outlier of chalk, also forms the summit of the hills rising above the southern cape of the Isle of Wight; this is separated from the central ridge by a broad valley occupied by the regular substrata, and the horizontal position is here resumed. The phœnomena of this formation are so beautifully displayed by the magnificent sections it exhibits on the southern coast, and they are often so highly interesting and important, especially in the Isle of Wight and Dorsetshire, that it appears desirable to consider them more in detail than the nature of this general article would allow: the concluding section of this chapter will therefore be dedicated to a particular review of them.

(*e*) *Height of hills, &c.* Chalk does not often bear the general character of a level or flat country, but on the contrary is subject to perpetual undulation of surface, the hills being remarkable for their smooth rounded outline, and for the deep hollows and indentations on their sides.

It may be affirmed that the most level tract of chalk in England is on the north-east and east of Cambridge. Near that place the Gogmagog hills stretch in the form of a full moon, and the flatness of the adjoining country gives an importance to their inconsiderable eminences. (G. Notes.)

The general character of the surface of the chalk, as will be gathered from the preceding observations on its range and extent, is hilly; the hills having on one side of them a precipitous escarpment, which in the long range extending from the coast of Yorkshire to that of Dorsetshire, is on their north-western side, while on the south-eastern they gently decline, and at the distance of several miles are lost beneath the superincumbent strata. (P.)

It also constitutes a remarkable character in the chalk ranges (of which numerous examples have been mentioned in describing its range and extent) to be frequently broken through by transverse vallies, giving vent to rivers often rising in districts of much less elevation beneath their escarpment, and among the subjacent strata. The low degree of consolidation possessed by this rock, has suffered it to be more easily acted on by the causes (whatever may have been their nature) which have excavated the vallies, and thus given rise to the above configuration of its surface. The phœnomena of these transverse vallies, and the nature of their connexion with the longitudinal vallies at the foot of the chain, is such as to overthrow completely the hypothesis which attributes the formation of vallies to the gradual but incessant action of the rivers now flowing through them; this point will be further examined in the chapter dedicated to this subject in the sixth book. (C.)

The highest summit of the range in Yorkshire is Wilton Beacon, which is 809 feet above the level of the sea; and there are several hills in the range traversing the more southern counties which are nearly equal, and even superior to Wilton Beacon, as well as several in the long ranges, traversing Hampshire, Kent, and Surrey in an easterly direction and forming the boundaries of the chalk basins of London and the Isle of Wight already described. The most elevated point of the whole is that of Inkpen beacon in Wiltshire, which is 1011 feet above the level of the sea.* (P.)

(*f*) *Thickness, &c.* The aggregate thickness of the upper and lower chalk is taken by the Rev. Mr. Buckland in his order of superposition, &c. as being 580 feet, and the two beds may be assumed without much error, as being of about equal thickness, but they may be supposed to differ considerably in different places.

Near Dover, the chalk with numerous flints and that with few flints, are together 480 feet thick, while that without

* The extremity of the Chilterns and of the North and South Downs is every where celebrated for the extent of their prospects. The boldness of the escarpment and the whiteness of the substance have given the idea of ornamenting the country in various parts by cutting away the turf. The white horse above Uffington in Berkshire, occupies about an acre of ground, and may be seen at some points of view at the distance of twelve miles. There is another land-mark of the same kind at Chervil near Calne in Wiltshire, and a third in the neighbourhood of Thetford. On the chalk hill that faces Weymouth is a representation of his late Majesty on horseback. Near Cerne is a figure of a giant holding a club in one hand and extending the other: this colossal figure is 180 feet in height. (G. Notes.)

flints is only 140 feet. (P.) At Handfast point on the coast of Dorsetshire, the flinty chalk is 600 feet thick, and that without 200 feet, but the peculiar position of this stratum at this place should make us cautious of relying on any estimate of its thickness, unless we are told in what manner it has been taken. (G. Notes from Middleton.)

Measuring the thickness of this formation in sections carefully constructed, of several parts of England in which it appears, and where the necessary data of the general level of the surface and the inclination of the beds are ascertained with tolerable precision, we may obtain an approximate result which gives between 600 and 1000 feet. (C.)

The best opportunity that has been afforded me of ascertaining the thickness of the chalk beds, is at Culver cliff in the Isle of Wight, where they are disposed vertically. A direct section of these beds seems to be about a quarter of a mile (or 1300 feet), and this is the general breadth of central ridge throughout the island except to the west of Newport, where it dilates and occupies a much greater superficial extent in consequence of the beds being inclined at a less considerable angle. (G. Notes.)

The chalk has been pierced by the well-digger in many places. In Lincolnshire near Rathby 300 feet; in Bedfordshire (Encyc. Brit.) 400 feet; in Kent, near Sittingborne, 363 feet; in Surrey, at Denbigh near Dorking 440 feet; in Hampshire, near Selborne (White) 300 feet, (G. Notes): but it is not mentioned whether these wells were sunk through the chalk, nor is any distinction made between chalk with flints, and without, and chalk-marle, into which these wells may have penetrated.

The flinty chalk, as before observed, is found forming the surface of this deposit almost every where. In two places however it has been found very thin. At Salcomb cliff in Devon, according to Polwhele, it is only three feet thick, and at Branscomb it varies from 12 to 40 feet (G. Notes): which may probably have been occasioned by the destruction of the upper beds, by the action of water. The flinty chalk has been penetrated beneath the beds of the London clay and Plastic clay formation at Stamford hill, three miles on the north of London, 100 feet, and 160 feet one mile on the east of London. (P.)

(*g*) *Inclination, &c.* The dip of the beds of chalk in the general is very inconsiderable. Near Hull in Yorkshire it is five yards in the mile towards the east. (G. Notes.)

On the north-west side of the *London basin*, as in Cambridgeshire and Bedfordshire, and about Devizes in Wiltshire, the dip is gentle to the south-east; while on its southern side, the dip is to the north and north-east. At High Clere in Wiltshire and Farnham in Surrey it is towards the north (G. Notes), but at Dover towards the north-east, less than one degree (P.); the general dip of the North Downs, extending from Dover to Guildford, varies from 10° to 15°; in the narrow ridge of chalk termed the Hog's back, extending from Guildford to Farnham in Surrey, the dip is very considerable, being above 45°. (G. T. vol. 2.)

The strata of which the range of hills called the South Downs are constituted, and extending from Beechy head on the coast of Sussex to Dorchester in Dorsetshire (and which therefore form the greater part of the external limit of the *Isle of Wight basin*) dip generally from 5° to 15° to the south; the inclination varying in different places (G. T. vol. ii. p. 171); and as the beds of chalk in the hills constituting the southern limits of the chalk basin of London dip towards the north, we might by analogy assume those of the southern limits of the Isle of Wight basin to have the same direction; but a remarkable deviation from that position occurs both in the Isle of Wight and on the coast of Dorsetshire, for there the beds are nearly vertical; but the phœnomena there exhibited will be better understood by referring this consideration to the detailed account of the sections presented by the chalk cliffs on the south coast in the following section. (P.)

(*h*) *Agricultural character.* Messrs. Cuvier and Brongniart represent sterility as one of the most decided characters of a chalk deposit, and mention Champagne as an instance of its soil being in some cases absolutely uninhabitable. In this country I should suppose the population of the chalk district less than of any other secondary rock in proportion to its size, but though a large part of the chalk land lies in common, I believe there is none absolutely unproductive; but Dunstable Downs and Luton Downs in Bedfordshire, and the Warden White hills form a tract of 4000 acres almost in a state of nature. The vallies are often extremely fertile, so much so that in Kent and Surrey many hop grounds are situated upon this description of soil, and the downs afford excellent pasturage for sheep. A chalk soil is favorable to the growth of sanfoin and clover, and if well manured, becomes good land for turnips, barley, and wheat. The red chalk of the Wolds north of Louth in Lincolnshire is considered excellent for turnips and barley. (G. Notes. Linc. Agr. Survey.) The beech is the tree best adapted to a chalky soil. It may be seen growing in great luxuriance at Knockholt, Tring, Henley, Fareham,, Norbury, &c. Hunmanby is well wooded notwithstanding its proximity to the sea.

The Chiltern hills in Oxfordshire were formerly covered with thickets and woods of beech, and afforded harbour to numerous banditti. Hence the office of steward of the Chiltern hundreds, now become a nominal office, the occupancy of which however, as it is held under the crown, enables a member to vacate his seat in parliament. (Capper's Dict.) Box hill in Surrey has received its name from the luxuriance of the box wood growing upon it, and which is to be met with all the way thence to Guildford. The excellence of the soil covering chalk is well known to the frequenters of Epsom and Newmarket. (G. Notes.)

(*i*) *Phœnomena of springs and wells in this formation.*— The following observations on this head are extracted from Mr. Middleton's memoir in the Monthly Magazine. (C.)

The lower beds of the chalk formation, and every fissure in them, are, with few exceptions, completely filled with water. All the rain and snow which fall upon chalk, percolate downwards to the base, where the water is stopped by a subsoil of blue clay, and that occasions it to accumulate in the chalk, until it rises to such a height as doth enable it to flow over the surface of the adjoining land. In this manner are formed the springs and rivulets which issue near the foot of every chalk hill. In the Cove at West-Lulworth, fine fresh water streams form the base of the adjoining mountain of chalk, just above the level of the sea. The water which issues from the chalk at Croydon, Beddington and Carshalton, forms the river Wandle, and the same thing happens at other places.

Mr. Hilton Joliffe made a culvert several hundred yards in length, from a level so low as to pass through his works in the chalk at Merstham in Surrey, by which a rivulet of water, sufficient to turn a mill, is constantly running off. It cost a considerable sum of money, all which it is believed might have been saved, by the easy operation of boring a few yards in depth, through the subsoil of blue clay, into the sand which lies under it. This culvert drains the water off in such a manner as to enable him to raise the lower beds of the chalk stratum. (Middleton. Monthly Mag. Nov. 1812.)

We have before observed that most of the rivers which traverse this formation, rise in the older rocks beyond its escarpment, and flow through valleys excavated across its chain; the fissured and porous character of this rock in fact prevents its giving rise to any considerable springs. (C.)

Two exceptions to this general rule have however already been mentioned, and will be found in the section presented by the sinking of Liptrap & Smith's well at their distillery one mile on the east of London, (page 45); by which it will be observed that, according to the report of the engineer who superintended the sinking of the well, the springs found in the chalk, which was bored into 160 feet, were more copious, and rose to the same height as the springs found in the beds of the Plastic clay formation overlying it. A considerable spring of pure water issues from the chalk at foot of the cliff on which Dover castle stands, discoverable only at low water, and judging from its situation, it may be considered as issuing from between the chalk with few flints, and that without flints, which are separated by a bed of soft chalk-marle. The well within the walls of the castle, said to be 400 feet deep, but in which the water stands a very inconsiderable height, is probably supplied from the same source. A copious spring of water, called Lidden spout, runs from the grey chalk between Dover and Folkstone. (P.)

Although two springs issued from the chalk as above recited, on the east of London, and within 125 feet from its surface, no spring was found by boring 100 feet into it at Stamford hill, three miles on the north of the metropolis; and it is said that at Royston in Hertfordshire, it has been penetrated to the depth of 400 feet without finding water. (P.)

The occasional absorption of the Mole in a part of its course, where running between the chalk hills of Surrey, perhaps deserves to be noticed, especially as Combden and Pope have exaggerated the circumstance into its occupying a constant subterraneous channel. The following account of this river is from Manning's history of Surrey. (C.)

'The Mole,' says Camden, 'coming to White hill (the same probably that now is called Box hill) hides itself, or is rather swallowed up, at the foot of the hill there; and, for that reason, the place is called the Swallow; but, about two miles below, it bubbles up and rises again; so that the inhabitants of this tract, no less than the Spaniards, may boast of having a bridge that feeds several flocks of sheep.' From this fabulous account, plainly founded on an idea suggested by common report, the reader might be led to imagine that the river actually disappears at this place—forms a channel beneath the surface of the earth, and at a certain distance rises again, and pursues its course above ground. But the truth of the matter seems to be this. The soil, as well under the bed of the river, as beneath the surface on each side, being of a spongy and porous texture, and by degrees probably become formed into caverns of different dimensions, admits, through certain passages in the banks and bottom, the water of the river. In ordinary seasons these receptacles being full, as not discharging their contents faster than they are supplied by the river, the water of the river does not subside, and the stream suffers no diminution. But in times of drought, the water within these caverns being gradually absorbed, that of the river is drawn off into them; and, in proportion to the degree of drought, the stream is diminished. In very dry seasons, the current is in certain places (particularly at Burford-bridge, near Box hill, and a little lower, between that and Norbury park gate, and at that gate and Norbury meadows) entirely exhausted, and the channel remains dry, except here and there a standing pool. By the bridge at Thorncraft it rises again in a strong spring, and after that the current is constant. At a place called the Way-pool, near the turnpike-gate, but on the side of the river next to Box hill, the method in which the water is thus drawn off, is visible by the observer. It hath here formed a kind of circular basin about 30 feet in diameter, which is supplied, in the ordinary state of the current, by an inlet from the river of about two feet in breadth, and one in depth. This inlet being stopped, the water in the basin is soon observed to subside; and, in less than an hour, totally disappears: when the chasms, through which it passes off, at different depths from the upper edge of the bank may be easily discerned. And, from this circumstance of betaking itself occasionally to these subterraneous passages, the river probably derived its present name of the Mole. In more ancient times it seems to have been called the Emlay. (Manning's Surrey, vol. I. iii.)

Near the bottom of Hawke's hill is a large pond, formed by several strong springs, which are seen in many places rising from its bed with strong ebullitions. It turns a mill, capable of grinding 20 loads of corn a week, and in a few yards runs into the river Mole. (Manning's Surrey, vol. I. p. 482.)

Section III.

Particular view of the Sections of this Formation in the cliffs of the southern coast.

We have already observed that the sections of this formation, exhibited in the many magnificent ranges of cliffs on the southern coast which result from them, and whence our island is supposed to have derived one of its earliest names, are so important as to demand a particular examination, and on this we now propose to enter. (C.)

(*a*) ISLE OF THANET.*

Commencing our examination on the east, these natural displays of the interior of this formation will be found to commence in the Isle of Thanet, which is entirely composed of this rock, the section of which forms continuous lines of low cliffs along the northern and eastern borders, rapidly worn away by the action of the sea, and often in consequence presenting fantastic appearances of detached pillars. The north-eastern cape, called the North Foreland, forms the loftiest point; the cliffs here, however, are only between 100 and 200 feet in height: between this point and Margate, the lowest strata are exhibited, the chalk without flints making its appearance: hence the strata gradually decline, though under an imperceptible angle, towards the south-west, in which direction the upper beds of the chalk sink and disappear beneath the more recent formations, which intervene, and separate this chalky tract from the main chain of the North downs of Kent.

On the south of the Isle of Thanet, a flat tract of this character extends beyond Deal, to Walmer castle, where the cliffs of the South Foreland emerge and gradually gain an imposing height, through a tract of 15 miles to the south-west, ranging by Dover towards Folkstone.

These cliffs afford one of the best opportunities for studying the chalk formation which can any where be found; they have been particularly described in a memoir communicated to the Geological Society; and as the observations there given appear to contain a more minute examination of the several constituent strata than has been made elsewhere, a great part of it is here inserted.

(*b*) CLIFFS EAST AND WEST OF DOVER.†

A natural section of the chalk is presented by the cliffs extending from Dover about eight miles eastward towards Deal, and five miles westward towards Folkstone.

The highest point of the range is in the immediate neighbourhood of Folkstone, about a mile north of the town, and is, according to the survey published by the Board of Ordnance, 575 feet above the level of the sea. From the signal-house above Folkstone, the depression of the cliff towards Dover is very gentle. The hill on which Dover Castle stands, is, at its highest part near the Turnpike, 390 feet high:—and if we suppose the cliff, where the castle walls terminate, to be

* By the Rev. W. D. Conybeare.
† From the Transactions of the Geological Society, vol. 5.

50 feet lower, which must be near the fact, we shall assume it to be in that place, 340 feet high. From this place, the cliff, generally speaking, declines in height gradually towards Deal, terminating about a mile from that place, and disappearing near Walmer Castle, beneath rubble and alluvial matter with which it is there covered to the depth of 15 or 20 feet.*

In this long range of cliffs, which in many places forms an immediate barrier to the sea, it is not to be expected that every part should be equally accessible to investigation. Between Deal and Dover there is but little difficulty, nor for nearly half the way from Dover towards Folkstone; but in the latter half of that distance, an immense fall, or rather, it should seem, repeated falls have taken place; so that that part of the cliff of which the beds remain *in situ*, is, at its extremity beneath the signal-house, nearly a mile from the shore. The ruin lying between this cliff and the sea, for about three miles in length, affords scenery inferior in beauty to the "Undercliff" of the Isle of Wight, only because from its want of soil, it is less susceptible of cultivation; while from the same cause its grandeur is more striking. The greater part of it, however, is sufficiently covered by herbage to have become a pasturage for cattle. The cliff, bounding this ruin towards the sea, is from its position, evidently not *in situ*, and it is equally clear that the enormous masses of which it is composed, have fallen forward from near the summit of the cliff *in situ*.

In the less precipitous parts of the cliff, and particularly along that part of it, between which and the sea the ruin lies, it is in a considerable degree covered by herbage; which however does not prevail so greatly as to prevent the observation, that it is, throughout its whole length from Walmer Castle to Folkstone, very distinctly stratified; that the strata are numerous, regular, and perfectly defined, although from the nature of the cliff it is not possible every where to trace

* Almost the whole line of these cliffs is more or less covered by alluvium, but of two sorts; one consisting of a red sand or sandy marle, occasionally containing spangles of mica and very considerable quantities of broken flint, detached doubtless by remote causes from the beds of the upper chalk which once enclosed them, and which have been destroyed. This alluvium prevails most near the signal-house on the summit of the cliff above Folkstone: and it is seen in many places filling up the gullies or deep indentations made below the surface of the chalk, most probably by the action of water. The other variety of alluvium consists of a greyish earth, enclosing small rounded portions of chalk and occasionally of flint; but between this latter and the chalk in situ, are occasionally visible considerable deposits of chalk rubble, as on the summit of the low cliff on the west of Dover, and near St. Margaret's bay on the east. (G. T. vol. v.)

the stratification. The chalk with numerous flints, that with few flints, that without flints, and the grey chalk, all appear in the course of this range; and with them are connected some beds consisting wholly of organic remains, that have not been described as occurring in other places. The strata lie in the following order, and are collectively about 820 feet thick.

1st. *The Chalk with numerous flints*; it is about 350 feet thick, and may be thus divided:

I. With few organic remains, (*a*) of the sketch beneath, lying upon

II. A bed consisting chiefly of organic remains; in which numerous flints of peculiar forms are interspersed; and a few beds of flint run along it. This bed (*b*) is termed, *the Chalk with interspersed flints.*

2d. *The Chalk with few flints*; This stratum (*c*) is about 130 feet thick.

3d. *The Chalk without flints* is 140 feet thick, and consists of

I. A stratum containing very numerous and thin beds of organic remains, (*d*) 90 feet thick.

II. A stratum about 50 thick, with few organic remains.(*e*)

4th. *The Grey Chalk*; this is estimated to be not less than 200 feet in thickness. (*f*). *

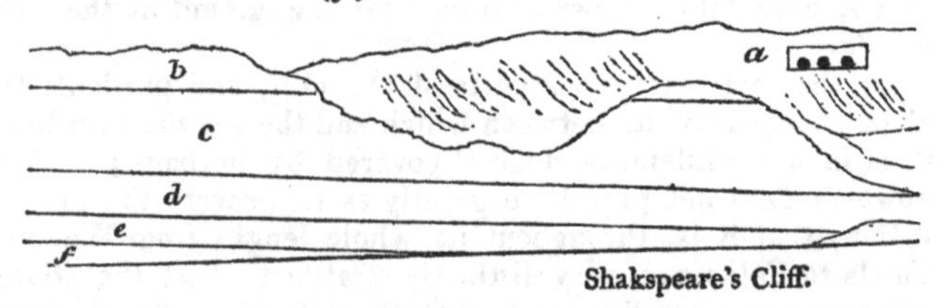

Shakspeare's Cliff.

The above sketch (which includes Shakspeare's cliff and about a mile on the west of it) exhibits the actual position of all the beds, but the bed (*a*) forms the principal part of the cliff at Dover Castle.

The cliffs do not run in the same direction throughout their whole length. From the signal-house above Folkstone to the South Foreland, their general bearing is north-east by east; from the latter place the cliff is somewhat curved to St. Margaret's bay; whence, to Walmer castle, the direction is nearly due north. Although the stratification is every where visible in a greater or less degree, it is best observed by tracing the junction of the chalk without flints, with the superincumbent bed

* Coast of France.

of organic remains. This is visible with little interruption for five miles, from the signal-house above Folkstone to the foot of Shakspeare's Cliff, in which distance it dips only about 300 feet, in the direction of north-east by east.

Proceeding from St. Margaret's Bay, the beds of flint appear to have nearly the same dip towards the north: whence it may be concluded, that the chalk strata in the neighbourhood of Dover dip somewhat less than a degree towards a point between north-east by east and north.

CHALK WITH NUMEROUS FLINTS.

I. *With few Organic remains.*

The low range of cliff between Walmer Castle and St. Margaret's Bay, being about five miles long from north to south, consists altogether of the chalk with numerous flints. The rise of the cliff is gradual; its highest part being that immediately contiguous to St. Margaret's Bay, where it scarcely attains 200 feet in elevation. In consequence of its nearly uniform hardness from the base to the summit, it is almost precipitous, and suffers but little from decay or casual fall; but it is so low in some places, as at Kingsdown Bay for instance, as scarcely to exceed 20 feet above the beach. The beds of flints are frequent and thin, being on an average scarcely two feet apart; and the flints are obviously separate from each other. But a remarkable deviation from this general rule is observable immediately on the west of St. Margaret's Bay; where a continuous stratum of flint, about an inch and a half thick, rises from the beach, and is readily traced at least two miles from that bay: soon afterwards another rises which is about half an inch thick, and is visible for nearly a mile about 20 feet below the former.

Quitting St. Margaret's Bay for Dover, the chalk with numerous flints appears to rise gradually; forming, at the latter place, when viewed from the shore, apparently about one-third the whole height of that part of the cliff upon which the castle stands. Soon after leaving St. Margaret's Bay, the beds of flint begin to increase in distance and in thickness; in the latter respect they go on increasing to that part of the cliff which is immediately beneath the castle; where the thick beds, protruding at least two feet beyond the chalk, give rise to the idea of their consisting each of one mass of flint, but which their rugged edges, as viewed from below, seem to render at least doubtful. Some of these beds exceed a foot in thickness.

The chalk with numerous flints is again visible on the height west of the castle, at a still greater elevation. That it forms the upper part of this height, was proved in making the extensive fortifications on its summit, as well as the openings a little beneath it, immediately below the citadel, which were intended for the reception of some part of the troops stationed at Dover during the war. These openings (*a* of the preceding sketch) are in an unfinished and ruinous state, but they served to evince an interesting fact. Each of these four chambers is driven into the hill about 100 feet, and is perhaps 20 wide and 15 feet high; and in each, the only roof left to support the superincumbent chalk is *a bed, not of flints, but of flint*; the whole is one nearly-continuous, though not regularly tabular and evenly-disposed mass. The belief of the existence of this fact, in regard to at least many of the beds of flint in the upper chalk near Dover, was previously entertained, by observing the workmen on the shore cleave several blocks of chalk, each eight or ten feet square, close to the beds of flint passing through them; and in every instance, examination proved that the flint of each bed so exposed, was connected together: not that it formed one plane surface; but, though varying in thickness from six to 18 inches, the flint would, if it could have been taken off whole, have exhibited occasional cavities, which, collectively, would have formed but a small proportion of the whole surface. A man who had been employed on this work during eighteen years, asserted that he had always observed the same fact.

The flint thus exposed for the first time, is sometimes cracked through in several places, from one cavity to the next: and the fractured surface always appears more or less white and opake. Such a fracture seems explicable only by the supposition of a contraction having taken place in the flint while in its natural position. Nor does it appear at all improbable that a contraction had actually taken place. Flint newly disengaged from its natural bed, is much more brittle, requires a much lighter blow to break it, than flint that has been long exposed.* This may perhaps be owing to the moisture or water belonging to the flint in its natural state, but which it loses in great measure by the joint action of the air and sun.

As the opake white substance, which in some cases only surrounded the edges of the fractured surface, did not project

* The flint gravel used for mending the roads round London, is in some places providently taken from the pit some time before it is wanted, and exposed to the action of the air and sun: for this practice the alleged reason is, that it hardens; which, probably is the fact.

beyond the central part, still retaining the black colour of the internal part of the flint, it is clear that it was not a coating; but on the contrary, this circumstance seems to furnish strong, if not conclusive evidence of its being the consequence of disintegration, proceeding from causes that have not hitherto been explained. The alluvium of the surrounding country corroborates the supposition; every where it includes a multitude of fragments of flint, the broken surfaces of which always exhibit the same appearance of disintegration.

It is also to the progress of disintegration that we are to ascribe the existence of the white opake coating by which the mass of every flinty stratum is more or less covered, while yet in its natural bed. In no instance did there appear any well defined line of separation between the flint and the opake coating, which not unfrequently is half an inch thick, and which by exposure to the sea becomes more compact, and hard enough to admit of a conchoidal fracture. Between its outer surface and the black flint, it is not uncommon to observe two or three thin bands of flint. If the white substance be the consequence of disintegration, it seems remarkable that these bands should have been thus left untouched. In order to ascertain the nature of the white substance surrounding the flint, three portions were selected with care. One from without the band—another from between the band and the flint—and the third of one in which there was no band. These my brother took with a view to determine what proportion of each consisted of lime, in case any should be found. The two first consisted wholly of siliceous matter; the last of 86 per cent. of silica and 14 of carbonate of lime. All the fragments were of a granular texture, and sufficiently hard to cut glass; each also left a whitish streak on the finger when rubbed with considerable pressure upon it; the last in the greatest degree.

Large fragments of striped flint of a grey colour are often discoverable among those which have been taken from the inferior beds in the upper chalk, but they frequently contain a nucleus of black flint, from which the grey stripes diverge as from a common centre.

Such flints as are interspersed in the chalk of the stratum with numerous flints, have usually some organic appearance. They are occasionally found in pear-shaped masses resembling the head of the alcyonium; sometimes in the form of nearly perfect spheres, which are solid, and do not commonly exceed half an inch in diameter, but are often much less. Others of no particular external form, have internal cellular or ramifying cavities which seem to indicate the same origin. Others again seem so decidedly to have been formed around sponges, that the flint has entered all the ramifications, the forms of which therefore remain.

It is not uncommon to find flints inclosing many of the shells observable in the chalk, and impressions of the few varieties of echinus common to that of Dover, the shells having been replaced by carbonate of lime, or the space they once occupied being left vacant; so that the internal cast of the shell, which is of flint, is in some cases connected with the surrounding mass by fine filaments of siliceous matter, arranged precisely in the order of the small perforations commonly visible in the shell, which therefore must have been formed while the shell was yet entire. Shells enclosed in flint are usually filled with the same substance; if only adhering to, or partially imbedded in it, they are generally filled with chalk.

About 40 feet below the summit of the cliff beneath the castle walls, lies a bed of a substance greatly resembling hard chalk marle, parallel to the beds of flint. It is about 18 inches thick, and is distinguishable from below by its being of a brownish yellow colour; and being harder than the chalk, it protrudes, presenting a rugged knotty surface. Such portions of it as fall, are collected by the lime burner, who can convert them into lime only by using coal, instead of the ashes usually employed in the burning of chalk. Between this bed and the summit, a horizontal crevice is visible in the chalk, indicating the presence of a bed of chalk marle. Several other beds of hard chalk marle are visible in the upper chalk east of the castle, between the beds of flint.

II. *Bed of Organic Remains with interspersed Flints.*

When viewing the middle part of the cliff from its base beneath Dover castle, a singular roughness is visible. By ascending the green slope of some ruin, it will be found to proceed in a great degree from its having inclosed a vast multitude of organic bodies of various kinds, amongst which the remains of a few varieties of the echinus, and the ochreous marks of some varieties of sponge, are extremely frequent. In part also the ruggedness proceeds from a vast number of small flints irregularly interspersed through the bed, but which are not visible from below, because such parts of them as are exposed partake of the colour of the bed; for being separate, and mostly small, they do not commonly shew any fracture.

The numerous knotty projections of this part of the cliff are much harder than chalk commonly is, and than that in which they are imbedded. These remains of organic bodies do not lie in thin or separate beds; but form one large bed, occupying about one-third part of the whole height of the cliff beneath the castle, and in the centre of it.

The beds of flint lying immediately above this bed of organic remains, are less frequent and much thinner than those which are nearer to the summit of the cliff, and some beds of flint are visible among the organic remains. Indeed there does not appear any decided line of separation between this bed and the superincumbent chalk with numerous flints; and hence we may consider this bed, though it contains comparatively very few beds of flint, as the lower part of the chalk with numerous flints.

The whole bed has, from below, a greyish appearance; and by this it may be traced by the eye for at least two miles, dipping gently in its course, which terminates at the foot of the cliff just at St. Margaret's bay, four miles on the east of Dover. It may be seen along the cliff at the back of the town of Dover, and is visible west of it as forming the upper part of Shakspeare's cliff, and terminating on the summit of the cliff about two miles beyond it. Its run is discoverable, not by the exterior roughness alone, but also by the presence, near the bottom of it, of two parallel and rather thick beds of flint, which are about four feet apart, and may be seen along the whole course of the bed, as it has been described. Between these beds of flint, lies a thin bed of soft marle, which, becoming friable and crumbling away by exposure, leaves a crevice which accompanies the beds of flint, and which is visible for the greater part of their run; and a similar bed is visible parallel to it, about three feet beneath the lower bed of flint.

The newly broken parts of such masses as fall from this stratum on the beach, shew that the chalk of it is yellower internally than the superincumbent chalk with numerous flints, which is very white; they are also extremely rugged, and the more prominent parts are much harder than chalk commonly is. It is impossible to detach any of the numerous inequalities on these masses, without discovering some organic appearance. Some resembled vegetable stems coated with chalk of a different colour. There often appears a cylindrical mass of whitish chalk, surrounded by concentric coatings of the same substance of a darker colour, which sometimes amounted to ten in number. Ochreous traces of several varieties of sponge are likewise visible; but by far the greater number of the projecting portions consist, when detached, of shapeless masses of chalk which are considerably hard, and which in some respect or other, either by exhibiting a slightly porous texture, or a striated surface, always induced the belief of organic origin. These striated portions are very hard within; the external striæ are sometimes very regular. Among other organic remains afforded by these masses, was extricated the cast, in chalk, of a small nautilus, which had not been previously discovered in the upper chalk.

Such is the general hardness of this bed, that the workmen employed in blasting the cliff, and squaring the chalk for the purposes of the harbour, always leave untouched such masses of it as fall, except they belong to the lower part of the bed; which, containing fewer organic remains, is readily squared. The axe, when struck upon the chalk of the upper or middle part of this bed, returns a sound so exactly similar to that of striking upon flint, that the workman could only convince me that no flint was there, by clearing away what he had struck.

Throughout this bed of organic remains, numerous thin veins of a grey colour run, generally speaking, parallel with the stratum. These veins however, are not straight, but undulate, terminating imperceptibly, being again renewed a little above or below. Some masses that had been split by the workmen along these veins, gave the opportunity of examining their nature, and it became very evident that they originated in the presence of some organized body. It was easy to detach from every part of the newly exposed surface, hard conical masses, striated from the summit to the base by lines of a dirty brown colour, which were glossy and moist: and where the continuity of the cone was accidently interrupted by fracture on the side, the same appearance was discoverable within. It was evident that the nearly horizontal part of these grey veins connected together the neighbouring conical masses. Wherever a flint or a shell was imbedded in contact with one of these veins, it exhibited superficially the same striated appearance as the conical masses of chalk.

The flints interspersed through this bed of organic remains are generally of remarkable forms, and shew either internal or external evidence of their having been formed in or upon some organized body. They are not uncommonly of a nearly pherical shape; and when solid, there is uniformly, as far as my observation goes, a small indented circle upon each: when not solid, they always contained a nucleus having the appearance of a sponge of the same shape as the flint: these rarely exceed an inch in diameter. Others are cylindrical, and inclose another flint of the same form; others (and they are numerous) are conical, having a flat base, around which is always indented an oval, within which there is sometimes the indented mark of a sponge: some of these are solid, others are lined with tuberculated chalcedony of a bluish aspect; these are about two inches high: a thin lining of blue chalcedony, which is extremely greedy of moisture, is by no means uncommon within some of the flints of this bed. There are others very common to it, whose external marks, consisting either of deep indentations or small rugged projections, bespeak the probability of their formation being in some way or other connected with organic matter. But there are other flints which it is not easy to describe. They inclose a cylindrical flint, resembling the stem or a branch of a vegetable, which, passing along the mass, is visible at each end, where sometimes it divides into numerous little branches: another of the same description crosses it, giving to the external flint a peculiar shape, and inducing the belief that it must have been deposited around some organic substance, of which the form is preserved by the internal ramifications. The whole of these flints are numerous in this bed of organic remains; but I did not discover any resembling them in form and character in any other part of the chalk.

The numerous shells of the echinus, or rather the calcareous spar which has replaced them, are almost always whole; rarely was one visible that had suffered depression; but the chalk with which they are filled, instead of being finer than that in which they lie, as is frequently the case in the echini of the upper part of the chalk with numerous flints, is on the contrary much coarser and of a somewhat sandy aspect.

This bed of organic remains with interspersed flints, is separated from the bed on which it lies, the chalk with few flints, by a bed of marle two or three inches thick, which lies about 15 feet below the two beds of flint before mentioned. The exterior roughness of the bed is however far less, and the interspersed flints are fewer, for 10 or 12 feet of its lowest part, than in the middle or the upper part of it.

CHALK WITH FEW FLINTS.

The chalk of this bed is soft and white, though not of so pure a white as that with numerous flints. It contains a few thin beds of organic remains, which, arguing from the ochreous characters that are frequently visible, may be considered as being chiefly of sponges: these beds are most frequent and determinate just below the thin bed of chalk marle forming the separation between this and the superincumbent bed. About 20 feet below that bed, two somewhat thicker beds of soft chalk marle run parallel with each other and with the line of separation, and at about three feet apart. As the marle shivers by exposure, these two beds may readily be traced along the cliff as crevices, rising from beneath the beach about two miles east of Dover, and pursuing their course without interruption, except such as is caused by the occasional falls of the cliff, quite to its termination beneath the castle: they are also very visible in some parts of the cliff above the town, where its surface is exposed. Traces of them may be seen on Shakspeare's cliff; but from their position in it, and in that further west. as well as from the nature of the cliff itself, which is too precipitous to be easy of access, they cannot readily be traced along it.

The marle of these beds commonly shivers by exposure in a direction parallel with the stratum; but undulating grey veins pass along it, and here and there may be extricated from them small conical masses in every respect similar to those which have already been mentioned as occurring in the superincumbent stratum.

Just beneath the thin bed of marle forming the line of stratification, two thin beds of separate flints are very visible, but not lying in the same manner as those belonging even to the thinnest beds in the chalk with numerous flints; their largest surfaces are not parallel with the stratum; but on the contrary, as they lie in every direction, they do not form an even line in regard to each other; and this is the general character of the few thin beds of flints occurring in this stratum, which do not continue for any considerable distance. Flints sometimes lie in the occasional thin beds of sponges which appear on the face of the cliff, and sometimes exhibit impressions of them on the surface. The flints interspersed in the chalk of this stratum are frequently cylindrical, and are sometimes in the form of the bulbous head of the alcyonium, or in shapes resembling vegetable stems: such flints I have observed here and there of more than two feet in length and scarcely exceeding half an inch in diameter, but they were always cracked across in several places.

The grey veins so numerous in the lower part of the superior stratum, are almost as frequent in this, but prevail most just above or below the thin beds of organic remains and of flints above mentioned, and in the neighbourhood of those numerous and nearly parallel crevices which are so many indications of regular stratification.

The ammonite has hitherto, I believe, been supposed to be first visible in the under chalk, or that without flints. A large one lies in the cliff about a mile eastward from Dover, nearly in a horizontal position, and just above a bed of flints which runs for some distance only a foot or two above the base of the cliff, and there are many interspersed flints within a few feet all around it. Two other ammonites from 12 to 18 inches in diameter, are visible at low water, in chalk, belonging evidently to that with few flints, but whether it be in situ or not, it is difficult to determine. One of them is distinctly oval.

About a mile and a half east of Dover, and near the place at which this stratum rises from the beach, I found the cast of a nautilus in it.

Iron pyrites is by no means uncommon ; it usually occurs in globular masses, coated by crystals having the form of the octohedron. which sometimes are attached to a flint : in one instance it was observed filling up the cracks in one. One mass had been formed around a terebratula, of which the shell, filled by pyrites, remained, but in a friable state. This stratum yielded to my search pectinites, terebratulæ, and the palates and vertebræ of fishes. A nearly perfect specimen of one species of the striated shell or inoceramus (G. T. vol. v. pl. 1. fig. 3.), perhaps the only one hitherto discovered, was found by the workmen employed in squaring the chalk : it was nearly filled with flint, and was partially imbedded in it.

Several excursions along the cliff between Dover and Folkstone, both at its base and on its summit, as well as the occasional opportunity of ascending or descending it, enabled me at length satisfactorily to discern the nature of the connexion of the chalk with few flints, with that on which it reposes, namely, a thick stratum without flints, enclosing numerous thin beds of organic remains, lying nearly close together.

A thin bed of soft marle lies between these strata. It may be readily traced along the cliff, as a crevice, for a considerable distance, but is most conveniently viewed while ascending Shakspeare's cliff from the town. Flints are here and there visible a few feet above this bed of marle, but not one was discoverable below it. Between this and a similar bed of marle nine or ten feet beneath it, are to be seen many of those thin beds of organic remains, which are characteristic of the stratum on which the chalk with few flint reposes. Both these beds of marle are also seen in the most elevated part of the low cliff, between Shakspeare's cliff and the town of Dover.

CHALK WITHOUT FLINTS.

I. *Stratum containing numerous thin beds of Organic Remains.*

This stratum, both internally and after exposure, is yellowish and without flints ; for not one appears, either in the cliff, or in the numerous masses lying at its base : in hardness, it exceeds the chalk with interspersed flints.

The low cliff immediately contiguous to Dover on the west,

consists wholly of the chalk of this stratum, except the summit of its most elevated part, which consists of the chalk with few flints (see preceding sketch); but in this place it does not so decidedly appear to consist of a number of thin beds of organic remains, as it does after rising from the beach at Shakspeare's cliff; nevertheless it manifestly consists of a vast accumulation of organic exuviæ. This stratum may be traced without interruption for nearly four miles; but the summit of the cliff between Dover and Folkstone, for the latter half of the way, quite to its termination near the signal-house above the latter place, decidedly belongs to the chalk with few flints: a close examination of it discovers here and there a few interspersed flints, and a single bed of them is visible about 40 feet below the summit, just half way between the two places. Wherever a path is practicable, the cliff is so sloping as to be covered with a verdure which prevents an accurate discovery of the stratification by ascending it, and it cannot always be seen from either above or below.

The external roughness occasioned by the numerous thin beds of organic remains in this stratum, is far less considerable than that of the chalk with interspersed flints. Still the same appearance of sponges is visible after long exposure, but they lie close together, and when detached are less; and they are not unfrequently separated by the remains of shells, so small as to be nearly in a state of comminution, a large proportion of them being varieties of the striated shell or inoceramus. The two varieties of the echinus so common in the chalk with interspersed flints are occasionally seen in this bed, but are less numerous. Ammonites from 12 to 18 inches in diameter and of a circular form are not uncommon: I saw several, all of which lay parallel to the strata. Though the thin beds of organic matter are nearly in contact in the lower part of the stratum, they are more separate in the upper part of it.

Through one of the many large masses lying on the shore, a bed about nine inches thick took its course, consisting of remains essentially differing from the rest. It consisted chiefly of ramose appearances about half an inch in diameter, and the masses detached from it greatly resembled those of the alcyonium visible in the sand of the Isle of Wight, described by fig. 12, pl. 29, vol 2, of the Geological Transactions.

In this stratum I did not perceive any traces of pyrites, so common in other parts of the cliff. It includes grey veins similar to those of the chalk with interspersed flints and of that with few flints, but they are far less numerous, and though the organic remains which occasion them are similar, they are much smaller.

II. *Chalk without Flints, and with few Organic remains.*

This stratum rises immediately at that part of the base of Shakspeare's cliff which is nearest to Dover, and is separated from the stratum containing numerous beds of organic remains, which reposes on it, by a bed of soft marle. As this marle, like all the others of the same nature that are visible in these cliffs, becomes friable and falls away by exposure, it serves as a certain guide to the stratification, and may be seen with little interruption for more than half the way to the signal-house above Folkstone, and at intervals for the other half; so that the connexion of the two strata may be traced for five miles without difficulty. Without this aid, however, there could only have existed such hindrance as naturally arises from occasional verdure on the face of the less precipitous parts of the cliff, being those above the undercliff: for the chalk of the two strata differ greatly in appearance. That of the stratum I am describing is soft, and even white in comparison of that which lies upon it, though not so white as the chalk with few flints; and for six feet beneath the bed of marle, is of so sandy a texture occasionally, as to be even friable, but here and there it assumes the appearance and compactness of a sand-stone.

The nearly horizontal crevices in other parts of the cliff appear to be nearly parallel with the stratification; but in this, the crevices differ from that position, and are even in a transverse direction; those that are nearly vertical are numerous, giving to the chalk in many places, an angular appearance, not common to any other parts of the cliff. It runs along the base of the cliff for somewhat less than half a mile, and in that space, affords, even on a close inspection, the traces of but few organic remains.

It incloses masses of pyrites, some of which are crystallised externally in the form of the octohedron, but their general form is spherical. Others, and they are not uncommon, are in the form of cylinders, rounded at each termination, to which there is frequently a short stem attached; the whole having the appearance of organic origin. When broken across, they are always found to radiate from the centre.

Here and there appears a small bed of sponges, of which the ochreous forms are visible; but this stratum contains none of the grey veins so numerous in the superincumbent strata. It is about 50 feet thick.

GREY CHALK.*

The grey chalk differs from the strata reposing on it, in being softer. It varies considerably in different places, in respect of colour and texture; being much more sandy and less compact than in others. A fair specimen of it yielded to my brother, by analysis, 82 per cent. of carbonate of lime, and 18 of silex and alumine, chiefly of the former, and a trace of the oxide of iron.

It first rises from the beach at the foot of the low cliff (see preceding sketch) contiguous on the west to that which is known by the name of Shakspeare's; but its separation from the chalk without flints is not at that place perfectly defined. In less than a quarter of a mile beyond the place at which it rises from the beach, the two strata are very distinct; the white being separated from the grey by some very thin beds of a sandy appearance and yellowish colour. As the grey chalk rises, its colour becomes deeper, and it is here and there so extremely soft, that the rain in descending the cliff, has carried down and deposited at its base considerable quantities. It is in these places particularly, that beds of sandstone from one to four or five inches thick, and extremely hard, take their course parallel with the stratum for a short distance, projecting beyond the face of the cliff from a few inches to two or three feet. A fan-shaped projection of this kind overhangs a copious stream of pure water issuing from the grey chalk, termed Lidden spout, protruding beyond the surface of the cliff, not less than five feet.

Not a flint is visible in this bed: its organic appearances are numerous, but do not differ considerably from those of the superior strata. Along the crevices running parallel with the stratification, the chalk is commonly of a deeper grey than the bed in general, and incloses some still darker appearances of ramification, resembling in their general form some varieties of broad-leaved fuci, which are somewhat softer than the chalk in which they are imbedded. The reverse of this is occasionally observable in the more solid parts of this chalk, and where its colour is of a lighter grey; for there it is sometimes traversed in every direction by very numerous and small ramifications of a colour still lighter. The remains of the echinus are numerous, and their shells are replaced by carbonate of lime

* This bed of grey chalk might more properly be designated chalk marle, and therefore its description ought strictly to be referred to the beds immediately beneath the chalk; for the convenience of the traveller however it is here inserted.

of a grey colour: every one I saw had not only suffered depression, but the shell was also in every instance broken; a circumstance which is the more remarkable, as it is rarely discoverable in any of the superior strata.

Small masses of pyrites are very common, and there are some of singular forms. Thin strings of it, sometimes ramified, the branches terminating in a point, are very numerous; but sometimes on the contrary they are terminated by a spherical bulb, or by one that is elongated and resembling the head of the alcyonium, but very small; these are generally hollow. A remarkable crystallization of pyrites is also very common. A string of octohedrons piled with considerable regularity on each other, and above an inch in length, is crossed by another similar to it,—the termination of each being the quadrangular pyramid of an octohedron:—these again are crossed at right angles by another, which is terminated in like manner, so that the three have one common centre: giving to the whole the appearance of the commencing crystallization, or the skeleton, of one large octohedron. Thin beds of sponges occasionally appear, their remains being either ochreous, or of a lighter colour than the chalk itself; but these are visible only when the stratum has risen considerably, and near to the beginning of the undercliff. Near the stream called Lidden Spout, I found the cast of a large nautilus in grey chalk, but much harder than that of the stratum.

(c) SUSSEX CLIFFS.*

After an interval of about 40 miles occupied by the formations below the chalk, and which, after quitting the green sand cliffs near Hythe, presents a flat and uninteresting coast, (excepting where the section of the central ridge of iron sand affords the picturesque cliffs on both sides of Hastings,) the chalky cliffs re-appear with much magnificence in the promontory of Beachy Head, which forms the termination of the range of the South Downs against the coast. These cliffs are said in Henshawe's Survey to be 575 feet high; they exhibit both the lower and upper chalk declining under a gentle angle to the S.W.: they extend westward without interruption about four miles, when they are broken through by the valley which gives issue to the Cuckmere river. On the west side of this valley the cliffs again rise, and are continued till a second similar interruption is occasioned by the mouth of the river Ouse, towards which a small outlying hummock of the plastic clay superstrata crowns the chalky cliff at Chimting castle; beyond

* By the Rev. W. D. Conybeare.

the Ouse the cliff is again resumed, having, at Newhaven castle hill, another outlier of plastic clay and sand reposing against it: hence, the chalky cliffs, though gradually declining in height in proceeding westward, are continued to the neighbourhood of Brighton, where they finally disappear, the beds of this formation having sunk beneath the superstrata of the Isle of Wight basin.*

West of Brighton, the coasts of Sussex and Hants present only a low uninteresting tract; but crossing to the Isle of Wight, the chalk re-emerges from the superstrata near its eastern point, and rises with its usual magnificence into Culver cliffs. Here, indeed, appearances of more than usual interest occur; for here we first enter upon that remarkable district in which these beds, so generally characterized by their horizontal position, assume that vertical arrangement which has been hastily assumed as peculiar to older and more chemical depositions, and as resulting in such from the circumstances of their original formation; but which, as we shall hereafter have occasion to shew, are limited to no single geological æra, and in the great majority of instances, if not in all, have been demonstrably produced by the mechanical force of subsequent convulsions. In our examination of this part of the coast, we have the further advantage of following an able and enlightened guide, the author of the excellent letters to Sir H. Englefield; to the correctness of whose description the writer of this article can bear the fullest testimony, having verified them all by a careful personal survey throughout the whole district.

* In the Royal Institution Journal, No. 8, p. 227, is an account of the cliffs at Brighton and on the east of it, by J. F. Daniell, Esq. who describes a bed consisting chiefly of flints, but containing rounded masses of granite, slate and porphyry, and resting near the town on chalk, but as being there covered by chalk rubble. This superincumbent rubble, as it proceeds eastward, is described by him as partaking more and more of a regular aspect, as having the appearance of chalk about half way between Brighton and Rottingdean, and as containing two contorted layers of flint, while others passing down it obliquely, also traverse the bed of alluvium, containing masses of flint, granite, porphyry, and slate. Having, as I conceive, some reasonable doubts as to the correctness of the preceding observations, which seemed to imply the opinion that the alluvial bed of rounded masses of primary rocks was interstratified with the chalk, I requested my friend Thomas Hodgkin, then a resident at Brighton, to re-examine the cliff, which he has done with attention, and the result is, that every oblique bed of flint perceived by him above the bed of granite, &c. stopped at it, resuming its direction beneath it; and he is of opinion that this bed of decayed alluvium, instead of being interstratified with the beds of chalk, was heretofore washed by the action of the sea, when at a higher level than it now is, into a crevice, either created by that action in the chalk, or by washing out a bed of soft chalk marle, once occupying the place of the masses of granite and of other rocks—an opinion extremely reasonable. (P.)

(*d*) ISLE OF WIGHT.*

The general configuration and relations of the series of inclined strata in this Isle, will be sufficiently understood by referring to the section of the Isle of Purbeck (see Pl. 2, fig. 5.) It will be there seen that the strata exhibit their greatest inclination in a part of the chalk range, and that on either side † of this, the deviation from an horizontal position gradually decreases. The ingenious hypothesis of Mr. Webster, which indicates how such an appearance may be presented by the partial section of strata, which (were it possible to trace their inferior prolongations beneath the surface, and to supply the portions which may have been stripped off from above), might be found to be disposed in parallel curves of a double flexure, is indicated by the dotted lines following the supposed directions of those curves.

That such cannot have been the original position of these strata is demonstrably evident; for among the vertical strata of the plastic clay formation is one composed of a thick layer of rolled chalk flints, the accumulation of which in such a posture is obviously contrary to the laws of gravitation; other beds, also containing decided fragments imbedded in them, occur in the substrata beneath the chalk (particularly in the iron sand), and the whole series is decidedly of mechanical formation.

It remains, therefore, that the phœnomena must be referred to some subsequent convulsion, which has violently produced either an elevation on one side, or a subsidence on the other, of these inclined masses; and from the form of flexure in the strata, the disturbing force (whatever may have been its nature) seems to have operated with the greatest energy in a lateral direction. The central line along which this force has acted, may be traced nearly east and west about 60 miles, from the eastern end of the Isle of Wight to Abbotsbury in Dorsetshire; for though the vertical chalk ends at Whitenose, an highly inclined saddle of the substrata may be traced in the continuation of the same line.

The first section of these inclined strata, in advancing east to west, is presented by Culver cliffs, a magnificent range of

* By the Rev. W. D. Conybeare, F.R.S. &c.

† In the Isle of Wight, this is only seen in the inferior strata; for the vertical portion of the superior strata being abruptly covered by horizontal beds of still later formation (see section of Alum bay, pl. 2, fig. 6), the gradual change from the vertical to the horizontal observable in Dorsetshire is here connected.

precipices near the eastern extremity of the Isle of Wight, forming a promontory separating White-cliff bay on the north-east from Sandown bay on the south-west. In the former, the superstrata of the plastic clay and sands may be observed in a position perfectly vertical, forming low cliffs; immediately on the south of which, the chalky strata tower to a stupendous height, their inclination being about 70°; rounding the cape towards the south, this inclination decreases to about 60°; the direction of the dip is north-north-east: all the upper beds of chalk contain alternating strata of flinty nodules, occasionally exhibiting that singular shivered texture described below;* the lower strata, as seen in Sandown bay, are destitute of flints, and the lowest consist of a yellowish white marle or argillaceous chalk; this rests on the green sand formation.

The southern extremity of the Isle of Wight, from Dunnose to St. Catharine's, exhibits a lofty range of downs, separated from the highly-inclined central ridge by a broad valley, and presenting an horizontal stratification. The upper region of these hills consists of chalk and chalk-marle, the central of the green sand, and the lower of the iron sand hereafter to be described; but on the side towards the coast, the inferior strata are concealed by vast masses of the superior, which have subsided in that direction, and form as it were a talus in front of them, constituting that most picturesque district so well known by the name of the Undercliff, which well describes its position; for a fracture which runs through the upper strata of

* All the flints, except those detached nodules in the body of the strata, are universally found in a most extraordinary state; they are broken in every direction into pieces of every size, from three inches diameter down to an absolutely impalpable powder. The flints thus shivered, as if by a blow of inconceivable force, retain their complete form and position in their bed. The chalk closely invests them on every side, and till removed, nothing different from other flints can be perceived, excepting fine lines indicating the fracture, as in a broken glass; but when moved they fall at once to pieces. The fragments are all as sharp as possible, and quite irregular, being certainly not the effect of any peculiar crystallization or internal arrangement of the materials, but merely of external violence. This new and most extraordinary appearance was first observed in a small pit on the Shorewell road, just beyond the parting of the road to Yarmouth, but no opportunity was afterward omitted of examining both the cliffs and the pits in many parts of the whole range, and the appearances were every where nearly similar, differing only in the circumstance that in some places the flint seemed to have been more generally and completely shattered than in others. It may not be improper to mention the places where these phœnomena were the most particularly investigated, as they may guide others in their researches, beginning at the eastern point, and proceeding westward; 1. Whitecliff bay; 2. Brading shute; 3. Pit on Brading down; 4. Hollow road at Knighton; 5. Arreton pit; 6. Pit above Shide bridge; 7. Pit just out of Carisbrook town; 8. Pit south of Carisbrook castle; 9. Freshwater cliffs; 10. Cliffs in Alum bay. (W. p. 20.)

the green sand formation, about half a mile from the coast, exhibits this section in a continued line of precipice; beneath which the whole space intervening, down to the beach, is occupied by a series of terraces formed by masses of strata (chalk and green sand) which have subsided from above, and generally settled in an highly inclined position dipping towards the interior. A wild scene of irregular confusion is thus produced; masses of the sandstone project in striking crags, combined in a thousand pleasing forms with the luxuriant foliage to which the deep dingles between the terraces afford a shelter. By these subsidences, vast masses of chalk have been brought down from their parent stratum, whose lower limit is between 500 and 600 feet above the beach, to the very edge of the sea; and near Ventnor a considerable cliff of chalk is thus seen on the coast, nearly adjoining to one of the iron sand on the same level.

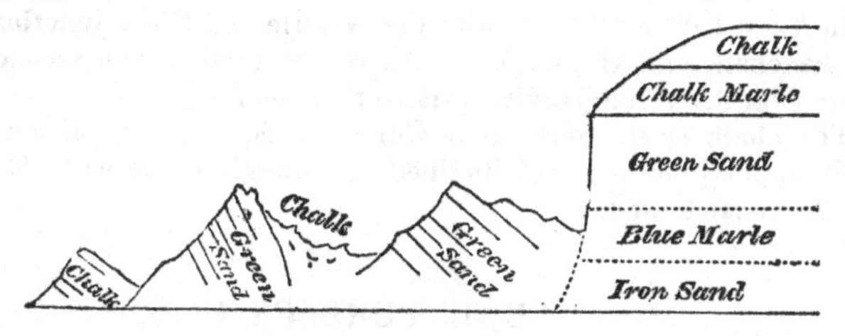

The preceding wood cut exhibits a sectional profile of this singular tract.

Proceeding towards the western extremity of the island, we again meet with a section of the central ridge of inclined chalk, corresponding to that which has been before noticed at its opposite extremity in Culver cliffs.

This section commences in the cliffs on the east of Freshwater bay, where the junction of the chalk and inferior strata affords appearances which are an exact repetition of those in Sandown bay. At Freshwater Gate, the whole range of this chain is broken through by a valley which separates the western extremity of the island into a distinct peninsula, and presents the remarkable phœnomenon of a spring rising almost within a stone's throw of the southern coast, and but little above the level of the sea, and yet flowing towards the opposite coast. This seeming paradox admits an easy explanation, by considering that the deep æstuary of the Yar in fact brings the tidal level almost as near this point, on the one side, as the other; the spring has its source in a diluvial deposit of clay and gravel, which partially occupies the lower part of this breach in the chalk, and may be seen hanging on the base the cliffs on either side : this clay, resting on the inclined substrata, would naturally turn the current in a northern direction.* Thence the chalky cliffs extend in a magnificent line of mural precipices, known by the name of the Main Bench, for two miles to the west; where, ending in the bold promontory of the Needles, they form the western cape of the island; beneath this cape lies that singular group of insulated pyramidal masses of chalk from whose figure it derives its name. These, having yet resisted the action of the surrounding waves, remain monuments of its ravages, which have destroyed the continuation of the chain of which they once formed a part. An idea may be formed of their size, when it is stated that the hull of the Pomona, a 50 gun frigate, (which in 1811, on her return from Persia, struck upon the point of the most western Needle) did not reach one-fourth of their height.

In Alum bay, on the north of the Needle point, the junction of the chalk and the upper strata represented in the second Plate (fig. 6.) accompanying this work is seen.

The chalk strata here, as in Culver cliffs, are vertical near their upper junction, and inclined in an angle of between 50° and 60° near their lower.

(*e*) ISLE OF PURBECK.†

In order to continue our examination of the sections presented by the chalk ranges, we must next cross the channel to the eastern extremity of the Isle of Purbeck, where in the prolongation of the same direct line with the central range of the Isle of Wight, and facing the Needle's point, another magnificent and interesting section of the vertical and inclined chalk is exhibited in the cape called Handfast-point.

This promontory forms the division between Studland bay on the north, and Swanwich bay on the south; the former exhibiting the superior, the latter the inferior junction of the chalk. The chalk strata first emerge under a very low and almost inappreciable angle, and thus continue round the extreme point of the cape; near which four or five detached pillars of chalk rise above the water, being evidently portions

* This point affords a good station whence we may expect to estimate hereafter the progress made by the encroachments of the sea, in wearing away the coast, since a very small advance will cut through the beach into the source of this rivulet, and joining the æstuary, render this peninsula a distinct island: the isthmus is at present protected by a shingle bank.

* By the Rev. W. D. Conybeare, F.R.S. &c.

which have resisted longer than the rest, the destroying action of the waves. A little to the south of this point, the vertical strata of chalk commence, and a configuration of the strata of a very remarkable kind takes place. The horizontal strata suddenly turn upwards into a curve forming near a quarter of a circle, and the vertical layers of flints meet the bent part of the chalk, as so many ordinates would meet a curve, decreasing in height as they get more under it. It is so impossible that description should do justice to this extraordinary arrangement, that we annex a sketch taken from one of the plates accompanying Mr. Webster's letters to Sir Henry Englefield, and published in his splendid work, the 'Description of the Isle of Wight.'

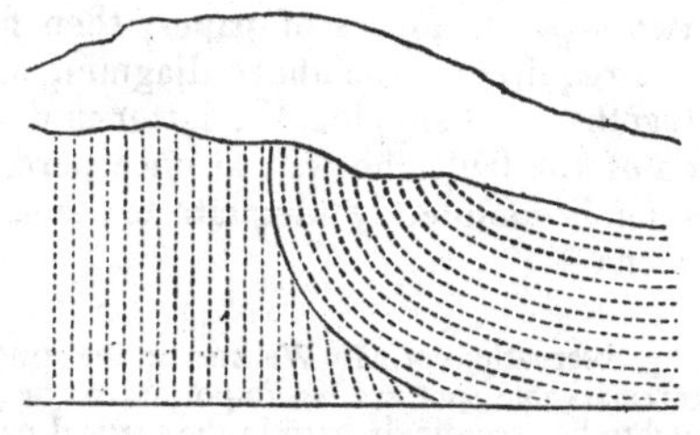

Mr. Webster has proposed an ingenious, but perhaps not altogether satisfactory solution of this remarkable position (see plate 2, fig. 5). Instead of repeating this, we shall attempt to shew that it may be simply and completely accounted for by the obvious supposition of such a fault, or dislocation of the strata, as is familiar in all geological inquiries; and considering the vast angular motions which these masses of strata must have undergone, might naturally be expected here: for this purpose we must have recourse to the subjoined diagram.

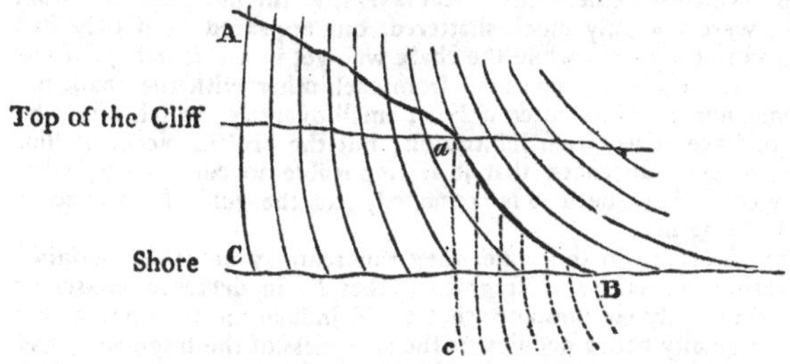

Let A B be the assumed line of fault; then let the mass of strata A B C, be moved along it in the direction from A to B, till the stratum A C be brought into the position indicated by the dotted line *a c*; and the same arrangement of strata which actually takes place in the cliff, will be the result. In order to produce this agreement, we have only to suppose that the line of fault, during a part of its course, coincides with the liner of stratification, which cannot be considered as an impossible or even improbable circumstance.

The flexure of the innermost strata of the vertical series (see plate 27 of Sir H. Englefield's Isle of Wight), as accurately represented in Mr. Webster's excellent view of this spot, affords an additional confirmation of this hypothesis; for we might naturally expect to find a portion of the strata in which such a flexure occurs, transferred to this very place by the fault assumed. If any person will take the trouble of copying the curved and vertical strata as represented in Mr. Webster's delineation on two separate pieces of paper, then filling up the vertical strata as required in the above diagram, and lastly applying them together and slipping the latter under the former in the direction of the fault, he will at once perceive that this simple supposition is completely adequate to the solution of the actual phœnomena.*

* The following observations of Mr. Webster on the strata of the chalk and flints at this remarkable spot are too important to be passed over.—'This chalk I found to be exceedingly hard in the vertical part, and also in the curved; so much so that it will not mark, nor can the nail make any impression upon it; but at some distance from the curved strata, where the chalk is horizontal, it resumes its softness.

In the vertical strata, the chalk is far from being uniform in its texture; appearing as if formed by the union of masses of chalk of different qualities; some parts being denser than others, and of rather a darker colour. When large masses fall down, they frequently separate into roundish fragments, which leave a lumpy and concreted appearance. It might be called a brecchia chalk, composed of roundish lumps of hard chalk, cemented by chalk somewhat softer. In it were numerous veins of calcareous spar, well formed crystals of which were in the cavities.

The flints which were here in vertical layers, at the usual distance from each other, were not only much shattered, but appeared as if they had been reduced to fragments while the chalk was yet in a soft state; for the fragments were in general separated from each other with the chalk between them: nor was this latter only in small quantity, which might be supposed to have arisen from infiltration; but the broken pieces of flint were often at such distances, that it is impossible to conceive by what means they could have been so far removed, had the chalk been solid at the instant of fracture.

When the flints are in this state, they can scarcely be called nodules; they are rather collections of fragments, that lie in detached masses or groups; and the only circumstance that could induce me to suppose that they were originally entire nodules, is the sharpness of the fragments, and the groups assuming the same regular situations with respect to each other, that the nodules do.

Yet although the relative position of the parts of each nodule are frequently so entirely changed, that it is necessary to suppose more motion than could have taken place had the chalk been quite solid, this is by no

Near the junction of the chalk and inferior formations in Swanwich bay, the inclination of the strata decreases, and here exactly the same series of phœnomena may be observed as in the Isle of Wight.

Near the east end of the Isle of Purbeck, many singular coves are excavated into the interior, some of them in a nearly circular form, the entrance from the sea being narrow, and opening into a wider basin: this form arises from the greater hardness of the inferior bed, the Portland limestone; which, having offered a stronger resistance to the action of the waves, projects in bold capes; while the inner strata have been more extensively worn away. The back of several of these coves, is cut into the chalk strata, and fine cliffs are thus exposed: this is the case at Worthbarrow Bay, Lulworth Cove, and Durdle Cove; in all these the chalk strata are nearly vertical, and the lower beds are seen in the same order so often alluded to.

From Durdle Cove to Whitenose point, about one mile and a half to the west, the chalky cliffs line the entire coast. This interval is interesting because it presents the termination of the vertical chalk strata, and because it is the last chalky cliff on this part of the coast, the hills of that formation here trending inland.

The termination of vertical beds appears to take place, in consequence of the course of the chalk and the line of the coast trending to the north of the axis of the disturbing force: the manner in which the transition from the vertical to a gently inclined position takes place, will be best understood by consulting the subjoined sketch.

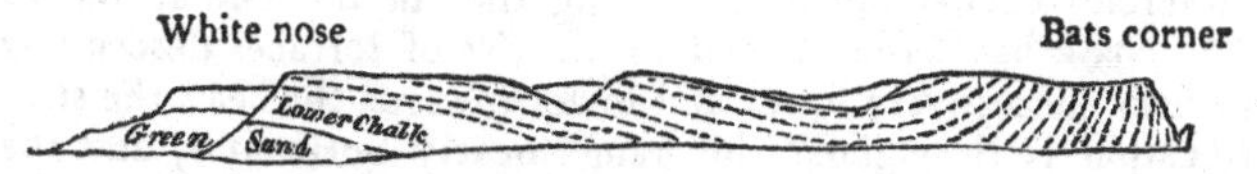

As at Handfast point, the vertical chalk is here hard, the horizontal in its usual state of softness; this seems to indicate some connexion between the inclination and consolidation of the strata.

All the members of the formation are the same as before.

means the case; every gradation being observable, from nodules of the usual shape, only much cracked, to those I have just described. These flints do not frequently fall into fragments in the hand, as those of the Isle of Wight, the parts being firmly imbedded in the chalk; but there is the same variety in the size of the fragments, from large pieces to the finest powder.

The nature of this singular chalk and flint is best examined on the south side of Handfast point, in Swanwich bay, where much of the cliff had fallen down; on the east face it is both more difficult to land, and to get access to it. (W. 166, 167.)

(*f*) BETWEEN LYME, AND SIDMOUTH IN DEVON.†

It has been stated in our account of the range and extent of this formation, that several insulated, and, as they are termed, outlying masses of it, occur considerably beyond the western termination of the principal and continuous range formed by it. The sections of the south-western of these detached groups against the coast in the east of Devon, afford the last chalk cliffs in this direction.*

These commence a little to the west of Lyme Regis (on the western border of Dorsetshire), and form a line extending between six and seven miles westward to the mouth of the Axe. The chalk does not however occupy the whole of the cliff, but is confined to its higher region; the central being possessed by the subjacent green sand, and the lower by the lias; for the chalk and green sand, in their extension westwards, overlie all the intermediate formations, and become immediately incumbent upon the lias, which in its turn terminates about two miles east from Axmouth; the green sand being thus brought into contact with the marle of the new red sandstone formation. The undulation of the upper line of the cliffs cuts in four places through the chalk to the subjacent green sand, forming five distinct summits which bear the names of 1. Ware; 2. Pinhay; 3. Whiteland; 4. (not particularly distinguished); 5. Dowlands. This enumeration proceeds from east to west: between Ware and Pinhay a fault occurs which throws down the strata to the west; beneath Pinhay and Whiteland an undercliff occurs much resembling that in the Isle of Wight. The chalk has here subsided in a series of terraces descending in beds successively lower and lower towards the sea; the stratification is throughout this range nearly horizontal, and the lower strata agree with those before described.

On the west of this range, the river Axe empties itself into the Channel through a wide valley occupied by the subjacent new red sandstone, beyond which are the precipices distinguished as the White cliffs. Here the strata dip rapidly to the west; the green sand and chalk reappear in the face of the cliff, and are brought down in succession by the inclination of the strata to the level of the beach. A portion of the upper part of this cliff was about fifteen years since detached, and subsided towards the beach. This point exhibits a striking

† By the Rev. W. D. Conybeare.

* Mr. De la Beche has communicated a detailed account of this part of the coast in a very able memoir laid before the Geological Society, from which the above brief abstract is principally taken.

scene of ruin, the shattered masses and pinnacles of the chalk being grouped together in the most picturesque manner. The village of Beer, notorious for the daring community of smugglers who inhabit it, and form as it were a peculiar race distinguished by many singular customs, lies in the recesses of a ravine traversing the white cliff, beyond which it is continued to the promontory of Beer head, the chalk here occupying the whole cliff. Round this point the precipices continue without interruption to Branscombe mouth westwards; the strata rise in this direction, and the green sand and new red sandstone, or rather marle, again crop out and form the base of the cliff: so that the strata from White cliff to this point, appear to lie in a kind of basin. The extent of this range is between three and four miles.

From Branscombe mouth the cliffs extend about five miles westwards to Sidmouth, being broken by two ravines into the separate groups of Branscombe, Dunscombe, and Salcombe: their base is uniformly constituted by the marle of the new red sand, upon which the strata of green sand repose. Branscombe (the easternmost summit), has a covering of chalk, as has the east end of Dunscombe (the middle summit), remarkable as being the western termination of this formation in England. The upper surface of the chalk is here furrowed into considerable inequalities, and large masses of it lie beyond its general line—the proofs of the abrasion and destruction which it has undergone. Salcombe, the western summit, no longer exhibits any traces of the chalk; but the abundance of the flints of this formation scattered over it, prove that it has formerly existed here also.

(*g*) COAST OF FRANCE. *

On the opposite coasts of France, a series of sections may be observed almost exactly answering both in character and position to those above described, demonstrating the former continuity of the constituent strata. †

* By the Rev. W. D. Conybeare.

† On reviewing the many remarkable points of agreement between the cliffs on each side the Straits of Dover, it seems a supposition too reasonable to be ranked among mere hypotheses, that they were once united, and that they were separated at some very remote period by an irruption of the sea, which in all probability washed away the connecting mass; for the unreasonableness of the popular notion that the two countries were simply rent asunder by some sudden convulsion, will become apparent, when it is further stated, that the chalk without flints on the west of Dover is not less than fifty feet in thickness, while that of Cape Blanc Nez is

Thus, from the low and marshy grounds near Calais, and between that town and Uissant, a range of chalky cliff rises directly facing those of Dover, and exhibiting the same subdivisions of this formation, viz. 1. Chalk with flints, forming the summit of the cliffs; then in succession, 2. Chalk with few flints; 3. bed with organic remains; 4. Chalk without flints; 5. Grey chalk. This range of cliffs is however much more limited than that on the English side, not exceeding three miles in extent; it commences about a mile west of Sangatte, forming the cape called Blanc Nez, and stretches towards St. Pot. The strata rise, as on the English side, under a low angle westward; and are, near St. Pot, as at Folkstone, succeeded by the substrata of blue marle and green sand.

From Cape Blanc Nez, the chalky downs recede inland, ranging in a semicircle round the district of Boulogne, and approaching the coast again near Etaples. The included area is occupied by the inferior strata, and corresponds in position with the similar denudation (as it is termed) of Kent and Sussex: exhibiting, together with the same formations, some which are lower in the series; a careful comparison of these two districts would be interesting and important. We shall subjoin to our account of the English denudation, the few particulars we possess as to the French.

Though, as we have said, the southern extremity of the semicircular escarpment of the chalk again draws near the coast on the north of Etaples (thus answering the termination of the South downs at Beachey head on the English side), yet no cliffs, we believe (for we have not personally inspected this point), occur on the French side; but a range of sandy dunes, accumulated at the foot of the hills, preserves them from the destructive agency of the waves.

scarcely thirty feet thick, and that each of the overlying strata at the latter place, is thinner than those near Dover, in about the same proportion; so that the height of the cliffs on the opposite shores is very different;—that immediately on the west of Shakspeare's being not less than 500 feet, while that of Cape Blanc Nez does not exceed 300 feet in height. Hence, supposing the two countries to have been once united, it may be assumed that the several chalk strata became gradually thinner in approaching that part which now constitutes cliffs on the coast of France: hence also, it may be concluded that the country in the neighbourhood of Calais, once constituted a part of that tract which is now termed the chalk basin of London. (G. T. vol. v.)

It is not necessary, in order to account for the agreement of the opposite coasts, to suppose that chains of uniform height with those occupying the land on either side, once traversed the space now possessed by the waves; for it is very probable that these chains may, from their original formation, have had a lower level,—such as might carry them beneath the surface of the ocean in some part of their course. (C.)

From the mouth of the river Cauche, near Etaples, past those of the rivers Authie and Somme, as far as Treport, the shore is formed all the way by a line of sandy dunes, and the land is low. South of the Cauche, and as far as the Authie, a series of argillaceous and sandy superstrata may be seen covering the chalk: these seem to indicate this tract as the eastern termination of the Isle of Wight basin.

At Treport, about twelve miles eastward from Dieppe, the chalky cliffs again rise, and here commences their principal range on the French coast; for they extend hence about sixty miles to Cape la Heve near the mouth of the Seine, presenting an unbroken barrier of a dazzling white, excepting where they dip into some creek or cove, or open to afford a passage to some river or streamlet. The cliffs of Dieppe, Cape L'Aailly, St. Valery en Caux, Fecamp, and Cape d'Antifer or de Caux, occur in this interval. Near the latter, are some remarkable insulated pyramids of chalk resembling in form and circumstances the Needles of the Isle of Wight, and denominated from the same analogy les Aiguilles. This long precipitous range may be regarded as the continuation of the chalky hills in the centre and south of the Isle of Wight, which lie directly over against it, following the line of bearing of the strata from north-west to south-east.

The cliffs are often of stupendous height, exceeding, it is said, in this respect, those of England, and sometimes attaining an elevation of nearly 700 feet, but this is probably exaggerated; Cuvier assigns between 300 and 400 only: their most lofty and striking point is near Fecamp. Their stratification is nearly horizontal, and they are almost entirely formed of the upper, or flint-bearing portion of the formation. In some points they are crowned by patches of the superior strata; this is the case near the light-house of St. Margaret's on the west of Dieppe, where a section of the plastic clay series is afforded, exactly corresponding to that of Newhaven in Sussex (G. T. vol. iv.); the former being perhaps an outlier on the south-west, as the latter is on the north-east of the Isle of Wight basin. *

Cape la Heve, where the chalk cliffs terminate, is two miles and a half east of Havre. The chalk without flints, and green sand, may be observed in this vicinity. Hence the hills run south, and their section against the Seine produces a cliff at Orcher three leagues above Havre; the intermediate tongue of land is low and consists of lias, containing crocodiles bones, &c.

* From the notices of fossils in different French writers, it seems probable that the upper strata also occur above the cliffs near Fecamp, and that the beds under the chalk are to be seen near the Cape de Caux.

On the south of the river the cliffs terminate at Tronville sur Mer, west of Honfleur, close to the mouth of the Toucques river; and here the inferior oolite is exhibited (as at Bridport) underlying the chalk and green sand. Further west, between the mouths of the Toucques and Dives, opposite the rocks called Vaches noirs, is a cliff which exhibits chalk resting on a blue marle containing oysters. To the west of this point the coast is entirely occupied by inferior strata. *

* The comparison of the interval between these chalk cliffs and the transition rocks of the Cotentin, with the sections of the Dorset and Devon coast, would prove very interesting: round Caen, extending nearly to the mouth of the Orne, oolite occurs; at Bayeux, inferior oolite: near the mouths of the Seuille and Virre, are cliffs of lias with gryphites and ammonites; they extend on the west of the latter river to Virreville near Carentan. Coal is worked at Litry, two leagues south-west of Bayeux. The junction of the transition slate runs by the banks of the Vire to St. Lo, and thence east to Aulnay.

CHAPTER II.

BEDS BETWEEN THE CHALK AND OOLITE SERIES.

Section I.

(*a*) *Introductory view.** Viewed on the large scale, the interval between the chalk and oolites may be described as occupied by a series principally consisting of beds of siliceous sand, which probably have an aggregate thickness in the greater part of their course of not less than 1000 feet, and form that extensive sand tract which is universally to be traced beneath the escarpment and inferior termination of the chalky ranges.

In the southern counties, where this series is exhibited in the most complete manner, its subdivisions are clearly ascertained; but in the midland and north-eastern, they have not as yet been so well determined.

Where the series is fully developed, the sandy beds are found to be divided into two groupes, separated by an intermediate bed of clay; the upper sandstone being distinguished by the general occurrence of numerous interspersed specks of a greenish substance, probably coloured by oxide of iron, and the lower by a deep ferruginous hue derived from the abundance of brown oxide of iron they contain: hence the former is usually called the Green Sand Formation, and the latter that of Iron Sand; occasionally, however, some of the upper beds assume the same ferruginous appearance.† Interposed be-

* By the Rev. W. D. Conybeare.

† The occasional occurrence of the green particles in other beds, particularly in the chalk marle, and the ferruginous character of some of the beds of the upper or green sand formation, has been an occasion of deceiving even experienced observers who have confined their examination to a single point; but when the field of observation is more extensive, and the great and continuous ranges of these upper and lower sands, are traced from their sections on the coast through the interior, particular attention being paid to the course and character of the intermediate clay, their family and relations will be clearly ascertained. These observations principally refer to the Weald countries.

tween the upper or green sand and the chalk, beds of an argillo-calcareous character are always found, occasionally themselves also mingled with sand; and into those beds the lower chalk seems to graduate by an almost insensible transition, by the admixture of argillaceous and siliceous matter through its substance. To these intermediate beds the name of Chalk marle has been given; hence we have the following subdivisions,—beginning with the highest member of the series and reckoning downwards.

A. Chalk marle.
B. Green sand.
C. Clay dividing the sands; which, as it occurs on the most extensive scale in the Wealds of Kent, Surrey, and Sussex, may be called the Weald clay.
D. Iron sand.*

In applying the term formations to these subdvisions, we wish to be understood as using it only as a convenient designation for a large assemblage of similar strata. Viewed under a more general aspect, the whole series perhaps might be considered as constituting but one formation; yet each of these subordinate members is in itself of sufficient importance to require a distinct and specific notice, and singly forms the mass of considerable ranges of hills.

All these formations are probably of marine origin. It should be noticed, however, that the Vivipara, a fluviatile shell, has

* It would have been considered desirable to subjoin a list of the synonymes under which Mr. Smith describes these formations, but he candidly acknowledges the difficulty of discriminating between them; and it can scarcely be considered as any impeachment of his general accuracy, to add that he has not succeeded in the attempt.

The great foundation of his errors in this respect, appears to have been an hasty identification of the limestone beds associated in the Kentish rag or green sand formation, with those of the Portland series:—an identification which is absolutely contradicted by a comparison of the series of formations as exhibited in the Weald of Kent, in the Isle of Wight, and in Dorsetshire. Hence the clay underlying the Kentish rag and green sand (our Weald clay) is in many instances confounded, under the name of Oaktree clay, with that underlying the Portland series (our Kimmeridge clay).

The iron sand, described by Mr. Smith, under the names of brick earth, and sand, and rock producing the Portland stone (because it immediately overlies that formation), is from the same cause confounded, in the Weald district, with the real green sand formation.

The writer of this article, when on a tour in the Isle of Wight and Purbeck in 1813, formed detailed lists of the several strata constituting the series as exhibited in the various points where their sections are exposed in that interesting district: these are unfortunately in the custody of a friend now on the continent, but they will be added if possible in an appendix at the end of this work.

been supposed to occur in the Weald clay; but the identification of the species cannot be considered as complete.

In treating of these formations, the method pursued will be, first, to give an account of their general characters, as they are seen where the series is most fully developed, under the usual heads, considering each separately; and secondly, to give a more detailed view of the different districts occupied by them collectively, noticing the peculiarities presented in each of these, and the points concerning which further information is required to render their comparison entirely satisfactory and complete. The article, range and extent, among the general heads will therefore be restricted to a brief notice, as all local and geographical particulars will be fully treated of in the second division.

(*b*) *Foreign localities.* Before proceeding to the detail of the separate, we shall subjoin a comparative sketch of the analagous formations on the continent.

As we have before traced the chalky tract from England into France, we shall also find these subjacent beds following its course (the limits of which were then assigned) through that country; thus, they are seen skirting the chalk of the Boulogne district, opposite the appearance in Kent and Sussex, at the western limit of the chalk cliffs about Havre and Honfleur, and the eastern boundary at Valenciennes, where the green sand assumes a conglomerate character and is known by the name of Turtia. Between these points, the beds in question form a broad zone of sandy country circling round the area of the chalk, on the east side forming only a narrow band, but on the south and south-west occupying a considerable space. Mr. Omalius d'Halloy has described the series under the title of the lower chalk, which seems very unfortunately chosen, since it is only mineralogically applicable to a very small part of it (that corresponding to our chalk marle), and has led to much confusion both as to the description of the chalk formation itself, and its constituent fossils. This author notices the following subdivisions. 1. Chalk; sometimes of a coarser texture, occasionally mixed with clay, sand, and chlorite. containing pale flints abundantly.* 2. Tuffeau; coarse sandy chalk mixed with chlorite. 3. Sands and sandstones; often mixed with calcareous matter. 4. Greyish clay; commonly of a marly character, sometimes mixed with chlorite: the passages of

* Although the constant abundance of pale flints may seem to distinguish this bed from the chalk marle of England, yet the latter is represented as containing beds of flint in some parts of Cambridgeshire; and beds of chert occur beneath the harder varieties of it near Reigate in Surrey.

these modifications into one another, and their alternations, prevent the decided determination of their order of superposition, further than the assigning the highest position in the series to No. 1. Such is the account given, in which we may clearly recognise the chalk marles and green sands of our own series. More precise information may be soon expected from a survey of the western termination of the chalky cliffs of France, on the eve of being undertaken by Mr. de la Beche: a translation of Mr. d'Halloy's memoir will be found in the Annals of Philosophy for Feb. and April 1818. The beds in question form the green coloured tract in the accompanying map. The writer has seen series of organic remains agreeing with those of the English green sand from near Havre, and lists of similar ones found at Valenciennes.

On the northern borders of the Alps, the highest beds of the exterior calcareous chains consist of a dark coloured limestone often mixed with sand and green particles, and agreeing in its fossils with this part of the English series, with the addition of nummulites, which are rare (although they do occasionally occur) in these beds in England. Similar beds are mentioned, and in a similar position, on the skirts of the Maritime Alps near Nice, in Mr. Allan's account of that neighbourhood. (Ed. Phil. Trans.) They form the second limestone of the memoir referred to.

The sandstone of Saxony, generally known under the Wernerian name of Quader sandstein, which forms such romantic scenery between Dresden and Pirna, and extends through Silesia, skirting the primitive chains of the Erze and Riesengeberge on the north towards Glatz, and is again found on the south in Bohemia, probably belongs to this series. Von Reaumer notices that, in Silesia. it occasionally contains green particles; but its general characters, and its few fossils, give it a nearer resemblance to the paler beds of the iron sand as they are presented near Hastings, with which it agrees also in containing vegetable remains. Sandy tracts, probably of similar age, occur in many other parts of Germany, particularly between Bamberg and Bayreuth, near the cavernous limestone, but their relations are not as yet ascertained; nor do we yet possess the means of pursuing these beds, if indeed they exist, in other parts of the world.

Section II.

*Chalk Marle.**

These beds, which occur immediately beneath the chalk, graduate into the lowest strata of that substance, in such a manner that very often no distinct line of separation can be traced. The harder beds at the bottom of the chalk series, described in the former chapter, afford an example of this transition, and perhaps might have been placed with equal propriety under this head; it may therefore be necessary again in some instances to allude to them. On the other hand, the lowest beds of the chalk marle often pass into those of the green sand, and it is somewhat difficult to catch any precise characters for a formation which is thus rather intermediate between two others, than possessed of independent features. The following descriptions will be taken (unless when it is otherwise stated) from the appearance of these beds in Kent, Surrey, and Sussex, where their relations are most clearly developed.

(*a*) *Chemical and external characters.* The composition of these beds consists apparently of three ingredients, intimately mixed, but in variable proportions; 1st, cretaceous matter; 2dly, argillaceous matter; 3dly, sand. In the upper beds, near its junction with the chalk, the cretaceous matter prevails; and these usually appear as chalky beds, distinguished from the true chalk by a greyish or mottled character, and by a more laminated texture, and by falling to pieces when wetted and dried again. It greatly varies in hardness, but will not usually mark like chalk, and often acquires sufficient consistency for architectural purposes; its aspect is also more gritty than the chalk usually is,† Where the argillaceous matter prevails in excess, a tenacious argillaceous marle of the ordinary characters, and of a bluish grey colour, is the result. Beds of this character often underlie the former, as at Folkstone in Kent. Thirdly, where the sand prevails, a fine-grained greyish sandstone of loose texture is produced, which forms a link

* By the Rev W. D. Conybeare.

† A more detailed account of the character of the harder beds which are worked in Surrey for architectural purposes, and afford a good fire-stone, will be found in the account of the Weald district in the second division of this chapter. That of the Isle of Wight (W.) is a good building stone, is soft when first taken out of its bed, but hardens by exposure All the ancient Gothic churches on the north side of the island, have been constructed of it.

graduating into the beds of green sand. Sometimes these modifications occur in distinct beds with well defined lines of separation, but more usually blended together, and passing into one another. The more argillaceous form which occurs at Folkstone was found by Mr. R. Phillips to lose 13 parts out of 100, on being submitted to the action of an acid, indicating the presence of nearly 30 per cent. of carbonate of lime: the more cretaceous forms, 82 per cent. carbonate of lime, 18 silex and alumine, chiefly the former, and a trace of oxide of iron.

The marly varieties occasionally contain concretional masses of the more cretaceous and siliceous forms; beds of chert also occasionally occur in these strata, as at Reigate, and flinty nodules in some parts of Cambridgeshire. Besides the above varieties, a singular appearance is assumed by the lower chalk beds in many parts of Yorkshire and Lincolnshire; which, as has been mentioned in the previous chapter, have often a dark red colour, (see particularly the note on Hanstanton cliff, page 78); these beds probably belong to this part of the series, and derive their colour from the mixture of oxide of iron.

(*b*) *Mineral contents.* Irregular nodules and radiated masses of iron pyrites are common both in the cretaceous and argillaceous varieties; in the latter, geodes and septaria, with calcareous spar and selenite, and occasionally sulphate of lime, also occur, if the marly beds, which will hereafter be mentioned as occurring in Yorkshire and Cambridgeshire beneath the chalk, really belong to this formation.

(*c*) *Organic remains.* The upper cretaceous beds, near the junction with the chalk, contain organic remains of a nearly similar character with those of the lower chalk; viz. *nautilus*, *inoceramus*, *echini*, *alcyonia* and *sponges*, but the lower and more argillaceous strata are distinguished by a rich variety of singular and peculiar fossils, especially in the order of multilocular shells. In the following list references to the figures in Sowerby's Mineral Conchology are given. *Ammonites*. *A*. Mantelli, plate 55. *A*. minutus, 53. fig. 3. *A*. planicosta, 73. *A*. rostratus, 173? *A*. splendens, 103. fig. 1. 2. *A*. varians, 176. *Nautilus*. *N*. inæqualis, 40. fig. 10. *N*. elegans, 116. *N*. Comptoni, 121. *Hamites*; the species of this singular genus are numerous and very abundant at Folkestone; many figures of these are given in plates 61 & 62. Plate 168 exhibits a remarkable variety armed with spines from beds probably of this formation at Roak, in Oxfordshire. Plate 215 shews the same from the Isle of Wight, and some other species. These fossils may be considered as peculiar to, and highly characteristic of, this formation.

Scaphites, plate 18. fig. 4 to 7.
Turrilites costata, 36. *T.* tuberculata, 74.

Belemnites—some varieties small and fusiform.

Of *Univalves not chambered*, have been found,
Dentalium decussatum, plate 70. fig. 5; *D.* ellipticum, ib. fig. 6 & 7.
Vermicularia umbonata, 57. fig. 6 & 7.
Cerithium melanoides, 147.
Euomphalus.
Patella laris, 139. fig. 3.

Of *Bivalves*, occur,
Terebratula biplicata, 91.
Arca subacuta, 44.
Nucula pectinata, 192. fig. 6 and 7.
Pecten Beaveri, 158.
Inoceramus.

Of the *Echinus*, a variety of Spatangus is found.

Among *Zoophytes*; *Madrepora*, a conical variety probably belonging to the genus Turbinolia of Lamarck.
The *Pentacrinite.*

Remains of the higher or vertebral animals are not common: but apparent fragments of bones have been found at Folkestone, and a fossil fish in the analogous beds of Cambridge; where, and at Malling in Kent, several species of *cancer* are also found, and mineral wood, near Cambridge, (G. T. vol. 5. p. 115,) and at Folkestone (ibid. p. 27,) in numerous fragments, often of several inches in diameter, and sometimes retaining the woody fibre; they mostly lie on the green sand, and are enveloped in, or the interstices of their outer part are filled with, pyrites, which is commonly in a state of decomposition, one of the consequences of which is the formation of selenite in crystals, which are often well defined. These are either attached to the fossil wood, or lie imbedded near it.

(*d*) *Range and extent.* As in treating of the several districts occupied by these formations in the second division of this chapter, we shall be enabled to give all the geographical particulars connected with it with greater distinctness, we shall here confine ourselves to observing generally, that the beds now described occupy a band extending beneath the foot of the north-western escarpment of the main chalk range, and in the south-eastern counties beneath the southern escarpment of the north Downs, and the northern escarpment of the south Downs. The breadth of this tract varies generally from one to two miles, but if the Cambridgeshire galt is properly referred to these beds, it is there considerably greater.

(*e*) *Thickness.* Near Folkstone the cretaceous varieties occupy about 200 feet, and the inferior argillaceous beds about the same. From 300 to 400 feet may probably be assumed as a fair average thickness for these beds; which, like all other strata, are very variable in this respect when compared in distant points.

(*f*) *Inclination.* The strata are always conformable to the subjacent chalk, and therefore generally approach the horizontal position; but in the disturbed ranges of the Isle of Wight and Purbeck, they become nearly vertical.

(*g*) *Height.* The cretaceous and siliceous varieties, especially when they assume an harder texture, frequently form an under terrace beneath the escarpment of the chalk hills: this may be particularly seen in Berkshire. The highest hills thus formed, are those on which the old Roman camp called Sinodunum hangs over the Thames opposite Dorchester; they are detached and bold, and are probably about 500 feet above the sea. The argillaceous beds, having offered less resistance to the causes which have modified the surface of our continents, form low grounds at the base of the escarpment of these ranges.

(*h*) *Agricultural characters.* The cretaceous varieties are, like the chalk itself, favorable to the growth of the beech. Where this and the argillaceous varieties are blended, a warm crumbling marly soil, very rich and valuable, is produced; but where the argillaceous forms exclusively occur, a deep stiff clay which requires the labour of years to render it mellow. Flint gravel, derived from the overhanging chalk ridges, is however often spread over this tract, and materially modifies its characters.

(*i*) *Phœnomena of Springs.* The waters which percolate through the rifty strata of the superjacent chalk, are usually thrown out by some of the more tenaceous beds in the upper part of this series; but having passed the line of these springs, it is necessary to sink the wells to a considerable depth, often 200 feet, in order to pierce the retentive argillaceous strata and reach the waters percolating through the subjacent sandy beds, and thrown up by their argillaceous partings.

Section III.

GREEN SAND.*

This is, in respect of its mass, and the number and beauty of its organic remains, one of the most important formations between the chalk and oolites. In all the southern counties its beds occupy a great thickness, and may be readily traced, as distinct from the other members which occupy this interval; but, as we have before remarked, in the Midland and Northern counties their course is much more obscure, and cannot yet be considered as fully elucidated.

(*a*) *Chemical and external characters.* The green sand consists of loose sand, and of sandstone. The sand is siliceous, but the cement, when in the form of sandstone, is generally calcareous. Both sand and sandstone mostly contain minute portions of a substance which has been termed Green Earth, † which has not been chemically examined, but very probably derives its colour from the suboxide of iron; and very commonly spangles of mica; subordinate beds and masses of chert, and veins of chalcedony usually occur; and also frequently alternating beds and nodules of limestone, which in the Isle of Wight is termed *Rag*, and which is identical with the Kentish Rag. Much obscurity has been produced by confounding the limestone of this formation with that of Portland; from which, in geological position, characters, and fossils, it is perfectly distinct. Beds of clay also form occasional separations in this formation. In the series of these beds, chert, flint, and chalcedony continually pass into each other by insensible gradation: owing to their hardness, they are often discoverable on the summits of hills, as at Leusden, Pilsden, &c. The flinty and chalcedonic varieties are much more frequent in the west of England than in the eastern part of Kent, but they occur very commonly in Surrey: the most beautiful are found near Charmouth in Dorsetshire.

The differences of appearance which characterise different beds of this formation, and often indeed affect all its beds in particular localities, arise, first, from a difference of texture, which passes from a very coarse-grained sandstone, and even a

* Chiefly by the Rev. W. D. Conybeare. Much information has also been extracted from Mr. Greenough's Notes.

† These green particles are not confined to this bed, but are seen also in the London clay above the chalk, in the lower chalk, in the Purbeck beds beneath the iron sand, and in the upper beds of the Kimmeridge clay when in contact with the Portland limestone.

distinctly conglomerated rock containing large rounded fragments of quartz, to the finely granular form: of which latter variety, the quarries of Blackdown in Devonshire, which principally supply England with whetstones, afford the best examples; and secondly, from the greater or less quantity of the green particles; these often prevail to such a degree, as to impart to the rock its predominating aspect; often they are so few, as to permit it to assume a grey or even buff colour. In some beds also, the sand is deeply coloured by brown oxide of iron; a circumstance which gives weight to the opinion, that the green particles owe their colour to the same metal. In the latter case, it is difficult if not impossible to distinguish the individual beds from those of the subjacent formation of iron sand; but if the formation be regarded as a whole, and the tract occupied by it traced continuously to some distance, all such partial difficulties will always be found to vanish, and the general truth will be elicited with the clearest evidence. Here, as throughout this science, extended and combined observation is the only safe guide.

It does not appear that any certain order of superposition can be traced to any distance in the varieties above described, but that they continually pass into one another, and are irregularly blended together; a constant uniformity of character can hardly, from the circumstances of the case, be expected in extensive depositions, so obviously mechanical in their origin. The Fuller's earth beds of Nutfield in Surrey form another variety subordinate to this formation.

(*b*) *Mineral contents.* Iron pyrites has been found in this rock at Folkstone in Kent and Caistor in Lincolnshire, and hæmatitic and stalactitic iron have been seen in the ferruginous beds. Near Nutfield in Surrey, it contains crystallized sulphate of barytes of a yellow colour, of which the interstices are often filled with opake quartz. The cherty nodules already mentioned, often afford chalcedonies and quartz crystals of great beauty. It is scarcely necessary to add that the calcareous matter dispersed through this formation, often yields crystallized varieties: the occurrence of spangles of mica has been already mentioned.

(*c*) *Organic remains.* The organic remains of this formation are extremely numerous; and often when, as at Blackdown, imbedded in the more siliceous varieties of its rocks, occur in a state of preservation equally singular and beautiful, the original calcareous matter of the fossil being entirely replaced by an infiltration of chalcedony. In this state, it is often easy to detach them completely from the loose sandy matrix; and they then appear, although having undergone a thorough conversion of substance, with all the sharpness and character of recent specimens.

Of the higher animals no remains have yet been found, except a few teeth of fishes, both of the conical and lanceolate figure.

Of the Testacea, the remains are so numerous, that the quarries of Blackdown alone afford 150 species. It would not be consistent with the plan of this work to enter into the detail of such a list; we shall limit ourselves therefore, in most instances, to specifying those figures of which can be referred to in Sowerby's Mineral Conchology.

UNIVALVES CHAMBERED.

Ammonites auritus, plate 134.† *A.* inflatus, 178. *A.* monile, 117. *A.* Nutfieldiensis, 108. *A.* Goodhalli, 225.
Nautilus undulatus, 40. *N.* simplex, 122.
Hamites spinalosus, 215.
Turrilites costata, 36. *T.* obliqua, 75, fig. 4.
Belemnites, elongated with a small furrow on the apex.

UNIVALVES NOT CHAMBERED.

Helix gentii, 145.
Trochus, *Solarium*, *Turritella* } these are figured in Smith's plate of green sand fossils.
Natica canrena? Parkinson, vol. 3. plate 6. fig. 2.
Murex; Smith's plate of green sand fossils.
Pleurotoma rostrata, 146.
Rostellaria; Parkinson, vol. 3. plate 5. fig. 11 & 12.
Auricula incrassata, 163.
Ampullaria?.
Planorbis euompalus, 140. fig. 8. *P.* radiatus, 140. fig. 5.
Turbo carinatus, 240. fig. 7.
Vivipara extensa, 31, fig. 2.

TUBULAR UNIVALVES.

Serpula.
Dentalium medium, 79, fig. 5.
Vermicularia concarra, 57, fig. 1 to 5.
Patella locris.

† All the references, unless where otherwise specified, are to the Plates of Sowerby's Mineral Conchology.

BIVALVES.

Arca carinata, pl. 44. fig. 11.
Cucullæa glabra, 67. *C.* carinata, 207. fig. 1. *C.* fibrosa, 207, fig. 2.
Nucula.
Trigonia dædalea, 88. *T.* spinosa, 80. *T.* eccentrica 208. fig. 2. *T.* affinis, 208. fig. 3.
Pecten quadricostatus, 56. fig. 1 & 2. *P.* quinquecostatus, 56. fig. 3 to 8. *P.* echinatus.
Pectunculus; Smith's green sand plate.
Terebratula biplicata, 90. *T.* intermedia, 15. fig. 8. *T.* ovata, 15. fig. 3. *T.* lyra, 138. fig. 2. *T.* pectita, 138. fig. 1.
Cardium Hillanum, 14. *C.* proboscideum, 156. fig. 1. *C.* umbonatum, 156. fig. 2 to 4.
Venus angulata, 65. *V.* equalis, 31. *V.* lincolata, 20. *V.* plana, 20.
Cardita tuberculata, 143.
Dianchora striata, 80.
Corbula gigantea, 209. fig. 56. *C.* lævigata 209. fig. 1. 2.
Chama canaliculata, 26. fig. 1. *C.* conica, 26. fig. 3. *C.* haliotidea, 25. *C.* plicata, 26. fig. 4. *C.* recurvata, 26. fig. 2. *C.* digitata, 174.
Ostrea crista galli, (not figured). *O.* gregarea, 111. fig. 1.
Inoceramus; same varieties as in the chalk marle.
Mya mandibula, 43.
Modiola pallida, 8.
Perna.

The family, *Echinus*, presents in this formation several species of the divisions Cidaris and Spatangus, and one small species of Conalus. In these there is a considerable resemblance to those of the chalk, but seldom a complete identity. This is the lowest formation in which the Spatangi have yet been found in England, and the only one besides the chalk which affords Conulus. Many of the species of Echinus in green sand are very small. The Encrinital remains are few and uninteresting; detached joints only of two species have been found. The Coralloid remains are neither numerous nor important; but a few small turbinated and porpital madrepores are found.

The Alcyonic remains in this formation, are more numerous and important than those of any other excepting the chalk; in the remaining strata indeed these fossils are comparatively few, and generally afford obscure traces only.

Among the numerous genera of these fossils which have never yet been accurately studied and classed, we may specify the following: 1. Ramose, much resembling that in the chalk; 2. Funnel-shaped; figured as the frontispiece to the 2d vol. of Parkinson's Organic Remains; its mass appears to have been formed by an irregular plexus of reticulated fibres, in which traces of a general radiation from the centre of the bottom can nevertheless be observed. 3. The tulip-shaped Alcyonia, described and figured by Mr. Webster in the 2d vol. of the Geological Transactions; these have, in perfect specimens, a ramifying root, and a stem, not very long, carrying a bulbous head, in the upper part of which is a funnel-shaped mouth. This head is formed of an irregular plexus of reticulated fibres, traversed longitudinally by rows of pores arranged in concentric circles round the axis; each of these pores has a central fibre running through it. 4. Having a very short stem, or rather neck, supporting a large head, divided into many lobes; of this form the varieties are many and grotesque. 5. Shaped like a cucumber.

The vegetable remains as yet found in this formation are confined to fragments of silicified wood.

(*d*) *Range and extent.* Traces of this formation may be observed underlying the escarpment of the chalk in Yorkshire and Lincolnshire, but those counties have not been examined with sufficient care to enable us to enter into any detail; and the same remark equally applies to the midland counties, in which also the alluvial debris at the foot of the chalk hills renders it difficult to trace this substratum. Following the escarpment of the chalk westwards, we find near Childrey and Wantage, in the vale of the White horse (Berkshire), very decided exhibitions of this formation, which may thence be traced continuously through Wiltshire and Dorsetshire underlying the chalk; in Wiltshire it often constitutes a secondary range of hills standing in advance of those of the chalk formation, and nearly rivalling them in height, as is the case at Warminster and Stourhead; and it forms the base of the vallies which deeply indent the great chalk escarpment, as at Pewsey and the vale of Wardour.

On the confines of Dorset and Devon, it presents many high and insulated masses, constituting what are called outliers; thus, it forms the summit of Leusden and Pilsdon hills and the extensive table-land of Blackdown, which stretches far to the west, covering great part of the eastern division of Devonshire, being subdivided by many vallies into several long ridges. In the course of its progress to the south-west, the green sand overlies in succession the terminations of the oolites and the lias, and becomes, in the western part of Blackdown, immediately incumbent on the newer red sandstone.

Still farther westwards, and beyond Exeter, we have another outlying mass of this formation capping the long range of Haldon hill, which is divided only by an intermediate valley from the granite of Dartmoor; a singular instance (in this island) of the near juxtaposition of primitive and very modern rocks.

In the south and centre of Dorsetshire, the green sand follows the curve which the chalk has been described as making through that district, and constitutes outlying conical summits which form well characterised land-marks near Abbotsbury.

Through the Isles of Purbeck and Wight it accompanies the elevated strata of the chalk, assuming itself a conformable inclination. We shall have occasion to return to this part of its course in a more particular manner, at the close of this chapter; as also to its range through the Weald of Kent and Sussex, where it forms as it were a ring of hills ranging parallel to those of the chalk, and forming an inner band circling round the centre of the denudation.

Besides its general range as above shortly sketched, Professor Buckland has noticed a singular and insulated occurrence of this rock near the south-west angle of the London basin; where the chalk, which borders that basin on the south, crops out and exhibits rather unexpectedly (from the low character of the ground between this and the general area of the London basin) this substratum at Kingsclere, and again near the northern foot of Inkpen beacon. It is seen, however, only through a very small space, sinking on all sides beneath the chalk.

(*e*) *Height of hills, &c.* In the Northern and Midland counties this formation occupies tracts comparatively low. In Wilts, the hills near Warminster, and those of Stourhead, masked by the conspicuous building called Alfred's Tower, rise to about 800 feet. In Devon, Blackdown rises 817, and Haldon 818 feet. In the Weald the greatest height of this formation is at Leith hill in Surrey, 993 feet.

(*f*) *Thickness, &c.* No natural section of this bed, affording a satisfactory opportunity of ascertaining its thickness, is known. It certainly however differs in different places. Through Yorkshire, Lincolnshire, Norfolk, and from Cambridge through Bedfordshire, along which counties it appears on the west of the chalk escarpment, this bed is but thin.

In the vale of Pewsey in Wiltshire and near Warminster, it occupies a greater breadth on the surface; whence it may be assumed as being thicker than it is higher north. At Pottern Parsonage near Devizes the well is 126 feet deep chiefly in the grey sand (T. 115), and one at Hagworthingham in Lincolnshire, in the same sand, is 110 feet deep. (G. Notes.) At Shaftesbury in Dorsetshire the wells (G. Notes) are very deep. It is 200 feet thick at Blackdown in Devonshire. But if the breadth of surface be an indication of thickness in a bed, we must look for the greatest thickness of this on the east of Alton in Hants, where its breadth is about 12 or 14 miles: but we have no account of sinkings in this neighbourhood. The greatest thickness of this stratum is given by Professor Buckland in his order of superposition, as 300 feet. In the Weald however it probably exceeds this.

(*g*) *Inclination, &c.* The dip of this bed may be considered to be nearly the same as that of the chalk which it underlies, usually dipping at a low and almost inappreciable angle beneath the escarpment of that formation; but in those districts in which the strata of chalk become highly elevated, these also assume a conformable position: to enter further into details, would be only to repeat without necessity what has been already stated in describing the inclination of the chalk.

(*h*) *Agricultural character.* This formation usually affords a light loamy soil; which, in the vallies formed by it in Wilts, has been found to possess considerable fertility. The elevated Downs covered by it in Devonshire and in Surrey, are still however mostly left in the sate of unreclaimed heath.

(*i*) *Phænomena of springs, &c.* The waters which rise in this formation, and are thrown out by the thin seams of clay which alternate with it, are frequently chalybeate. The general character of the formation being porous, it is often necessary to pierce it to a considerable depth in order to reach the water thrown out by the great retentive substratum of the Weald clay.

Section IV.

WEALD CLAY.*

This formation has hitherto attracted a very slight degree of attention, and a very few words will be found to comprise all that is at present known concerning it.

It is exhibited on the largest scale in the Wealds of Kent, Surrey, and Sussex, where it separates the central nucleus of ferruginous sand from the encircling ranges of the green sand formation. Mr. Smith has designated this district by the name of the *Oaktree clay*, but as that name was originally applied to

* By the Rev. W. D. Conybeare.

an argillaceous stratum considerably beneath the iron sand (the same which is marked by the appellation of Kimmeridge clay in our arrangement), and which is consequently of distinct and anterior formation to this; † since also Mr. Smith has applied the term to the Kimmeridge beds in his Stratigraphical arrangement of organic remains, it would manifestly only perpetuate a source of confusion to retain it in this place.

(*a*) *Chemical and external characters.* Its chemical character and aspect varies from that of a dark tenacious clay, to that of a blue or grey calcareous marle of an earthy and friable texture. It contains occasionally layers of argillo-calcareous concretions replete with shells of the genus Vivipara fluviorum. The interior of these is usually filled by calcareous spar; and as the cement has firmness enough to admit a slight polish, masses of this description are occasionally wrought for ornamental purposes, and form what is well known in many of our gothic buildings by the name of Petworth marble. It should be remarked, however, that the Purbeck marble has been in many instances miscalled by this name; but the latter is marked by the shells contained in it being of smaller size, and by a greater delicacy and elegance of texture and appearance. The taper shafts which cluster round the columns of the early gothic style, are generally formed of the Purbeck marble, the Petworth being used chiefly for the slabs of altar tombs, and other coarser works. It is little worked at present, being so liable to break that it is difficult to make it into any thing that is not bulky; under the saw it will frequently break off and fall to pieces. Some quarries are still however wrought near Kirdford and North Chapel in Surrey. It is said also that beds of fullers' earth occasionally occur in this formation.

(*b*) *Mineral contents.* These appear to be limited to the specks of mica which occur in many of its beds, nodules of iron pyrites, and selenite.

(*c*) *Organic remains.* Its organic remains have not as yet been enumerated by any writer; those given under the head Oaktree clay in Mr. Smith's Stratigraphical arrangement, being in reality those of the Kimmeridge clay. The most characteristic shell is, however, undoubtedly that which marks the Petworth marble, namely, the Vivipara fluviorum, which is found not only where the quarries of that stone are now worked, but in many other localities through the vale occupied by this formation, in Kent, Surrey, and Sussex; they are mingled with minute bivalves, supposed to belong to the genus

† Full proof of the distinction of these formations will be given in treating of the Kimmeridge clay.

Cypris of Lamarck, and figured in Sowerby's Min. Conc. Plate 31. fig. 1.

Since the recent Vivipara is a fluviatile shell, it is argued that this formation, at least that part of it which contains the marble in question, must have been deposited from fresh water; but the identification of species does not appear to be made out with such decided evidence, as to render it proper hastily to admit this supposition, until the other organic remains of the formation shall have been examined and ascertained. The work of Mr. Mantell on the fossils of Sussex, will doubtless ere long supply this deficiency.

(*d*) *Range and extent.* With regard to the range and extent of this formation, we must begin by observing, that the difficulty of discriminating between this and the Folkestone clay in the Northern and Midland counties prevents our speaking positively as to its course in those quarters, neither are we as yet in possession of materials which will enable us to trace it in the Western counties.

The Southern counties can alone, in the present state of our information, be cited as affording undoubted localities of this bed. In these, however, several distant points in Dorsetshire, the Isle of Wight, and the Weald of the South-eastern counties, enable us to assign to it a considerable extent, and to speak of it with the lights derived from comparative observations. In Dorsetshire it may be traced in the Isle of Purbeck, dividing the green and iron sands, but it thins off to nothing in the western part of that peninsula. In the Isle of Wight it may be observed holding a similar position, and having obtained a greater thickness, it may be traced following the green sand under the central chalky range—and again underlying the same formation in the southern range of the undercliff.*

Its position in the Weald is sufficiently indicated in the head of this article: it there possesses a thickness many times greater than in either of the preceding localities.

* In the neighbourhood of Black-gang chine this stratum is very visible. It was the cause of a great land-slip which took place in the year 1799, and of another fall eight or ten years ago; springs of water, to which both may be attributed, still continue to make their way through the blue marle on the east of Black-gang chine.

Whenever the land springs act on this stratum, it becomes of the consistence of mud, runs out, and leaves the sandstone without support; which, being deprived of its foundation, of course tumbles down. (W. p. 134.) To this circumstance is to be attributed, in part at least, that immense ruin which forms the striking feature of the Isle of Wight; and which being now variegated by rocks, and woods, and cottages, and corn fields, is become the principal attraction of the visitors to the island.

(*e*) *Elevation.* This formation never rises into hills of any height. In the Weald, where alone it possesses sufficient thickness and extent to give any character to the general form of the surface, that surface is an uniform broad vale.

(*f*) *Thickness.* We possess no data for estimating its thickness, beyond those which may be deduced from the consideration of its superficial extent. On this ground we may safely assign to it in the Weald of Kent more than 300 feet. In the Isle of Wight it appears not to exceed 100.

(*g*) *Inclination.* We have no particular observations to make concerning the inclination of its strata, which appears to be always parallel to the superstrata of green sand and chalk.

(*h*) *Agricultural character.* This formation presents to the agriculturist a soil of pale, cold, and retentive clay, naturally covered with woods, and requiring a large capital, combined with superior skill, attention, and activity, to convert it to any other purpose.

(*i*) *Phænomena of springs, &c.* The aluminous chalybeate spring, analysed by Dr. Marcet, in the Isle of Wight, appears to have its origin in this formation.

Section V.

IRON SAND.*

In arriving at this, the lowest of the formations which intervene between the chalk and oolites, we become released from the difficulties which still partially obscure the history of some of those strata of which we have lately treated. We are able to trace and identify the present series throughout the island.

This formation may be best studied in the neighbourhood of Hastings. In different counties it has received the name of *Carstone* and *Quern stone.* It is not possible to assign the synonyme employed in the geological works of Mr. Smith, since from an erroneous identification of the strata in different districts, he has attached different names, and attributed different geological positions, to this formation. It forms the micaceous brick earth of the midland counties in his description; but in the Weald, he ascribes this to the Folkstone clay, and confounds this sand with that which underlies the coral rag.

(*a*) *Chemical and external characters.* This formation is composed of a series of strata, in which sand and sandstone

* Chiefly by the Rev. W. D. Conybeare.

prevail, occasionally alternating with subordinate beds of clay, loam, marle, fullers' earth, and ochre.

The sand and sandstone are entirely siliceous, and generally contain brown oxide of iron in a considerable proportion ; often indeed in such quantity as to have rendered many of its beds worth the working as an ore of that metal, while the forests of the county were still in a state to afford a ready supply of fuel on the spot. Hence the tracts occupied by this formation, at once strike the eye from the brownish red aspect of their soil. Some of the sandstone beds, however, in which the iron is less abundant, are of a yellow, and even of a light buff colour.

The texture of these sandstones is evidently mechanical; they often indeed form coarse grained conglomerates, consisting of pebbles (principally quartzose), from the size of a pigeon's egg to that of a pin's head, imbedded in a ferrugino-siliceous cement ; hence a regular gradation may be traced into a very fine-grained sandstone.*

This formation often contains (especially in Bedfordshire, Dorsetshire, and near Hastings) a considerable quantity of fossil wood, and even regular beds of wood coal. The sands alternating with these beds also much resemble, in some places, those occurring in the great coal formation ; this is particularly the case at Lulworth Cove in Dorsetshire, where the strata of this series completely assume the character of an imperfect coal formation. These circumstances have led to expensive but abortive attempts to procure this combustible from these beds near Bexhill in Sussex.

The following additional particulars are extracted from Mr. Greenough's notes. Ferns, charred wood, and other supposed associates of coal, occur in the white and grey sandstones of this series, but rarely in the ferruginous. The sinkings at Bexhill in Sussex, attended with so great an expense, were conducted in these beds. It is said that a kind of cannel coal occurs on the banks of the stream dividing Heathfield and Waldron parishes in the same county, extending for a quarter of a mile in beds from two to ten inches thick near the surface,

* About Horsham in Sussex this formation yields flags for pavements: at Battle Abbey the groin work which is in good preservation, is of freestone belonging to this series; it forms a good coping stone: near Lynn in Norfolk the iron clinkers are much esteemed as a building stone, and are in common use about Tunbridge in Kent; being little subject to injury by exposure. At Faringdon in Berkshire they are made into mill-stones. The want of materials however for the roads is severely felt in the Weald of Kent and Sussex, and generally wherever the ferruginous sand appears on the surface, as the sandstone is more usually of a friable texture. (G. Notes.)

at the bottom of Geer's Wood, and on the skirts of Filmoor. (Morn. Chron. Dec. 1810. G. Notes.)

Silicified wood occurs in it at Aspley, Crawley, and other places in Bedfordshire; and jet is said to have been worked in the parish of Wotton on the borders of Surrey and Sussex. (G. Notes.)

Coaly matter occurs in the parishes of Crawley and Warplesdon in Surrey; and at Flimwell the sand contains carbonaceous matter and a thin bed of vegetable. (G. Notes.)

The beds of fullers' earth which occasionally alternate in this formation, have been extensively worked in Bedfordshire. An account of them is given in the Philosophical Transactions for 1723, by Mr. Holloway; as this memoir is very short, possesses considerable merit, well illustrates the disposition of the beds of this formation in that part of their course, and is moreover highly interesting as a very early specimen of accurate geological observation, it is reprinted in the note below.*

* An account of the pits for Fullers' earth in Bedfordshire; in a letter from the Rev. Mr. B. Holloway, F.L S. to Dr. Woodward, &c. From the Transactions of the Royal Society, vol. xxxii. p. 419.

Bedford, 6th July, 1723.

"I went a few days ago to the Fullers' earth pits at Wavendon near Woburn, where there are several pits now open; but, as the men were only at work in one, and I understood the earth was disposed in much the same manner in all, I did not trouble myself to go down into more than that wherein they were digging; in which I found things disposed thus.

From the surface, for about six yards depth, there are several layers of sands, all reddish, but some lighter coloured than others, under which there is a thin stratum of réd sandstone, which they break through; and then for the depth of about seven or eight yards more, you have sand again, and after that come to fullers' earth; the upper layer of which, being about a foot deep, they call *cledge*; and this is by the diggers thrown by as useless, by reason of its great mixture with the neighbouring sand, which covers, and has insinuated itself among it; afterwards they dig up the earth for use to the depth of about eight feet more, the matter whereof is distinguished into several layers, there being commonly a foot and a half between one horizontal fissure and another. Of these layers of fullers' earth, the upper half, where the earth breaks itself, is tinged red, as it seems by the running of water from the sandy strata above, and this part they call the *crop*, betwixt which, and the cledge above mentioned, is a thin layer of matter not an inch in depth, in taste, colour and consistency not unlike to Terra Japonica. The lower half of the layers of fullers' earth, they call *wall-earth*; this is untinged with the red above mentioned, and sesms to be the more pure and fitter for fulling; and underneath all is a stratum of white rough stone, of about two feet thick, which, if they dig through, as they very seldom do, they find sand again, and then is an end of their works.

One thing is observable in the site of this earth, which is, that it seems to have every where a pretty horizontal level, because they say that when the sand ridges at the surface are higher, the fullers' earth lies proportionately deeper.

The ochre of Shotover hill occupies a similar geological position; a list of the accompanying strata will be found in the appended note.*

The workings for iron in the Weald of Sussex are described in the agricultural survey for that county, from which an extract will also be found below.†

In these works they seldom undermine the ground, but as they dig away the earth below, others are employed to dig and carry off the surface; otherwise, the matter above, being of so light and flitting a nature, would fall in and endanger the workmen; for, as was observed before, the stratum of sandstone which occurs before they come to the fullers' earth, does not lie, as in coal-pits, immediately over the matter they dig for, like a ceiling, but even in the midst of the superjacent strata of sand, and therefore can be no security to them if they undermine them.

The perpendicular fissures are frequent, and the earth in the strata, besides its apparent distinction into layers, like all other kinds of matter, by reason of its peculiar unctuousness, or the running of the adjacent sand imperceptibly among it, breaks itself into pieces of all angles and sizes.

For the geographical situation of these pits, they are digged in that ridge of sand-hills by Woburn; which near Oxford is called Shotover; on which lies Newmarket heath by Cambridge, and which extends itself from east to west, every where, at about the distance of eight or ten miles from the Chiltern hills, which in Cambridgeshire are called Gog-Magog; in Bucks and Oxon, the Chiltern hills, from the chalky matter of which they chiefly consist; which two ridges you always pass, in going from London into the north, north-east, or north-west counties in the manner before mentioned: after which you come into that vast vale, which makes the greater part of the midland counties of Cambridge, Bedford, Bucks, Northampton, Oxford, and Gloucester, and in which are the rivers Cam, Ouse, Nen, Avon, Isis, and others, which I take notice of because it confirms what you say of the regular disposition of the earth into like strata or layers of matter, commonly through vast tracts, and from whence I make a question, whether fullers' earth may not be found in other parts of the same ridge of sand hills, among other like matter."

We must observe however that the author has fallen into an error in supposing Newmarket heath, which in truth is occupied by the chalk formation. to be the continuation of this chain of hills.

* Sections of the Ochre pits, Shotover hill.

	feet	inches
Beds of highly ferruginous grit, forming the summit of the hill	6	
Grey sand	3	
Ferruginous concretions	1	
Yellow sand	6	
Cream-coloured loam	4	
Ochre	0	6

Beneath this is a second bed of ochre separated by a thin bed of clay; then succeeds an interval of nearly 40 feet occupied by various alternations of ferruginous and sometimes cherty and argillaceous loams of a deep cream colour; beneath these is the formation of Aylesbury limestone.

† The soil of Penhurst is gravelly to an intermediate depth: at the bottom of the Earl of Ashburnham's park, sandstone is found, solid enough for the purposes of masonry. Advancing up the hill, the sand rock is 21

(*b*) *Mineral contents.* The occurrence of beds of iron-stone and ochre in this formation, has been already mentioned. Specimens of hæmatitic and stalactitic iron may be found among the former. Nodular concretions, having an outer shell of iron ore containing a nucleus of loose sand, and others in which bands of a darker and lighter shade, are so disposed as to resemble the aspect of the Egyptian pebble, are common. Chert and quartz crystals are found in some of the beds, but never in any considerable quantity; nor do they ever rival those of the green sand in beauty.

(*c*) *Organic remains.* The organic remains of this formation have yet received very little illustration. They appear to be very sparingly dispersed through it, generally, although abundant in some particular spots. Of these, Faringdon in Berkshire is the most productive: we have collected from the pits near that town, a large nautilus, fragments of ammonites, belemnites, ostreæ, terebratulæ, and spines of an echinus cidaris; but the most abundant and interesting fossils of this spot are the spongitæ, of which many varieties, tubular, funnel-shaped, and palmated, occur: very beautiful minute corallines also occur, and a very singular ramose tubular fossil divided like a chambered shell by transverse septa.

Insulated casts in the septa of the ammonite are common

feet in thickness, but so friable as easily to be reduced to powder. On this immediately a marle sets on, in the different depths of which the ironstone comes on regularly in all the various sorts as follows:

1. Small balls.—Provincially *twelve foots:* because so many feet distant from the first to the last bed.
2. Gray limestone. Used as a flux.
3. Foxes.
4. Riggit.
5. Bulls.
6. Caballa balls.
7. Whiteburn. — What tripoli, properly calcined and treated, is made of.
8. Clouts.
9. Pity.

This is the order in which the different ores are found.

Advancing on, there is a valley where the mineral bed seems entirely broken, and the sandstone sets on. At the distance of something above a mile, the ironstone is again seen—another intervention of sand, and then at low water when the tide goes out, the beds of ironstone appear regularly on the shore; an indisputable proof that, however the appearance of the surface may vary, the substrata continue the same.

In taking the range northwardly, from the bottom of Ashburnham Park for 12 miles at least, the strata are nearly the same, there being no material inequality of surface that does not partake of sandstone, marle, ironstone, and sand again at the top.

throughout this stratum in Oxfordshire, and a telliniform shell occurs in the ironstone of Shotover in Oxfordshire.

The fossils given by Smith in his stratigraphical arrangement under the head Brick Earth, are partly from this formation, partly from the Folkstone marle; those which have the locality, Steppingley park, Bedfordshire, are from the iron-sand.

At Shanklin Chine in the Isle of Wight, are nodules of ferruginous marle in this formation, containing obscure casts of turbinated univalve and bivalve shells. The lowest beds also near Sandown fort in that island, contain various shells.

The occurrence of vegetable remains in this formation has been already noticed in connexion with the coal beds occasionally traversing it: fragments of silicified wood are sometimes found.

(*d*) *Range and extent.* The tract extending along the foot of the chalk hills in Yorkshire, has not as yet received sufficient examination to enable us at present to offer any particulars concerning the course of this formation in that quarter, and the marshes and alluvium on the west of that escarpment, in Lincolnshire and Norfolk, generally conceal it. In the west of Cambridgeshire, however, we find it well exhibited, and having here once gained a full view of it, we may trace it hence continuously through the remainder of the island. Here it forms a band nearly four miles broad, between Fenny Stanton near St. Ives and Huntingdon, rising into low hills: this band, still holding the same breadth, preserves a direction nearly south-west through Bedfordshire, crossing that county from Potton to Woburn, and entering Buckinghamshire at Bowbrick hills, near Fenny Stratford. Throughout this part of its course, it constitutes the principal mass of a well defined range of hills, based on the Oxford clay. This chain of hills is continued through Buckinghamshire in the same direction, from Brick hill on the east to Brill hill on the western border; but the range is here traversed and broken through by numerous broad vallies, and thus separated into insulated groups. Throughout all that part of Buckinghamshire which lies west of the Grand Junction Canal, the Aylesbury or Portland limestone, which here makes its first appearance underlying the iron sand, forms the lower and central regions of those hills, and the iron sand is confined to their summits. In the valley to the south, separating these hills from the chalk range, the principal mass of this formation must hold its course; but the whole of this tract is so much concealed by debris of flint gravel, derived from the latter, that little can be seen.

In the adjoining county of Oxford, the iron sand and its

subordinate beds of ochre cover the Aylesbury limestone beds on the summit of Shotover hill, which forms a corresponding elevated group to that of Brill, over against which it stands; and it may be traced hence throughout the southern vale, as the dark red soil, prevalent on the road to Tetsworth, at once indicates to the eye.

On the opposite side of the Thames, the system of hills of which Shotover forms a part, is continued through the north of Berkshire by the range of which Cumnor hurst forms the highest point on the east, and Faringdon clump on the west. Both these summits are capped by the iron sand, but it has been denuded off from a great part of the intervening ridge, and occurs in these points only in the form of insulated and outlying groups. The main course of the formation lies through the southern valley, but is mostly concealed by alluvia from the overhanging ranges of chalky downs.

In Wiltshire, we are informed by Mr. Townshend, that this sand occurs at the foot of Beacon hill in the road from Pewsey to Chippenham, and may thence be traced rising up all the way to Lockswell heath and Dring hill. That it also emerges from beneath the clay at Foxhanger, west of Devizes, and thence rises up to Seend. It is here described as being a pudding-stone, composed of rounded quartz, whose cement is siliceous with a red calx of iron, containing ore formerly in much request for the furnace and the forge, and forming the material whence the ancient Britons wrought their Quern stones.

It seems probable that in proceeding farther west, this formation thins out and expires, since we do not meet with any traces of it in the west of Dorsetshire; but in the south-east of that county it is very strikingly displayed throughout the Isle of Purbeck, where it consists of many beds of quartzose conglomerate, and of coarse and fine-grained sandstones containing seams of wood coal. It here forms a narrow stripe (the high inclination of its strata reducing the space they occupy in their superficial appearance) running parallel to the chalky downs, and along their southern foot.

In the Isle of Wight it forms the substratum of all the southern half of the island; cliffs of it may be traced from Sandown bay on the east to Freshwater bay on the west, excepting where concealed by the subsided ruins of the undercliff.

In the south-eastern counties, this formation constitutes the central chain of hills which traverses the Weald of Sussex from Hastings to Horsham, and sends off several branches, one of which extends to Tunbridge Wells. It is separated on the north from the Ragston hills by a broad valley of deep clay through which flow the two streams which unite to form the Medway. On the south, a similar broad valley of clay separates the range from the Sussex chalk downs. (G. Notes.)

(*e*) *Elevation, &c.* In the Midland counties, as has been already observed, this formation constitutes the mass of a chain of hills extending through Bedfordshire, and forms the summits of the continuation of the same chain which ranges through Bucks, Oxon, Berks, and Wilts. This chain however is much broken. Its highest point appears to be at Brill hill on the borders of Buckinghamshire and Oxfordshire. The greatest height of the hills of this formation, which constitute the central nucleus of the Weald in the south-eastern counties, is at Crowborough Beacon.

(*f*) *Thickness.* The greatest thickness of this formation appears to be in the Weald country, where it cannot be estimated at less than 500 feet; but we have no precise data to assist us in estimating it either here or elsewhere.

(*g*) *Inclination.* The position of the strata of this formation is generally conformable to that of those which we have previously described, always keeping a parallelism to the great chalk ranges, and dipping in the same direction, and under a similar angle. This must however be understood with the following limitation;—that towards the south-west this stratum thins off and terminates much sooner than those which succeed it, not extending beyond the eastern division of Dorsetshire; whereas the green sand stretches into the very heart of Devon. This line agrees very nearly with that of the subjacent oolites.

(*h*) *Agricultural character.* The soil produced by the decay of the beds of this formation is esteemed for early crops, and is in some places exceedingly fertile; but a vast proportion of its surface is covered by heath, as is the case at Disingham Heath in Norfolk, which is high and spacious. In Cambridgeshire it forms excellent garden ground: in Bedfordshire it is considered excellent for the growth of potatoes, and there are extensive plantations upon it near Woburn; and some hop-grounds in Kent. It is very favorable to the growth of wood. The Weald of Kent and Sussex, consisting of it, is so named from the forest which formerly covered it. This sand, until the art of manufacturing iron by coal was discovered, was in great request for the furnace and the forge. (G. Notes.)

(*i*) *Phænomena of springs, &c.* This formation resembles in this respect all other aggregations of loose and porous strata, divided by a few others of more tenacious quality: the wells are generally deep: the waters are, very frequently, from the abundance of iron in the rocks they traverse, chalybeate; of which Tunbridge wells is a known instance.

Section VI.

*Particular account of the distribution of the formations between the Chalk and Oolites, in the several districts occupied by them.**

The reasons which appeared to render it desirable to subjoin to the more general view of these formations already given, a more precise account of their local phœnomena, have been already stated in the introduction to this chapter. We shall commence this article with a description of the structure of the Weald of Kent, Sussex, and Surrey; since it is in this tract that the formations in question acquire their greatest thickness and importance.

A. *The Weald.*

This district may be generally described as stretching along the coast of the channel from Folkstone to Beachy Head, and thence extending westwards into the interior as far as the confines of Sussex and Hampshire.

A lofty range of chalk hills, already described in treating of that formation, forms the general boundary of this district, excepting where it is open to the sea; and as the formations within the area make their appearance by emerging from beneath the surrounding chalk, it has been frequently described as a great denudation. It would, however, be highly rash to assume that the chalk at any period actually covered the whole space in which the inferior strata are now exposed, although the truncated form of its escarpment evidently shews it to have once extended much farther than at present.

To the encircling hills of chalk a valley of variable breadth, occupied by the chalk marle and its subjacent clay, succeeds. This valley, through a great part of Kent and Surrey, was anciently known by the name of the Vale of Holmesdale; a name celebrated by the traditionary lines which recorded the successful stand said to have been made in defence of the ancient laws and liberties of the country against the Norman invader.

> 'The Winding vale of Holmesdale
> Was never won nor ever shall.'

The inner border of this valley is formed by a second range of hills running round in a line nearly parallel with those of the chalk, which they rival and even surpass in height: these are composed of the green sand formation. At the foot of this second amphitheatre of hills, is a second valley much broader

* By the Rev. W. D. Conybeare.

than the former, and occupied by the argillàceous beds which we have called the Weald clay.

A central and third range of hills, still nearly of the same height, forms the nucleus of the whole district. It is composed of the thick strata of the iron sand formation; but in some places near the valley of the Rother, which traverses the centre of this range, a series of argillo-calcareous beds has been dug into for the sake of the limestones they afford: these form the base on which the iron sand reposes, and must be considered as introductory to, if not identical with, the upper part of the Purbeck limestone series.

A very interesting geological phœnomenon is presented by the course of the rivers watering this district, and the arrangement of the vallies which convey them. We have already noticed that the two grand vallies of this district—that of Holmesdale, and that of the Weald clay—are parallel to the direction of the strata; but these do not form the channels through which any of the more important streams seek the sea, for these generally have their source in the central ridge of iron sand; and flowing thence both to the north and to the south, in directions nearly at right angles both to these vallies and the strata, traverse the ranges of green sand and chalk, through gorges opened across them, in their way to join the Thames on one side and the Channel on the other; instead of being turned by their escarpments into the great subjacent vallies, as they would be if the fractures in those escarpments were repaired, and forced to empty themselves into Romney marsh and Pevensey level. In no place perhaps is the important fact of a double system of vallies crossing each other transversely—(a fact which we shall hereafter see to be of the greatest consequence with reference to theories on the origin of the present inequalities of the earth's surface) more strikingly displayed. In treating of the chalk ranges, we have already noticed the rivers which thus break through their line.

In order to trace the several formations we have above generally indicated, with the greater certainty in their course through the interior, we shall derive the clearest elucidation of the structure of the ranges composed by them, in beginning with the sections presented along the line of the coast. Without indeed attending to those, we should be in some danger, from the resemblance of some of the beds, to confound the great formations of the green sand and iron sand, or to mistake the hard and sandy varieties of the chalk marle for members of the green sand; but if we first attend to the structure of the great ranges of hills, and the clayey vallies which separate them, as developed on the coast, and thence trace them continuously through the interior, it will be found impossible to arrive but at one conclusion.

The termination of the chalk range, which forms the northern boundary of the Weald, against the coast on the north-east of Folkestone, has been already described at page 104; where an account is given of the grey chalk, or cretaceous form of the chalk-marle which underlies them. This reposes on the argillaceous form of the chalk marle, which constitutes the mass of a crumbling bank of no great elevation, stretching along the beach for nearly a mile from the foot of the chalk cliffs towards the village of Folkestone, where it is succeeded by the green sand. It is replete with various organic remains, of which a list has been already given in treating of the chalk marle. It will be found important in our subsequent remarks to trace this argillaceous marle through the interior. The green sand, emerging from beneath this marle, rises near Folkestone into cliffs which continue, with an height of from 100 to 150 feet, to line the coast as far as Hythe; where the alluvial flat of Romney marsh, evidently gradually gained from the sea, at present keeps it off from the foot of the hills.

The green sand, as exhibited in this section, is a coarsely-granular aggregation of rounded fragments of quartz, cemented by calcareous matter, and interspersed with very numerous particles, and sometimes large kernels, of the green earth which characterises the formation.

Where the intervention of Romney marsh protects the continuation of the hills belonging to this formation from the wasting action of the waves, a sloping talus extends from their base to about two-thirds of their height; and here a long low line of precipice, nearly resembling both in structure and situation that of the undercliff in the Isle of Wight, hangs over it. This precipice continues about three miles through the parishes of Lympne and Aldington; * it presents finer varieties of the sandstone alternating with beds of limestone, which appear generally through Kent to mark the lower part of this formation. The organic remains occurring in this range are Nautilus, Ammonites, Trochi and other turbinated univalves, Ostreæ, Pinnæ, Terebratulæ both plain and plicated, Cardia, Pectines, Pectunculi, Arcæ, Echinitæ spatangi, and E. conuli.

---

* In Mr. Greenough's Geological map this line of cliff is represented, but owing to an accident, probably in the colouring, as included in the clay district of the Weald, instead of that of the green sand.

Near Orlestone, the range of the green sand hills trends inwards, and is succeeded by lower eminences composed of the Weald clay, which skirts the central part of the marsh, and forms a broad tract of flatter country.

Towards the south-west extremity of the marsh, the strata of iron sand, forming the central chain of hills, emerge from beneath the Weald clay near Oxney Isle and Rye. At Cliff End near Winchelsea, the barrier of marsh land terminates; and the sea, attacking the foot of this central chain, exposes a bold section of it in a line of cliffs extending thence past Hastings, and terminating at the distance of about seven miles, near Bexhill.

The strata of this formation, near Rye, are coarse-grained and highly ferruginous; near Hastings they are less ferruginous, finer, very loose and friable, frequently crumbling down from the summit of the cliffs in a shower of loose sand, which accumulates in a talus at their base. Traces of vegetable impressions may be frequently here observed.

Taking Winchelsea as a centre, the succession of the formations on either side, intervening between this point and the chalk hills, will be found to correspond. Hence, in proceeding south-westwards we have again to traverse in an ascending order the formations we have already crossed in descent. The alluvial flat of Pevensey marsh, however, here interposed, as was Romney marsh on the other side, has prevented any sections from being formed on the coast; but the Weald clay may be seen forming a broad flat tract from Bexhill beyond Pevensey; and the green sand and chalk marle, though here occupying a less breadth than on the northern side, and not forming a distinct range of hills, may be observed underlying the chalk near the foot of the lofty cliffs of Beachy head.

Having thus followed the order of these formations as exhibited on the coast, we have next to trace their course through the interior of the Weald; and here we shall depart from our usual arrangement, and commencing with the central chain which forms the nucleus of the whole series, trace the succeeding formations intervening on either side between this and the great boundary of the chalk hills in an ascending, instead of a descending order; since, in this instance, this change of method will be found attended with many advantages.

The area occupied by the central range of iron sand may be generally represented by a long narrow triangle of which the northern end of the base is in Oxney Isle, the southern at Bexhill, and the apex four or five miles west of Horsham. The whole tract is elevated, and forms the ranges of Ashdown and St. Leonard's forests; Crowborough Beacon, 804 feet above the sea, is the highest point.

Near Battle, the substrata on which this great arenaceous formation reposes, have been pierced in the lime-works on the east of Ashburnham, and been found to consist of a series of fifteen thin beds of limestone, alternating with shale; the seven upper beds are of a grey colour, and are said to be nearly composed of an aggregate of shells; the eight lower beds are described as of a blue colour, and as being a perfectly indurated calcareous marle. This argillo-calcareous formation will probably be found, on an attentive examination of all its beds, especially the lowest, to coincide with that of Purbeck, hereafter to be described; the upper beds are, however, said to present different fossils. Until it shall have been more scientifically observed, however, it is impossible to speak with absolute certainty; and therefore the present general description of the Weald appears to afford the most proper place for noticing them. An account of the section presented by these works, extracted from the Agricultural Survey, will be found in the note below.*

* GREYS.

	Thickness. feet	inches	
The First Limestone	3	3	8 feet Shale
Second	0	9	9 ditto
Third	4	0	39
Fourth	1	1	3
Fifth	0	8	3
Sixth	0	8	2
Seventh	8	3	4

BLUES.

The great blues by far the best.

	Thickness. feet	inch.	feet	inch.
Eighth	2	0	1	6
Ninth	0	6	0	4
Tenth	0	9	1	3
Eleventh	1	2	0	4
Twelfth	0	8	1	1
Thirteenth	1	1	1	6
Fourteenth	0	6	8	0
Fifteenth	2	3		

The last stone is fine enough to set a razor.

These inferior argillo-calcareous strata are covered by that great series of beds of ferruginous sand which constitute the mass of this central group of hills. In the general description of the formation before given, an account of the beds of iron ore worked in this part of the Weald will be found.

The great zone formed by the Weald clay, constituting a flat tract about five miles in average breadth, circles round this central group, ranging on the north from the Isle of Oxney by Tunbridge towards Haslemere, and thence returning south-east to Pevensey. It is generally characterised by its calcareous concretions containing the Vivipara, and forming what is called the Petworth marble. On this subject we have nothing to add to the statements given in the general account of these formations. It is of great importance, in order to correct all misapprehensions which have arisen concerning the structure of this district, to trace correctly the line of the next formation—that of the green sand; and this fortunately may be done without the possibility of error, by following its course continuously from its section on the coast between Folkestone and Lympne; for its upper limit being marked by the course of the argillaceous chalk marle of Folkestone through the vale of Holmesdale, and its lower limit by the great clay valley of the Weald, it is effectually cut off from the formations on either side, with which it may form occasional similarity of character, and has indeed been confounded by very able observers, who have examined it only partially and without a due attention to this circumstance.

From the sectional cliffs before described between Folkestone and Lympne, this formation may be uninterruptedly traced to the borders of Surrey, constituting a well marked ridge of hills from 600 to 800 feet in height, and about two miles broad, ranging immediately on the south of Maidstone, Sevenoaks, and Westerham, although occasionally broken through by transverse vallies, particularly that of the Medway. Although, however, the mass of this formation may be thus uninterruptedly traced, it undergoes some change in the composition and external characters of its constituent beds as it advances westwards. East of the Medway these present, as on the coast, sandstones of a loose texture, full of large green particles, alternating towards the lower part of the series with beds of limestone; but in proceeding towards Sevenoaks and Westerham, the green particles decrease in quantity; the prevailing aspect of the sandstone becomes more ferruginous, and its grain finer; cherty beds also appear more commonly dispersed through the mass.

On the north of this range lies the vale of Holmesdale, throughout the bottom of which the argillaceous marle of Folkestone may be traced, without interruption, from the sea coast into Surrey. Above this, on the opposite side, rises the escarpment of the chalk hills, the bottom of which is formed by the grey chalk or cretaceous chalk marle, often containing beds of fire-stone, occasionally marked by spots of green earth.

Having thus traced these formations uninterruptedly from the coast into Surrey, it will be our next object to describe their appearance on the east of that county, near Merstham and Reigate; and this we shall do somewhat more minutely, since on the ground above stated—namely the continuous course of each formation from the coast—we feel ourselves compelled to dissent from the opinions advanced by a writer concerning whose eminent services to English geology one estimate only can be formed; and who, from the inspection of this single spot, has pronounced the fire-stone beds, which we assign to the chalk marle formation, to belong to that of green sand, and the range which we consider as the true green sand, to be iron sand. A particular account of the observations made by Mr. Phillips on this district in Feb. 1821, is therefore subjoined.

Section from Merstham Chalk pits to Nutfield.

The Merstham pits are situated in the chalk marle, the colour of which here is very different to that of the grey chalk or chalk marle of Folkestone, which is much darker, owing perhaps to its being in immediate contact with the subjacent blue marle, but which is not the case at Merstham.

The kilns are situated at the lowest level at which the marle is fit for the kiln, beneath which it becomes hard and partakes largely of the nature of stone for 30 or 40 feet in depth; the lowest bed being, to use the term of a very intelligent miner, 'ratchelly'—rubbly, very loose, and easily cleared away.

This 'ratchelly' bed rests immediately upon the series of fire-stone beds, which have during some years been worked as a quarry, affording the chief supply of fire-stone for the London market. At this time however, the quarry was full of water, but an account of it by T. Webster will be found in the fifth volume of the Transactions of the Geological Society. In addition to the information there given, it may be useful to annex a description of these beds, by the intelligent miner of whom mention has already been made.

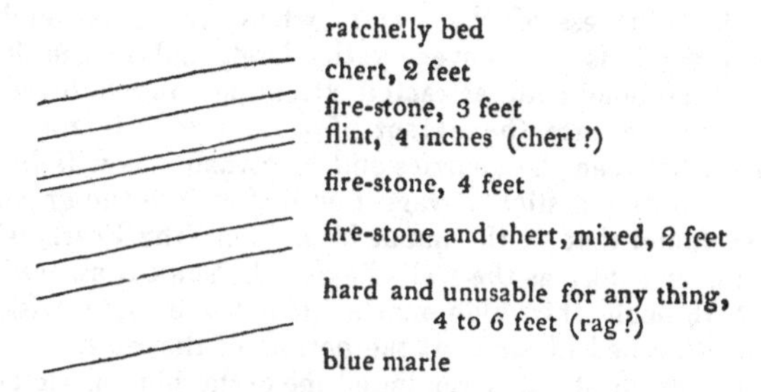

These beds dip on an average nine inches in every yard to the north, and together furnish a never-failing supply of water, which then filled the quarry, owing to a circumstance presently to be detailed. The water finds its way not only between the beds, but also in great quantity through nearly vertical fissures in the beds themselves, wide enough for the admission of the hand.

The following diagram describes the whole series of beds near Merstham.

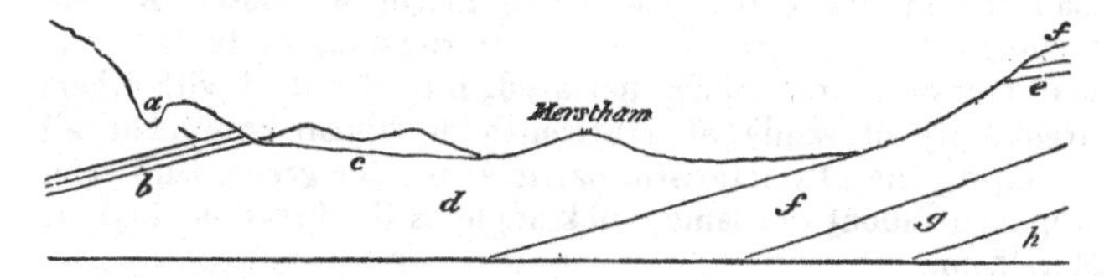

a. hollow in which the limekilns are situated; *b.* fire-stone beds; *c.* tunnel; *d.* blue marle; *e.* fullers' earth beds; *ff.* green sand; *g.* probably Weald clay; *h.* probably iron sand.

To drain the quarry at the upper part of the fire-stone beds *b*, a tunnel, *c*, was some years ago driven up from the lowest land near Merstham, and in a very nearly horizontal direction, through the blue marle *d*; which marle immediately underlies the fire-stone beds, but at the actual contact is of a yellowish colour.

The draining of the beds by means of the tunnel, diverted a stream of water, which formerly issued from another place, and turned a mill; but as the tunnel has lately been stopped, the water has resumed its old course, and issuing from beneath the picturesque mound on which Merstham church is built, and flowing into the lower land, it again supplies the mill. The water appears to find its way through the upper part of the blue marle, where it contains layers of somewhat compact calcareous sandstone, inclosing green particles and mica.

The blueness of the water, where it issues, and indeed wherever it is seen covering the land, and in the ditches of the neighbourhood, attracted attention, and induced a close examination into the extent of the marle. It appears every where between the quarries and Merstham, as will be evinced by thrusting a stick a very few inches into the ground; and we* found that at the inn at Merstham (the Feathers) a well had been sunk, as the well-digger who sunk it assured us, 260 feet through this blue marle alone; the water rising from beneath a bed of stone at the bottom of the well.

The marle strongly reminded me of the blue marle at Folkestone, but we saw no organic remains.

From Merstham we walked to Nutfield, looking every where for the continuation of the blue marle, and we found it for about two miles in the ditches and fields on both sides of the common road. In Nutfield marsh, the water appeared here and there a little ferruginous, giving an external tinge to the marle.

So soon, however, as we began to rise towards Nutfield church (situated in a lane below Nutfield and considerably lower than the fullers' earth beds,) the marle was lost. A sand enclosing layers, often so highly ferruginous as to bear the character of an iron-stone, appeared, interstratified with others (frequently of sand) of a greenish or bluish cast; but all exhibiting the characteristic particles of the green sand, and dipping at about the same quick angle as the fire-stone beds at Merstham.

I suspect these ferruginous appearances to have been mistaken for the iron sand, which is described by T. Webster as appearing on the surface south of Merstham.

Our principal object at Nutfield was to see the fullers' earth beds, as regards their position; they are described by the before-mentioned author as lying between the green sand and the iron sand.

Not one of the three fullers' earth pits now working was opened to the bottom of the fullers' earth. In the middle, and most important one, however, we procured it to be done, and found the fullers' earth resting on a sand and sandstone in part highly ferruginous, but no where without the green particles belonging to the green sand; on which therefore we conclude it rests, and not on the iron sand: in a word, that the fullers' earth is enclosed in, and subordinate to, the green sand formation. The latter is often ferruginous in places; the masses of stone overlying the fullers' earth are often completely so,

* I was accompanied by my friend S. L. Kent, M.G.S.

and masses and layers even of iron-stone were seen in the banks of the lane leading up to Nutfield church as before recited.

In pursuing our walk to Red Hill near Reigate, we found nothing but green sand. The nature of Red hill itself we had not time for exploring, being anxious to see whether the low grounds between it and Merstham consist, as between Merstham and Nutfield, of the blue marle. We found the marle again half a mile north of Red hill, and about a mile south of French's turnpike, which is situated on a little rise of green sand. At this place we met with a farmer, who told us that the blue marle occupies all the low grounds (or much of the low ground) south of Merstham—that it is locally termed 'black land'—is in many places very *holding* (produces good crops) and where it is suffered to lie fallow every third year wants little or no manure; a strong proof of its marly nature.

We entered the Baron's cave, the sand of which is referred by T. Webster to the iron sand. The opening to this cave is on a mount north of Reigate and overlooking it completely. The descent of the cave is towards the town, but is not in a direct line, at an average angle perhaps of 30 or 35 degrees for 200 or more feet, with one branching vault near the centre towards the west about 150 feet long, from 10 to 12 feet high, and about 12 feet wide. The sand is very white, except where it is traversed by ochreous veins; mica is found in it sparingly and in small fragments as well as extremely minute green particles, visible only by the help of a glass. It seems therefore to belong to the green sand formation.

I pointed out to a farmer some fragments of Petworth marble lying on the side of the road, and employed in mending it, near French's turnpike; he informed us that it is found in layers in blue clay, (the Weald clay *g* of the preceding section) south of Red hill, and also of the range on the summit of which the fullers' earth pits are situated: an additional reason for assuming the sand and sandstone on which the fullers' earth lies to belong to the green sand, since the Weald clay lies between the green and iron sand.

From * the points just described, the green sand ridge, after being cut through by the valley of the mole, ranges on the south of Dorking; Leith hill, which is about 990 feet above the level of the sea, and possesses a remarkably bold and mountainous character, forming a part of it. Dorking itself

* The remainder of this article is by the Rev. W. D. Conybeare

appears to stand upon the Folkestone marle. The upper beds of the sand formation crossed in proceeding thence to Leith hill, are coarse-grained and very ferruginous, and alternate with beds of clay; so that they appear like a second formation of iron sand resting upon the green sand. These beds continue to intervene between the chalk marle and more decided green sand throughout the western part of Surrey. The strata constituting the mass of Leith hill have the green particles very sparingly interspersed, and are of a brownish buff colour; they are nevertheless well characterised as belonging to the formation usually denominated green sand, by their alternating beds of chert passing into chalcedony, and by the abundance of the same variety of alcyonium which Mr. Webster has described in the same formation in the Isle of Wight. The sections of these form white rings, traversing in every direction the darker mass of the rock.

Immediately south of Guildford, the ferruginous upper beds are strikingly displayed at St. Catherine's chapel. The more decided green sand may be observed beneath it on the south of Godalming, whence it ranges south-west to Hind Head, a summit of this ridge nearly rivalling Leith hill in height, and rising 923 feet above the sea. Between Leith hill and Hind head, the chain of this formation is broken through by the valley of the Wey, but is otherwise continuous, and breaks down the whole way with a very bold escarpment towards the great valley of the Weald clay on the south.

The upper ferruginous beds intervening between this ridge and the chalk marle, occupy the whole extent of Alice Holt and Woolmer forests; and the chalk marle containing, as at Merstham, beds of firestone, is well displayed between Alton and Selborne.

A good sketch of the geology of this district will be found in the commencement of that most elegant and entertaining of all works of the kind, White's Natural History of Selborne; hence we extract the following particulars.

The chalk rises immediately on the south-west of Selborne into a considerable eminence called Selborne Hanger. In the pits on its sides, the Ostrea crista galli, Cornua ammonis and Nautili (of the species probably common in the lower chalk) have been found. At the foot of this hill, is a thin bed of clay which divides the chalk from the subjacent strata of firestone on which the village stands. The roads to Alton, and that to the forest, are also deeply worn into this rock. It consists of a white firestone, alternating with thin beds of a blue rag. It is extensively quarried for the same purposes as the similar strata at Merstham, and it appears to rest on beds of clay which divide it from the sands of Woolmer forest.

To the south of Haslemere, the green sand range, following the mantle-shaped disposition of all the strata round the central nucleus of iron sand, becomes as it were reflected towards the south-east. It may be traced by Blackdown hill and Brinksole heath on the north of Petworth; but the hills in this direction become lower; there is a quarry of this rock well characterised in the grounds of Lord Egremont's park at Petworth.

Between Petworth and the chalk escarpment on the south, the upper ferruginous sands, before noticed, prevail; and at the very foot of the escarpment at Duncton hill, the chalk-marle formation is well displayed.

We have not traced these formations through the remaining and south-eastern part of Sussex to the sea, but that district will shortly receive full illustration from the publication of a work expressly dedicated to it by Mr. Mantell of Lewes.

The continuation of this denudation on the opposite coast of France, where it occupies a semicircle of about twelve miles radius, ranging round Boulogne as a centre, and skirted by a lofty amphitheatre of chalky hills, exhibits a series of formations lower than any of those which appear on the English side, and which can therefore be only spoken of in this place by an anticipation of the general method pursued in this work. Yet, as it is only in consequence of its connection with the denudation of the Weald, that this tract acquires importance in the eyes of the English geologist, we shall throw together in this place the imperfect materials we possess concerning it, trusting that they may have the effect of leading to a more full examination of a district connected in so interesting a manner with the geology of our own coast.

The escarpment of the chalk pursues the semicircular sweep above described, round this district; the green sand may also be traced on the north-east of Uissant, and the hills close to Boulogne, on which Buonaparte's tower is built, are capped with strata of ferruginous sand, possibly our own iron sand. But instead of these formations occupying, as in England, the whole of the denuded area, we find the coast, through the greater part of its breadth, lined with a calcareo-argillaceous formation which underlies the sand last mentioned; this may be studied in the cliffs on either side of Boulogne, which however are, from the nature of the material, in a very crumbling state.

One of the upper beds of this formation consists of a remarkable calcareo-siliceous grit. Boulogne is principally built of this rock. It often forms the upper stratum of the cliffs on the north-east of that town, and is particularly abundant at a place called le Creche, between it and Uissant. This rests on some beds of argillaceous limestone, separated by clay; and along the bottom of this cliff is found a coarse limestone of a brown colour, full of cornua ammonis, turbinated univalves, &c.

These alternations of argillaceous and calcareous beds continue to extend into the interior towards the east of Boulogne. In a deep pit sunk at Souverain Moulin, about five miles from the town in that direction, in a fruitless attempt to procure coal, the workmen passed through twelve such alternations, and then pierced a solid calcareous rock 100 feet in thickness, containing ammonites. Below this, occurred a thin seam of wood coal, and then 20 feet of a shelly limestone full of turbinated univalves, small oysters, serpulæ, &c. together with impressions of ferns and other vegetables; and lastly, another thin seam of carbonized wood, resting on coarse limestone.

We have considered, in colouring the Map annexed to this work, the alternating beds of marle and limestone as belonging to the same series which, near Battle and in the Isle of Purbeck, underlie the iron sand: but it should be added that Mr. Buckland rather inclines to refer them to the older formation of the Oxford clay, believing the calcareo-siliceous grit which covers them, to be the same with that associated with the coral rag of England.

The coral rag is exhibited with well marked characters about nine miles south-east of Boulogne, near Samers, and the great oolite may be seen in the country round Marquise at the same distance on the north-east. Both these points approach very closely to the chalk escarpment; so that if the green sand, iron sand, &c. exist at all in these directions, they must be greatly reduced in extent.

Near Marquise, the oolite comes in contact even with the older rocks of the coal formation, which shew themselves in this corner of the denudation almost immediately beneath the chalk. This coal district presents a band of mountain limestone accompanied by another of regular coal-measures. The principal marble quarries are at Ferques, and the principal coal mines at Hardingen. The stratification is extremely confused and contorted.

Section VII.

*Isle of Wight.**

As we have already observed, the formations which it is the object of this chapter to describe, form the substrata throughout the southern half of this island. They may be most effectually studied at their junctions with the range of vertical chalk which crosses the island from east to west; on the former side at Sandown bay, on the latter at Freshwater bay.

Of this appearance in Sandown bay Mr. Webster gives the following account.

'Immediately below the chalk without flints, is a stratum of yellowish white marle, or argillaceous chalk. This is also wholly without flinty nodules. A thick stratum of sandstone succeeds, consisting of siliceous sand united by calcareous matter, and containing also mica and green earth. It is often very friable, and being here of a whitish colour, is not readily distinguished, at a distance, from the chalk.

The next stratum consists of a dark blue or grey marle, which readily falls to powder when exposed to air and moisture; accordingly its place is marked by a considerable hollow, the cliff having mouldered away, so as to form a slope which may be climbed up without much difficulty.

The blue marle is followed by a very thick stratum of dark red and highly ferruginous sand, sometimes containing beds firmly cemented into rock. To this succeeds a yellow sand stratum; then shale; and a very irregular succession of dark red and yellow ferruginous sands with clay, and shale, extending nearly to the middle of the bay.

The section of these ferruginous strata forms a very lofty perpendicular cliff, distinguished by the name of Red cliff; which bears a striking contrast to those of the neighbouring chalk. Vast masses of it are constantly falling down; on which account it is rather dangerous to walk underneath.

Near the termination of these cliffs towards the middle of the bay, are several thin strata of a stone composed wholly of bivalve shells in a calcareous matrix, much resembling Purbeck stone; but the shells are larger. These strata are from three inches to one inch in thickness, are separated from each other by beds of shale and fibrous carbonate of lime, and have the same inclination as the strata lying immediately above them. The greatest part of what could easily be got at, has

* By the Rev. W. D. Conybeare.

been carried away, and used as paving, for which it answers very well, being of so regular a thickness, as not to require any preparation.'

The blue marle of the above description corresponds with the formation denominated in this work the Weald clay.

The junction on the east of Freshwater bay presents an exact repetition of the above.

Advancing about a mile south-east of this junction, Mr. Webster observed the argillaceous strata on which the ferruginous sandstone reposes, make their appearance near Brook point. His account of these is very interesting.

'The ferruginous sand cliffs continued some way farther, preserving nearly the same inclination. But the strata succeeding to it, and which dipped with a gradually decreasing angle, until they were at last nearly horizontal near Brook, were very different. They consisted of a succession of beautifully coloured plastic clays alternating with beds of red and yellow sand, sandstone, slate clay with fossil shells, and also limestone containing veins of calcareous spar.

At Brook point the cliffs interested me much. They were about thirty feet in height; and were composed chiefly of clay resting upon a bed of soft sandstone, which contained a considerable quantity of sulphur, arising from the decomposition of pyrites.

At this place I observed many masses of a coaly blackness, bearing the exact form and resemblance of trunks of trees that had been charred, lying on the beach, and imbedded in the clay cliffs, and also in the rock.

In some parts, the ligneous fibre was still evident. In other parts, the wood had been converted into a substance much resembling jet; its blackness being intense, its cross fracture conchoidal, and its lustre very great. Other parts of the trees were entirely penetrated by pyrites: and considerable groups of crystals of this substance were frequently attached to the outside.

They were imbedded in clay of various colours, white, grey, yellow and red; and lay in irregular horizontal strata of several inches in thickness, being often pressed flat, by the incumbent weight. Over this stratum of clay, which is about eight or ten feet thick, there is another, of the same depth, of sand and gravel highly ferruginous; and the water which filters from it is strongly impregnated with sulphate of iron.

On lifting up some of the sea weeds which grew upon the shore between high and low water mark, I was surprised to find almost all the rocks below them composed of petrified trees, which still retained their original forms. They were of various sizes, from eight or ten feet long and two feet in diameter, to the size of small branches. The knotty bark and the ligneous fibre, were very distinct; and they were frequently imbedded in masses of clay now indurated and in the state of an argillaceous rock.

Some parts of these trees were converted into iron-stone; and other parts consisted of a great variety of substances, being partly calcareous, ferruginous, pyritous, bituminous, and ligneous; and the whole exhibited a beautiful example of the astonishing processes of nature in converting vegetables into coal, and in filling their substance with solid rock. It is obvious that the lowest strata of the island are exhibited here and at Sandown foot.

Where the strata become horizontal along the line of the undercliff on the south of the island, the same succession of chalk, chalk marle, green sand, blue marle, and iron sand may be traced. The best section is afforded by the precipitous sides of Black gang chine. In the undercliff, the green sand contains distinct alternations of calcareous and cherty beds.'

Section VIII.

Isle of Purbeck. *

On the eastern promontory of this peninsula, on the north of Swanwich, and directly opposite the western extremity of the Isle of Wight, to which we have already traced them, these formations re-appear together with, and underlying, the vertical chalk.

Mr. Webster thus describes the section presented at this point. 'The order was exactly the same as what I before observed in the Isle of Wight. Under the flinty chalk, was chalk without flints; then chalk-marle; green sandstone, with limestone and chert; but a slight trace of blue marle; and, lastly, ferruginous sand, sandstone slate, shale, &c. forming the bottom of the bay.

The lower chalk and chalk marle were, however, much indurated; and the limestone and chert, with many fossils, were mixed among the green sand in the most confused state imaginable; exhibiting the same knotted and lumpy appearance as I have before mentioned with respect to the chalk, but still more remarkably.

* By the Rev. W. D. Conybeare.

The inclination of the strata, at the bottom of the clay, lessened as they approached the town of Swanwich; in these cliffs, I observed a stratum of fossil wood, similar to what has been described at Compton in the Isle of Wight, and occupying the same situation.'

The section fig. 5. in the plates accompanying this work, exhibits a representation of the series at this spot. It will there be farther seen that the substrata of the Purbeck and Portland limestone, &c. emerge from beneath these arenaceous formations, and constitute the southern promontory of this peninsula. On the western side of this, the coast again trends northward. At the point where it again approaches, in consequence of this inflection, the line of the chalk hills, is a tract about four miles long, rendered in the highest degree remarkable and interesting by a series of nearly circular bays, guarded in front by projecting capes and reefs of the Portland rock, and excavated backwards into the interior as far as the chalk; thus forming regular oval basins, not having more than one-third of their circumference open towards the outer sea. The precipitous sides of these basins afford the finest sections of all the formations intervening between the Portland beds and the chalk. The outer capes of these coves, which exhibit the Portland beds, are lofty, being transvere sections of a ridge of hills of that formation traversing the peninsula longitudinally from east to west. The inner cliffs, which exhibit the arenaceous formations, are low; being cut into a valley, which, throughout this tract, separates the above ridge from that of chalk. Cliffs, and towering eminences of the latter, form the back of each cove.

There are six of these coves, which occur in this order beginning from the east; 1. Worthbarrow cove; 2. Lulworth cove; 3. Stare cove (a very small excavation); 4. a larger cove unnamed; 5. Man-of-war cove, and Durdle or Barn-door cove; remarkable from a perforated and arched rock forming its eastern cape, whence it has acquired the latter name.

The sections in all these correspond to those before described in the Isle of Wight, and at Swanwich bay. At Lulworth cove, which is perhaps the most remarkable of these on account of the regularity of its circular form, the beds of wood coal contained in the iron sand are very strikingly exhibited.

Mr. Webster has well observed that these coves are equally remarkable for their singular form and picturesque character, as from their comprising in so short a space the epitome of so considerable a portion of the English strata.

Section IX.

Attempt to trace the arenaceous formations through the Midland counties. *

The iron sand cannot, it is believed, be pursued in Dorsetshire much to the west of the point to which it has been traced in the last section, its line of basset being entirely overlaid in that direction by the green sand, which thus becomes immediately incumbent upon older formations; nor is it again to be distinctly recognised until we arrive at the neighbourhood of Devizes in Wiltshire. The course of the green sand through this interval has been already sufficiently indicated in the section dedicated to that formation; but at this point, where both the sands are again distinctly to be traced, it becomes important to bestow on them a more particular attention, in order to connect the history of these formations in the southern and midland counties. We regret, however, that we can do little more at present than point out the subject for future examination.

The green sand here extends throughout the vale of Pewsey; and its lowest beds may be seen in the canal locks near Devizes, resting on the Weald clay. The iron sand may be traced farther to the west, rising on the banks of the hills of the coral rag range towards Bowood.

Proceeding to the north-east, under the northern foot of the Marlborough downs, the traces of these formations become much obscured.

Beneath the great escarpment of the chalky range, is a sub-escarpment, or lower terrace, occupied by beds very analagous to those which occur in the chalk marle at Merstham and at Selborne. Beneath this sub-escarpment may be traced beds of green sand, well characterised both by their texture and fossils, but unimportant in thickness and extent. In the valley beneath, where we should naturally look for the course of the iron sand, almost every thing is concealed by alluvial debris, and a few insulated points of this formation are alone visible.

The same condition of things extends through the whole vale of the White horse; but the traces of decided green sand cannot be observed much farther east than Wantage. Here the sub-escarpment of the indurated chalk marle beds assumes a greater height and importance, rising on the banks of the Thames opposite Dorchester into two conspicuous conical hills,

* By the Rev. W. D. Conybeare.

one of which is crowned by the entrenchments of a Roman camp. It seems not improbable that the green sand and chalk marle may run together in those hills, and pass so gradually into each other as not to be distinguishable. At their foot lies a broad argillaceous flat; but in the absence of more decided evidence of the presence and position of the green sand formation, it must remain somewhat doubtful whether this tract belongs to the clay of Folkstone or that of the Weald; although from its relation to the green sand of Wantage, which, at the distance of about six miles westward, seems to be placed above the same clay, we are inclined to identify it with the latter.

About two miles north of these hills, the iron sand, emerging from beneath the argillaceous strata last mentioned, exhibits itself on the banks of the Isis in the low cliff which supports the small church, and gives name to the village of Clifton.

The same inferior terrace of indurated chalk marle, which we have thus traced through Berkshire, continues to range beneath the chalk hills of Oxfordshire, between one and two miles in advance of this great escarpment. It is extensively quarried for building, &c. in the parish of Roak. It here contains the following organic remains.

Turrilite.		
Scaphite.		
Hamites plicabilis,	Pl. 234,	Sowerby.
Hamites armatus,	Pl. 168,	ditto.
Pecten Beaveri,	Pl. 158,	ditto.
Ammonites varians,	Pl. 176,	ditto.
Ammonites rostratus,	Pl. 173,	ditto.

It contains also an Echinus and some other fossils of the Folkstone clay.

The traces of green sand have not yet been observed through this district, and perhaps do not exist. The chalk marle seems to repose immediately on a deep blue clay which forms a tract about two miles broad, succeeded by the iron sand, the limits of which in this and the adjacent counties have been already sufficiently traced in the general article on that formation.

Tetsworth stands on the above clay; which, from the absence of decided traces of green sand, and our imperfect acquaintance with its fossils, we are not able to identify positively, either with that of Folkstone or the Weald. Possibly from the absence of the intermediate green sand, both these formations may have come into contact and be confounded together in one general clayey tract.

The same constitution of the district immediately succeeding the chalky range, may be traced through the other midland counties (Buckinghamshire and Bedfordshire) into Cambridgeshire, but is much covered and concealed by the great accumulation of the diluvial debris of the chalk hills, which converts much of this interval into a vast plain of flint gravel.

The indurated chalk marle is extensively quarried at Totternhoe in Bedfordshire and Reach in Cambridgeshire. It is there known by the name of *clunch.* It affords by burning a good lime. Some of the beds resemble those of Riegate, and are used for the like purpose, as a firestone. Reniform masses of radiated pyrites are common, and one singular bed is full of similar masses of yellow indurated marle, externally of a green colour, and of all sizes from a hazel nut to an ordinary potatoe.

This clunch, or indurated chalk marle, forms the Castle hill at Cambridge, and most of the heights in that neighbourhood. Its line of junction with the upper chalk is said to range by Royston, Balsham, and Newmarket, between the chalk marle and the iron sand, which occurs on the west of this county. Near Gamlingay may be traced, as in Oxfordshire, a broad argillaceous tract, the clay composing which is locally known by the name of *galt.*

The chalk marle which reposes on this bed is said not to be separated from it by any strongly decided demarcation, but rather to pass into it by an insensible gradation.* Some beds of green sand occur near the junction, and others at the bottom of the galt, and near its junction with the iron sand. Its organic remains seem rather to identify the galt with the clay of Folkstone, than with that of the Weald. It is possible, however, that both may be blended together, almost without the chance of discriminating them, in a country where little is to be seen but fen and marsh. The Isle of Ely presents a

* I have never been able to observe, says Professor Hailstone, any strong line of separation betwixt the clunch and the succeeding stratum of gault on which it rests. I believe they pass into each other. The lower beds of clunch become more sandy, and gradually assume the nature of an argillaceous loam. In the next observable stage of transition, the mass assumes a greenish-grey colour, and a plentiful admixture of dark green sand is uniformly dispersed through its substance. At the same time it contains numerous irregular dark brown nodules of a ferruginous indurated marle. At length these foreign matters disappear, the mass becomes uniform, and ends in a bluish clay or argillaceous marle called gault. This occurrence of green sand in the confines of the two beds, was first noticed by Mr. Warburton at the brick-pits near the Castle hill, from which he inferred that it always takes place under the same circumstances; an inference which is borne out by the testimony of the most experienced brick-makers about Cambridge. (G. T. vol. iii. p. 249.)

hillock of sand (perhaps of the green sand) resting upon a part of the galt.

The adjacent parts of Norfolk have never been carefully examined. The beds immediately underlying the chalk at Hunstanton cliff have been already mentioned, when tracing the course of the chalk formation in that quarter (page 78.) They remind us most of the ferruginous sands immediately underlying the chalk marle near Guildford in Surrey.

In Lincolnshire, as has already been observed, the lower beds of the chalk range are of a reddish hue, derived from the intermixture of iron. Ought these to be referred to the true chalk or rather to the chalk marle? Beneath these the following strata have been observed.

a. Quartzose ferruginous pebbly sand ..	from 6 to 10 yards.
b. Calcareous clay, containing beds and concretions of oolitic limestone..	— 12 to 14 —
c. Granular quartzose sandstone and sand, varying from dark brown to light grey, and containing shells	considerably thicker than the two former beds.

It is very desirable that the organic remains of the stratum *a.* should be ascertained, and until this is done it is impossible to identify it with certainty; mean while, it may be referred with the greatest probability to the green sand.

These beds rest on strata of argillaceous shale, which have been pierced to the depth of more than 100 yards. They appear to belong, in part at least, to the Oxford clay, the intermediate beds being here deficient; but it is impossible in the present state of our information to hazard more than a conjecture on the subject.

An examination of the foot of the Yorkshire Wolds, and more especially of the coast where their lower strata meet the sea on the north-west of Flamborough head, would very probably clear up all difficulties on this subject, and enable us to institute a satisfactory comparison between those formations in the northern and southern counties, and we may hope that even the present suggestion may contribute to further this object.

CHAPTER III.

OOLITIC SERIES;

Including all the Strata between the Iron Sand and Red Marle.

Section I. *General view.**

(*a*) *Of these formations in England.* We now enter upon the second of the three great subdivisions of the formations classed by us under the Supermedial Order; a series on many accounts highly interesting and important, since it is the great repository of the best architectural materials which the Island produces,—and since the history and relations of its several members, although they occur extensively on the continent, have in this country alone been fully developed and illustrated.

It would here be unjust to suppress honorable mention of the name of Mr. Smith, well known from his Geological Map and Sections. It is his great and real merit to have been the first person who clearly ascertained and established the true order, position, and course of these formations; and superseded the vague and general denomination of secondary or shell limestones, by introducing the determination of a full, precise, and particular knowledge.

The system of formations occupying the interval of which we have above determined the limits, may be generally described as consisting of a series of oolitic limestones, of calcareo-siliceous sands and sandstones, and of argillaceous and argillo-calcareous beds, alternating together, and generally repeated in the same order; i. e. a formation consisting of many beds of oolitic limestone, resting upon one of calcareo-siliceous sand, and that again upon an argillo-calcareous formation.

Neglecting for the present minor subdivisions, we may with sufficient accuracy represent three of these systems as comprehending all the beds which intervene between the iron sand, and the saliferous or new red sandstone. Each of these systems is based on an argillo-calcareous formation of great thickness,

* By the Rev. W. D. Conybeare.

which always constitutes a well marked line of demarcation, preventing the possibility of mistaking or confounding them together; the oolitic rocks of each system generally forming a distinct range of hills, separated from those of the other systems by a broad argillaceous valley. Hence we may give a synoptical view of the whole series, as divided into the upper, middle, and lower oolitic systems. *

1. Upper Oolitic system.	Argillo-calcareous Purbeck strata, separating the iron sand and oolitic series. *Oolitic* Strata of Portland, Tisbury, and Aylesbury. Calcareous sand and concretions (Shotover hill and Thame.) Argillo-calcareous formation of Kimmeridge and the vale of Berks, separating the oolites of this and the next system—Oaktree clay of Smith (generally).
2. Middle Oolitic system.	*Oolitic* Strata associated with the Coral rag (Pisolite of Smith). Calcareous sand and grit. Great Oxford clay (Clunch clay of Smith) separating the oolites of this and the next system.
3. Lower Oolitic system.	Numerous *Oolitic* strata, occasionally subdivided by thin argillaceous beds; including the Cornbrash, Forest marble, Schistose oolite and sand of Stonesfield and Hinton, Great oolite, and Inferior oolite. Calcareo-siliceous sand, supporting and passing into the inferior oolite. Great argillo-calcareous formation of Lias and Lias marle, constituting the base of the whole series.

The difficulties more than once already alluded to, as to the precise and proper use of the word *formation* will here present

* In the former editions of Professor Buckland's Synoptical view, the principle of adhering to the great natural division of these systems by the principal argillaceous formations, was sacrificed to the convenience of subdividing the numerous beds of the lower system; so that the two upper systems are here classed together, and the lower divided into two. He is now, however, inclined to admit the propriety of employing the more natural divisions adopted in the text.

themselves. It seems applicable with sufficient propriety to designate the distinction between these three oolitic systems, separated as they are by vast intervening argillaceous deposits; and even to the distinction between those argillaceous deposits and the oolites they support—but scarcely so to the minor subdivisions which take place in the lower oolitic system. And we must observe generally, that all the systems appear to have resulted from the gradual and successive action, through a long period, of similar causes, uninterrupted by any violent change of circumstances. Whence these causes produced, at one time argillaceous, at a second arenaceous, and at a third calcareous and oolitic deposits; or why these alternations are several times repeated, are questions which it does not belong to the present state of science to answer.

Each of these beds is characterised by its peculiar organic remains, and very often even the minutest subdivisions may be thus identified,—remains of many extinct genera of oviparous quadrupeds, allied more or less nearly to crocodiles and monitors, but apparently inhăbitants of salt water only,—various vertebral fishes,—testacea of every description,—corolloid zoophytes,—encrinites, &c.

These formations, in England, occupy a zone having nearly thirty miles in average breadth, extending across the island from Yorkshire on the north-east to Dorsetshire on the south-west.

(*b*) *Foreign localities.* The British Islands present no traces of these formations beyond the limits above assigned, with the exception of the lias, which occurs in Ireland in the county of Antrim, near the south-east border of the basaltic district; in the Isle of Sky, and some other of the Hebrides.

In France, as we have before stated, some of the oolites of this series may be traced in the denuded tract surrounding Boulogne, particularly at Marquise; they may be seen also succeeding the chalk on the west on the coast of Normandy, beyond the mouth of the Seine, and between that point and the transition district of the Cotentin.

A recent examination of that coast by Mr. De la Beche, who has kindly communicated to us the general results of his observations, enables us to add to the notice already given (see the note on the account of the chalk cliffs of the French coast) the following more detailed and corrected particulars.

Along the mouth of the Seine, on both sides, the chalk and green sand repose on a blue marle and marle-stone. At Trouville sur Mer, the oolites of the upper and middle formation, i. e. the Portland stone and Coral rag, emerge from beneath this marle. Between Villers sur Mer and Dives, the clay separating the second and third system (the Oxford clay) forms the base of the cliffs, which are capped by the lower beds of the coral rag, and an overlying mass of green sand. Still further west by St. Comme, Arromarche, St. Honorine, Virreville, and Grandcamp, the cliffs present the inferior oolite resting on lias.

Hence, a zone of these formations extends, circling round the chalky and arenaceous border of the basin of Paris, by Caen, Alençon, Poitiers, Bourges, Auxerre, Bar le Duc, and Mezieres, the oolites stretching beyond all these places, succeeded at a wider distance by the subjacent lias;* and still further by the red marle, which, with a few occasional interventions of coal-measures, reposes on the transition and primitive chains of Bretagne and la Vendée on the west; of Limousin, Auvergne, Forez, Beaujolais, and Moroan on the south, and of the Vosges on the east.

Within the interior area of the basin of Paris, there is also a small denudation exposing the oolites, in a district called the Pays de Bray, a little north-west of Beauvais.

The extensive chains of the Jura mountains is principally composed of lias and the oolitic series. There is a very interesting description of the part of this chain near Lons de Saulnier by Mr. Charbant, in which it appears that its base is composed of red and variegated marle containing gypsum; that on this reposes an extensive formation of beds of gryphite limestone (lias), alternating with marles containing fossils exactly similar to those which characterise this formation in England; that these are covered by a series of oolitic beds, the lowest of which (like our inferior oolite) abounds in ferruginous particles, and is surmounted by a fine granular oolite and freestone, forming the escarpment of the first terrace of these hills, and lastly that, above this first terrace, are others each presenting an oolitic series based on argillaceous beds which separate it from that beneath. Nothing can possibly present closer analogies to the arrangement of these formations in our own island.

Professor Buckland has the following observations on the oolite and lias of the Alps.† The two principal varieties of the oolite or jura limestone, are: 1. A compact grey marble; 2. A granular oolite; the latter occurs abundantly in the Tyrol, in the valley of the Adige below Trent, and occasionally in the Salzburg mountains; the former prevails in Switzerland, and

* Mr. Omalius d'Halloy, or at least his translator in Thomson's Annals, calls all these formations Lias, including the oolites under that appellation.

† Annals of Philosophy, June 1821.

generally through the Alps; near Aigle, on the south-east of Vevey, it assumes the character of red compact marble similar to that of Salzburg; and at Roche, in the same neighbourhood, it is full of organic remains resembling those of the English coral rag; but from the compact nature of the matrix in which they are imbedded, these are visible only on the surface of the weathered blocks. This last observation may be applied also to a large proportion of the younger alpine limestone beds in the Tyrol and Salzburg, in which the organic remains are for the most part concealed by the extreme compactness of the stone; but, when apparent, are referable to the same classes with the oolite fossils of England. Such remains are distinctly visible at Nafels, near Glarus, in Switzerland, and at Halstad, in Salzburg; where also the limestone becomes partially oolitic.

The lias (like all the other formations in the Alps) is destitute of its alternating beds of clay, but maintains its position between the oolite and new red sandstone. At the salt mines of Bex, it reposes immediately on the upper bed of saliferous gypsum, where it is a dark-blue compact limestone, and contains ammonites, nautilites, terebratulites, and many bivalves identical with those of the lias of England.

At Halstad, it occupies a similar position between the oolite and red marly sandstone that covers the saliferous limestone, and is full of ammonites, belemnites, and other lias fossils. At Seefulden, near Inspruck, it contains fish similar to those which occur in the English lias at Lyme Regis. And at Mischelle, near Trent, it may be seen dividing the oolite from the red marle and new red sandstone.

In the central parts of Germany, the lias stretches from the Raue Alp (a continuation of the Jura chain) through Wurtembourg to Nurembourg, Gotha Wurtzbourg and Cobourg, occupying the greater part of the lower country between the mountains of the Black forest, Bergstrasse, and Vosges on the west, and the Bohemer Wald and Thuringer Wald. It does not appear that any decided oolites occur in this tract, but its details have never been fully examined. In the north of Germany, the lias may be traced surrounding the Hartz, and stretching on the north of the Erzgeberge, reposing on the red marle; it is described by Mr. Freisleben under the appellation of *muschel kalkstein*. Oolites here occur associated with it, but in no greater quantity.

In Russia, the lias certainly occurs near Moscow, but we have no distinct particulars concerning the secondary formations of that vast country.

The tracts of these formations in England, France, Germany, and Russia, may be considered as parts of a continuous series of deposits occupying the great central basin of Europe. That on the Italian side of the Alps, although necessarily mentioned in connection with that chain, belongs to a distinct and southern basin; of which a great part of Spain also probably forms a portion. Much of the limestone of that country may perhaps prove to be lias, and the oolites are distinctly described by Mr. Townshend as extending round Anchuela del Campo, about half way between Saragossa and Madrid.

No observations on these formations beyond the limits of Europe have yet been published, which can authorise us in attempting to identify them.

Section II.

Upper division of the Oolitic series, including 1. The Argillaceous Limestone beds of Purbeck; 2. Portland Oolite; 3. Kimmeridge clay.

The natural and geographical relations of these formations having induced us to refer them to a single section, we shall pursue the following order in describing them.

Each formation will be first separately treated of under all the usual heads, with the exception of those assigned to range and extent, height of hills, and inclination, which will be postponed to a general article placed at the end of the particular account of these three formations, as viewed separately, since in all these respects they are so intimately associated together, that confusion and needless repetition would be produced by any other mode of considering them.

As the particular description of the three formations will thus form paragraphs in the same section; the usual letters (*a*) to (*i*) will be employed to mark those devoted to the Purbeck beds; they will be doubled for the Portland (*aa*) to (*ii*), and trebled for the Kimmeridge clay (*aaa*) to (*iii*). (C.)

1. PURBECK BEDS.

(*a*) *Chemical and external characters.* The Purbeck beds, which occupy the highest place in this series, consist of many thin strata of argillaceous limestone, alternating with schistose marles, and forming an aggregate more than 300 feet in thickness. Mr. Webster thus describes them.

The Purbeck stone consists chiefly of shells (principally the Helix vivipara), partly whole, and partly in a state of comminution, imbedded in a calcareous cement, which is sometimes very pure and crystallized, and sometimes in a state approaching to an indurated marle. These beds are separated by others entirely without shells; and also by layers of shale and marle, the shivery nature of which allows the stone to be quarried with much ease. It is thus obtained of various degrees of thickness, according to that of the bed, and the whole hill consists of many alternations of these strata * (W. 172.)

The stone, well known by the name of Purbeck marble, and which was formerly much used in our gothic churches for columns and monuments, was nearly the uppermost of these beds; and differs from the common Purbeck stone, only in the purity of the calcareous matter, and in the shells being more entire. It agrees exactly, or very nearly so, with the Petworth marble; but is now out of use, and the quarries are filled up, and scarcely known. (W. 172.)

(*b*) *Mineral contents.* The only mineral contents of these beds appears to be the pyrites occurring in the marle, and the gypsum consequent to its decomposition. Gypsum, both striated and amorphous, in a bed of brownish or reddish clay, is worked in Durlestone bay on the coast of the Isle of Purbeck: its geological position is between the *Portland* and *Purbeck* beds. (Mr. Webster, G. Notes.) In the interstices of the limestone the strata about Peverel point, which forms the northern extremity of Durlestone bay, are found crystals of selenite in fibrous marle: the surface of this marle is covered with farinaceous gypsum, which also appears in an indurated state, forming alternate strata with limestone. Pyrites abounds

* The following Section of the *Purbeck strata* are by Mr. Middleton, (Monthly Mag. Dec. 1812, p. 395.)

1.	Various beds of stone brash, black shiver, and compact rock, in alternate layers These are supposed to rise and form the surface near Handcross, and in a line east and west of that place in Sussex, and also from Swanage to Durlstone bay, in Dorsetshire; at the latter place, I had a full view of them in the cliffs, and estimated the thickness of them at	80 feet
2.	*Leaning vein*, a bed of very penetrable stone, abounding with small shells raised and shipped at Swanage and Durlstone bay, to London, for flag pavement	7
3.	Stone not esteemed of any value, and black shiver in many beds	60
4.	*Freestone*, raised and shipped at Swanage and Durlstone bay, for London, in Purbeck squares	5
5.	Various beds of stone, in low estimation and black shiver..	20
8.	*New vein*, a bed of good and free working stone, raised and shipped at Durlstone bay for London	5
	As the strata rise toward Tillywhim quarry, we there meet with	
9.	Many beds of stone in low estimation, and black shiver lying under the *New vein*. These the quarry men call Rag-stone, and they are in thickness about	100

in the latter, and hence the sulphuric acid concerned in the formation of the selenite and gypsum seems to be obtained. (Maton. G. Notes.)

(*c*) *Organic remains.* Beautiful impressions of fish are frequently met with by the quarry men between the laminæ of the limestone, and abundance of fragments of bones, some of which belong to the turtle. Complete fossil turtles have also been found, and one extremely perfect. (W. p. 163.)

Mr. Johnson of Bristol* possesses a very perfect head of a crocodile found in Purbeck; but the character of the matrix is not quite decisive, as to whether it belongs to these or the Portland beds.

The shells of this formation have not been accurately examined; the most common is a small and elegant species of Vivipara, supposed to be a freshwater shell, but the subject requires further investigation.

(*d*) *Range and extent.* (*e*) *Height of hills.* These subjects will, for the reasons primarily stated, be treated of in a common article with reference to these beds—those of Portland and the subjacent Kimmeridge clay.

(*f*) *Thickness.* According to the measurements of Mr. Middleton (the only ones published), as given in the note to (*a*), the thickness of these beds in Purbeck is 291 feet; the higher numbers that have been sometimes given, arise from taking the Portland beds into the account.

(*g*) *Inclination.* The inclination of the strata of the three associated formations, will be most conveniently treated of together in the article at the end of this section.

(*h*) *Agricultural character.* A marly clay.

(*i*) *Water.* A retentive stratum; the wells must be sunk through it to the bottom of the Portland beds.

2. PORTLAND OOLITE.

(*aa*) *Chemical and external characters.* This formation consists of several beds of a coarse earthy limestone. The different beds often vary in their characters, nor are the same beds of an uniform texture in different localities. The calcareous rocks, through all the three divisions of the oolite series (where they do not, as in the case of the Purbeck beds and lias, assume new features from the admixture of argil) are not easily to be distinguished (excepting by the aid of their organic remains) from one another, though very readily dis-

* The remainder of this article is by the Rev. W. D. Conybeare.

tinguishable from those of other formations. The character of yellowish white calcareous freestone, generally mixed with a small quantity of siliceous sand, being common to all their varieties, and nearly all of them becoming occasionally oolitic.

The varieties to be observed in the Upper series, or that of Aylesbury and Portland, are, a fine grained white oolite, a loose granular limestone of earthy aspect and of various shades of yellowish grey; and more rarely a compact cretaceous limestone having a conchoidal fracture. In Wiltshire and Dorsetshire, many of the beds contain layers of chert alternating with them like the flints in the chalk formation; the lower beds are very sandy, and often very abundant in green particles resembling those of the green sand. Nodular concretions of calcareo-siliceous grit occur in the sandy beds in Oxfordshire. (C.)

The more oolitic varieties (principally quarried in the Isles of Purbeck and Portland) afford the great part of the stone used for architectural purposes in the vicinity of London.

Mr. Webster gives the following account of the Portland quarries. The uppermost beds of the isle of Portland consist of an oolitic rock, and they are numerous. That which appears on the summit, and is called the Cap, is of a yellowish colour, and porcellanous character; it is only burnt for lime. The next bed is worked for sale, being the best building stone. Those below this bed contain numerous casts of shells that injure the stone, which is only used for coarser purposes; and with these beds alternate others consisting of chert. (W. p. 197.) *

* A more detailed account of the several quarries in the Isle of Portland may be seen in the Monthly Magazine for Jan. 1813, p. 481, whence the two following Sections are copied.

Section of Waycroft and other Quarries on the East side of Portland.

	feet.
1. Vegetable mould, less than	1
2. *Stone brash*, a cream-coloured limestone	3
3. Parting of ditto and black-blue clay	1
4. *Cap*, a cream-coloured stone in three layers, with partings of clay, and so hard as to turn the steel points of chisels and pick-axes	10
5. The *White bed*, or the highest layer of marketable stone: two feet off the top of this stratum is egg-shaped	5
Parting, abounding with grey flint, of no use	2
6. The *Middle bed* of marketable stone, containing few marine impressions, and occasionally either in one or two beds	5
Parting stone, containing many shells, of no value	2
7. The third bed with few or no shells; this is the best stone, and it varies in thickness in the several quarries from	7 to 14
	43

(*bb*) *Mineral contents.* The beds at Portland and Tisbury contain beautiful yellow sulphate of barytes (sugar candy stone) and crystallized calcareous spar. (G. Notes.)

In the next quarry, the third bed (7) is in two layers; the lower of these is free from shells, except about one foot of its top, and the upper one contains very few shells.

We visited four other quarries on this side of the island; but as they do not differ materially from the above, it would be improper to repeat my observations.

Under the foregoing marketable stone, many layers of flint and unserviceable stone, to the depth of about 50 feet.

On the east side of Portland, the whole thickness of the stony strata, is about 93 feet, and beneath that is black blue shiver of great thickness.

Sections of Quarries on the West side of Portland.

In the first quarry we visited, I found the vegetable mould, the stone brash, and the beds of hard stone cap, similar to what they are on the east side of the island. But in this place the marketable stone is about 11½ feet thick, in one bed; whereof about 2½ feet of the top is egg-shaped.

In the adjoining quarry immediately under the Cap (4) is Roach (a mass of fragments of oyster shells compressed and cemented in a very hard stone) six feet thick, upon a bed of the best saleable stone nine feet thick.

The third quarry is similar to the two former, down to the Roach, which in this case is five feet thick, and united to the best stone without any parting. In this quarry the Roach was cut into blocks and sent to Millbank, Westminster, for building the abutment of Vauxhall bridge. And many others of these blocks were prepared for the same purpose, but they remain in the quarry at this time. (Aug 1812.)

In the fourth quarry the Roach is in the same state as in the third quarry, but reduced to three feet thick, and at this place the best stone is 11 feet.

		feet.
	In Gosling's quarry, the stone brash and two beds of the Cap, are increased in thickness to	20
	Roach in one bed 4 feet, and two other feet of it are united to the top of the *White bed*, together	6
5.	*White bed*, a marketable stone, exclusive of the two feet of Roach which are united to it	8
	Many layers of flint and stony rubbish, including one bed of a white tender stone, which is not fit for great exposure	6
6.	Two other beds of Roach in the place of the middle bed of saleable stone	6
7.	The *Third bed* (7) of saleable stone, contrary to the usual order, is not near such good stone as the White bed (5) in this quarry	6
	The above marketable and other stone lies upon many layers of flints, and beds of unserviceable stone to the depth of about	55 or 60
	The whole of the stony strata in this place is about	112

And under that is black blue shiver several hundred feet thick, whereof there are about 100 feet above the level of the sea.

Middleton, Mo. Mag. Jan. 1813, p. 481.

(*cc*)· *Organic remains.** (The references to figures where not otherwise specified, are to Sowerby's Mineral conchology.)

* By the Rev. W. D. Conybeare.

A section of a more descriptive character, drawn up by Miss Bennett from the quarries excavated in this formation at Chicksgrove, in the vale of Tisbury, Wilts, is printed in Sowerby's Mineral Conchology, vol 2. p 58; we here subjoin it, presuming that the beds here called chalk are the cretaceous limestones mentioned in the text.

A Section of Chicksgrove Quarry, South of the village and of the river, in the parish of Tisbury, in Wiltshire.

1. Top of quarry. Rubble, fourteen feet. No shells in this bed. (Impure chalk.)
2. Stone not good, two feet. The lower part of this bed contains the same shells as the chalk below it.
3. Chalk, two feet. Trigonias three species; Pectens like those of Thame, Oxfordshire; Ostrea several species; a thick equivalve; bivalve which is common in the rubble beds of freestone; a small bivalve, perhaps Unio; two other small bivalves, and a Trochus like those of the flinty chalk. (Hard chalk.)
4. Flint, four inches. (Approaching chert.)
5. Chalk, eleven feet. A rubbly chalk without shells. (Hard chalk.)
6. Spangle bed, five feet six inches. Contains Ammonites, Oysters, and various other shells changed into the spar. (Limestone containing some white, but no green sand.)
7. Walling Rag, two feet six inches. Fragments of shells changed into spar. (Like No. 6, only coarser and harder.)
8. Devil's bed, two feet. Fragments of shells changed into spar, smaller shells than the Walling Rag. (Like No. 6.)
9. Great Rag, three feet. No shells, or only small fragments. (A compact sandy limestone, with minute grains of green sand.)
10. Brown bed, three feet. Contains Ammonites. (ess compact than the last, with more green sand, some parts of a loose texture.)
11. Trough stone, three feet four inches Trigonias, the shell changed into spar, and Ammonites. (Similar to some parts of the last.)
12. White bed, two feet eight inches. Contains Ammonites. (Between 10 and 13)
13. Hard bed, three feet six inches. Trigonias, the shell changed into spar, and Ammonites. This bed is very like No. 11. (Rather less green sand than No. 10.)
14. Fretting stone, two feet. A soft stone and no shells. (A loose sandy limestone with green sand)
15. Under bed, two feet. Fragments of shells changed into spar. (More compact and finer grained than the last, and holding less green sand.)
16. Under bed, two feet six inches. Contains Trigonias, the cast of the outside of the shell a soft stone. (Like the last, except that it contains no spar.)

The whole depth of Chicksgrove Quarry to the bottom of the stone is 61 feet four inches, measured by John Montague, foreman of the quarry.

The scales of fish, erroneously supposed to have been found in this

We believe that remains of fish are occasionally met with in this formation as well as in the preceding.

The shells which occur are principally the following:

CHAMBERED UNIVALVES.

Ammonites triplicatus, *T.* 92, fig. 2.
gigauteus. *T.* 126.
Lamberti. *T.* 242.
Nutfieldiensis. *T.* 107.

UNIVALVES NOT CHAMBERED.

Turritella; Smith, fig. 2.
Natica; Smith, fig. 1.
Solarium Conoideum. *T.* 11. m.
Trochus.
Ostrea expansa. *T.* 238.
Crenatula; Parkinson, plate 15. fig. 5.
Pecten lamellosus. *T.* 239.
Trigonia clavellata, *T.* 87. u.
gibbosa. *T.* 235. 236.
Astarte cuneata. *T.* 137, fig. 2.
Lutraria ovalis. *T.* 226.
Nerita sinuosa. *T.* 217, fig. 2.
Unio.
Cardita.
Cyclas; Smith, fig. 3.
Venus; Smith, fig. 5.

Of these shells, the Ammonites triplicatus and Pecten lamellosus are most characteristic. In the section of Chicksgrove quarry (see page 175), other particulars concerning the shells of this formation will be found.

No other Zoophytal remains are mentioned than those of a beautiful aggregated madrepore, specimens of which, imbedded in a semitransparent chert, occur at Tisbury in Wilts.

Large fragments of wood are common.

quarry, were from a tile-stone quarry on Lady-Down, in the parish of Tisbury, and about one mile north-west from Chicksgrove quarry.

The above are the names by which the different beds are known by the people who work the quarry.

Most of the stone contains calcareous spar, in the place of the fragments of shells dispersed through it, but No. 14 and 16 are without it; the spangle bed contains most. The rare stratum called by geologists 'White freestone,' and here called chalk, but from which it differs in its situation, occurs also at Brill in Buckinghamshire, and at Upway in Dorsetshire.

The sections presented at Brill hill in Buckinghamshire exhibit a remarkable coincidence with the above.

(*dd*) *Range and extent.* (*ee*) *Height of hills.* Referred to the general article at the end of this section.

(*ff*) *Thickness.* According to Mr. Middleton's measurements, these beds, in the Isle of Purbeck, exhibit altogether a thickness of about 120 feet; and this appears the estimate on which the greatest reliance may be placed.

(*gg*) *Inclination.* Referred to the general article at the end of this section.

(*hh*) *Agricultural character.* A poor stone-brash soil.

(*ii*) *Water.* The water issues abundantly from the bottom of this rock, where it is thrown out by the subjacent Kimmeridge clay.

3. KIMMERIDGE CLAY.

(*aaa*) *Chemical and external characters.* These beds consist of a blue slaty or greyish yellow clay (the *Oaktree clay* of Smith,) containing selenite; but it sometimes contains beds of highly bituminous shale, as near Kimmeridge on the coast of the Isle of Purbeck, where these are used as fuel, whence they have obtained the name of *Kimmeridge coal*; and hence the name of Kimmeridge clay as applied to the whole formation. The beds are most instructively displayed (G. Map) near that place. They are also finely exposed (G. Notes) on the coast of the Isle of Portland.

On the east of little Kimmeridge, where the cliffs are abrupt, they are composed of a slate-clay of a greyish yellow colour, finely slaty, containing both animal and vegetable impressions. The plates of which the rock is composed, become much more evident after it has undergone some decay; or, when sound, after it has been exposed to the fire. It divides spontaneously into large tabular masses. The fracture of the rock is earthy, with many small specks and nodules of indurated clay. The outside of the rock is covered by a thin layer of calcareous spar. The mass effervesces with acids, but the nodules of indurated clay do not. This rock passes gradually into a bituminous shale; but the first transition is into a slate-clay of a lighter or darker colour, the joints of which are covered with iron pyrites. It burns with a yellowish flame, giving out a sulphureous smell, and becomes afterwards of a light grey colour. The second transition is to a bituminous shale called *Stony Coal* [*Kimmeridge Coal* *] the specific

* Near Smedmore in the parish of Great Kimmeridge, is found what the country-people call '*Coal-money*', generally on the top of the cliffs, two or three feet below the surface, enclosed between two stones set edgeways

gravity of which is 1.319. Its colour is dark brown without any lustre: it effervesces slightly with acids, contains no iron pyrites, and burns readily with a yellowish rather smoky and heavy flame. The smell is bituminous but not sulphureous. The Kimmeridge coal, however, rests upon, and is covered by, the slate-clay first described. (G. T. vol. i. p. 263.)

Several useless attempts have been made for coal, by sinking into the stratum, as at Sunning Well near Oxford; on the edge of Bagley wood near Farringdon in Berkshire: near Whiting's farm, between Shaftesbury and Hargrove in Dorsetshire they sunk by subscription upwards of 100 feet in search of coal through shale and clay. (G. Notes.)

It has sometimes been erroneously supposed that these beds were the same with the aluminous shale of Whitby in Yorkshire (G. Notes), which belong to the lias beds.

(*bbb*) *Mineral contents.* These beds furnish the selenites of Oxfordshire, which are daily formed by the action of decomposing pyrites upon the oyster shells and other fossils contained in them. (G. Notes.)

Alum was formerly manufactured in the parish of Kimmeridge. (Wilson's Mountains. G. Notes.)

(*ccc*) *Organic remains.** The most interesting remains contained in this stratum are doubtless those of the extinct genera allied to the order Lacerta, but evidently calculated for a marine abode; the vertebræ, paddles, &c. of a species of Icthyosaurus differing from those in the lias; the vertebræ, phalanges, and head of another Saurian animal, perhaps a variety of Plesiosaurus, have been found at Kimmeridge and Headington; bones, apparently of Cetacea, likewise occur.

The shells are given according to Smith and Sowerby.

CHAMBERED UNIVALVES.

Nautilus.
Ammonites, five species; Smith, fig. 7.
Belemnites.

and covered with a third, together with the bones of some animal. They are from two to three and a half inches in diameter, and a quarter of an inch thick, round, on one side flat and plain, on the other convex with mouldings. On the flat side are two, sometimes four, small round holes, perhaps the centre holes by which they were fixed to the turning-press: they are supposed to have been either amulets or money. In support of the latter opinion, it may be observed, that 'down with your coal', is in some counties a cant expression for pay your money.' (G. Notes.) There has also been found in the neighbourhood a shallow bowl of Kimmeridge coal, six inches high and as many in diameter, containing coal-money. (Hutchings Hist. Dorset. G. Notes.)

* The remainder of this article is by the Rev. W. D. Conybeare.

UNIVALVES NOT CHAMBERED.

Trochus; Smith, fig. 3.
Turbo; Smith, fig. 2.
Melania Headingtoniensis. *T.* 39.

BIVALVES.

Ostrea deltoidea. *T.* 148.
crista galli, and three other species.
Astarte lineata. *T.* 179.
ovata; Smith.
Trigonia costata. *T.*
clavellata. *T.*
Venus; Smith, fig. 8.
Modiola.
Cardita.
Cardium.
Mactra.
Tellina.
Chama, two species; Smith, fig. 2.
Avicula.
Pecten.
Terebratula; Smith, fig. 9.

Serpulæ.

Of these shells, the Ostrea deltoidea appears to be the most characteristic.

(*ddd*) *Range and extent.* (*eee*) *Height of hills.* Referred to the general article at the end.

(*fff*) *Thickness.* Mr. Middleton assigns upwards of 700 feet as the thickness of this formation in the Isle of Purbeck, but this seems excessive. Mr. Buckland, in his 'Order of Superposition', gives only 600. Near Oxford, where the beds thin off, the thickness cannot exceed 100 feet. In the pit at Sunningwell, on the north edge of Bagley wood, it was only 70 feet.

(*ggg*) *Inclination.* Referred to the general article.

(*hhh*) *Agricultural character.* A tenaceous clay often covered with oak woods; whence, in Wiltshire, this formation has been called the Oaktree clay.

(*iii*) *Water.* 'The water,' says Mr. Smith, 'in this formation, is deficient and of bad quality"; in fact, from its retentive nature, it must generally be pierced before any copions supply can be found.

General account of Range and Extent, Height of Hills and Inclination of Strata of the Purbeck Beds, Portland Beds, and Kimmeridge Clay. *

(*d*) *Range and extent of the three formations above described.* None of these formations have as yet been observed to the north of Buckinghamshire, where the Portland beds first make their appearance underlying the iron sand, (whose course through that county has been traced in a preceding article,) and forming a constituent part of the same chain of hills. It may thus be traced from a mile north of Stukeley, a village about five miles on the west of Leighton Buzzard, Bedfordshire, to the hills between Winslow and Aylesbury, and dipping thence to the south, it underlies a great part of the vale of Aylesbury. Further west, on the borders of Oxon and Bucks, it rises from Thame, and culminates on the insulated group of Brill hills, where it has only a very thin covering of iron sand. The section here displayed corresponds very precisely with that of Chicksgrove as given in the note to page 173. The quantity of green earth in the lowest sandy strata is here very remarkable. Pursuing the same direction through Oxfordshire, to the south-west, it may be traced through the parishes of Milton and Hazely on the south-east of the Thame; and rising thence to the north-west, culminates in Shotover, circling round the more elevated ridge of that hill, immediately beneath the iron sand and ochre which form its summit: it is here however quarried in the village of Garsington only. It exhibits, round the slope of this ridge, inferior sandy beds, some of which are very full of particles of green earth; while the lowest contain those large nodules of calcareo-siliceous grit, which from their size and grotesque appearance will attract the notice of every one who ascends Shotover in following the old London road.

At the foot of the high ridge of Shotover, the Kimmeridge clay makes (as far as has been traced) its first appearance in the village of Headington, where it is seen in many of the quarries overlying the oolites of the next division. In its previous course through Bucks, the Portland beds seem to rest immediately on the Oxford clay (both the Kimmeridge clay and the coral rag being wanting)

On the Berkshire side of the valley of Oxford, the Kimmeridge clay may be traced on the hills answering to Shotover, in the pit sunk by Sir George Bowyer in a fruitless search for coal near the northern edge of Bagley wood; but the strata

* By the Rev. W. D Conybeare.

thin off in this direction, the Portland beds are no longer to be seen, and the clay itself soon terminates, the iron sand overlying its outgoing and coming into contact with the subjacent coral rag belonging to the next division.

Immediately on the south of Abingdon the Kimmeridge clay pursues its course westwards through the vale of Berkshire, following the line of the Berks and Wilts canal as far as Wotton Basset, where it turns more to the south, and ranges on the east of Calne and of Seend,* reposing all the way on the inner edge of the coral rag. Near Seend, the chalk and subjacent green sand overlie and conceal it, advancing in a sort of projecting cape even into the district of the great oolite. In the interval above mentioned (from Abingdon to Seend) the Portland beds, reposing on the Kimmeridge clay, are only seen at one point; namely, at Swindon, where they are extensively quarried: elsewhere they are either concealed by the alluvial debris scattered over this valley, or have their basset edge overlaid and hidden by the superior strata.

On the south of the projecting cape of chalk and green sand before mentioned, which extends from Seend to Stourhead, a wide valley, opening in the ranges of these formations between Stourhead and Shaftesbury, again exposes the subjacent strata. This denudation, which extends on the east nearly to Wilton (being traversed by the river Nadder and occupying the vale of that name, also known by that of the Vale of Wardour) is skirted by an escarpment of green sand, ranging from Mere by Chilmark towards Wilton on the north, and along the right bank of the Nadder on the south. The iron sand being here wanting, the green sand reposes towards the south of Chilmark, on a zone of argilleo-calcareous beds belonging to the Purbeck series (which we here notice for the first time) ranging thence to Lady down. The thin slabs of this formation are here raised for tile stones. Within this zone, the subjacent strata of the Portland series occupy the interior of the denudation, being extensively quarried at Fonthill, Tisbury, and Chicksgrove. On the west of Fonthill (along what may be termed the mouth of this denudation) a clay appears beneath these beds (probably that of Kimmeridge, although coloured in the Geological Map as the Oxford clay.)

Much disturbance seems to have been experienced by the strata within this denudation; since, although nearly horizontal at Chicksgrove, yet at Chilmark, Fonthill, and Tisbury, they are inclined nearly 40 degrees, dipping towards the north and east.

* Mr. Greenough's Map represents its course accurately as far as Calne, but errs in terminating it abruptly at that point.

Through the north of Dorsetshire, the continuation of these formations is concealed by the projection westwards of the vast overlying platforms of chalk and green sand extending over their basset edges. Beyond the escarpment of the chalk towards the southern coast of that county, however, we again meet with them under the following circumstances.

1st. Near Upway, on the road from Dorchester to Weymouth, a zone of the Purbeck beds may be observed for some distance immediately beneath the escarpment of the chalk; they are here however very imperfectly exhibited; but they may, nevertheless, be obscurely traced to the point where the chalk hills meet the coast, six miles east of Weymouth; where we shall presently return to them.

The country intermediate between Upway and Weymouth exhibits what is geologically termed a Saddle of the two inferior division of oolites; presenting in succession, on either side of a central point between these two places, the coral rag, Oxford clay, and forest marble, dipping on either side from this central point.

On the southern flank of this saddle, close to the passage from Weymouth to Portland, the Kimmeridge clay may be seen resting against the coral rag.

The Kimmeridge clay also forms the substratum of the whole Isle of Portland, and rises high on its northern face, where it is capped by an abrupt escarpment of the superior oolitic beds. All the strata sensibly decline, though not under a very rapid angle, to the south; thus giving the profile of the island, as seen either from the east or west, that appearance of an insulated inclined plane which at once distinguishes it from the other headlands. This declination brings the line of junction between the Kimmeridge clay and oolite to the level of the sea, near the south extremity of the island, which is formed by low calcareous cliffs, worked by the action of the sea into numerous caverns, some of which communicate with funnel-shaped craters on the surface of the island, through which the waves may be seen boiling within the narrow limits in which they are pent. All the coasts of this island are steep, the base of Kimmeridge clay forming a sloping talus surmounted by crags of the oolite, scarred by numerous quarries. Near Bow and Arrow castle is a remarkable rocky defile on the top of the cliffs, imitating in miniature one of the Derbyshire dales.

2. We have now (returning to the coast where the chalk hills meet it east of Weymouth) to trace the formations which constitute the subject of this section through the Isle of Purbeck. The inclined position of the strata occupying the district thus denominated, and the deeply excavated caves which

form so remarkable a feature near its western extremity, have been already described. (see the Chapter on Chalk, p. 110, and that on Green sand and Iron sand p. 160). We there observed that the Portland beds, dipping inland in an angle of from 45 to 60 degrees, formed the exterior barriers and capes at the mouth of th e coves, while the vertical strata of the lofty chalk downs ranged along their bottom; their sides exhibiting in section all the intermediate formations. The represensation beneath will convey a sufficiently accurate idea of the

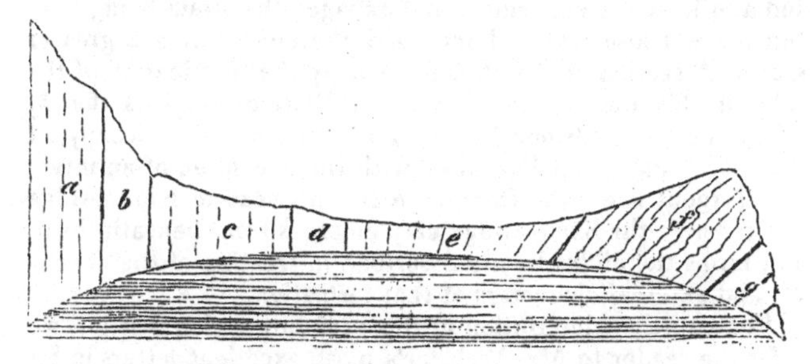

a. Chalk. *b.* Chalk marle. *c.* Green sand. *d.* Blue marle.
e. Iron sand. *f.* Purbeck beds. *g.* Portland beds.

appearances which prevail in all of them. The more solid masses of the Portland rock, having resisted most strongly the action of the destroying causes which appear to have excavated these coves, often form a reef of rocks in front of them; in one instance a projecting crag of it has been worn into a remarkable arch through which a boat can pass; this is called the Barn-door.

While the convulsions which have here acted on the strata, have raised the solid beds of Portland rock in mass, in regular inclined planes, the softer superincumbent argillaceous beds have been bent by the lateral pressure (to which they have more readily yielded) into many singular contortions, exactly similar to those which have formed the subject of so many observations and so much theory in the transition slate districts. It may indeed be stated as a general fact, that when in a series of inclined strata, solid, thick, and compact beds of stone are found associated with thin argillaceous and yielding beds, the former, however elevated, usually pres nt regular planes of stratification; while the latter are bent and twisted into the most irregular curves. It is scarcely possible to conceive a stronger proof that this disposition is not the result of original formation, but of subsequent mechanical violence.

Lulworth Cove is the most remarkable of these singular bays, the names of the others will be found in the pages before referred to.

From the east of Worthbarrow, the last of these coves in that direction, the Purbeck and Portland beds form a ridge of hills extending longitudinally from west to east across the Isle of Portland, ranging parallel to the vertical chalk ridge, (from which they are separated by a valley occupied by the intermediate sands), about a quarter of a mile broad at the west, and a mile at the east end near Swanage; the strata being here thicker and also less inclined, and therefore filling a greater space. A section of the cliffs formed by the termination of the ridge in this direction, is seen in the Plate of Sections, fig. 5, extending to Durlstone head. The Portland beds occupy all the southern coast of Purbeck, with the exception of an interval of about five miles from St. Adhelm's head to Kimmeridge bay; where the Kimmeridge clay emerges from beneath them, and forms the cliffs already described in treating of the chemical and external characters of that formation.

For a further description of this very interesting district, we refer the reader to Mr. Webster's most excellent letters in Sir Henry Englefield's Description of the Isle of Wight. The personal observations of the present writer may serve as a further confirmation of their perfect accuracy.

The argillaceo-calcareous beds underlying the iron sand at Sandown bay in the Isle of Wight, mentioned before in page 157, and those underlying the iron sand of the Weald in Sussex (see page 148), must probably be referred to a part of the Purbeck series.

(*e*) *Height of hills.* Brill hill on the borders of Buckinghamshire and Oxfordshire, exhibits probably the highest point attained by any of these formations; the Portland beds are there near the summit, which has only a very thin cap of iron sand. In Shotover hill, where the cap of iron sand is thicker, they range about 50 feet below the summit, which is 559 feet above the sea. In Berkshire and North Wilts, they are confined to low ground; in the denudation of the valley of the Nadder, however, the surface is raised, and the Purbeck beds there acquire some elevation at Lady Down. Near Upway in Dorsetshire, they are near the foot of the chalk escarpment, but still at some height above the sea. The height of the Isle of Portland is about 300 feet; and that of the ridge of Purbeck and Portland stone, traversing the Isle of Thanet, about the same.

(*i*) *Inclination.* Through Buckinghamshire, Oxfordshire, Berkshire, and North Wilts, these beds are nearly horizontal, dipping in an almost inappreciable angle south-south-east. In the denudation of the vale of Nadder, as has been observed, they are elevated sometimes to an angle of 45°; at Upway and Portland they decline in opposite directions, forming a portion of what may be termed the Weymouth saddle (see p. 192); and in Purbeck they form a part of the system of highly inclined strata constituting that district.

The remarkable contortions of the Purbeck beds have been mentioned in the preceding article: they are accurately represented in the plates of Sir H. Englefield's Isle of Wight.

Although these formations do not, when examined in any particular spot, exhibit any want of conformity with one another, it is evident that they cannot be considered as strictly conformable throughout; since in Buckinghamshire we find the Kimmeridge clay absent, and the Portland beds reposing on the Oxford clay. ‡ The superior strata of the sands and chalk exhibit a like want of conformity with regard to these, since they overlie their basset edges completely in the north-eastern counties, and in several instances in the south-western.

Section III.

Middle division of Oolites.

Synonymes.—The Superior Oolite, Oxford Oolite, or Pisolite of other authors, is subordinate to this formation; as are also the Calcareous grit and sandy strata which form the lowest beds.

I. CORAL RAG.*

(*a*) *Chemical and external characters.*† This formation comprises a series of beds occupying a thickness of from one to two hundred feet; in the upper part of which, containing the oolite, the calcareous matter; and in the lower, the siliceous prevails. The coral rag (properly so called), which as character-

‡ Mr. Smith has inaccurately represented the Kimmeridge clay (his *Oak-tree clay*) as here intervening, since the gryphæa dilatata and other characteristic shells of the Oxford clay may be traced up to the very junction with the Aylesbury lime at Waddesdon hill, Bucks.

* By the Rev. W. D. Conybeare.

† We have confined ourselves to the characters of this formation in the Midland and South-western counties, entertaining doubts as to the propriety of referring to it the Yorkshire district so allotted in Mr. Greenough's Map.

ising, gives name to the whole formation, occurs principally towards the middle of the series. Such at least is the disposition of the beds near Oxford, where they have been most attentively examined; and in Wiltshire (where the confused account of Mr. Townshend indicates the reverse), it has been ascertained that the same order prevails.

The general relations of the fossils in these beds, and their association, as constituting a single and well marked range of hills between the vallies of the Kimmeridge and Oxford clay, are sufficient to indicate the propriety of considering them as a single formation.

We proceed briefly to state the character of the various constituent beds of this formation.

The Upper calcareous beds are;—a calcareous freestone of tolerably close texture, full of shells comminuted into fragments generally too small to ascertain their species; more or less oolitic, frequently very indistinctly so, but occasionally passing into beds in which the oviform grains are much larger than in any of the other oolites. This variety, which though by no means universal, is yet, where it occurs, characteristic of the formation, has given rise to the name *Pisolite* employed by Mr. Smith. The colour of all these beds is of a yellowish white, becoming palest in the most oolitic, and passing occasionally into shades of light grey. It rises in thick beds often traversed by lines of division oblique to the plane of stratification; which hence, unless where the stone is exposed to a considerable depth, becomes obscure. It affords a tolerable material for building, but far inferior to the oolites of Portland already described, and to those which will be hereafter described as belonging to the third division, scaling off in large flakes after a few years exposure to the weather. Oxford has to regret its vicinity to this formation. It is a tolerably good limestone, but probably contains from one-tenth to one-third of sand.

The coral rag, which lies under this freestone, consists, as its name denotes, of a loose rubbly limestone, mingled with and often almost entirely made up of a congeries of several species of aggregated and branching madrepores. Two or three irregular courses of this rock intervene between the freestones and the inferior sandy beds; they often assume a marly character and grey colour; they are used for lime and the repair of roads.

The sandy, or rather siliceo-calcareous beds, consist of a thick deposit of yellow coloured quartzose sand, usually containing about one-third of calcareous matter, and traversed by irregular strata and concretions of indurated calcareo-siliceous gritstone. These rest immediately on the subjacent Oxford clay, and may be traced through the whole escarpment of the hills composed of this formation supporting the rag and freestones. It is in the calcareous grit of this sand that the fossils of this formation occur in the greatest quantity, and especially in those immediately beneath the coral rag beds. The ostrea gregarea characterises the sand generally.

Iron is more or less abundant throughout the sand; sometimes so much so as to give it the appearance of the iron sand described in the former chapter.

In the vicinity of Weymouth, the beds, at the junction of the Kimmeridge clay and the freestones of this formation, are also very sandy and ferruginous.

(*b*) *Mineral contents.* This formation affords scarcely any thing which deserves notice under this head; calcareous spar is of course common, but does not present any beautiful varieties. Crystals of quartz have also been found, though very rarely, in the Headington quarries near Oxford.

(*e*) *Organic remains.** The remains of Vertebral animals are scarce in this formation; but vertebræ of the Ichthyosaurus have been found in its beds of calcareous grit.

The shells are numerous, especially in the beds of calcareous grit, and chiefly of the following species; for figures of which we may refer to Sowerby's Mineral Conchology.

CHAMBERED UNIVALVES.

Ammonites excavatus, *T.* 105. *A.* giganteus, *T.* 126. *A.* plicatilis, *T.* 166. *A.* vertebralis, *T.* 165. *A.* splendens, *T.* 103, fig. 3.
Nautilus.
Belemnites.

UNIVALVES NOT CHAMBERED.

Melania Headingtoniensis, *T.* 39. *M.* striata, *T.* 47.
Turbo muricata. *T.* 240. fig. 4.
Helix.
Trochus bicoronatus. *T.* 221, fig. 2.
Ampullaria; Smith, fig. 2.
Turritella?

Serpulites.

---

* The plates given by Townshend of the fossils of this formation require the following corrections. The Echinites copied from Plott's Oxfordshire, plate 5, fig. 5—9, are not from the coral rag, but from the great oolite; there are however species nearly similar in the coral rag. In Plate 6, figs. 1, 2 & 12, are from other formations. With these exceptions they may be consulted usefully. Smith's catalogue and figures are both good.

BIVALVES.

Ostrea gregarea, *T.* 111, fig. 3. *O.* crista galli, Smith, fig. 4.
Pecten fibrosus, *T.* 136, fig. 2. *P.* lens, *T.* 205, fig. 2. 3. *P.* arcuatus, *T.* 205, fig. 7. *P.* similis, *T.* 205, fig. 6.
Chama, same as in the Oaktree clay.
Trigonia, casts of several species.
Lima rudis. *T.* 214, fig. 1.
Lithophaga.
Mytilus.
Modiola.

Fragments of a fibrous shell are common, but not sufficiently perfect to ascertain whether they belong to the same genus with the Inoceramus of the chalk formation.

Many beautiful *Echinites* occur in this formation; viz. of the division *Cidaris*, three species; 1. *C.* papillata, much resembling that of the chalk, (Parkinson, vol. 3, pl. 1, fig. 9.); 2. *C.* intermedia (same plate, fig. 6.); 3. *C.* diadema (same plate, fig. 4.): and of the genus *Clypeus*, two species; viz. 1. Clypeus sinuatus (the same which occurs in the Great and Inferior oolite) pl. 2, fig. 1; and 2. the Clypeus clunicularis (a small oval variety) see Smith's plate of the fossils of this formation, fig. 6., where it is rightly restored to this genus, having been confounded by other writers with genera to which it did not belong. Here we may observe that a new genus, the Clypeus, makes its first appearance, and an old one formerly noticed, the Spatangus, is no longer found, not being known in any formation below the green sand.

This also is the first formation (in descending the series) in which in this country * any considerable number or variety of madrepores occur in a fossil state; the species occurring in the formations we have before described being few and scarce. We here notice several species of the divisions *Caryophillia* and *Astrea*, following the Lamarckian arrangement of this family, viz. 1. a Caryophyllia approaching to *C.* Carduus, but not muricated (Parkinson, vol. 2, pl. 5, fig. 5.); Caryophyllia cespitosa? a smaller branching madrepore clustering in groupes; and of the division Astrea one species, approaching to Astrea favosa (Smith's plate of coral rag fossils, fig. 1.); a second (Parkinson, pl. 7, fig. 11.), and a third approaching to Astrea annularis.

It must be stated, however, that the subject of fossil madre-

* The Maestricht beds, which repose immediately on the chalk in the Netherlands, are however rich in madrepores.

pores has never been treated in this country with the accuracy it requires; the superficial and external characters of this species have alone attracted attention, their internal structure and mode of increase have been scarcely noticed : and we have consequently little or no real and precise knowledge concerning them.

Fossil wood is often found in the calcareous grit of this formation.

(*d*) *Range and extent.* Considering it as yet doubtful whether the tract on the edge of the eastern moorlands of Yorkshire, coloured in the Geological Map of Mr. Greenough as belonging to this formation, does not rather belong to the great oolite, we shall add the few notes we possess concerning it as an appendix to the whole suite of oolites, without venturing decidedly to assign its geological place among them.

Passing then over this tract, we certainly do not find any traces of the coral rag formation in proceeding to the south-west, until we come to the middle of the island.

Its earliest appearance in this direction is on the east of Oxford, where it forms the elevated platform rising on the south-west of Otmoor, and occupying the interval between the confluence of the Charwell and Thame with the Isis. This platform supports the still higher ridge which (exhibiting the Portland beds and iron sand) constitutes the summit of Shotover hill. The whole of its surface, which extends about five miles from east to west, and seven from north to south, is covered with quarries of which the principal are those of Headington two miles east of Oxford, at the foot of the high ridge of Shotover, in which the junction of the beds of this formation and the Kimmeridge clay, which lies above them, is well displayed.*

As the strata here dip nearly to the south, the platform occupied by them subsides in the same direction, Beckley on the north being its most lofty point, whence it sinks towards Sandford, being there very little above the level of the Isis. Near this last point, the sandy beds occupy its escarpment, the whole way, the basset of the coral rag crosses the river Isis, and the plane of the formation, rising northwards, occupies the middle of the range of hills on the west of that river,

* The upper surface of the freestone beds was in the year 1812 laid open in one of the quarries to a considerable extent, by stripping off the superjacent clay; it had the appearance of having been marked by the action of water upon it before the deposition of the clay, and presented occasionally small round cup-shaped cavities which seemed to have been worn into it the stony strata were traversed also by many perpendicular rents of various breadth, into which the clay had insinuated itself.

crowned with Bagley wood. This ridge corresponds with the opposite platform, and must have been originally continuous with it, the intervening valley of the Isis near Oxford appearing to be a breach in the chain of hills opened by subsequent denudation.

The Bagley wood hills are, like Shotover, crowned with iron sand, which constitutes the summits of Foxcombe hill and Cumnor clump; but the Portland beds do not extend so far north, although the Kimmeridge clay may be traced along the base of the iron sand of Foxcombe hill, interposed between it and the coral rag. This however also thins out before it gains Cumnor hill, in which the iron sand rests immediately on the coral rag. The same thing also happens on the opposite side near Shotover hill, where the clay likewise thins out beneath Forest hill, an insulated summit of iron sand just on the north of Shotover.

North of Cumnor hill is Whiteham hill, an insulated mass of the coral rag and its subjacent sand, hanging over Ensham, and constituting what is called an outlier. This is the highest point of the coral rag.*

From Cumnor hill the platform of the coral rag, extending westwards, forms a range of hills about 200 feet high (skirting the north of Berkshire) between the rivers Isis and Ock. The escarpment of these hills, which exhibits the inferior sandy beds, is towards the former; the gradual slope on the back of the strata, towards the latter. The superficial breadth of the coral rag is here about four miles.

At Farringdon are two summits of iron sand, resting on the coral rag, one on the east of the town marked by a conspicuous clump of firs, the other on the south-west.

* The most interesting circumstance with regard to these hills remains to be mentioned. A large accumulation of alluvial pebbles and blocks, often of considerable size, and derived apparently from some transition district on the one hand and from chalk on the other (comprising quartz, sandstone like that of the Lickey, hard black flinty slate, porphyry, and in addition to these chalk flints), covers a great part of Bagley wood, and pebbles of the same kinds are scattered, though more sparingly, over the summit of Whiteham hill. Now since Bagley wood is considerably elevated above the neighbouring district, and Whiteham hill is completely insulated, steep, and at least 300 feet above the vallies which surround it, we have here a most decisive proof of the excavation of the vallies at a period long subsequent to the formation of those hills; since when these blocks and pebbles were transported hither, there must have been uniformly inclined planes from their native sites to their present locality. That they should have rolled up the present escarpments is a physical impossibility. The phœnomenon is of exactly the same kind with that of the granitic blocks of the Alps, transported to the sides of the Jura chain; but the inferences are here more direct, in as much as it is impossible to call in the imaginary aid of ice-bergs to float the transported materials in this instance.

Hence, the same platform and ridge continue to extend to the west-south-west, traversing Wiltshire on the north of Highworth and Wooton' Basset, as far as the hills of Bow-wood and Bromham near Chippenham. They here acquire a southern direction. Near Seend there is a breach in the chain, through which the Berks and Wilts canal passes; but further south it again rises and becomes conspicuous at Steeple Ashton. Here, however, the extension of the green sand and chalk westwards, which has before been noticed as overlying the Kimmeridge clay, covers up the basset of this formation likewise.

To the south of this projecting cape of green sand, near the place where Alfred's tower marks its highest point, a range of low oolitic hills may be traced running east of Wincanton, in a southern direction, to Stourminster, and continuing the line of the coral rag hills. These are indeed coloured in Mr. Greenough's Map as a part of the great oolite, but we believe they are in reality separated from that formation by a valley of clay following the line of the intended Dorset and Somerset canal. Mr. Smith has assigned this range, both in his map and sections, to the coral rag; and we are inclined in this respect to follow his authority, confirmed as it appears to be by the general line of bearing of the formation in question.

On the south of Stourminster, the western projection of the green sand and chalk through Dorsetshire again conceals the basset of this, and indeed of all the oolitic formations; but on the south of this projection, where the inferior strata reappear skirting the coast, the coral rag again comes in exactly where, pursuing the same line of bearing, we might expect it —at Abbotsbury.

In order to understand its course from this point, it will be necessary to repeat the observation which has been before made when treating of the Portland beds; namely, that the country between the chalk hills and Weymouth forms, geologically speaking, a saddle, the axis of which (consisting of a nucleus of our third division of oolites) ranges east and west, having successive collateral zones of the second and first division of oolites on either side. Hence the coral rag of Abbotsbury (forming one of the northern collateral zones), ranges thence in a western direction, near Broadway to the coast, between Weymouth and Ringstead bays*, forming a well marked ridge

* All this part of Mr. Greenough's Map requires correction, having been founded on a supposed identification of the ferruginous sandy beds above the coral rag in this part of its course, with those of the inferior oolite, which has since proved to be erroneous; the line there assigned to the inferior oolite is really that of the coral rag.

with vallies of Kimmeridge clay on the north and the Oxford clay on the south. Professors Buckland and Sedgwick have minutely examined this district, and will soon communicate their observations to the public; the beds at the junction of the Kimmeridge clay and coral rag have here a peculiarly sandy and ferruginous character, which has occasioned them to be mistaken for the inferior oolite.

The southern collateral zone of the coral rag occupies what may be called the Peninsula of Weymouth; ranging from Wyke regis to Weymouth fort, having a valley of the Oxford clay on the north and the Kimmeridge clay hanging on its southern slope at Portland ferry.

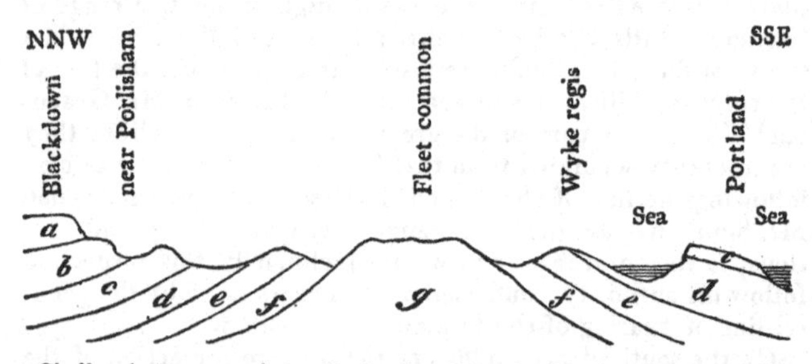

a. Chalk. *b.* Green sand. *cc.* Purbeck and Portland beds. *dd.* Kimmeridge clay. *ee.* Coral rag. *ff.* Oxford clay. *g.* Forest marble and great oolite.

The above rough section will give a general idea of the district.

(*e*) *Height.* Whiteham hill in Berkshire, 576 feet above the level of the sea, is probably the highest point attained by this formation; the average height of the low chain of hills occupied by it, seems to be about 400 feet.

(*f*) *Thickness.* The thickness of the coral rag and calcareous sand together may be taken at from 100 to 150 feet; each of these divisions occupying an equal moiety of this total thickness.

(*g*) *Inclination.* It is generally very little, averaging about 1 in 150, and therefore less than one degree; but the strata are often traversed by parallel lines of cleavage, cutting their planes at various angles; care must be taken to distinguish these from the true lines of stratification. The appearances resulting from the mixture of these lines are often singular and puzzling, presenting the phœnomenon of beds nearly horizontal, associated with others which seem to be highly inclined.

(*h*) *Agricultural character.* The outer slope of the hills of this formation towards their escarpment being occupied by the calcareous grit, affords a light sandy soil. Their surface and back, where the substratum is the coral rag, presents a loose stone-brash well adapted for turnips, barley, &c.

(*i*) *Water.* From the rifty and porous character of these strata, it is necessary to penetrate to their junction with the subjacent clay, in order to procure any considerable supply of water. The springs which flow over the superior clay, frequently sink into the rifts of this formation; and, after being for a time engulphed and concealed, are again thrown out by the basset of this subjacent clay; examples of such swallow-holes, as they are commonly called, may be seen at Headington near Oxford.

II. OXFORD CLAY. ‡

This is the Clunch Clay *of Smith, and forms the separation between the middle and inferior assemblage of oolites, including subordinate beds of limestone called the Kelloway Rock.*

(*a*) *Chemical and external characters.* This formation consists of beds of immense thickness of a tenacious and adhesive clay of a dark blue colour, becoming brown on exposure, and containing argillo-calcareous geodes and septaria These latter are frequently termed, from an obvious analogy, turtle stones. Some of these geodes are employed at Melbury in Dorsetshire as a coarse marble. The argillaceous strata are frequently mixed with calcareous, and sometimes with bituminous matter, affording in the latter case an inflammable shale;* the section appended in the note will give a general idea of the manner in which these beds occur.† The lower part of

‡ Chiefly by the Rev. W. D. Conybeare.

* The appearances of coal in this bed has given rise to numerous trials, encouraged by ignorance or fraud. Among these was one at the expense of Sir Edward Baynton and the Marquis of Lansdowne, on the south-east of Tetherton. T. 127. Search for coal has been made in various parts of this deposite, as near Elston in Bedfordshire: and an unsuccessful trial was made on the south-east of Tetherton in the parish of Goddington at the expense of Sir Edward Baynton and the Marquis of Lansdowne. (Agr. Sur.) Other trials have been made at Bruham in Somerset, at Pyrton in Wiltshire, and Enshan heath in Oxfordshire. (G. Notes.)

† Near the village of Donnington, which is about six miles on the west of Louth in Lincolnshire, and near the termination of the western escarpment of the chalk range, this formation has been bored into to the depth of upwards of 300 feet without passing through it. It affords the following list of alternating beds of clays, stones, and bituminous slate-clays. (G. T. v. iii. p. 396.)

this formation occasionally contains irregular beds of limestone; which, from their partial occurrence, can only be con-

Section afforded by boring at Donnington in Lincolnshire.

		feet	in.
1.	A clay soil	3	—
2.	Dark coloured clay	9	—
3.	Soft grey slate with marine impressions	1	—
4.	Blue argillaceous stone	—	6
5.	Dark-coloured clay	3	1
6.	The same as No. 3	1	—
7.	Laminated clay slightly indurated	23	—
8.	Soft grey slate slightly inflammable	5	3
9.	Same as No. 8, but darker coloured	5	3
10.	Indurated clay with white marine organic remains	37	6
11.	Same as No. 10, but harder and blacker	7	3
12.	Dark coloured bituminous inflammable schist	6	—
13.	A dark blue coloured ironstone	—	3
14.	Laminated indurated clay with white marine organic remains	33	—
15.	Same as No. 14, but harder, with marine impressions of thin leafy pyrites	10	4
16.	Dark blue argillaceous stone	—	4
17.	Hard indurated laminated clay with impressions of thin leafy pyrites	18	4
18.	Laminated bituminous schist, with white marine organic remains, and inflammable	1	10
19.	Dark blue ironstone	—	2
20.	Laminated bituminous schist, same as No. 18	11	—
21.	Dark blue ironstone	—	1½
22.	Laminated bituminous schist, same as Nos. 18 & 20	18	10½
23.	Dark indurated clay, with some white marine organic remains	3	6
24.	Laminated bituminous schist, same as Nos. 18, 20, & 22	8	—
25.	Dark indurated clay, same as No. 23	3	—
26.	Laminated bituminous schist, same as Nos. 18, 20, 22 & 24	4	6
27.	Dark indurated clay, same as Nos. 23 & 25, intermixed with thin seams of laminated schist	30	3
28.	Grit	—	2
29.	Brown laminated schist	—	2
30.	Hardstone, bind or argillaceous stone	2	10
31.	Hard laminated bituminous schist	1	2
32.	Same as No. 30	2	—
33.	Hard laminated bituminous inflammable schist	2	4
34.	Inflammable slaty bind	3	—
35.	Hard laminated bituminous schist, very inflammable	3	7½
36.	Hard dark blue bind, interlaid with thin strata of bituminous schist	13	9½
37.	Very inflammable schist	—	2
38.	Hard dark blue bind, same as No. 36	3	8
39.	Argillaceous stone	1	—
40.	Same as 39, but not so hard	1	—
41.	Hard dark blue bind, same as Nos. 36 & 38, in which the boring was discontinued	22	10
		309	

sidered as subordinate to it. These have been noticed chiefly at Christian Malford and Kelloway bridge near Chippenham in Wiltshire, and have been denominated, from the latter locality, the Kelloway rock. This stone occurs in irregular concretions, the exterior aspect of which is brown and sandy, the interior being harder and of a blueish colour. It consists almost entirely of a congeries of organic remains, among which several varieties of ammonites are predominant. The beds of clay (says Mr. Smith) which immediately cover this rock abound in selenite, and below it are found a brown aluminous earth and bituminous wood. Beds of clay separate the Kelloway rock from the oolites of the next division.

This limestone is only used for mending the roads, and as there are very few excavations for this purpose, it is difficult to trace its course; Mr. Smith, however, mentions the following localities in addition to those which have been already stated; Thames and Severn Canal near South Cerney; Kennet and Avon Canal near Trowbridge; Wilts and Bucks Canal near Chippenham; a Pit sunk in a fruitless search for coal at Bruham near Bruton, Somersetshire.

(*b*) *Mineral contents.* Iron pyrites and selenite occur abundantly in this, as indeed in all argillaceous formations. The association of sulphur with the clay strata seems to afford an interesting subject of enquiry. Mr. Smith believes the mineral waters of Melksham, &c. to be derived from the beds of this formation immediately contiguous to the Kelloway rock.*

Every variety of this stratum in the preceding section, agrees in the two following properties—the presence of calcareous matter, which is manifested by a brisk effervescence when any part of it is submitted to the action of acids; and secondly, a more or less abundant admixture of pyrites; all the bituminous slates, when exposed to the action of fire, burn with a very strong offensive smell, but those found below the depth of 80 yards were not so disagreeable in that respect, as those which were higher in the stratum. Varieties 33 and 35 are remarkable for their inflammability, and burn with a thick bituminous flame, appearing nearly equal in this respect to common coal; but after the bitumen was exhausted, the remainder was left undiminished in size. The organic remains observed in these beds, as their general characteristics, are impressions of ammonites and some small bivalve shells. (G. T. vol. iii. p. 327.)

In the Philosophical Transactions for 1787, is an account of the strata penetrated in sinking 478 feet for water in this formation near Boston; these consisted of clay mingled with marly concretions (called in that account chalk) and what is there termed gravel—probably loose and rubbly marle-stones—at 470 feet from the surface, a thin bed of stone (perhaps the Kelloways rock) was met with.

An interesting account of the sinking through the strata at Melksham will be found in the 'Guide' for that watering place.

* See further particulars of these springs in the article on the waters of this formation.

(c) *Organic remains.* Since the organic remains which occur in the Kelloway rock appear to be peculiar and characteristic, we shall give a double list of the remains of this formation; first, those which occur in the argillaceous beds generally; and secondly, those which distinguish the Kelloway rock.

1. Organic remains of the *Clay.*

Bones of the Ichthyosaurus occur, but are rare, and of a different species from those in the lias.

The following shells may be specified.

Chambered Univalves.

Ammonites Duncani, *T.* 157.

A. armatus; Smith, fig. 3. The ammonites in this formation are generally so much compressed, that it is difficult to ascertain their specific characters, and often still preserve their pearly shell; these circumstances are common to most of the argillaceous beds.

Nautili and *Belemnites* also occur.

Univalves not chambered.

Rostellaria.

tubular ditto.

Serpula, a peculiar variety; Smith, fig. 5.

Patella latissima. *T.* 139, fig. 1.

Bivalves.

Ostrea palmetta. *T.* 111, fig. 2.

Gryphæa dilatata. *T.* 149.

Perna aviculoides. *T.* 66.

The occurrence of fossil wood has already been noticed.

2. Shells of the *Kelloway rock,*

Chambered univalves.

Ammonites calloviensis. *T.* 104.

sublœvis. *T.* 154.

Kœnigi. *T.* 203; and several species not figured.

Nautili and *Belemnites.*

Univalves not chambered.

Rostellaria; Smith, fig. 1.

Bivalves.

Cardita deltoidea. *T.* 197, fig. 4.
Chama digitata. *T.* 174.
Gryphæa incurva. *T.* 112, fig. 2.
Pecten fibrosus. *T.* 136.
Plagiostoma obscura. *T.* 114, fig. 2.
Avicula inequivalvis. *T.* 244, fig. 2.
Terebratula ornithocephala. *T.* 101.*

(*d*) *Range and extent.* The uncertainty which as yet prevails concerning the true divisions of the oolitic formations which appear in Yorkshire,† prevents our being able to speak with absolute decision concerning the course of this formation at the north-eastern termination of the diagonal zone which it forms across the country. It should seem, however, that it certainly occurs on the coast of that county at Yew Nab near Filey bridge, underlying the calcareous grit of the coral rag formation; and it is probable that the greater part of the Clay vale, lying along the Yorkshire Derwent as far as New Malton, and intervening between the chalk and oolite hills, is of this formation. On the south of New Malton, the chalk advances close to the oolite hills and overlies this argillaceous tract.

South of the Humber, it may be traced in a southern direction through Lincolnshire, following the course of the fens along the Ancholme navigation, and the Witham river, to Boston; proceeding in the same direction, it forms the substratum of the western portion of the Cambridgeshire fens and those which border on Huntingdonshire; hence it has sometimes been called the *Fen Clay.* Where it enters Lincolnshire on the north, it forms a very narrow tract, not exceeding three or four miles across; but, from the drawing away of the chalk hills and their intermediate substrata which cover it, towards the south-east, a great breadth of it becomes exposed on the south of this county; a line taken at right

* The references, unless where otherwise specified, are to Sowerby's Min. Conchology.

This list would undoubtedly be much encreased were these strata more explored: but affording few materials applicable to œconomical uses, they have been very little laid open or examined. The most characteristic shell is the Gryphæa dilatata.

† Mr. Smith in his general Map, and more lately in his particular Map of Yorkshire, has represented the *Alum shale* of that county (near Whitby, &c.) as belonging to this formation; but the continuity of this tract with the great zone of the lias formation, the whole list of its organic remains, and its general characters, prove beyond the possibility of doubt that it really belongs to the lias, and not to the Oxford or clunch clay; it is correctly represented in Mr. Greenough's Geological Map. This subject will be more fully discussed in treating of the Lias formation.

angles to its course being not less than 15 miles; and this may be stated as its maximum of its superficial extent in this island.

It is here bounded by the low range of oolitic hills on which Lincoln stands on the west, and by the chalk hills on the east; the sands beneath the chalk appearing (from the absence of the two first divisions of oolites) to rest immediately on its upper edge; but the fenny state of this tract renders it difficult to speak with decision concerning its structure.

Proceeding to the south-west, a line drawn from Peterborough* to Bedford, Buckingham, and Bicester, will nearly mark its inferior junction with the subjacent oolites—and another from Huntingdon to Oxford its superior junction. Near Huntingdon, its upper edge appears to be in immediate contact with the iron sand; but on entering Buckinghamshire,† the Aylesbury limestone becomes interposed, and on the borders of Oxfordshire the coral rag, the whole assemblage and suite of strata being here full and complete. The breadth of the tract occupied by this formation is greatly reduced in proceeding from the north; being at Huntingdon more than twelve miles, and between Bedford and Oxford not more on an average than five or six, and often still less. West of Oxford it ranges along the valley of the Isis westwards as far as Cricklade; thence bending on the east of Malmesbury, it traverses Wilts in a south direction, following the course of the Avon past Chippenham to Melksham, having a breadth of from five to eight miles, and being always bounded by the escarpment of the coral rag on the south and east; and by the rise of the hills formed by the subjacent oolites on the north and west.

From Melksham, its course continues south through Somersetshire and Dorsetshire passing by Wincanton and Sturminster, on the south of which it is overlaid by the great western extension of the chalk and green sand. In this part of its course it is not more than three miles across, and has always the subjacent oolites on the west, and the coral rag escarpment on the east, except near Wincanton, where the green sand (overlying the basset of the coral rag) advances upon this formation.

This formation reappears in the south of Dorsetshire, in the

* In Mr. Greenough's Map the Oxford clay is carried a little too far to the west between Peterborough and Higham Ferrers, the escarpment which hangs over the right bank of the fen being really formed of the subjacent forest marble.

† Mr. Smith has divided this clay district between the Oak-tree clay (Kimmeridge) clay, and the Clunch (Oxford) clay, allowing the traces of the former to be obscure; the author of these remarks has however, by by tracing the characteristic fossils of the latter only, to the very foot of the Aylesbury limestone hills, convinced himself, that the former is entirely wanting.

Weymouth denudation, where its course will be sufficiently indicated by referring back to the description of that pursued by the coral rag in that district, since it occurs interposed between the central nucleus of the subjacent oolites and the collateral chains of that rock.

(*e*) *Height, &c.* This stratum must be considered as being for the most part extremely low, as is the case throughout the considerable tract occupied by it in Lincolnshire, where it scarcely exceeds the height of the fens which penetrate it, and which separate it from the point of its reappearance on the surface in Huntingdonshire; whence to Bedford, this stratum forms the vale of the Ouse. On the west and south-west of Bedford, and on the east of Buckingham, occur some exceptions to the general flatness of the country formed by this stratum: but the eminences are not considerable. Thence to Oxford it is low and flat, and on the south-west of that place it forms the vale of the Isis for several miles. Gentle eminences again appear near Malmsbury and occasionally on the south of it, as near Trowbridge and on the north and west of the vale of Blackmoor. (G. Map.)

(*f*) *Thickness.* The thickness of this formation must be very great, probably exceeding 500 feet. At Boston, as has been already stated, it has been penetrated to the depth of 478 feet; from comparing its breadth taken horizontally, and its inclination, it may be estimated at about 700 feet in the midland counties, and cannot be much less in the south-western.

(*g*) *Inclination.* It appears to be very nearly conformable to the formations before described, although a slight difference in this respect, between the different members of the series, must be supposed, in order to account for the occasional appearances of beds in some places which are wanting in others, and for the upper edge of this formation being thus at one time in contact with the iron sand, at another with the Aylesbury lime, and at a third with the coral rag. A very slight deviation however is sufficient to account for this, and no more probably takes place. When indeed it is said that beds are conformable to each other, the circumstances under which they appear to have been deposited would scarcely warrant us in expecting to find an absolute and mathematical parallelism, but only an approximation to it.

The general dip of this formation appears to be to the east and south-east under an angle scarcely if at all exceeding 1°, except in the elevated tract north of Weymouth, where the direction and dip change, and the angle is often as high as 30°.

(*h*) *Agricultural character.* The vale which extends from Melbury to Shaftesbury in Dorsetshire is famous as a butter country; but this clay is generally barren and untractable, unless where covered, as in Lincolnshire, by the White clay of that country, which is chalk breccia. (G. Notes.)

The tract occupied by this formation is indeed so frequently covered by alluvial debris that its agricultural character is very variable from this cause.

(*i*) *Phænomena of springs.* It is generally necessary to sink through this retentive formation in search of water. The well sunk at Boston to the depth of 478 feet has been already mentioned, and even then the attempt was abandoned without success.

The following mineral springs occur in the course of this formation, but perhaps their true seat is the upper beds of the subjacent oolites. Waters containing purgative salts; at Stanfield (in Lincolnshire); Kingscliff, (Northamptonshire); below Cumner (Berkshire); Melksham and Holt (Wiltshire). Iron and carbonic acid at Seend (Wilts), and green vitriol at Somersham, Huntingdonshire. It is to be regretted that no correct analysis of the mineral waters of this island, with the exception of a few of those of most distinguished resort, has yet been made; the analysis of the Melksham water will be found in the 'Guide' before alluded to.

Section V.

Lower Division of Oolites. *

A. *Upper beds*, associated with the Great Oolite and containing the subdivisions *Cornbrash*, *Stonesfield slate*, *Forest Marble*, and *Great Oolite*.

Introduction.—In the general view of the Oolitic series (p. 166,) a synoptical table of this lower system of oolites (which far exceeds in thickness and importance the two former,) has been already given. Since the subdivisions into which it may (occasionally at least) with propriety be resolved, are so numerous that it would introduce some confusion to treat of them all in a single section, at the same time that many of them cannot be considered as either sufficiently universal or extensive to be entitled each to a separate section; and since an arrangement so minutely divided would both have perplexed the memory and given rise to much unavoidable repetition, we shall, led by the great principle of convenience, attempt to reconcile as far as may be consistent with that precision and generalization of method which appear desirable, pursue a

* Chiefly by the Rev. W. D. Conybeare.

middle course, and assign to this system two sections; viz. the present, to its upper, and the ensuing, to its lower beds; and this division will be sufficiently natural, since all its upper beds may fairly be regarded as subordinate to the great oolite, and all the lower to the calcareo-siliceous sand which forms the base on which it rests, although the line at the junction of the upper and lower beds must still be drawn in a somewhat arbitrary manner. In the present section then it will be our object to treat under the several usual heads, of all the circumstances connected with the upper beds subordinate to the great oolite, noticing in order under each head the characters of the several subdivisions.

(*a*) *Chemical and external characters.* Viewed generally, the chain of hills composed of this oolitic system will be found to consist of one great oolitic mass, resting upon the beds of calcareo-siliceous sand (itself containing some beds of coarser oolite) which we have referred to the next section; but on more minute examination it will be found that the upper part of this great oolitic mass, forming the acclivity of the hills where they rise from the valley occupied by the Oxford clay, present strata of a character sufficiently distinct from the great body of the oolite to entitle them to a separate description; these, instead of rising in thick masses, are generally either fissile or rubbly; are much mingled with clay, forming as it were a link between the principal deposit of purely oolitic beds, and the succeeding argillaceous beds; in place of the yellowish tinge of the oolite, they have very generally a blue colour, or in some beds a pasty appearance and a dead white colour not unlike chalk. As far as our observations at present extend, it should appear to be impossible, or nearly so, to trace any divisions of these upper beds resting on the great body of the oolite, which may be certainly applied to every part of the course pursued by this system of rocks through the island; since they appear rather as accidental varieties of this great oolitic deposit, where, as we have already observed, a mixture of argillaceous precipitates led the way to the great mass of the latter character which prevailed in the next succeeding period. In some instances, however, the precipitates of argillaceous and calcareous matter, during this intermediate period, appear to have followed one another alternately, at sufficient intervals to allow the formation of tolerably thick beds of either kind, which may be traced with regular order through tolerably extensive districts; and even where this regularity does not absolutely exist, still a general uniformity of character will be found to distinguish the upper members of the series from its other portions.

In the neighbourhood of Bath, and throughout an extensive tract in the adjacent counties of Gloucester, Somerset, and Wilts, this regularity is the most observable; and we shall therefore proceed to draw our descriptions principally from this district, premising that our materials will be in great measure derived from the various publications of Mr. Smith, to the great accuracy of whose observations, as relating to the district in question, we can from personal examination bear the highest testimony, although we cannot agree with him in believing that these minute divisions can be traced or indeed exist throughout the whole course of the great system of oolites of which we are now treating, as regular beds, decidedly distinct from each other.

A general idea of the stratification of this district will be afforded from the section in the note,* the result of a careful examination of it under the guidance of the Revd. Mr. Richardson: we shall proceed, therefore, more particularly to state the chemical and external characters of the upper oolitic beds in the order here exhibited.

The *Cornbrash* (No. 2), (No. 1 being the lowest beds of the clay described in the preceding article), is a loose rubbly limestone, of a grey or bluish colour, especially near the superincumbent clay, but on the exterior brown and earthy; it rises in flattish masses rarely more than six inches thick. The upper beds of the third oolitic system in Oxfordshire, which are with probability referred to this rock, are often of a pasty or chalky consistency and colour. In Wiltshire it is known by the name of the cornbrash or corn-grit. The latter appel-

* *Section in the neighbourhood of Tellisford and Farley Castle, ten miles south of Bath.*

1. Lower part of the Oxford or clunch clay full of selenite.
2. Cornbrash .... from 8 to 16 feet.
3. Clay.................... ... .8 to 14
4. Calcareo-siliceous sand and gritstone ...... usually about 10
   but near Norton swelling to 40
5. Forest marble.. 18
6. Sand beneath the forest marble............................ 2
   but sometimes swells like No. 4 to a greater thickness.
7. Clay ....from 40 to 60
   sometimes however thinning off to 20
8. Great oolite 130

Mr. Smith's Section of the strata through Hampshire and Wilts to Bath may also be advantageously consulted with reference to this district. The only inaccuracy which deserves mention is, that the cap of the lower beds of the great oolite on the hill above Mitford Castle is not distinctly represented.

lation however is improper, because it is not a grit; being here rubbly, it is not fit for any purpose, excepting for lime and the repair of the highways; but at Malmesbury (T. p. 267,) where it is thick and solid, it is much quarried for building. Its fragments, rounded at the edges, may be noticed as they are turned up by the plough, either round Atford, and at Wraxall, or at Chippenham, and the interjacent country. It may readily be discovered by the superincumbent red soil which constantly attends it. (T. 103.)

The joints of the rock are usually separated by clay, which makes it tenacious, and occasions the excavations formed in it to hold water.

No. 3. The *clay* beneath is generally white near its junction with the cornbrash, and afterwards blue.

No. 4, No. 5, and No. 6, are so intimately associated, that they require to be described together. This assemblage consists of beds of limestone generally fissile and divided by argillaceous partings, lying between two strata of calcareo-siliceous sand and gritstone; the lower of these is indeed generally insignificant, but often swells into greater thickness and importance: these sandy strata appear to contain about one-third of calcareous matter: the grit-stone contained in them is hard enough however to scratch glass, and forms irregular slate-like concretions. The limestone lying between these sands is that known by the name of *Forest marble*; its beds, generally speaking, are thin and slaty; sometimes however beds of two or three feet thick may be found. The colour of the stone is generally grey or bluish, externally brownish, appearing on examination to be frequently composed of a congeries of dark coloured shells, interspersed with white oolitic particles: bivalve shells are most common in the thick beds—univalve in the thin. Decomposed pyritical wood often gives a partial redness, and some of the joints have a reddish tinge: the texture of the stone is coarse-grained; the structure of its masses fissile, whence coarse roofing slates and flag stones are in general use in the villages on the course of this rock; the more solid beds have also been occasionally worked as a coarse marble, being susceptible of a tolerable polish, and variegated by the contour of its imbedded shells; from this circumstance, and the occurrence of these strata in Whichwood forest, Oxfordshire, it has derived its name,

The partings of clay between the beds of this rock vary in thickness from less than an inch to more than a foot. There can be little doubt that the *Calcareous slate of Stonesfield* near

Woodstock, Oxfordshire, so remarkable (as we shall hereafter have occasion to observe) for the singular variety of its organic remains, among which the spoils of birds, land animals, and amphibia, occur mingled with vegetables and sea shells, belongs to the same part of the series with the forest marble. The assemblage of beds here worked, consists of two fissile beds of a buff coloured or grey oolitic limestone called *pendle*, each about two feet thick, separated by a bed of loose calcareo-siliceous sandstone called *race*, about the same thickness. Concretions are frequent in the latter (G. Notes), and are called *whim-stones* or *potlids*; they are partially oolitic, sometimes blue in the centre, and vary from six inches to two feet in diameter: their form is generally that of a flattened sphere; they do not break concentrically, but into parallel planes; and they often contain shells. The pendle, after being quarried, is suffered to lie exposed to the action of a winter's frosts,* and the blocks being then struck on their edge with a mallet, freely separate into slates sufficiently thin to afford a light material for roofing. The quarries are principally situated in the valley immediately on the south of Stonesfield village, which branches off eastwards from that of the Evenlode. The mode of working is by driving horizontal galleries about six feet high into the side of the hill, and then extracting the two strata of pendle laterally, piling up the refuse masses of the intermediate bed of race, so as to support the roof: deep perpendicular shafts communicate with these galleries. These workings have been carried on from remote times to a considerable extent, so that both sides of the valley are completely honey-combed by them. Beautifully plumose stalactites are often found in the fissures of the rock, and are called by the workmen, from an obvious though coarse analogy, tallow.

Beds of calcareous slate also occur in the oolites of this system at Easton and Collyweston in Northamptonshire, a few miles south of Stamford, and, it seems probable, belong to the same part of the series with those above described; we have however no particular description of these quarries. The slate of Stonesfield is much more calcareous than that of Collyweston, which is stated by Bishop Watson (Chem. Ess. vol. iv.

* At Stonesfield, the pendle is raised in blocks about a foot thick, and only between Michaelmas and January. As soon as they are raised, the blocks are wetted, and covered with earth, until there is a prospect of frost, when they are uncovered; but if the frost goes, are immediately covered again, or they would not split. The action of frost opens the joints of the slates sufficiently for a blow with a mallet to complete the operation. The price of slates at the pit is about 40s. per 1000 of the ordinary size, but they have been obtained so large as 21 feet square. (G. Notes)

p. 316) to imbibe more water, and retain it for a longer time than the Westmoreland slate; but it does not imbibe half so much, nor retain it one-fourth the time that a common tile does. (G. Notes.) In this part of the country also, the escarpment on the right bank of the Nen, between Higham Ferrers and Peterborough, exhibits beds resembling forest marble at Raunds and Stanwick.

The whole mass of this oolitic system in Dorsetshire (excepting the inferior oolite and its sand) presents the fissile character of the forest marble; but it seems more probable that the great oolite here passes into this structure (as it undoubtedly does occasionally in other places), than that the forest marble, generally a suborbinate bed only, should here swell to such a disproportionate thickness, and the great oolite itself be wanting.

No. 7. *Clay over the upper oolite*, or *Bradford clay*, consists of a common blue marly clay, which, at the point of its contact with the great oolite, is replete with the peculiar organic remains hereafter to be described. It is sometimes wanting, and it then becomes impossible to distinguish the upper beds of the great oolite from those of the forest marble.

No. 8. *Great Oolite.** This is, both in thickness and utility, by far the most important of the British oolites: it consists of a stratified calcareous mass, varying in thickness from 130 to more than 200 feet; softer and harder beds, (the former characterized by those distinct oviform concretions which give name to this series of rocks, the latter exhibiting them more rarely and obscurely) alternate in this mass of strata. The former afford the freestone which renders this rock so valuable; but these strata vary much, both in thickness and quality, even in quarries in the same neighbourhood. The Kettering freestone of Northamptonshire is rendered extremely beautiful by the distinctness of its oolitic structure: that of Bath has generally a finer grain: this has been employed in the late repairs of Henry the Seventh's chapel at Westminster. St. Paul's was built principally from the quarries about a mile north of Burford in Oxfordshire. Fragments of comminuted shells may be observed in all the varieties mingled with the ova, but so completely broken down that it is generally impossible to ascertain their species; hence arises the rarity of such specimens from this rock, and our consequently imperfect knowledge of its fossils. The colour of the freestone beds is generally white with a light cast of yellow. Of the other beds, some are grey, some almost blue in the middle: sometimes

* We have principally copied this description from Mr. Smith.

also, beds of a brown ferruginous cast are interposed, especially at the bottom of the series near its junction with the fullers' earth. The upper beds, in which the shells are more distinct, and which afford indifferent freestone, cannot be easily distinguished from the forest marble. The structure of the freestone is in thickly bedded masses, which, if traced to a distance, will often be found to thin out. Many of the other beds exhibit a laminated cleavage, not parallel to the greater lines of stratification, for which they have sometimes been mistaken, and thus given rise to accounts of highly inclined beds alternating with horizontal ones in this rock: many appearances of this kind may be seen in the quarries near Badmington park, the Duke of Beaufort's seat in Gloucestershire.

Traces of magnesia have been discovered in some of these beds on the Cotteswold hills, otherwise they appear to be a tolerably pure carbonate of lime, dissolving in acids with very little residuum. Thin partings of clay, and sometimes large irregular interposed masses of that substance, may be observed between the upper strata.

(*b*) *Mineral contents.* Scarcely any thing occurs in this series which deserves to be noticed under this head. Calcareous spar is almost the only substance which presents itself. Of this, the finest specimens are those from Stonesfield, which are of a bright transparent yellow; they occur as stalactitic masses, forming sometimes fine slender tubes, and sometimes beautifully radiated and plumose crystallizations. Quartz crystals have been found, but are extremely rare; and even that universal mineral, iron pyrites, is scarcely to be seen.

(*c*) *Organic remains.* The variety and abundance of these compensates, as is usual in the secondary rocks, for the barrenness in their mineralogical list: here we have also in one bed, the most singular assemblage of organic remains presented perhaps in any formation, comprising land animals, birds, insects, amphibia, fishes, shells, and vegetables. The organic remains are however chiefly to be sought for in the upper beds, the great oolite itself affording but few perfect specimens, although replete with comminuted fragments. We shall premise some general observations on the distribution of these remains in the different beds, and then give lists in parallel columns of the shells which have been figured from each of them.

The *Cornbrash*, as Mr. Smith observes,* though altogether but a thin rock, has not its organized fossils equally diffused or

* The following notices are principally copied from Mr. Smith, as far as the beginning of the notice on the particular families of organic remains.

promiscuously distributed: the upper beds of stone which compose the rock, contain fossils materially different from those in the under; the clusters of small oyster shells and the stems of the pentacrinus lie near together, and not many others are found near the bottom of the rock.

In the *Forest marble*, though its various beds are composed of little else but a mass of shells, loose and whole specimens are rare, and extracted with great difficulty. A few however are occasionally found in the clay between the stone; bones, teeth, and wood, firmly imbedded in the rock, are some of its most characteristic indications. Pickwick and Atford quarries, a few miles east of Bath, used to be most famous for these; but since it has been generally understood that the same stratum may uniformly be expected to produce the same organized fossils, other quarries of the same stone have been searched and found to contain them. Small turbinated shells are frequent in this rock: the univalves are more common in the thin, the bivalves in the thicker beds.

If the *Calcareous slate of Stonesfield* be correctly assigned to this part of the series (which is rendered still more probable by the occurrence of the same teeth and palates in both instances), we here find the only known instance in which the remains of birds and terrestrial animals have been found in beds of antiquity at all approaching to these; they here occur mingled with *winged* insects, amphibia, sea shells, and vegetables, presenting at once the most interesting and difficult of problems connected with the distribution of organic remains.

The *Clay above the great oolite* contains few fossils, except in its lowest bed, where in immediate contact with the upper surface of the subjacent oölites; but in this point it abounds with remains of the pear encrinus, with many small coralloids, and several peculiar terebratula.

The *Great oolite.* In the great mass of this rock, as has been before observed, perfect remains are rare, from the generally comminuted state in which they occur; in some of the upper beds however they are more easily distinguished. Many small turbinated shells, like those of the forest marble, here occur in the stone, and a bed containing numerous madrepores, several of which appear identical with those of the coral rag, is also found in part of the series. Most of the fossils of the subjacent clay are also common to the upper beds of the oolite.

We proceed to a more particular enumeration of the families of organic remains occurring in these beds.

VERTEBRAL ANIMALS.—*Mammalia.* The calcareous slate of Stonesfield presents bones, believed by Mr. Cuvier to belong to a species of Didelphys, one of the opossum tribe; they are absolutely imbedded in the slate, together with various marine remains, and not subsequently intruded into its fissures; they present therefore an unique instance of the occurrence of beings of such an order, in strata older than the youngest members of the superior or tertiary class.

Oviparous quadrupeds.—*Sauri.* A well characterized crocodile, but of a species distinct both from those now known to exist, from those found in a fossil state in Germany, and from one at least of the French fossil species, has been dug up at Gibraltar near Oxford, and is now in the collection of that University; it is from a bed towards the upper part of this oolitic system, perhaps the Cornbrash.

An immense animal, approaching in its dentition and characters to the Monitor, occurs at Stonesfield in the calcareous slate; specimens have been found which must have belonged to an animal 40 feet long and 12 high: the lower jaw vertebræ and extremities are preserved in the Oxford collection. In many respects this animal bears a great analogy to the Lacerta gigantea, described by Sœmmering in the Munich Transactions as being found in Bavaria. See a translation of his Memoir in the Annals of Philosophy for September 1821.

The Ichthyosaurus has not yet been noticed, but as it occurs in the beds both above and below these, it is probably to be found in this part of the series also.

Testudines.—Remains of two or three species of Tortoise occur in the Stonesfield calcareous slate.

Fishes.—Teeth,* palates, and vertebræ of fishes of several varieties are found both in the Stonesfield slate and in the forest marble of Atford near Bath: the same varieties seem to be common to both places.

Birds.—Leg and thigh bones, apparently belonging to birds, are imbedded in the Stonesfield slate.

Insects.—*Coleoptera.* Specimens which have been decidedly pronounced by Dr. Leach to be the Elytra of Coleopterous insects, occur in the Stonesfield slate; they are of two or three different species.

Crustacea.—Two or three varieties of the Crab or Lobster tribe occur also in the Stonesfield slate.

---

* These little fossil productions were, a century since, in common use with the ladies, as ornaments; and, what is a still more singular circumstance connected with their history, they seem to have been applied to the same purpose by our ancient British ancestors: as several strings of them were discovered in the Wiltshire Barrows, among other ornaments, opened by the late Mr. Cunnington, of Heytesbury; and now preserved by his widow, in her exquisite collection of Barrow antiquities.

It is impossible to close this list of the remains of so many tribes of the rarest occurrence in a fossil state, and mostly unknown in the older rocks, thus singularly assembled and mingled together in a single and insignificant formation, and that in one limited spot, without remark. Those acqnainted with the general distribution of organic remains, but not with the locality in question, would naturally be led to suspect that some of them, for instance the remains of the didelphys, the birds, and the coleopterous insects, must have been introduced into fissures of the regular strata at a subsequent and comparatively recent period, in the same manner as the bones discovered in the rock of Gibraltar: but the slightest examination of the spot, or even of the slabs brought from thence and containing these remains, will at once remove this suspicion, since they are found intimately associated with the shells which characterize this part of the oolitic series. The beds themselves are also most clearly to be traced holding a regular course together with the superior and inferior beds of this oolitic system, and cannot therefore be considered as a local, overlying, and recent deposit. Several varieties of shells, the most characteristic of which is a small studded trigonia, and several vegetables, principally flags, ferns, and mosses, occur in the same beds. Specimens of all these remains are preserved in the collection at Oxford; and Professor Buckland has obtained beautiful and accurate drawings of the whole series. It is greatly to be desired that these may shortly be submitted to the public.

We must account for the presence of the Didelphys, birds, and Coleopterous insects, in the same manner as we do for the wood and remains of land vegetables not unfrequent in the strata: the amphibia may have belonged to species principally marine. It is evident from peculiarities in their structure, that many of the fossil animals, generally resembling the amphibia, lived entirely or almost so in the sea, and were to the now existing amphibia what the cetacea are to mammalia.

Testacea.

Cornbrash.	Forest Marble.	Clay over oolite.	Great oolite.
MULTILOCULAR UNIVALVES.			
Ammonites.			
A. Discus. *T.* 12.	specimens in these strata scarce and ill defined		
Nautilus.			
	Nautilus; one variety in the Stonesfield slate.		
Belemnites.			
	a fusiform belemnite in the Stonesfield slate.	a small slender belemnite.	
UNIVALVES NOT CHAMBERED.			
	Patella rugosa.		
Voluta.			ill defined traces and casts, perhaps of *Turbo*, *Melania*, & *Ancilla*.
Turbo.			
Turritella.	*Turritella.*	*Turritella.*	
Rostellaria, same as in Kelloways stone.	*Rost ellaria*; Smith, fig. 3.		
Ampullaria.			
Natica? Smith, fig. 1.			
	Ancilla; Smith, fig. 3.		
		Trochus.	
		(Annelidæ of Lamarck.)	
TUBULAR IRREGULAR UNIVALVES.			
Serpula, several species.	*Serpula.*	*Serpula*, several species.	*Serpula.*

Bivalves equivalved.

Cornbrash.	*Forest marble.*	*Clay above the oolite.*	*Great oolite.*
Modiola.		*Modiola*, several species.	The fossils of this bed having (from the difficulty of procuring detached specimens) been little examined, this list is very imperfect. The upper beds contain most of the shells in the preceding column.
M. imbricata. *T.* 212. f. 1. 3.			
M. aspera. *T.* 212. f. 4.			
M. plicata. *T.* 248. f. 1.			
Unio.			
U. acuta. *T.* 33. f. 5. 6. 7.			
U. ..... Smith, f. 7.			
Trigonia.			
T. clavellata. *T.* 87.			
T. costata. *T.* 85.	*T.* costata. *T.* 85.	*T.* costata. *T.* 85.	
Cardium.			
C....... Smith, f. 6.			
Cardita?			
C. deltoidea. *T.* 197. f. 4.			
C. lirata. *T.* 197. f. 3.			
C. producta. *T.* 197. f. 1.			
Mya?	*Mya.*		
Venus?	*Venus?*	*Venus?*	
V... Smith, f. 5.			

Inequivalved.

Cornbrash.	*Forest marble.*	*Clay above the oolite.*	*Great oolite.*
Ostrea.			
O. Marshii. *T.* 48.	*O.* crista galli.	*O.* crista galli.	*O.* crista galli.
	Ostrea ..... Smith. f. 4.	*O.* acuminata. *T.* 135. 2.	
Pecten.			
P. fibrosus. *T.* 136. 2.	*P.* fibrosus. *T.* 136. 2.	*P.* fibrosus. *T.* 136. 2.	*Pecten.*
P. laminatus. *T.* 205. f. 4.	*P.* .... Smith, f. 5.		

Cornbrash.	*Forest marble.*	*Clay above the oolite.*	*Great oolite.*
Avicula.			
A. **echinata.** ***T.*** **243.**			
A. **costata.** ***T.*** **244. f. 1.**	....	***A.*** **costata.** ***T.*** **244. f. 1.**	
Lima.			
L. **gibbosa.** ***T.*** **152.**			
Terebratula **(not plicated.)**	....	***Terebratula*****; a nonplicated species not figured.**	***Terebratula,*** **not figured; same as in the superincumbent clay.**
T. **subrotunda.** ***T.*** **15. f. 1 & 2.**			
T. **intermedia.** ***T.*** **15. f. 8.**			
T. **digona.** ***T.*** **96.**	....	***T.*** **digona.** ***T.*** **96.**	
T. **ornithoce-phala.** ***T.*** **101. f. 12. 4.**			
T. **obovata.** ***T.*** **101. f. 5.**			
Terebratula **(plicated.)**			
T. **obsoleta.** ***T.*** **83. 7.**	....	***T.*** **obsoleta,** ***T.*** **83. 7.**	***T.*** **obsoleta.** ***T.*** **83. 7.**
		T. **reticulata. Smith, f. 10.**	
		Chama.	
		C. **crassa. Smith, f. 6.**	
		Plagiostoma, **Smith, f. 7.**	***Plagiostoma*****; same as in the superincumbent clay.**

From the Testacea, we may proceed to the *Echinital family.* Here we have several species of the family Cidaris, which do not appear to be strongly distinguished from those of the coral rag, and occur principally in the Cornbrash, in the clay over the great oolite, and in the upper beds of that rock. Of the family Clypeus, the Clypeus sinuatus and clunicularis appear to be found in all the beds; and in the Cornbrash, and clay over the great oolite, is a very depressed species of Conulus, which might at first sight be mistaken for Clypeus sinuatus with its sinus filled up by the stony matter of its matrix.

Of the *Encrinital family*, the Cornbrash, forest marble, and great oolite present some species of Pentacrinites, and the clay over the great oolite, together with the upper beds of that rock, is distinguished by that beautiful and well known species the Pear Encrinite of Bradford, the Apiocrinites of Mr. Miller's Monograph of this class, in which full details of the anatomical structure of these singular beings, and engravings on the most ample scale are given. We refer for particulars to the table of the geological distribution of the Encrinites accompanying that work.*

The class of *Madrepores*, or lamelliferous Polyparia of Lamarck, presents several species in a bed near the top of the great oolite. Sufficient attention has scarcely been given to them to distinguish between the species of this series and the coral rag; it appears, however, on the whole, that a turbinated Caryophyllia, a large ramose one approaching Caryophyllia carduus, and small clustered ramose one approaching C. flexuosa or cespitosa, are hardly to be distinguished from those in the coral rag: that the same remark will apply to a variety of Astrea approaching A. favosa, but that there are some other varieties of Astrea distinguished by the nodular form of their mass, and the delicacy of their stars, which seem peculiar to the great oolite; and that a branching species of oculina covered with stars laterally disposed, is also peculiar to the great oolite, and some of its immediately superjacent beds.

These remarks, however, are offered only in the conviction of their being very imperfect; and with the intention of leading to further examination of this subject. Figures of some of these may be seen in Smith's plate of the fossils of the oolite. The forest marble contains a few species of astrea.

The clay above the great oolite contains a species of *cyclolite* (madrepora porpites), a *tubipore*, and a small but beautiful variety of branching *millepore*, with the pores most elegantly arranged in spiral lines round the branches. Lamouroux, in his additions to his republication of Ellis's work on corals, has

* Among the extraneous fossils imbedded in the white clay, the most interesting are the Encrinites, first noticed by the Rev. Benjamin Richardson at Burfield, near the summit of the hill, on the southern hanging of which Bradford stands. They were next discovered south of the river, on the surface of the rock, in the same bed of white clay, but more than 150 feet lower than Burfield, and a little elevated above the level of the river. Finally, they were traced on the summit of the opposite hill, yet always deposited in their proper bed. (T. 268.)

given correct figures of this fossil under the name *Terebellaria* ramosissima, Pl. 82. fig. 1.; it is also coarsely figured in Smith's plate.

From the figures given by Lamouroux, a flustriform coralloid, Berenicca diluviana Pl. 80. fig. 3, and one of the Cellariæ, Alecto dichotoma Pl. 81. fig. 12. Two small varieties may be cited as often found in the clay above the great oolite investing its shells.

Irregular cylindrical branches often occur in all these beds, which appear to have derived their origin from *alcyonia*; and in the great oolite, well characterized fragments of these zoophytes, exhibiting distinctly their spongy and cellular texture, may be frequently traced. In the upper beds of the great oolite we may observe congeries of minute millepores and cellepores, and the small varieties of tubercular, ramose, and perforated sponges, figured by Lamouroux, tab. 84. f. 5 to 10. as distinct species, but which are more probably only individuals in different stages of growth and states of contraction.

In all these beds fragments of *fossil wood* may be traced, and more particularly in the forest marble.

The Stonesfield calcareous slate exhibits many beautiful vegetable impressions, chiefly ferns, flags, and mosses, many of which nearly resemble those of the coal formation.

(*d*) *Range and extent.* The formations from the cornbrash to the great oolite inclusively, form the mass of a well defined range of hills traversing the island diagonally from Yorkshire to Dorsetshire, and rivalling or surpassing the great chain of the chalk hills, in continuity, extent, and elevation. To this range Mr. Smith has given the appropriate name of the *Stonebrash hills*, from the stony fragments mixed with the superficial soil, which are commonly known among agriculturists by this denomination. It will facilitate our conception of the position of these various beds, then, to state that, where they can be distinguished from each other, the Cornbrash is generally found forming the first acclivity of these hills where they begin to rise from the valley occupied by the Oxford clay, which accompanies them on the east and south-east; that the forest marble and calcareous slate extend still further on their rise; that the great oolite, emerging from beneath these, forms their most elevated region and brow; and finally, that the subjacent beds, associated with the inferior oolite, and to be described in the next section, are displayed in the escarpment and slope of these hills towards the west and north-west; the lias occupying the plains at their foot.

Occasionally, however, especially in Northamptonshire and Oxfordshire, the causes (whatever they may have been) which have evidently produced great degradation and wearing away of parts of this chain, have denuded the cap of the great oolite near the escarpment of the main chain of hills, and thus exposed a very considerable breadth of the subjacent sandy beds, causing the oolite to terminate in a low terrace (yet sensibly elevated above the denuded tract), and considerably within what appear to have been its ancient limits. This fact is clearly indicated by the occurrence of numerous, scattered, and insulated masses of this rock dispersed throughout the denuded space. These masses, which often occupy more than a square mile in extent (sometimes several miles) occur not only as regular outliers (i. e. forming the caps of insulated hills in the regular plane of the strata), but also in the lower grounds and vallies surrounded by hills of the subjacent sands, being in this case always disposed in considerably inclined planes, as if they had been precipitated into their present situation from the summits of the adjacent hills over which they appear to have once extended. We may conjecture with much probability that the denuding cause which has manifestly excavated the valleys of this tract, (and which may possibly have been the currents of the retiring waters beneath which all the secondary beds of our present continents have obviously been formed) undermined, in the course of its action, these masses, by removing the loose materials of the subjacent sand; while the more solid beds of the incumbent oolite partially resisted its destructive agency, and left behind these relics of its former extent.

These particulars may indeed seem more closely connected with some of the following heads of description, than with the range and extent of these strata; but they will be found, as we proceed, so essential to the full elucidation of the present subject, that they could not have been omitted in this place without manifest inconvenience: and, having thus obtained a clear view of the general circumstances connected with the position of these rocks, we shall be enabled to trace more readily and rapidly the particulars of their local distribution.

Although these beds form (with very partial interruptions arising from the overlying of the chalk and green sand formations near the two extremities of the line in Yorkshire and Dorsetshire) a continuous zone extending across the island from north-east to south-west, and have their track marked by a line of hills which, considered in a general light, must be regarded as a single chain; yet it seems necessary to break this extended line into shorter intervals for the convenience of description, since we have to attend both to the upper beds and superior edge as well as the inferior edge of this important system: and were we first to trace the former, and then return to trace the latter, through the whole island, the eye and attention would be fatigued and perplexed by the necessity of recurring twice to geographical positions essentially connected with each other; but, by dividing the line into shorter intervals, and tracing the superior and inferior limits of this system through each of these in succession, we keep the attention fixed on the same district. As an additional argument in favour of this method, it may be stated that this long chain of hills. having suffered more or less from denudation, in different parts of its course, naturally subdivides itself into distinct groupes, generally corresponding with the divisions we shall assume. These will be, I. Yorkshire; II. from the Humber to the borders of Northamptonshire and Oxfordshire; III. Oxfordshire; IV. the Cotteswold or Gloucestershire District; V. Somerset and North Dorset; VI. South Dorset.

I. *Course through Yorkshire.* In pursuing this course we shall for the present pass slightly over Yorkshire, reserving to a separate article the considerations which influence our views as to the identification of the rocks of that district. We shall now therefore simply state that we are inclined to consider the oolitic series which stretches from Scarborough to the Howardian hills, forming an escarpment on the south of the sandy hills of the Eastern Moorlands, as rather belonging to that system which we are now considering, than to that of the coral rag; to which, on account of its corals, it has been referred both by Mr. Greenough and Mr. Smith. Bending round from the Howardian hills towards the south-east, it becomes overlaid and concealed by the chalk and green sand hills near Pocklington; but according to Mr. Smith's late map, reappears in the same line at North and South Cave, on the banks of the Humber; on the opposite side of which, in the direct continuation of this line, the great oolite does undoubtedly make its appearance: so that we have almost the evidence of a direct continuity of the strata in favor of the view here proposed.

II. *Course from the Humber to Oxfordshire.* From the Humber, the Stonebrash hills, which are here possessed of a very small elevation, stretch due south through Lincolnshire, the metropolis of which stands on their course. They at first divide the vallies of the Trent and of the Ancholme navigation, and afterwards have the valley of the upper part of the river Witham on the west, and that of the lower part of the same river, which trends round in a singular manner, on the east. Their breadth to the north of Lincoln is inconsiderable, and here the great oolite almost exclusively prevails; but to the south of Lincoln they gradually expand, and it is said that near Sleaford the cornbrash can be traced on the eastern slope of the hills, separated from the great oolite by a thick bed of clay. From Sleaford we may trace the upper beds, which will of course indicate the inner edge of the tract occupied by this series, due south to Peterborough:* then, skirting the confines of Huntingdon and Northamptonshire, and forming the escarpment overhanging the Nen at Raunds and Stanwick† where shelly beds of a blue colour, sufficiently compact to take a tolerable polish, and apparently agreeing with the forest marble, are worked for ornamental purposes. They proceed hence south to Bedford,‡ and then skirt the Ouse to Buckingham and the borders of Oxfordshire; where, after examining the middle and outer edge of this part of the oolitic tract, we will resume our account of their course.

Through this tract, the great oolite forms a zone to the west of the places above mentioned. It does not appear that the line of demarcation between the oolite which is worked in the quarries at Stamford and Kettering on this line, and the upper beds of this series, can be traced with precision. If the calcareous slate of Collyweston (south of Stamford) be justly referred to the forest marble, a line drawn hence by Raunds and Stanwick will however ascertain a part of this division.

The western boundary of the great oolite, and its junction with the subjacent sands, also requires to be more carefully examined through the south of Lincolnshire and Rutlandshire. It appears to extend some way over the summits of the sandy

* The following are the memoranda made by the author of this article in crossing the range of this system of beds from Peterborough by Stamford towards Leicester. "Between Peterborough and Castor a stonebrash soil prevails; at Castor, beds like those of the forest marble may be observed; at Sutton near Castor fine oolite is worked; from Wansford to Stamford a closer-grained buff coloured limestone clouded with blue prevails: the same character extends quite to the edge of the great oolite about two miles east of Uppingham, where it is succeeded by the red coloured beds of the subjacent sands."

† Improperly included by an extension (though a very slight one) of the colour assigned to the Oxford clay, under that formation in Mr. Greenough's Map.

‡ These beds of limestone are worked in the north of Bedfordshire, at Bletsoe, at Felmersham, near Bedford, and at Turvey; and on the north of Buckinghamshire, at Olney, Newport Pagnell, Thornton, and Buckingham; at Skerrington field near Newport Pagnell, the palates and teeth which have been considered as characteristic of the forest marble, have been found together with the jaw of a species of lacerta.

hills which overhang the vale of Belvoir. In Rutlandshire, it extends within two miles on the east of Uppingham,* a broad interval between the edge of the great oolite and the lias, being occupied by the subjacent sands.

From the neighbourhood of Uppingham, the junction of the great oolite and sands appears to range by Rockingham to Kettering: hence a narrow tongue of the great oolite proceeds south-west towards Northampton, on the north-east of which town it forms some well defined hills between Weston Favel, Kingsthorpe, and Boughton, but in many of the intermediate points its course is very obscure; appearing, as it were, entangled among the subjacent sands which occupy a broad denuded tract on both banks of the Nen, but particularly on the north, between Peterborough and Higham Ferrers. On the south of this denudation, the great oolite, skirting round Higham Ferrers, proceeds along the summit of the hills hanging over the right bank of the Nen (not however quite extending to their escarpment), and ranges about a mile south of Northampton. It thence passes a little on the south of Blisworth Tunnel on the grand junction canal; but there occurs a detached portion of it considerably to the north of this line on the east of the road from Northampton to Althorp, extending from Duston to the edge of Lord Spencer's park. This oolitic tract is not in the regular and elevated plane of that stratum, but appears to be a vast subsided mass, being situated at a considerably lower level than the hills of the red sand (inferior to the oolite in regular geological position) which overhang the town of Northampton; particularly Hunsborough hill, distinguished by its ancient circular camp: yet it is clearly identified by its characters and fossils with the oolitic beds which occur on the summits and backs of these very hills, in their regular and undisturbed position. The exact boundaries of this insulated oolitic tract are often obscure; as far as it was possible to trace them in a single examination, they are correctly indicated in Mr. Greenough's Map.

West of Blisworth Tunnel, the oolite advances to the north as far as Stowe-nine-churches; but a considerable denudation again exposes the subjacent sandy strata in the neighbourhood of Towcester, round which town the inferior edge of the great

* The colour of the inferior oolite and sands in Mr. Greenough's Map is here in some copies carried too far to the east; but the dotted lines are correct, or nearly so.

† The lines of junction given in this part of Mr. Smith's Map are purely imaginary. Mr. Greenough's exhibits a near approximation to correctness; absolute correctness could, from the obscurity of the ground, be attained only by an observer long resident in the county.

oolite skirts on the south and west, about two miles distant from it, proceeding westwards, by Blakesly to Culworth (the exact line of junction being however in this part of its course frequently obscure) : hence it trends to the south, forming the summit of the range of hills which skirt the valley of the Cherwell about five miles distant from the course of that river, which it approaches still nearer at Aynhoe.*

III. *Course through Oxfordshire.* Having thus traced the inferior edge of this oolitic system into Oxfordshire, as we have previously done its upper edge, we shall in the next place trace both the edges through that county, which will on many accounts form a convenient division. Resuming then our account of the upper beds on the borders of Buckinghamshire and Oxfordshire, whither we have already pursued them, we find the same blue and shelly beds which we had noticed at Buckingham, worked at Marsh Gibbon, Ambrosden, and Merton, † forming a very low swelling ground on the north of Otmoor, and separated by a tolerably thick bed of clay from the great oolite; at Bletchingdon about three miles on the east of Merton, a rock appearing to possess similar characters with the forest marble has been worked for tomb-stones, &c. The pillars in the inner quadrangle of St. John's college Oxford, are from these quarries, and the chimney-pieces in some of the neighbouring villas built in Charles the Second's time, appear

* The residence of the author of this article in this part of the country has enabled him to trace this portion of the course of the oolites more in the detail. Culworth village stands on a brow formed of the subjacent sandy strata, against which the oolite here appears to terminate abruptly, ranging immediately on the south of the village: hence, the line of junction ranges south, passing just above the brook on the east of Thorpe Mandeville church, where the whole summit of the hill is occupied by the subjacent sand There is however a subsided mass of the oolite immediately at the foot of the hill on the north-east of the church; and about five miles to the north of this place and of the general line of junction, there is another subsided mass, near Woodford, half a mile east-south-east of the village; the beds here dip under an angle of 30° to the south, and are worked as limestone quarries. Proceeding south, the oolite attains the brow of the escarpment looking down on Marston St Lawrence, on which the villages of Gretworth and Farthinghoe are built; thence continuing above the Spa at Astrop, to Aynhoe. Strictly speaking, the beds which have been here denominated oolite, are, at their line of junction with the sands, rather to be considered as referable to the fullers' earth rock of the neighbourhood of Bath (that which stands at the head of the next section) with which they closely agree in their characters and fossils.

† Mr. Smith has extended the colour of the clunch clay over this ridge; and the same error is committed in most copies of Mr. Greenough's Map; but the dotted lines placed to direct the artist who coloured the maps, indicate it, though they have not been attended to. The lines drawn to indicate the course of the forest marble in this part of Mr. Smithis Map, are purely imaginary.

to have been formed of it. The rock on the south of Bletchingdon, as far as Islip, and on the first hill beyond this village on crossing the little river Ray, agree sufficiently both in character and position with the cornbrash. At the point last mentioned they approach within half a mile of the coral rag series; so that the intervening clay must be here comparatively thin.

If any regular division of the upper beds of this series from the great oolite can be here traced, it must probably cross the Cherwell a little to the north of Enslow bridge; since the beds at Gibraltar, close to this place, agree most nearly with the Cornbrash. Here the remains of a well characterised crocodile have been found. The section afforded by a well sunk at Blenheim appear to confirm this line; it is as follows; limestone rock 70 feet; blue clay succeeded by a lighter clay in which was the water, 10 feet; total depth 80 feet: the water was probably in this instance carried by the clay above the great oolite. The crop and course of this clay on the surface, has not been clearly ascertained, but may probably be traced along the ravine between Ditchley and Blenheim parks, and then along that on the north of Stonesfield which opens at the village of Fawler (south of Charlbury) into the valley of the Evenlode. Crossing the Evenlode, the forest marble beds appear to crown the hills through the forest of Whichwood, whence its further course will be most conveniently treated of when we proceed to trace the course of these beds through Gloucestershire, since the Evenlode seems to form the natural boundary between the system of hills connected with the uplands of North Oxfordshire, and those dependent on the Cotteswold hills.

The boundary of the Oxford clay and Cornbrash may be sufficiently assigned by a line drawn from Kidlington on the River Cherwell to the north of Lechlade on the borders of Gloucestershire.

Having thus traced the inner edge and upper strata of this oolitic series through Oxfordshire, we may proceed to trace the inferior and exterior edge through the same district, since the natural divisions of the country coincide in this instance with the political, inasmuch as the range of these beds through Gloucestershire (on which we shall next enter) forms the continuous range of the Cotteswold hills, which, having suffered less from the denuding causes that appear to have ravaged the tract now under consideration, present less complicated phœnomena.

We have before traced the outline of the great oolite at its junction with the subjacent sandy rocks through Northamptonshire to Aynhoe on the borders of Oxfordshire, and near the banks of the river Cherwell. It crosses that river between the villages of Steeple Ashton and North Aston.* Close to the brow of North Aston hill, near the road from Oxford to Banbury, may be seen a junction of the fullers' earth rock or lowest bed of the oolite with the subjacent sand, and below this, near the bottom of the valley between it and Deddington, a subsided mass of the same rock.

From North Aston the junction passes by Duns Tew, and then crosses the valley on the south to Worton Wood on the north of Sandford. The oolitic beds then keep the summit of this hill to Great Tew,† and thence range about half a mile on the north-west of Swerford, forming a terrace above the general ridge of the sand hills near that town. In the valley, however, between Swerford and Great Rolwright, they are thrown down by a considerable subsidence, pitching in a steep angle towards its bottom, which they cross close to the entrance of Swerford park; and lying here, as in a hollow trough, again rise towards Great Rolwright. The ridge of Rolwright hill, as far as Cornwell on the confines of Worcester, Gloucester, and Oxfordshire, forms the extreme line of the great oolite; but its lower beds are here very sandy, and not to be distinguished without difficulty from the substrata. The same construction prevails in Chipping Norton hill; the valley separating these hills, and those intersecting their branches, exhibiting the sandy substrata. The junction passes by the hamlet of Churchill, and crosses the Evenlode near Ascot d'Oiley on the east of Shipton near Whichwood.‡

On the north of the line above described, between the Cherwell and the Evenlode, is an extensive district occupied by the brown ferruginous sands so often mentioned as forming the substratum on which the oolites repose in this part of their course. Several long and somewhat lofty ridges, connected on the north with the table land of Edge hill, are constituted by these beds; but in the lower situations throughout this tract, are scattered several of those remarkable insu-

* Mr. Smith's representation of the line of this junction through Oxfordshire, in all his Maps, is purely imaginary; it can indeed only be traced by continued residence on the spot, which is the authority on which the above particulars are given. The representation in Mr. Greenough's Map is perfectly correct: we have been particular in stating the details, as they have never yet been presented to the public.

† The vallies within this tract, however, occasionally expose the subjacent sandy beds by denudation, as may be seen at Gagingwell near Euston.

‡ Mr. Greenough's Map carries the inferior oolite too far eastwards along this valley.

lated and subsided portions of the rocks belonging to the great oolite (the original superstratum which probably extended over this district before the excavations of its vallies) to which we have already alluded in the beginning of this article: of these a full account will be found in the subjoined note.*

* Since these insulated and subsided masses of the oolitic beds, distant as they often are from the general line of its present limit, present an interesting geological phœnomenon, and are also important to the agriculture of this district, as affording the only limestone quarries scattered through an extensive sandy tract, it may not be unacceptable to add a more detailed account of their localities, in which we shall refer to the great and accurate survey of this county executed by Mr. Davis of Lewknor, and published by Mr. Cary. Mr. Greenough's Geological Map exhibits all the great features of this district with as much accuracy as can be attained on a scale so reduced.

If then we begin this description, commencing from the valley of the Cherwell, we find the first tract of this nature about three-fourths of a mile west of Banbury, in the fields along the summit of the low hill rising above Neithrop, and immediately on the east of Withecombe farm. This patch does not contain above a few acres: the strata are considerably inclined, and well identified, by their organic remains and general character, with the lower beds of the great oolite in this district (probably those analogous to the fullers' earth rock of Bath), although the general line of these beds is seven miles to the south. The second patch (which exhibits exactly similar beds) occurs about three miles west-south-west of the last, where the most northerly of the two roads marked in Mr. Davis's Map from Broughton to Lower Tadmarton crosses a small brook about a mile west of the former village; where the road descends the brow towards this brook, quarries of these limestone beds may be seen extending on the right towards Page's farm; and where, having crossed the brook, it ascends the opposite brow, similar quarries may be seen. It is scarcely possible to ascertain the exact limits of this patch, but they are probably not very extensive. 3. On the left of the lane leading from Lower Tadmarton to the Danish camp on the adjoining hill on the south, about a furlong from the village, and just before another lane leading from Upper Tadmarton joins it on the right, is another insulated quarry of these beds. 4. Descending westwards from the hill on which this camp, supposed to have been the station of the Danish army before the battle of Hooknorton, is situated, and which commands all the neighbouring country, if we proceed along the lane by Hooknorton lodge and about a mile beyond it we find a more extensive patch of these limestone strata extending along the lower platform on the north of Hooknorton. 5. About a mile and a half west of these quarries, on the very borders of Oxford and Warwick shires, a little beyond the foot of the north-east branch of Great Rolwright hill, and close to the point where the old road leading hence to Edgehill (apparently an ancient trackway, skirting the edge of the escarpment, since it here forms the county boundary) crosses the Stour near its head at a place called Traitor's ford, is a subsided portion of the same beds rapidly changing their dip from south to north. 6. About two miles east of Epwell, close to the point where the lane from Epwell to a farm called Lower Lays, crosses the lane from Shutford to Brails, and on the west side of this cross, is a similar patch extending on both sides the brook. 7. About a quarter of a mile west of Epwell, and a little on the north of the Warren house, where the two lanes cross each other, is another subsided portion of these limestone beds, dipping about 50° to the north. 8. On the summit

IV. *Course through Gloucestershire, from the Evenlode to the Avon.* Having thus pursued these beds to the confines of Oxfordshire and Gloucestershire, we find them, in the latter county, attaining their greatest thickness and most decided character, and the range of hills constituted by them assuming greater height and bolder features, and forming a more continuous chain: this is well known under the name of the Cotteswolds.

Following the *upper edge of the series* along the first rise of this chain, we find the junction of the Oxford clay and the highest beds, or Cornbrash, passing a little north of Lechlade and Cricklade, with a westerly course, and then turning southwards by Malmsbury, Chippenham, and west of Melksham: a little on the north and west of this line, the hills of Whichwood forest exhibit on their summits beds agreeing in aspect with the forest marble, which cross the Windrush a little north of Witney, and thence range immediately to the south of Tetbury. In this part of Gloucestershire, the clay separating the forest marble from the great oolite is stated (see Agricultural Survey) to be 80 feet thick. At Tetbury, the upper beds turn to the south, cross the London road from Bath on the west of Atford, and ranging on the inner slope of the hills which skirt the right bank of the Avon, reach that river immediately on the east of Bradford; where the clay covering the great oolite is well displayed, and its fossils most abundantly found.

The *inferior edge* of the great oolite, having crossed the Evenlode a little east of Shipton under Whichwood, follows the escarpment of the Whichwood hills on the right bank of the above river, following the branch of those hills which extends north towards Stow on the Wold, but is separated from the high ground crowned by that village, by a breach communicating with the valley of the Windrush; this valley, forming a deep denudation and cutting into the inferior beds, as far south as Burford, makes this branch of the Whichwood hills a long peninsulated ridge crowned by the great oolite, which has its junction with the inferior and sandy beds immediately above Rissington on the west, and Idbury on the east.

At Stow on the Wold, the continuous range of the Cotteswold commences, although the Whichwood groupe, already

of Overbrails hill, a very conspicuous outlier of the sandy beds, rising from the lias in the vale of Shipston upon Stour (marked with the letter *v* in Mr. Greenough's Map, but inadvertently coloured as lias) is a small patch of the oolite confined to a single field; but this is not a subsided mass, like those above described, but merely an instance of a part of the stratum catching the summit of an insulated hill in its regular plane.

described, must certainly be considered as an appendage to it. The inferior junction of the great oolite is on the brow of the hill about a mile east of Stow, but is indistinct; it thence follows the escarpment of the chain, which here forms a bold cape extending far to the north, into the confines of Worcestershire; the vale of Shipston on Stour lying at their foot as a broad bay included between this range on the west, and the lower range of Edge hill (whence as we before observed the oolite cap has been denuded) on the east. The Ilmingdon hills (almost separated from the main chain by the valley of Campden) form the northern point of this Cape, at the very extremity of which lies the insulated and lofty summit of Meon hill (marked l in Mr. Greenough's Map). This, standing in advance of the whole Cotteswold chain, and looking down on the great central plain of England, commands one of the most extensive prospects in the island. The sienitic summits of Charnwood forest may be faintly distinguished in the north-eastern horizon, and Caer Caradoc and Clee hills on the north-west; the long ridge of Abberley, marked by its three conical summits, and the nearer range of Malvern, illustrating by its abrupt forms and serrated outline the idea of a Spanish sierra, constitute the western boundary; while on the south-west the eye follows the escarpment of the Cotteswolds, and insulated and outlying groupes connected with it, among which Breedon stands preeminent, catching in the distance between them the high grounds of the forest of Dean: on the south-east, and east, the escarpment of the great chain of the oolite and subjacent sandy strata is seen circling round the vale of Shipston to Edge hill and Arbury hill.

The Ilmingdon hills and Meon hill have a cap of the great oolite; but the beds here displayed, being near their inferior limit, are coarse and sandy.

Hence the general outline of the Cotteswolds turns south * towards Winchcombe, the high platform between this place and Cheltenham being cut off by surrounding vallies from the main chain; the inferior junction of the great oolite always keeps near the brow of these hills. On the north of Winchcombe hill, the outlyers of Tredington † and Alderton, however, though lofty, appear only to exhibit the inferior oolite,

* Mr. Smith makes a long branch of the oolitic hills run off near this point from Aston Sub Edge in a north-westerly direction. He has been led into this error by an incorrect depth of shading in the great County Survey (which is old and very inaccurate), by which a low range of lias at the foot of the escarpment is represented as rivalling the main chain in height.

† Marked *i* in Mr. Greenough's Map, but erroneously covered with the colour denoting lias.

which is probably also the case with the more extensive insulated mass lying still further north between Evesham and Tewkesbury, and constituting the Breedon hills (marked *k* in Mr. Greenough's Map).*

From Cheltenham the escarpment of the hills and inferior junction of the great oolite passes about five miles east of Gloucester (having two outlyers of inferior oolite, Churchill and Robinswood hill, on the north-east and south-east of that city), and pursuing its course to the south, is deeply indented by the vale of Stroud; beyond which it projects in a bold cape hanging over Wotton under Edge, and then continues in a nearly straight line almost due south, ranging immediately west of the road from Gloucester to Bath, to within four miles of the latter city, around which all the streamlets flowing into the Avon run through vallies of denudation deeply furrowing the high platform of the oolitic hills: † thus the brook flowing

* In Drayton's fanciful poem, the Polyolbion, are some very characteristic lines descriptive of Breedon hill; they are put into the mouth of one of his singular local personifications, the vale of Evesham.

> " Yet more, what lofty hills to humble vallies owe,
> And what high grace they have which near to us are plac'd,
> In Breedon may be seen, being amorously embrac'd
> In cincture of my arms. Who though he do not vaunt
> His head like those who look as they would heaven supplant,
> Yet let them wisely note in what excessive pride
> He in my bosom sits, while him on every side
> With my delicious sweets and delicates I trim;
> And when great Malvern looks most terrible and grim,
> He with a pleasant brow continually doth smile."

† The position of all the constituent beds in each of the hills hanging over these denuded vallies round Bath, may be at once determined from the marked features of their outline and profile, by a spectator from a distant view, provided he be acquainted with the general structure of the range. The great oolite forms a flat table land on their summits, ending with an abrupt edge; this is succeeded by a gentle slope which marks the subjacent fullers' earth, a greener verdure and rushy grounds arising from the discharge of the oolite springs thrown out by these clayey beds are here seen: beneath is the lower terrace of the inferior oolite, which breaks down with a steep and almost precipitous escarpment to the vale.

In many instances (as at the north-east of Lansdown near the monument) large broken masses of the great oolite, having been precipitated from its escarpment, are spread over the slope of the fullers' earth, presenting a scene of rocky ruin resembling the undercliff in the Isle of Wight. This may be seen particularly at Warley rocks above Bathford. Similar subsided masses of the inferior oolite are also frequently piled against the foot of its escarpment.

Mr. Townsend gives a particular account of these dislocations; but since they affect the inferior oolite as well as the beds of which we are now treating, we shall postpone our extracts from it to the end of the article on stratification and inclination in the section assigned to the inferior oolite.

from Langridge by Swanswick entirely cuts the ridge of Lansdown from the main chain; and that coming down from Catharine's to Bath Easton, nearly affects Charmey down and Salisbury hill (an insulated summit of the great oolite hanging over Bath Easton on the west) in the same manner. Bathford stands at the mouth of another similar valley extending upwards to Box; and the main valley of the Avon (justly celebrated for its picturesque character) continues to form a denudation in the inferior strata as high up as Bradford, where the crop of the great oolite crosses the river.

The whole range of the Cotteswold escarpment throughout Gloucestershire, has in ancient military operations afforded a strong and commanding line, and is occupied by numerous camps; partly at least belonging to the line of defence fortified by Ostorius; a map of this line may be seen in the Archæologia for 1820.

V. *Course through Somerset and the North of Dorset.* From Melksham, whither we had before traced it, the *superior junction* of these beds with the Oxford clay proceeds in a slightly undulating line southwards,* a little east of Froome, Wincanton, and Stockbridge; beyond which, turning slightly westwards, it meets, and is overlaid and concealed by, the great south-western extension of the chalk.

The clay separating the upper beds from the great oolite (properly so called) has at first a very tortuous line (which is very correctly given in Mr. Smith's Map), occasioned by the configuration and inequalities of the surface cut by its plane, which is varied by lofty hills and deep vallies. Crossing the northern branch of the Avon close to Bradford, it ascends the back of the hill between it and the southern branch of the same river, which it crosses a little south of Farley, and then ranges over the summit of the elevated platform on its left or western bank, on which stand Charterhouse Hinton, and Norton St. Philip. The Cornbrash ranges along the escarpment on the right bank of the stream by Tellisford,† having a slight cap of the lowest beds of the Oxford clay abounding with selenite, and appears on the left bank as an insulated cap to the hill on that side between Tellisford and Norton. There is another insulated cap of this rock, not exceeding a few acres, close to

* The outline in Mr. Smith's Geological Map is here preferable to Mr. Greenough's, which is distorted by throwing the limestone of Stourminster (probably belonging to the coral rag) into the great oolite.

† This is the point to which the section at the beginning of the paragraph on the chemical and external characters of the great oolite particularly refers.

Pipe-house on the north of Hinton.* The calcareous sand and forest marble circulate round this tract by Hinton and Norton, the subjacent clay appears on the slope descending from Hinton to the great oolitic platform on the west of that village; hence, it follows the outline of that platform, keeping about a mile within its escarpment by Norton, Hardington, and Buckland Denham, ranging towards Froome. South of that town, the junction of the upper beds and great oolite appears to range towards Dorsetshire in a slightl west lulating line, parallel to that before assigned to the junction of these beds with the Oxford clay, and from two to four miles west of it, but it has probably never been accurately traced.

The *inferior junction* of the great oolite (that along which it rests on the fullers' earth) crosses the northern branch of the Avon a little on the west of Bradford, ranges round the upper part of the escarpment of the hill between this and the southern branch, crosses the latter close to Farley mill, and keeps the escarpment of the great platform of Hinton and Norton, passing along the brow of the higher range of hills above Mitford, Writhlington, Kilmersdon, and Mells.

The ridge of Hampton,† Claverton, and Odd Downs, hang-

* The hill between Mitford and Freshford affords the best opportunity near Bath for examining the whole of this series of strata; it is within three miles of that city; the vallies are cut down into the upper beds of the lias marle, and in ascending the hill, the inferior oolite, fullers' earth, great oolite, Bradford clay, forest marble and sandstone are crossed in passing towards the insulated patch of cornbrash, which, at Pipe-house, crowns the hill. The summit of this hill is covered with transported chalk flints.

† The quarries at the head of the inclined plane above Bathhampton, afford one of the best examples of the great oolite which can possibly be studied. The upper beds are here thin, and separated by clay seams (at one point a large imbedded mass of clay also occurs), and by a loose oolitic sand; the surfaces of these slabs present a congeries of organic remains, chiefly terebratulæ, small coralloid bodies, spines of several species of Echinus, and occasionally joints of the Bradford pear Encrinite (Apiocrinites of Miller). Fragments of the large fibrous shell which Mr. Sowerby considers as now ascertained to be a species of oyster, also occur. These beds occupy about six feet from the summit; then follows a bed about two feet thick, containing several immense masses of an undulating madrepore, and several tubipores; beneath this is the solid oolite which is quarried to the depth of about 30 feet. The lowest bed worked contains several large bivalves. Beneath the floor of the quarries, the beds of the great oolite extend on the slope of the hill, between 60 and 100 feet, forming a steep escarpment; beneath which, the black beds of the fullers' earth clay (full of fibrous calcareous spar) may be observed throwing out numerous springs. Vast blocks, precipitated in every direction from the great oolite, are however spread over the whole slope of the fullers' earth; forming a confused scene of projecting crags, over which it is scarcely possible to walk. It is remarkable that chalk flints are scattered pretty abundantly over this tract: they are also found on the summit of

ing over Bath on the south, forms a long insulated range, capped with the great oolite, advanced on the north of this platform, encircled along its scarped sides by successive zones of the inferior beds; and the ridge north of Wellow, between the above and the Hinton platform, is similarly situated.†

Most of the rivulets which in this part join the Avon, rise beyond the escarpment of the whole oolitic series, and flow through vallies which traverse the chain and divide it into many insulated groupes: the fact is similar to that which was formerly noticed with regard to many portions of the chalk range, and the whole configuration of these vallies bears the same evidence that it involves an absolute physical impossibility to attribute their formation to the streams which now employ them as a channel.

The junction of the great oolite and fullers' earth may be most distinctly traced through the whole of this district. But near Froome, the horizontal beds of the oolitic series abut abruptly* against the inclined strata of the mountain limestone connected with the lofty chain of Mendip, which here begins to rise, although as yet its peculiar strata are only exposed by denudation in the lower part of the vallies between Froome and Mells. It is the inferior oolite which is generally in contact with the mountain limestone, and therefore these vallies will

many of the adjacent downs, warranting the same geological inferences which were formerly drawn from the accumulation of transported pebbles on the top of the coral rag hills near Oxford. See note, p. 190.

† There are other smaller insulated summits of the great oolite resting on the platforms of the inferior oolite near the above ridges; such is Duncom hill north of Dunkerton, and Newbury hill above Mells.

* In order to explain how beds of such distant geological ages as the mountain limestone and oolite, come into immediate contact, in consequence of the difference of inclination through which their planes cut each other, we subjoin the following diagram.

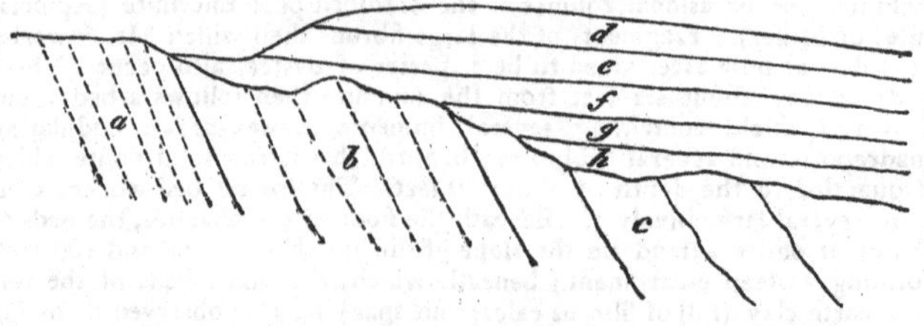

a. Old red sandstone. *b.* Mountain limestone. *c.* Coal measures. *d.* Great oolite. *e.* Fullers's earth. *f.* Inferior oolite. *g.* Lias. *h.* Red marle.

All these circumstances are illustrated in the district of the east end of the Mendips between Froome and Mells.

be more particularly described hereafter. Near Frocme, however, the great oolite approaches very near to, if it does not quite touch, the mountain limestone, and the line of bearing of these beds being thus thrown out of their regular course, it becomes doubly difficult to trace them: a bed of clay in the valley a mile west of Froome appears, however, to be that associated with the fullers' earth, and hence the inferior junction of the great oolite probably ranges by Cloford to Bruton; whence it seems to proceed along the edge of the hills on the west of Charlton, Horethorn, Milford, Port, and Sherborne, a few miles south of which the crop of this oolite is, like that of the other oolitic systems, overspread by the great western extension of the chalk and green sand through Dorsetshire.

VI. *Course in the South of Dorsetshire.* On the south of this overlying extension of the chalk and green sand, the beds of the present oolitic system re-emerge from beneath it near Bridport.* The rock here assumes the fissile character of forest marble, and it has been sometimes exclusively referred to that subdivision; but since it is in immediate contact with the fullers' earth clay and inferior oolite, and the thickness of the series seems also inconsistent with this supposition, we are rather inclined to believe that the whole oolitic series of the present formation here puts on this form.

This rock ranges to Bothenhampton, on the east of Bridport valley, having however an insulated portion capping the cliff on the west of Bridport harbour; but, on the east of the harbour, the cliffs exhibit only inferior oolite for more than two miles, where the second cliff in that direction (Burton cliff) subsides, and is succeeded by a low bank of the fullers' earth marle, characterized by its fibrous calcareous spar. Above this marle are quarries of the fissile oolite, which proceeds hither from Bothenhampton. The cliffs here terminate, a shingle bed (the commencement of the great Chesil bank) being interposed between the hills and the sea. The oolite is plainly to be traced through the base of these hills from Swyre to Abbotsbury, forming their general platform, which supports insulated summits of the overlying green sand at Swyre beacon and Abbotsbury camp.

Thus far the structure of the district is sufficiently regular, and the strata horizontal or nearly so; but beyond Abbotsbury

* All this district, and indeed the whole south of Dorsetshire, is erroneously represented in Mr. Smith's Map. Mr. Greenough's is correct in the neighbourhood of Bridport, but requires, between Abbotsbury and Weymouth, the alterations mentioned in the article on the course of the coral rag.

(where there appears to be a fault), we enter on the elevated Weymouth district, which we have before had occasion to describe, in treating of the Portland beds and coral rag. At Abbotsbury the coral rag advances close to the curve of the coast, so that we lose the beds of the present oolitic system for a time; but they reappear a little to the south in Fleet down, where they form a saddle extending across the head of the Weymouth æstuary near which it terminates, being encircled (as was formerly stated) by collateral zones of Oxford clay, coral rag, &c.

(*e*) *Height of hills.* As has been stated in the preceding article, the whole course of this system across the island is marked by an almost continuous range of hills. Sometimes, however, the whole escarpment of these hills is formed by the inferior oolite, and the great oolite forms a slight upper terrace ranging at a distance inland. In Yorkshire, we are inclined to attribute the eastern moorlands to the inferior sands, and the escarpment which skirts them on the south to the great oolite; but this, being a point open to controversy, must be considered accordingly. Crossing the Humber through Lincolnshire, the hills are low; they gain greater height in Rutlandshire, Northamptonshire, and Oxfordshire, but the highest points in this district are the summits rising near the edge of the escarpment, (as Arbury and Epwell hills) which belong entirely to the inferior sands, the great oolite ranging considerably on the south-east. In Gloucestershire, however, the great oolite, always crowns the brow of the escarpment, and gains its greatest height: the loftiest point of these hills is Cleeve hill, near Cheltenham, which is 1134 feet above the level of the sea; the next, Broadway, is 1086; Stow in the Wold is 883, and Landsdown near Bath is 813 feet.

These hills, thus breaking down with an abrupt escarpment to the north-west, fall with a very gradual declivity to the south-east. This circumstance is common to the chains formed by all the beds above the new red sandstone, and arises from the cropping out of their strata in the former direction against the older rocks of the north and west. An observation of Gilpin's, in his tour to the Wye, strikingly illustrates this inequality of fall on the opposite sides of the watershed of this chain. Standing on the high grounds between Cirencester and Cheltenham, where the head waters of the Thames take their rise on the back of the Cotteswolds, he remarks that while on the one hand this river wanders to the east through a course of more than 150 miles before it reaches the level of the tide at Richmond, on the other the Severn, flowing within 12 miles of this point at the western foot of the hills, is already a tidal river.

This range, like that of the chalk, is entirely broken through by many vallies; such are those of the Yorkshire Derwent, of the Humber, of the Witham in Lincolnshire, of the Charwell and Evenlode in Oxfordshire, and of most of the streams which join the Somersetshire Avon on the south of Bath; especially the branch of that river which flows from Emborrow Mere at the foot of the Mendips by Mells and Froome.

The insulated hills advanced beyond the chain in Gloucestershire have been already mentioned, but they generally belong rather to the inferior oolite than to the beds of which we are now treating. Many of the ranges near Bath, as Lansdown, Claverton Down, &c. which are capped with the great oolite, are, from the course of the last mentioned vallies through them, completely insulated. (See the particulars in the account of the range of these beds.)

(*f*) *Thickness*. The thickness of these beds is given in the list at the beginning of the present section for the neighbourhood of Bath: altogether they may be taken in round numbers at about 250 feet in this district.

On Farley Down near Bath, on Coombe Down, and in several other places near Bath, wells have been sunk through the great oolite to the depth of 100 to 130 feet. (T. 129.) The actual thickness of this bed alone seems here to vary from 130 to 150 feet, but it must be considerably thicker in the northern parts of the Cotteswolds, as must the superior beds on the south of Cirencester: 400 feet would probably be not an excessive allowance for the whole series in their thickest part.

(*g*) *Inclination*. The great oolite has a gentle dip towards the south-east, as have all the beds of this formation; there are however some apparent local exceptions which merit notice, more perhaps from their singularity, than from possessing any real importance. The following sections, afforded by quarries between Cross Hands and Petty France in Gloucestershire, are communicated by H. J. Brooke, Esq. M.G.S. who also referred us to a somewhat similar section of a quarry at Coombe Down near Bath, in the Monthly Magazine, with a description by J. H. Moggridge. dated March 17, 1796.*

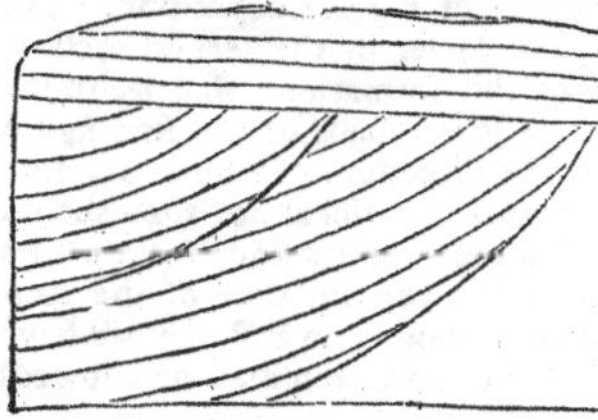

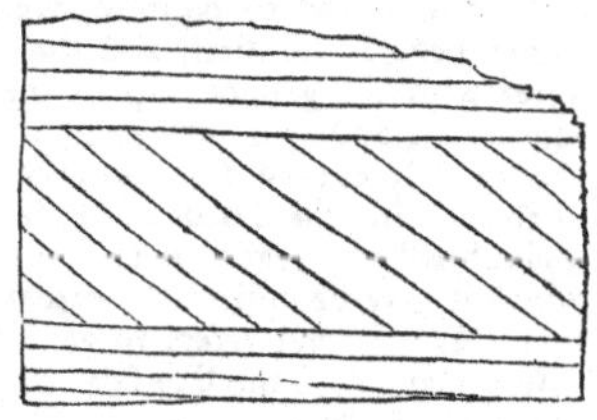

* It would be difficult to imagine that these beds, so differently circum-

In the Weymouth elevated district, the beds of this formation, which form its saddle, are sometimes considerably inclined.

The most remarkable phœnomena which require notice under this head, are those of the large subsided portions of these beds now found pitched at the bottoms of vallies traversing the subjacent rocks and at a considerable distance from the present line of extension of the formation to which they belong. We have already particularised many instances of this kind in the north of Oxfordshire. They must be considered as the remnants of the strata, once generally covering the hills at whose base they are now found; which, having been undermined by the causes that have excavated the vallies, have slid down into their present position, where they are naturally thrown into a direction inclined in general at a considerable angle to the horizon.

The precipitated masses of the great oolite scattered over the slope of the subjacent fullers' earth on the hanging of many of the hills round Bath, are phœnomena of the same kind.

The phœnomena of these subsided masses which have assumed their present position in consequence of having been undermined by the excavation of the subjacent soft strata, and thus slidden downwards, must not be confounded with those cases of subsidence which are accompanied by vertical fissures or faults traversing the strata to a great depth, and in which the whole series of strata affected by them (soft as well as hard) is let down together. Both the circumstances and causes of the latter are clearly distinct from the former; no partial undermining action can account for them; but they must be referred to causes seated at a great depth and acting in the same manner as the force producing earthquakes.

Further particulars of these undermined and precipitated masses in the neighbourhood of Bath, will be found at the end of the article corresponding to this in the next section; since, as the inferior oolite partakes of these disturbances, it will be more convenient to keep the description of the whole of these cases together.

stanced in regard to position, were the effect of regular deposition. The upper beds in the first, and the upper and lower in the second section, agree with the general dip of the beds of this formation; that is, to the south-east, and at a very small angle. The lower beds of the first figure are not only curved, but also take the reverse direction. Instead therefore of considering these beds as the result of regular stratification, they should undoubtedly be regarded as the consequence of contraction, from causes which it may be difficult to explain, during the consolidation of the stratum. Townshend refers to several similar instances near Bath: we have before mentioned the like false cleavage (if the expression may be allowed) in the coral rag.

(*h*) *Agricultural character.* The Cornbrash, says Mr. Smith, is chiefly an arable superior in quality to much of the Stonebrash hills; and, when otherwise appropriated to pasture, produces grass of a good quality. Indeed the mixture of calcareous, argillaceous, and arenaceous beds, in this part of the series, is favorable to their agricultural quality. The general course of the forest marble through the Cotteswold district, is said to be distinguished by the prevalence of timber trees, woods, and pasture. The soil over the great oolite is a loose stonebrash, absorbent, and any thing but rich; it is said however to answer fairly for turnips, and tolerably for wheat; but it never averages half the value of the subjacent sandy soils.

(*i*) *Phænomena of water.* The clays underlying the cornbrash and the forest marble generally hold up the water beneath these strata, so that this indispensable article may be readily procured; hence a more dense population (as Mr. Smith has very ingeniously remarked) distinguishes the course of these beds from the great oolite, where water can be obtained only in deep wells and at a great expense. These wells have often been sunk 130 feet through the rock to its junction with the fullers' earth, which throws out its springs and forms a weeping ground round the escarpments of the oolitic hills, as may be particularly noticed near Bath. Occasionally even the springs of the upper beds sink through this rock also, in consequence of some failure in the intervening clay; this is particularly the case in the forest marble, which has numerous swallow-holes, thus absorbing the springs of the cornbrash; thirty of these may be noticed in the space of half a mile round Hinton.

It is probable that the cornbrash or forest marble may be the true seat of the mineral waters found in sinking through the Oxford clay, and enumerated in treating on that formation. If the Pickering hills in Yorkshire really belong to this part of the series, the curious phœnomena of the several branches of the Rye near Helmsly, which flow through that escarpment by subterraneous channels, should be here noticed.

Section VI.*

Inferior members of the Third or Lower system of Oolites; including—the Fullers' earth, Inferior oolite, Sand and Marlstone of Smith.

The Oolitic system of which it was the object of the preceding section to trace the superior portion, is continued (or rather, considering the order of formation, introduced) by the series of beds which we are now to consider. These, although in some localities presenting well marked lines of distinction, in other instances run so much into one another by insensible gradations, that it is far from easy to assign any exact point in which any given member of the series can be strictly said to commence, or accurately to distinguish the contiguous members from each other.

Speaking generally, the great mass of calcareo-siliceous sands on which the rock called the inferior oolite reposes, may be said to form the most universal and characteristic feature of this series. These sands pass almost insensibly, by the mixture of various loamy and marly beds towards their lower limit, into the argillaceous formation which covers the lias; and towards their upper limit by an increase of calcareous matter, into the oolitic beds called, from their position, inferior. Between this inferior oolite, and that which we have already described in the last section under the name of the great oolite, a thick calcareo-argillaceous formation carrying beds of fullers' earth (whence it is denominated), and sometimes also beds of coarse oolite, is interposed, often forming a very conspicuous division in this part of the series. But, in other instances, this calcareo-argillaceous mass is either wanting, or by the prevalence of its calcareous matter passes into the form of an oolitic limestone; so that the limit of the great and inferior oolites can only be ascertained (if at all) by an accurate examination of their organic remains.

It must at once appear from the above observations, that we cannot as yet regard the exact demarcation of all these subordinate beds as being completely ascertained throughout the island; but this acknowledgment ought rather to be regarded as a proof of the high advance of geological information which has rendered the adjustment of points so minute an interesting object of enquiry, than as affording any ground for questioning its progress, in the consideration of the few and immaterial deficiencies still left unsupplied.

* By the Rev. W. D. Conybeare.

In treating of this part of the series, we shall again select the neighbourhood of Bath as affording the best type of its arrangement. The separate beds appear in this district most strongly characterized and most distinctly divided; they have also been most attentively examined. We proceed to consider them in the regular descending order.*

(*a*) *Chemical and external characters. Fullers' earth.* In the district above referred to, and as it should appear, very generally elsewhere, the great oolite of the preceding section reposes on a thick bed or series of beds of calcareo-argillaceous formation, which usually carries one or more indurated and rocky strata, besides frequent courses of a soft rubble-stone. In these the calcareous matter predominates. The hardest bed of this stone is blue in the interior, and used for mending

* The following may be considered as a continuation of the section referred to in the note at the head of the articles on the upper members of this system (page 202), and is numbered accordingly. It enumerates these beds as they are exhibited in the neighbourhood of Bath, and may be compared with those subsequently given from other localities, as notes to the inferior oolite: it is extracted from the list in Mr. Warner's Bath Guide.

	No.		feet.
Fullers' earth.	9.	Yellow clay.................................. Rather loose and porous; visible all round Bath by the slips on the declivities of the hills, occasioned by the springs flowing on the surface of the next stratum: it contains no fossils and is applied to no use.	12 or 15
	10.	Blue clay .... This stratum being compact throws out the upper springs round Bath; it contains terebratulæ.	12 or 15
	11.	Good Fullers' earth	8
	12.	Bastard Fullers' earth .... Having in its middle a thin rock which is abundantly furnished with organic remains (see the list under its proper head). Numerous corals are also found at its foot. This rock makes the best road-stone in the neighbourhood.	100 and upwards
Inferior oolite and sand.	13.	Inferior oolite or bastard freestone .... The upper part of the stratum is hard and is used for the roads; the lower is cut into tolerably good ashler near Froome: its fossils are numerous as will be seen in the list.	30
	14.	Calcareous sand .... At the foot of it is a bed of large Pectens and masses of coral, also confused masses of serpulæ mingled with belemnites. This constitutes the marle-stone of Smith, and is used in mending the roads.	50 and upwards

roads. Blue and yellow clays alternate with the above substances, and generally contain varieties applicable to the purposes of fullers' earth. The disposition of these beds near Bath, will be seen in the general section in the note below.

Inferior oolite and Sand. The inferior oolite is very generally distinguished from the great oolite by the larger proportion of brown oxide of iron disseminated through its mass, which sometimes occurs in the form of numerous minute globular particles occupying the same situation which the oviform particles hold in the great oolite. This variety is found in Dundry hill, and corresponds exactly in character with specimens from Bayeux in Normandy. From this admixture of iron the beds have generally a brownish cast, passing in the interior into a blue or grey tinge: the texture is coarse, and they are usually applied to mend the roads: occasionally, however, they yield a very good freestone for agricultural purposes, as particularly at Dundry hill near Bristol and Doulting hill near Shepton Mallet, in both which places the quarries are very extensive. A larger proportion of siliceous sand is usually mixed with the calcareous matter than in the great oolite, and therefore renders it a less profitable limestone. The fossils also, which are numerous, serve to distinguish it.

Near Bath, several regular and continuous strata of this rock rest on thick beds of a slightly calcareous sand, containing courses of irregular calcareous concretions; and near the bottom, where the proportion of calcareous matter is less, strata of soft sandstone.

These strata preserve a great uniformity of character through the south-western counties, and may be very advantageously studied in the fine sections on the coast on either side of Bridport harbour. An account of these will be found in the note, and may serve as a general type of their disposition in this quarter.*

* *Section of Down cliff between Seaton and Thorncombe, two miles west of Bridport harbour.*

The hill rising above the cliff exhibits a cap of forest marble, or rather of the great oolite assuming the character of that rock; beneath this the cliff exhibits

		feet.
1.	Inferior oolite and sand alternating, the sand towards the top passing into marle, about	80
2.	Sandy marle	50
3.	Rusty sand with ferro-argillaceous concretions whose cavities are filled with sand	50
4.	Greenish blue micaceous sandy marle, containing indurated concretions of similar constitution	80

This passes into the lias marle, on which it rests.

Immediately on the east of Burton cliff (about three miles east of Bridport harbour) the fullers' earth may be seen resting on the above beds.

In the midland counties the calcareous matter is less abundant in these beds, and a ferruginous sand and sandstone, containing a very small proportion of lime, predominates. The first beds, as exhibited in the section last referred to, appear to be wanting, or at least are so blended with the fullers' earth rock and great oolite as not to be distinguishable; while strata analogous in character to No. 2, 3, and 4 of that section are here exhibited on a much larger scale, and prevail exclusively over extensive districts, constituting that broad tract of red, or rather reddish-brown ferruginous sands, so well known in the north of Oxfordshire, Northamptonshire, and Rutlandshire.

In Oxfordshire, the series below the fullers' earth, as far as they can be ascertained in the neighbourhood of Banbury, appears to be,

1. Sand and sandstone with a slight calcareous mixture, highly ferruginous and frequently micaceous, sometimes containing large scales of mica in great abundance; the sandstone sometimes forming hard flagstones, but more generally soft. Few fossils, and those chiefly belemnites, occur, but towards the bottom are some beds containing rather more calcareous matter and more abundant fossils; these are externally of a rusty brown, and internally of a greenish grey colour, derived from the suboxide of iron. The white pearly shells of numerous plicated terebratulæ, scattered through this dark ground, and having their interior coated with crystallizations of calcareous spar, give a striking and pleasing character to specimens of these beds. Ammonites, belemnites, and gigantic limas also occur in them.

The calcareous sandstones in the upper part of this series, afford an indifferent and very unsightly material for architectural purposes, as the rusty looking buildings which characterise the neighbourhoods of Banbury and Northampton sufficiently evince.

The whole of this sandstone series is generally separated from the great oolite by a thick clay, corresponding to the fullers' earth clay; this may be seen particularly in the descent of Rolwright hill towards the east. Beds of micaceous loam also alternate with the sandy strata.

2. Marle and marly sandstone corresponding with No. 4 of the Down cliff section (page 236), and probably the marlestone of Smith. This marle is sandy, gritty, micaceous, and generally derives a green colour from a copious admixture of suboxide of iron. At first sight the brown ferruginous sands much resemble the iron sand, and the green varieties the green sand, beneath the chalk formation. The micaceons loams,

when more than usually indurated, wear an aspect very like the loose shillot afforded by decomposing greywacke. It contains regular courses of concretions, and sometimes continuous beds of an indurated calcareo-siliceous gritstone of similar character and constitution. This grit contains casts of many shells, but the shelly matter itself is rarely preserved; nor are the terebratulæ, which characterise the lower beds of No. 1, to be met with here. The characteristic shells will be mentioned under the proper head. Much of this gritstone readily decomposes on exposure to the atmosphere, but some of the beds, or rather some portions of those beds, possess a very high degree of induration: they are occasionally quarried and squared for flag-stones or excavated for troughs. Some of the most extensive quarries on these beds will be found at Fenny Compton hill, a little west of the tunnel of the Oxford canal, on the borders of Oxford and Warwick shires.

The hill above these quarries is crowned with ferruginous sandstone, beneath which is a thick bed of blue clay: then succeed some thin alternations of marly rocks and clay, and lastly comes the great mass of the green coloured calcareous grit, which is quarried to the depth of 30 or 40 feet.

Thus we have seen the lowest beds of this series to consist of a green sandy marle, containing concretions and rock masses of similar character, both in Oxfordshire and Dorsetshire. In the intermediate district the same concretions may be traced, as particularly near Stinchcombe in Gloucestershire, and Hinton St. George in Somersetshire: they are there called sand-burs or clay-burs, and are, we believe, the beds designated by Mr. Smith as marle-stone. These beds form a gradual transition into the lias marles.

If we are correct in assigning the sandy beds of the eastern moorlands of Yorkshire to this formation, we must add to the above account thin seams of imperfect coal, as occurring among its members: this subject will be however considered hereafter.

(*b*) *Mineral contents.* The beds included in this section present little deserving notice under this head. The fullers' earth contains fibrous calcareous spar; and the inferior oolite at Dundry, occasionally, though very rarely, affords quartz crystals. Common calcareous spar of course occurs abundantly in most of the beds.

(*c*) *Organic remains.* Those of vertebral animals are very rare in all these beds: a series of vertebræ were however discovered a few years back in the marly sandstone of Warkworth, Northamptonshire. They probably belonged to some large marine lacerta, but were dispersed soon after their discovery, and none of them could be recovered. Fragments of the claws, &c. of marine crustacea, of the crab or lobster families, occur in the inferior oolite at Dundry, and in the marly sandstone in the north of Oxfordshire. The Testaceous remains (as far as they have been hitherto figured) of these beds will be found in the following list. The species, in the column assigned to the inferior oolite, which have an asterisk prefixed, have all been found in the quarries of Dundry hill near Bristol.

The disposition of these remains in the inferior oolite is thus stated by Townshend. The lowest bed is distinguished by its abundant casts of ribbed and studded Trigoniæ; immediately over this is a hard and compact coral bed containing large specimens of Madrepora cinerascens; and then succeeds the superior bed, abundantly charged with the other fossils of the following list.

We have not distinguished the organic remains of the sand next below the inferior oolite, since they are few, and mostly the same with those of the oolite.

The fossils of the lowest members of this sand and sandstone, where they touch and gradually pass into the upper lias marle, will be found under the column, Marly Sandstone; we have not however incorporated those assigned to the marlestone by Mr. Smith, since he states that there exists great difference between the various beds of this marle in this respect, and we incline to believe that he includes under this name our marly sandstone and upper lias marle. This point cannot be determined till the publication of his "Strata Identified, &c." has proceeded further; the fossils placed by us in this column have all been found in the north of Oxfordshire and the adjacent district, where these beds are displayed in the most striking manner. They mostly come from the bottom beds of the green micaceous sandstones which rests on a cream-coloured marle, introducing the upper members of the lias clays: the terebratulæ however characterise the upper beds of this green sandstone at the foot of the browner sands.

CHAMBERED UNIVALVES.

Fullers' Earth.	*Inferior Oolite.*	*Marly Sandstone.*
Ammonites	**A.* discus. *T.* 12.	
A. modiolaris.	*A.* concavus. *T.* 94.	
Smith, fig. 2.	f. 1.	
It may be doubted	**A.* elegans. *T.* 94.	*A.* elegans. *T.* 94.
however whether	*A.* jugosus. *T.* 92. f. 1.	
this does not rather	**A.* Banksii. *T.* 200.	

Fullers' Earth.	*Inferior Oolite.*	*Marly Sandstone.*
belong to the inferior oolite.	**A.* Blagdeni. *T.* 201. **A.* Brackenridgii. *T.* 184. **A.* Brochii. *T.* 202. *A.* Brogniarti (see p. 190. vol. 2.) *A.* Gervillii (see p. 189. vol. 2.) *A.* Herveyi. *T.* 195. *A.* Stokesi. *T.* 191. *A.* Walcottii. *T.* 106. **A.* Sowerbii. *T.* 213. *A.* Annulatus. *T.* 222. *A.* Strangewaysii. *T.* 254. f. 1. 3. *A.* falcatus. *T.* 254. f. 2.	
	**A.* falcifer. *T.* 256. f. 2.	*A.* Brookii. *T.* 190.
	**A.* Browni. *T.* 263. f. 4.	*A.* Beechii. *T.* 280.
Nautilites. Smith, f. 1.	**Nautilus* lineatus. *T.* 41. *N.* obesus. *T.* 124. *N.* sinuatus. *T.* 194.	*Nautilus.*
Belemnites (slender with a deep groove from the apex.	*Belemnites* — slender with a deep groove from the apex. — long slender without any groove. *— very thick & short, the alveolar cavity occupying nearly the whole of the shell. The Belemnites figured by De Montfort (Conchyliologie Systematique, tom. 1) under the names Cetocis glaber, 93 genus, & Hibolites hastatus, 97 genus, occur at Dundry.	*Belemnites* — many with very large alveoli & exactly resembling those in the upper lias marle.

Fullers' Earth.	*Inferior Oolite.*	*Marly Sandstone.*
Trochus (casts of them.)	**Trochus* similis. *T.* 142. upper figs.	*Trochus?* a small species.
	T. concavus. *T.* 181. f. 3.	
	T. dimidiatus. *T.* 181. f. 1. 4.	
	T. duplicatus. *T.* 181. f. 5.	
	**T.* elongatus. *T.* 193. f. 2 to 4.	
	**T.* punctatus. *T.* 193. f. 1.	
	**T.* abbreviatus. *T.* 193. f. 5.	
	**T.* fasciatus. *T.* 220. f. 1.	
	†*T.* granulatus. *T.* 220. f. 2.	
	**T.* sulcatus. *T.* 220. f. 3.	
	**T.* ornatus. *T.* 221. f. 1.	
	**T.* bicarinatus. *T.* 221. f. 2.	
	**Nerita* lævigata. *T.* 217. f. 1.	
	**Cirrus* nodosus. *T.* 141. f. 2. & 219. f. 12. 4.	
	**C.* Leachi. *T.* 219. f. 3.	
	Planorbis euomphalus. *T.* 140. f. 8. 9.	
	Melanea lineata. *T.* 218. f. 1.	
	**M.* Heddingtonensis. *T.* 39.	
	Turbo ornatus. *T.* 240. f. 1. 2.	*Turbo ?*
	**Rostellaria.* 3 species not figured.	
	Turritella? 3 species.	
	Ampullaria. 3 species.	
	Conus? (casts of).	*Helicina* polita. *T.* 285.

Tubular irregular Univalves. *Annelidæ* of Lamarck.

Fullers' Earth.	*Inferior Oolite.*	*Marly Sandstone.*
	Serpula triquetra.	*Serpula.*

Equivalved Bivalves.

Fullers' Earth.	*Inferior Oolite.*	*Marly Sandstone.*
	* *Trigonia* costata. *T.* 85.	
Trigonia clavellata (small variety.) *T.* 87. and some other species.	* *T.* clavellata. *T.* 87. * *T.* striata. *T.* 237. f. 1. 2. 3. *T.* duplicata. *T.* 237. f. 4. 5. *Arca?*	
	* *Cucullæa* oblonga. *T.* 206. f. 1. 2.	*Cucullæa.*
	* *Nucula* margartacea? larger than the recent shell.	
Cardium, (Smith, fig. 5.)	* *Cardium?*	
Cardita, (Smith, fig. 4.)	* *Cardita* obtusa. *T.* 197. f. 2.	*Cardita.*
	* *C.* lunulata. *T.* 232. f. 1. 2.	
	* *C.* similis. *T.* 232. f. 3.	
	C. producta. *T.* 197. This approaches much to some of the following species of Lutraria; and the hinge has not been ascertained.	*Cardita* producta. *T.* 197.
Lutraria gibbosa. *T.* 42.	* *Lutraria* gibbosa. *T.* 42.	*Lutraria* gibbosa. *T.* 42.
	* *L.* lirata. *T.* 225.	
	* *L.* ambigua. *T.* 227.	
	* *Astarte* excavata. *T.* 233.	
	* *A.* lurida? a variety. *T.* 137. f. 1.	
	* *A.* Cuneata? a variety. *T.* 137. f. 3.	
Astarte ovata?	*A.* ovata.	

Fullers' Earth.	*Inferior Oolite.*	*Marly Sandstone.*
Unio.	*Unio* Listeri. *T.* 154. f. 3. 4.	*Unio* Listeri. *T.*154. f. 3. 4.
	U. concinna ? *T.* 223. f. 1. 2. This rather belongs perhaps to the lower sandy beds.	*U.* concinna. *T.*223. f. 1. 2.
Mya ?	**Mya* intermedia. *T.* 76. f. 1.	
	**M.* V. scripta. *T.* 224	*Mya* V. scripta. *T.* 224.
	**Fistulana* ampullaria. Parkinson, vol. 3. *T.* 14. f. 2. 4. 6. 7.	
	**Mytilus*, figured in Townsend's Character of Moses. Pl. 14. f. 4.	
Modiola anatina, (Smith, fig. 3.) & some other species.	**Modiola* plicata. *T.* 248. f. 1.	
	M. cuneata. *T.* 211. f. 1. & 248. f. 2.	*Modiola* cuneata. *T.* 248.
	* *Donax.*	
Venus ? two or three species.		
Tellina, (Smith, fig. 6.)	**Pinna* lanceolata. *T.* 281.	
Terebratula (nonplicated varieties.)	*Terebratula* (nonplicated varieties.)	*Terebratula* (not plicated.)
		T. subrotunda. *T.* 15. f. 1. 2.
		T. punctata. *T.* 15. f. 4.
T. intermedia. *T.* 15. f. 8.	**T.* intermedia. *T.* 15. f. 8.	*T.* ovata. *T.* 15. f. 3.
	**T.* carnea. *T.* 15. f. 5 & 6.	
	**T.* semigloba. *T.* 15. f. 9.	
	**T.* digona. *T.* 96.	
T. ornithocephala. *T.* 101. f. 1.	**T.* ornithocephala. *T.* 101. f. 1.	
	T. acuta. *T.* 150. f. 1. 2.	
	T. resupinata. *T.*150. f. 3. 4.	

Fullers' Earth.	_Inferior Oolite._	_Marly Sandstone._
Terebratula (plicated varieties.)	_Terebratula_ (plicated varieties.)	_Terebratula_ (plicated.)
_T._ media. _T._ 83. f. 5.	*_T._ media. _T._ 83. f. 5.	_T._ tetraedra. _T._83. f. 4.
	_T._ obsoleta. _T._ 83. f. 7.	_T._ crumena. _T._83. f. 2. 3.
	_T._ spinosa, figured in Townsend's character of Moses. Plate 14. f. 8 & 9.	
_T._ lateralis. _T._ 83. f. 1.		
_T._ concinna. _T._ 83. f. 6.		
	Ostrea trichites? the fibrous shell of which fragments are common in this formation.	
Ostrea rugosa.	_Ostrea_ rugosa.	_Ostrea_ several species.
_O._ acuminata. _T._ 135. f. 2.	_O._ acuminata. _T._ 135. f. 2.	_Gryphæa_ dilatata, _T._ 149.
_O._ Marshii. _T._ 48.		
	*_O_ gregarea. _T._ 111. f. 1.	
	*_O._ palmata. _T._ 111. f. 3.	
	_O._ resembling gregarea, but finely tuberculated.	
Pecten. Walcot's Bath fossils. f. 37.	*_Pecten_ lens. _T._ 105. f. 2. 3.	_Pecten_ several species.
	*_P._ barbatus. _T._ 281.	
	_P._ fibrosus. _T._ 136. f. 4.	
	_P._ equivalvis. _T._ 136. f. 1.	
	*_Lima_ proboscidea. _T._ 2. f. 64.	_Lima_ proboscidea. _T._ 264.
	_L._ gibbosa.	
	*_Avicula_ costata. _T._ 244. f. 1. also a variety with tuberculate ribs.	
	*_Perna_ Aviculoides. _T._ 66.	
	Crenatula?	

Fullers' Earth.	*Inferior Oolite.*	*Marly Sandstone.*
Plagiostoma ovalis. *T.* 114. f. 5.	**Plagiostoma* punctata *T.* 114. f. 1.	
	**P.* rigida. *T.* 113. f. 1.	
	**P.* gigantea? *T.* 77. a flatter variety.	

These beds present the following species of the family *Echinus.*

The fullers' earth and rock contains the same depressed species of Conulus with the cornbrash, Parkinson, vol. 3. T. 2. f. 2. The inferior oolite contains Cidaris subangularis (Parkinson, vol. 3. T. 1. fig. 4), another Cidaris not described, a third species resembling the Echinus angulosus of Lamarck, Encycl. Method. Pl. 133. f. 7., and a fourth belonging to the Lamarckian genus Echinus, but not described. The Clypeus sinuatus, and a smaller species called by Lhwydd clunicularis, also occur in the inferior oolite.

We have not as yet met with any Echinites in the sandy beds beneath the inferior oolite.

Of the *Encrinital* family, the Pentacrinites caput medusæ and Pentacrinites subangularis of Miller, occur in the inferior oolite and likewise in the marly sandstones.

Of the *Coralloid* order several genera occur in the inferior oolite; viz. an Astrea resembling A. siderea, a species intermediate between Astrea and Mœandrina, a Caryophyllia, a Fungia, a Cyclolites (the C. elliptica of Lamarck), the Alecto of Lamouroux (see his new edition of Ellis, T. 81. f. 12.), and a Cellepora.

Townsend also mentions the Madrepora cinerascens (Explanaria mesenterina of Lamarck), as found in the inferior oolite. None of these have yet occurred in the inferior sands.

Traces of *Alcyonia* are observable both in the inferior oolite and marly sandstones.

(*d*) *Range and extent.* These beds stretch diagonally across the island from north-east to south-west, sometimes forming the north-western portion of the great chain of Stonebrash hills described in the preceding section, and at others entirely confined to their escarpment in that direction. The former case occurs when the hills are low, as in the midland counties; the latter where they are lofty and steep, so as to comprise the aggregate thickness of these beds in the interval between their base and summit. This generally takes place in the Cotteswolds of Gloucestershire, and in the part of the chain which traverses the north of Somersetshire.

In the north-eastern counties (in which direction we shall begin to trace their course) these beds constitute the sandy district of the eastern moorlands in Yorkshire, if our views of the geological structure of that country be correct; but for the reasons so often already given, the particular examination of that district must be referred to a distinct article.

Crossing the Humber into Lincolnshire, we find these beds occupying the western portion of the Stonebrash hills, which (as we have before seen) traverse that county near its western border from north to south. The ferruginous freestones attendant on this part of the series, are well displayed on the borders of Lincolnshire and Leicestershire, in the hills above Grantham and the vale of Belvoir.

Proceeding southwards, they occupy all the western half of Rutland and the adjoining border of Leicestershire, extending from about two miles east of Uppingham as far nearly as Houghton on the Leicester road. They are however covered in many places with vast accumulations of transported blocks of gravel near their junction with the lias, which is therefore often obscure. Burrough hill, crowned with an extensive Roman camp (about five miles north-east of Houghton), forms a prominent summit terminating and commanding the range, and abounds with fossils. It does not appear that this district has as yet been examined with sufficient attention to ascertain the lines of junction of the beds of which we are now treating, either with the incumbent oolite or subjacent lias, in a manner fully accurate and satisfactory.

On entering Northamptonshire, the upper junction of these beds with the great oolite has already been traced with sufficient minuteness in the preceding section; the inferior junction with the lias clays, from Rutland to the confines of Oxfordshire, Gloucestershire, and Warwickshire, pursues, generally speaking, the lower region of an escarpment of hills ranging from north-east to south-west. This escarpment is however broken into by many transverse vallies, which form, as it were, deep and broad bays in it, and of course produce similar indentations in the junction which follows them. The first of these, reckoning from east to west, is that which extends from Market Harborough to the south of Maidwell, but this junction is much concealed by alluvial debris throughout this tract.*

* It would be easy to collect from the transported blocks in the vicinity of Harborough alone, a tolerably complete suite of English rocks from chalk to transition slate. A vast tract of these debris stretches over the lias clay at the foot of the escarpment described in the text along the borders of Leicestershire and Warwickshire from Houghton by Harborough, Lutterworth, and Rugby, towards Daventry: the alluvium must be of great

On the west of Market Harborough the lias clays are seen at Oxendon and Marston Trussel, and the escarpment of the ferruginous sands ranges above these places towards the eventful plains of Naseby heath, and thence by Walford, and near which it approaches and accompanies on the east the line of the Grand Union Canal, which enters and undermines that escarpment at Crick tunnel,—the tunnel being excavated in the subjacent lias clays beneath the marly sandstone which crowns the escarpment. Having passed the ridge, the canal traverses a denuded valley on the lias clays as far as its junction with the branch communicating with the Oxford Canal, which passes the same ridge under circumstances exactly similar at Braunston tunnel.

The breadth of the district of ferruginous sands in this part of its course, is very considerable; it is difficult however to state it correctly, on account of the very tortuous lines formed both by its superior and inferior junction, but if we reduce these to lines expressing their mean course, the distance between them will average about ten miles. The account given under the head "(a) Chemical and external characters" of the varieties presented by these beds in the midland counties, will sufficiently explain the structure of this tract.

Near the tunnels above mentioned at Crick and Braunston, the scarped edge of these beds forms a long level ridge of inconsiderable height; but near Daventry (about a mile west of which town the escarpment ranges) it becomes broken, and varied with loftier summits rising above the general surface of the platform.

Such are Burrow hill near Daventry, crowned with one of the most extensive ancient camps in the island; Arbury hill, similarly fortified and distinguished as a station in the trigonometrical survey;* this rises 804 feet above the level of the

depth,—it often forms low ridges of hills, and in some instances (as near Braunston) caps the escarpment.

* The accuracy of the astronomical observations made at Arbury has been called in question in the controversy respecting the measure of the meridianal arc deduced from this survey. One writer, Captain Kater, has supposed that a deflection of the plum-line may have been produced by the attraction of some rock of greater density at no great depth beneath the surface, arguing to the probability of such a circumstance from the occurrence of the sienitic group of Charnwood forest, about 25 miles north of this hill. An exact knowledge, however, of the structure of this district, demonstrates that the probability is every way against this supposition. The lias and new red sandstone formations are in this quarter very regular, and cannot be together estimated at less than 1000 feet in thickness; and as nothing in any of the neighbouring denudations indicates the protrusion of any ridge of the older and elevated rock strata through this mass, there is every appearance that it would be necessary to sink to this depth before any such rock could be met with.

sea; the surrounding and adjacent summits of Studbury, Staverton, Rydon, Badby, and those a little farther south at Cherwelton, Hellidon, and Byfield, nearly rival Arbury in height: they have the general appearance of a series of rounded conical knolls scattered confusedly over a platform of lower elevation, and constitute by far the most varied and picturesque scenery in this part of the country: they are entirely composed of the ferruginous sands, which also crown the insulated and detached (or outlying) hills of Shuckborough and Napton, rising from the lias plains at the foot of the escarpment on the west:† about four miles south of these is another similar outlier on which stands the village of Boddington. This group of eminences round Arbury, inconsiderable as is their absolute height, are yet the most elevated ground in this part of the island, and shed their waters to three different quarters of its circumference, a distinction in which they have few rivals; for, as Moreton observes, "from Hellidon downs there springs forth the Leame (which flows westwards by Leamington into the Warwickshire Avon, and thus is finally emptied through the Severn into the Bristol Channel); from Studbury hill, or very near it, the Nyne or Nen (which flows eastward till it disembogues in the æstuary of the Wash on the German Ocean); and lastly, from the hills on the north-west of Cherwelton,‡ the Cherwell (which flows south into the Thames); and all these springs are within an equilateral triangle whose sides do not exceed about a mile in length. From Studbury hill alone, indeed, and the grounds at the foot of it, the rain water that falls there runs down to three different points; a part of it westward to the Leame; another part eastward to the Nen; and the rest southward to the Cherwell."

These scattered summits, evidently the fragments of strata which have been once continuous, demonstrate to what an extent the action of the denuding causes has taken place in this quarter. The same causes appear to have formed that deep and broad valley indenting the course of this chain like a deep bay, through which the Cherwell flows to the south (the western branch of that river rising within it, and the eastern soon entering it): the Oxford canal avails itself of the level tract of the same valley, which completely traverses and bisects the course of the Stonebrash hills; this valley is excavated in the lias clays as far as Banbury and a little to the south of that

† These are erroneously coloured as if they were entirely composed of lias in Mr. Greenough's map.

‡ Near Cherwelton there is a small accumulation of chalk flint gravel, and other alluvial debris in the bottom among the hills.

town, but the ferruginous sands cross it between Adderbury and Astrop.

On the west of this valley, an elongated and insulated ridge of hills, capped by the strata belonging to this part of the series, extends from the north-east angle of Oxfordshire into the contiguous parts of Warwickshire, ranging on the west of Fenny Compton (the quarries of which place have been already mentioned in describing the general characters of the formation), to North End in Burton Dasset parish. This ridge forms a projecting head-land advancing over the subjacent lias plains, and commanding extensive views towards the Malvern hills at the distance of 40 miles.

A breadth of one mile, exposing a valley of denudation in the lias clay, and conveying a streamlet which joins the Cherwell at Banbury, separates this ridge from the range of Edge hill, an elevated platform terminating in a sharp angular point (near which is a very perfect Roman camp) on the north, and forming what may be considered as the eastern promontory of the great lias bay of Shipston on Stour, which has been before mentioned as deeply indenting the general line of the hills of this formation. The escarpment of Edge hill, which commands extensive views towards the chain of Abberley and Malvern, is very abrupt and almost precipitous towards the north and west, and nearly coincides with the boundary line of Oxford and Warwickshire. In the latter direction a little south of the inn called Sunrising (where the road to Stratford branches off from that to Warwick) a colossal figure of a horse, similar in design to that of the Berkshire downs, deeply excavated in the ferruginous sands, formerly gave from its colour the name of the vale of the Red horse to the plains of Shipston beneath. The original figure has been destroyed by recent enclosures; and modern art has only replaced it by a miserable colt; the breed appears to have become degenerate;

> " *Venale pecus Corythæ posteritas et*
> *Hirpini.*"

As far as the Red horse, the platform of Edgehill is varied only by long vallies, furrowing it to convey tributary streamlets towards the Cherwell in a south-east direction; but on the south of this point a groupe of rounded conical summits, (exactly resembling those already described near Arbury hill in Northamptonshire) rises above, and breaks its long level outline. Shenlow hill above the village of Shennington (which stands on the brink of a picturesque valley) is the most northerly and conspicuous of these summits; Epwell hill is another point of the same groupe, and is estimated in the Trigonometrical Survey, for which it furnished one of the stations, at 836 feet above the sea. The adjacent country affords the finest, or indeed (with the exception of some parts of the vale of Oxford) the only good scenery in Oxfordshire. These hills are entirely formed of the ferruginous sands; the subjacent marly sandstones occupy the escarpment of the general platform.

Near the foot of the Epwell hills, is one of the subsided portions of oolite which has been described in the preceding section.

On the west towards Brailes (stretching in a bold range by Compton Winyate), the descent of the escarpment presents many beds of ferruginous sandstones alternating with loamy marles, resting on a thick bed of dark blue clay, beneath which is a lower terrace of the marly sandstones resting on the lias clay. Brailes hill or rather hills consist of two detached and lofty summits rising like islands from the great lias bay of Shipston (or Vale of Red horse) one on the north-west, and the other on the south-west of Brailes;* these agree in composition with the opposite escarpment, but the greater of them (the southern hill) exhibits in one field on its summit a patch of oolite.

South of Brailes, the escarpment ranges by Whichford and Long Compton under the Rolwright hills, which exhibit oolite on their summit. These constitute what may be considered as the southern boundary of the lias bay of Shipston (or vale of Red horse); but the Evenlode, which rises in this quarter, flows southward through a valley which forms a prolongation, or to preserve the metaphor a creek, opening into that bay, and traversing the oolite hills; this continues tolerably broad (more than a mile across) till it approaches Shipston under Whichwood.

The ridge noticed in the preceding section as running on the north-west from Whichwood forest, and being covered with a cap of the great oolite, forms the western boundary of this vale. The strata beneath this oolitic cap, as seen on both sides the ridge, at Idbury on its eastern escarpment, and the denuded valley of Great Rissington on the west, are first, clay throwing out a series of springs, then ferruginous sandstones containing belemnites and terebratulæ (great Rissington stands on these beds) and beneath these clay again, containing large concretions of the marly sandstone of extraordinary induration. The hills proceeding north from Stow in the Wold above More-

* Inadvertently coloured as lias in Mr. Greenough's Map; where they are marked by the letter *u*.

ton in the Marsh (though separated near the former place by a breach opening into the denuded vale of Rissington) form the continuation of the same ridge with which they correspond in composition. These constitute the western border of the broad part of the great lias bay, and after circling round a sort of creek which forms the vale of Campden, terminate in a bold promontory at the Ilmingdon hills (the north-west cape of this bay), immediately on the north of which lies the insulated summit of Meon hill: all these points have been described already in the preceding section. The strata on the escarpment of the Ilmingdon hills, in ascending from the village, are first, a thick bed of marly sandstone, above this a series of clays traversed by some thin beds containing Ammonites, &c., then a very ferruginous rock: above this the slope of the hill is more gradual, and the road is not sufficiently deeply cut to exhibit the strata, which however are seen to be decidedly oolitic on the summit.

From this promontory, the Cotteswolds range in the line already sufficiently described in the preceding section, south-south-west by Cheltenham and Stroud towards Bath: they appear to be generally capped by the great oolite, and to exhibit the strata we are now describing in their escarpment: most of the insulated (or outlying) hills situated in advance of the chain and rising from the lias plains on the west, appear however to be wholly composed of them. Breedon hill is however so lofty, that it may be expected (for we are not aware that it has yet been attentively examined) to exhibit a cap of the great oolite; on this account it has been mentioned in the preceding section. This is the most important of these outliers, and forms a considerable group of hills lying north-east of Tewkesbury; it is marked *k* in Mr. Greenough's Map. The Tredington hills* a little south of the above (marked *i*) are capped by the sandy beds of the inferior oolite, as are Churchill (marked *h*) in the north-east, and Robinswood hill on the south of Gloucester.

Mr. Halifax has furnished us with the following memorandum of the section on the western escarpment of Painswick hill (north-west of Stroud), which well illustrates the general structure of this part of the chain.

* Tredington hills are inadvertently coloured as lias in Mr. Greenough's Map.

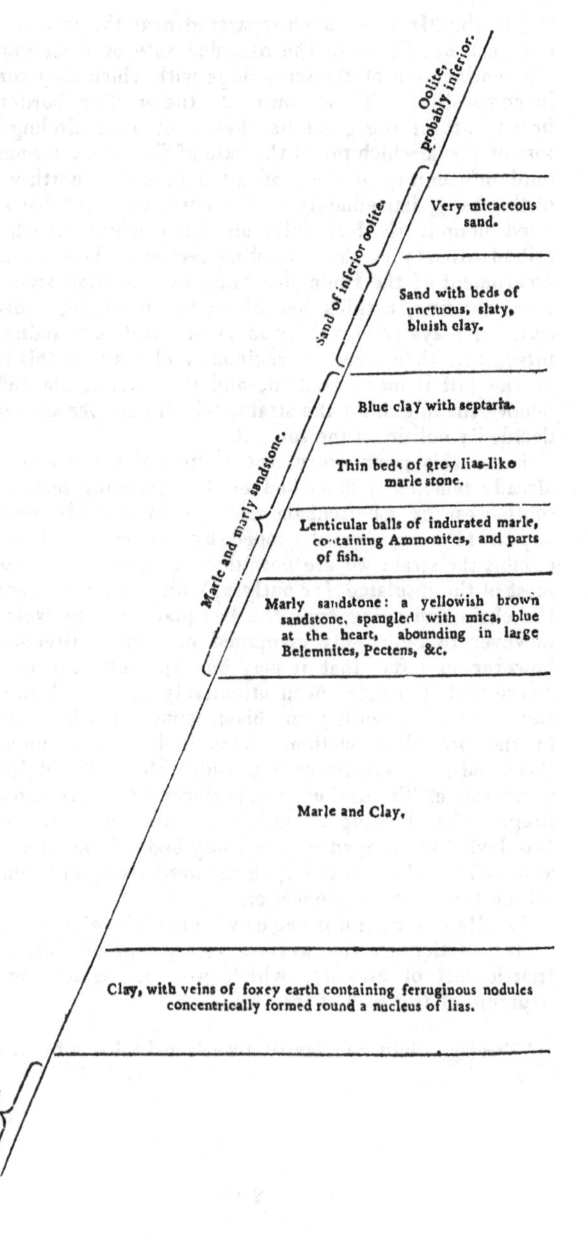
Oolite, probably inferior.
Very micaceous sand.
Sand of inferior oolite.
Sand with beds of unctuous, slaty, bluish clay.
Blue clay with septaria.
Thin beds of grey lias-like marle stone.
Marle and marly sandstone.
Lenticular balls of indurated marle, containing Ammonites, and parts of fish.
Marly sandstone: a yellowish brown sandstone, spangled with mica, blue at the heart, abounding in large Belemnites, Pectens, &c.
Marle and Clay.
Clay, with veins of foxey earth containing ferruginous nodules concentrically formed round a nucleus of lias.
Lias.

The course of the oolitic hills in the vicinity of Bath has already been sufficiently noticed in the preceding section, and it is now only necessary to add that, beneath the cap of the great oolite which crowns them, the series of strata now described occupy their middle and lower regions, occurring in the order indicated in the list inserted in the note on their chemical and external characters (p. 202), which may serve as a type of structure of all the escarpments surrounding Bath.* The inferior oolite generally forms lower terraces advanced in front of the higher table-land capped by the great oolite; and to the west of the general line of the escarpment, often constitutes detached summits or outliers (as may be particularly observed south of the line between Bath and Bristol): such are the ridge on which Newton park stands, the conspicuous hummock of Stantonbury hill (crowned with an extensive and perfect ancient British camp, connected with and strengthening the Belgic boundary line of Wansdike), the three conical summits of the Barrow hills and the adjacent ridge called the sleight in Timsbury parish; but Dundry hill (three miles south of Bristol) is the largest and most conspicuous of these detached outlying hills. This presents a long narrow ridge nearly four miles from east to west, and sending off a short branch from its western extremity towards the south-east; this ridge rises about 300 feet above the lias platform which supports it, and 700 feet above the level of the sea; it has a thick cap of the inferior oolite throughout, which has been very extensively quarried as a freestone near the western end; the eastern end exhibits a large camp corresponding with that on Stantonbury, and forming a part of the same defensive line.

The principal escarpment of the inferior oolite (ranging on the east of these outlying summits) proceeds from Bath to the south-south-east above English combe, Priston, and Paulton, being intersected however by several deep vallies of denudation which cut through the subjacent beds into the new red sandstone on which it rests. Clandown above Paulton is remarkable for the deep pits sunk through most of the intermediate beds into the regular coal measures; one of these exceeds 200 fathoms in depth, but this begins in the lias; another, just on the edge of the hill towards Paulton, commences in the inferior oolite: of this pit (peculiarly interesting as verifying the order of all the intermediate beds and exhibiting an instance unique, as far as this island is concerned, of any successful experiment

* Some interesting facts connected with the subsidences and slopes of the strata in this neighbourhood will be found under the head *inclination and stratification.*

to reach the coal from beds so far above it in the regular series) a section will be hereafter given. No inference however can be drawn from this district in favor of similar trials elsewhere, since two peculiar circumstances here concur; 1st, the manner in which these more recent beds here overlie the coal field, resting unconformably in horizontal planes on the truncated ends of the highly inclined strata belonging to the coal measures; and 2ndly, the thinning out of many of the beds in this direction; in consequence of which the sands of the inferior oolites, and the clay of the lias, are greatly reduced, and have almost vanished in many places, leaving the freestone beds of the inferior oolite almost in contact with the lower stony beds of the lias; the new red sandstone also being greatly diminished in thickness, so that a geological interval, equal in many other districts probably to 2000 feet is here reduced to less than a quarter of that depth.

South of Paulton, in Kilmersdon and Babington parishes, the inferior oolite spreads over the same platforms with the diminished lias, and it becomes somewhat difficult in many places to trace the latter as a distinct formation. In the south of the latter parish, and in Mells, the horizontal planes of the inferior oolite come in contact (in the manner represented in the wood-cut, page 228) with the inclined coal-measures and mountain limestone constituting the eastern portion of the Mendip hills, which expire in this direction by the lowering of their strata; so that they become buried beneath the level of these more recent formations, and are here exhibited only in the bottom and sides of the vallies of denudation. Such is the character of the district lying between Mells on the north, Frome on the west, and the two Cranmores on the south; an uniform and elevated plain of the inferior oolite spreads over its whole surface, furrowed by vallies about 150 or 200 feet deep, which expose the mountain limestone. The character of many of these vallies (particularly of that between Mells and Frome and its lateral branches) is highly romantic; the streamlets that flow through them being skirted by bold and rocky banks overgrown by feathering woods; while the geologist observes, as a feature of peculiar interest in their precipitous escarpment, the actual contact of the horizontal bed of inferior oolite resting on the truncated edges of strata of mountain lime, thrown up in an angle of from 50 to 60 degrees. This line of contact is sometimes perfectly level for a considerable distance (as if the edges of the mountain limestone strata had been rendered smooth by some mechanical force abrading them previously to the deposition of the inferior oolite), but in other instances it is rugged and irregular; sometimes the contact is marked by a breccia of fragments of the older, cemented by the newer rock, but this is by no means constant. On the north of Doulting hill, between West Cranmore and Shepton Mallet, the inferior oolite abuts against the old red sandstone which forms the nucleus of the Mendips, and here begins to display itself in considerable eminences at Downhead common. Doulting hill is celebrated for its quarries of the inferior oolite, which here affords a valuable freestone. Between Doulting and Shepton Mallet, the inferior oolite and lias abut against the inclined strata of Mendip, under such circumstances and with such considerable variations from their usual aspect, that it is not easy to pronounce to which formation many of these anomalous beds belong: similar beds continue to hang on the southern slope of the Mendips, on the north of the line between Shepton Mallet and Wells.

It is only however where immediately in contact with the Mendip range, that this obscurity prevails; for, from Doulting, the general line of the inferior oolite is continued south by Cannard's grave towards Bruton, forming a conspicuous escarpment rising above the lias plains which extend westwards: over these some outlying hills of the same formation are scattered; viz. the long ridge of Pennard hills, and the more striking and far seen cone of Glastonbury Tor. Brent knoll, a lofty lias hill rising out of the marshes on the border of the Bristol channel to the height of between 400 and 500 feet, also exhibits on its summit a cap of inferior oolite nearly coextensive with the ancient encampment which crowns it.

But to return to the main escarpment; this ranges from Bruton south-west to Castle Cary, and thence south by the two Cadburys, near the most southerly of which a detached summit of this formation gives position to one of the boldest and strongest ancient encampments extant in the island, girding its sides with triple fosses and stupendous mounds. Near this point the range circles westwards round the vale of Marston Magna, pursuing a line between Ivelchester and Yeovil to Ilminster.

Between Ilminster and the coast, the junction of the inferior oolite and sands with the lias ranges due south, but is very generally concealed by overlying hills of the green sand formation supporting occasionally chalky summits, and connected with the great western extension of these beds towards the Blackdown hills.

On the coast, the series of beds of which we are now treating is finely displayed in the sections presented by the cliffs between Charmouth and the head of the Chesil bank.

Proceeding from west to east, we first find the inferior oolite resting on the lias-marle on the summit of a steep pyramidal cliff about half way beneath Charmouth and Bridport harbour, called the Golden Cup, (perhaps a corruption of the Golden Cap, as derived from the reddish yellow colour of the summit strongly contrasted with the dark blue of its base): Down cliff succeeds, in which the dip of the strata towards the east has brought them somewhat lower: a third cliff in which the inferior oolite and sands occupy the middle region, (the hill above being crowned by the fissile variety of the great oolite), intervenes before we arrive at the mouth of Bridport harbour. The first cliff west of Bridport harbour no longer exhibits the lias clays, (these having here been carried by their dip below the sea level), and the lowest visible beds, being those which contain the concretions of green marly sandstone, intermediate between the sands and lias: in Burton cliff, which follows this, the inferior oolite itself sinks in like manner, being succeeded on the east by a low cliff of the fuller's earth clay full of fibrous calcareous spar. The average dip of the strata in these cliffs is about 1 in 50: their order and nature have been already given in the notes on the chemical and external character of this part of the series, page 236.

(*e*) *Height of hills.* This series of strata is in many instances confined to the escarpment of the hills capped by the great oolite; but occasionally, as in Rutlandshire, Northamptonshire, and Oxfordshire, constitutes the entire mass of considerable hilly tracts, and very generally caps the outlying hills advanced on the north and west of the great oolitic range. These hills have already been noticed in tracing the range and extent of these strata.

If we are correct in referring the sandy district of the eastern moorlands in Yorkshire to this series, we shall there find the most considerable heights constituted by it. The most lofty of these, Roseburry Topping, is stated in the Trigonometrical survey to rise 1022 feet above the level of the sea. To collect the facts belonging to this head together we will here repeat, on the same authority, that Arbury hill, Northamptonshire attains 804 feet; Epwell hill, Oxfordshire, 836 feet; and Dundry hill, Somersetshire, 700 feet above the same level.

(*f*) *Thickness.* The thickness of these beds near Bath has been already given in the section, page 202, which, with those forming its continuation page 235, will be found to amount together to 460 feet; but, in the midland counties, the thickness of the sands connected with the inferior oolite must be much more considerable: if we were to take 400 feet for the average, it would probably be far from excessive.

(*g*) *Inclination and disturbances of the strata.* These strata are in their general position strictly conformable to all the other secondary formations which we have as yet described, dipping beneath them towards the south-east under an almost inappreciable angle.

Some instances of those dislocations accompanied by subsidences, which are usually termed faults, may be observed in these strata, especially in that portion of them which overlies the south Gloucester and Somerset coal-field; where, although the greater part of the faults affecting the coal-field appear to have been produced by convulsions which have acted upon it before the deposition of these strata, and therefore do not derange the latter, yet in some cases the contrary takes place. An instance of this may be seen on the edge of Clandown hill, Somersetshire, where it hangs over the village of Paulton; here a fault traversing the subjacent colliery also throws down the inferior oolite 20 fathoms to the north, bringing it to the level of the lias; the hill on the north-west of Bitton (half way between Bath and Bristol on the north bank of the river Avon) exhibits a similar case; the southern point of that hill being formed of the inferior oolite sands, made to abut abruptly against the lias and subjacent new red sandstone of which all the northern part of the hill consists.

Instances of derangement on the slopes of the hills occupied by these strata, from masses which have been undermined either at the period when the vallies were originally excavated, or by the subsequent agency of the springs percolating through them, &c. and have been precipitated over the escarpments, such as have been already mentioned when treating of the great oolite, are much more common than the true cases of faults: we subjoin the particulars of several observed by Mr. Townsend in the neighbourhood of Bath; the two first cases indeed rather relate to the great oolite than to these beds, but we have been induced to keep the whole together from the convenience of presenting a more connected view of the phœnomena of this nature which the geological traveller may have an opportunity of examining in that district.

On the northern extremity of Lansdown, near the monument, we meet with, not the mouldering of a scarp, but its dislocation; for here the crop of the great oolite has fallen down on the back of the bastard free-stone; and the intermediate bed of clay, with its fullers' earth, is wanting. At this very time, (March 1803,) Mr. Bush is removing the rubbish, and has laid bare some of the best beds of the oolite, which, instead of dipping gently to the south-east, here fall to the north in a regular succession of fragments, between which are considerable chasms. The bastard free-stone, on which these fragments rest, is visible, reclining on its bed of sand, as that does on the blue marl and lyas, which here dip north, towards the Severn. (T. 192.)

Now the same kind of dislocation already noticed in the crop of the great oolite, near the monument on Lansdown, may be observed in the inferior termination of the same rock at Murhill, south of Winsley, opposite to the conflux of the Frome and Avon rivers. For the rock, which here forms an elevated cliff of about one hundred and forty feet, being unsupported by its subjacent clay, has slipped back and subsided on the bastard free-stone, leaving very extensive chasms, some of which are empty, but others have been occupied by fragments of the rock. Some of the fragments, precipitated to a considerable distance from the cliff, and almost buried in its ruins, measure more than forty feet by twenty. One of these, which dips at 40° south towards the river, contains more than five thousand cubic feet of free-stone. (T. 195.)

In this spot, the vast fragments I have mentioned point towards the vale beneath, as if they were precipitated from the cliff. But near the Dundas aqueducts, and at Dunkerton, I have observed a more remarkable phænomenon; for there, the inferior termination of the bastard free-stone has subsided, fractures have ensued, wide chasms appear, and numerous faults have been created: but these enormous fragments of the stratum have not been precipitated, nor do they point towards the vale beneath, but seem to be sliding back, and dip into the hill from which they proceed, and in which the whole rock appears. (T. 195.)

In such a position, the inferior surface of the lowest bed, which should repose on sand and be concealed, is brought to light, and may be distinguished by its peculiar fossils. (T. 196.)

So likewise to the east of Bath, as we ascend the hill towards Hampton Down, we pass over several visible subsidences of the bastard free-stone, and proceeding in the same direction, when we have passed the subsummit of the hill, and descend towards the canal by the rail-road, where the cliff appears, we count five depressions of this rock, with its sand bed, and at a lower level we find the blue marl, on which the river glides, and in the marl we find its peculiar rock. This subsidence of the bastard free-stone continues all the way to the Dundas aqueduct, and in it the canal is formed. (T. 196.)

South of this line, in the road from Widcombe to Claverton, we observe this rock has fallen down to a considerable depth, leaving chasms now occupied by clay and gravel. Here also we find ponderous masses, which dip into the hill. (T. 196.)

Near Bath, our most remarkable dislocation of the bastard

free-stone is on Mount Sion, where the summit of the rock is visible. From this elevated region, looking down upon all the hills by which Bath is immediately surrounded, we command, in the distant offscape, a view of Mendip. On this delightful spot, Dr. Parry has made extensive excavations, for the foundation of his magnificent structure, and exposed numerous dislocated fragments of the rock. At a lower level in the hollow way by which carriages will ascend to this high hill, he has sunk (A. D. 1811) pits to a considerable depth. These exhibit enormous blocks heaped up confusedly, some much inclined to the horizon, others nearly vertical, and falling headlong into the vale, through which the river flows. (T. 197)

Beneath this chaos we have the blue marl bed, which may be traced on the same high level, by Park-street to Camden Place, and from thence to Bath Easton. (T. 197)

To the south of Bath, under Beechen Cliff, the same marl appears, with its rock and springs: but in the intermediate space, between Mount Beacon and this cliff, the bastard free-stone, with its marl bed, has sunk down to such a depth, that the former is quarried near Widcombe Crescent, in the road to Claverton, and the latter is to be seen fifteen feet under sand, in the well of Caroline Buildings, that is nearly on a level with the river. By Hetlin Court, when the hot springs had failed to supply the usual quantity of water in a given time, the Corporation employed Mr. William Smith to remedy the evil. He laid open the ground, detected the cause of failure, and restored the springs. At that time I took notice of his operations, and at a great depth saw the springs through the blue marl. (T. 197.)

Subsequent to this operation, Mr. Palmer, then Mayor, sunk in the sand of the King's bath, as deep as he could venture to proceed, without endangering the pump-room, yet he did not arrive at the blue marl. From the bottom of his sinking he sent me a quantity of sand. This was alluvial, not calcareous, but siliceous, and in this sand I ascertained the green quartz, with iron, such as we find beneath our chalk, in Wiltshire, and from thence it came. (T. 197.)

Extensive portions of the bastard free-stone bed, disrupted and fallen down below their native level, may be observed in Collier's Lane, going up to Lansdown. But should any young geologist, occasionally residing, or a transient visitor in Bath, wish to see some dislocation of this rock, without extending his walk to Collier's Lane, or climbing the steep ascent to Claverton and Hampton, he may easily gratify his curiosity by walking on the canal bank to the boundary of Bathwick, and then crossing the Folly bridge to examine at his leisure the five successive depressions of the rock, which form as many steep ridges. At the foot of each, thus placed on different levels, the marl springs issue, always following the dislocations of the rock. (T. 198.)

The same appearances have been noticed in numerous places, but are no where so distinctly seen as in the immediate vicinity of Bath. Indeed no country can exhibit more interesting scenes for the geologist, or more numerous examples of dislocation, than this; for here, not merely the bastard free-stone, but all the strata, disrupted to a great extent, have fallen down towards the subjacent vallies, and after reiterated fractures, forming steps, have sent down enormous blocks, which are either piled up in heaps, or scattered on the declivities. These phœnomena are striking; but near Bath we have others abundantly more so. To the south of Prior Park, on the southern hanging of Coombe Down, where we look down on the wide expanse in which a little streamlet flows, we find the bottom bed of the great free-stone rock, and, to the east of this we have, nearly on the same horizontal level, the superior beds of the same rock. (T. 198.)

(*h*) *Agricultural character.* The different beds of this series are of necessity very variable in this respect: the fullers' earth, says Mr. Smith, presents a clayey soil distinguishable by less cultivation than upon the dryer soils of the oolitic rocks above and below it, which are more genial to the growth of corn. Wood and timber trees are also common to it on the slopes of the hills, of which it so often forms a part.

The more calcareous beds of the inferior oolite agree with the other stonebrash soils in agricultural character.

The sands below the inferior oolite afford in those districts where they are most extensively exhibited, viz. in the midland counties, a soil the fertility of which is spoken of in very high terms by agricultural writers. "This red district," says the author of the report on Oxfordshire, "may be considered as the glory of the county. It is deep, sound, friable yet capable of tenacity, and adapted to every plant that can be trusted to it by the industry of the cultivator."

(*i*) *Phœnomena of springs, &c.* The upper member of this series (the fullers' earth clay) throws out copiously the waters which have percolated through the great oolite: those of the inferior oolite and sands are thrown out by the subjacent marles; so that this series exhibits two lines of springs, one near its superior, and one near its inferior extremity At Chadlington and Deddington in Oxfordshire, and at Astcot, Northamptonshire, are mineral waters said to contain iron, sulphur, and sea salt; and at Clifton, Oxfordshire, one containing soda, selenite, and lime: these are probably situated in the lower marly beds. Is not the Scarborough water, which contains carbonate of lime, sulphur, sea salt, and iron, situated in the upper part of this series?

Section VII.

LIAS.*

(*a*) *Chemical and external characters.* This formation consists of thick argillaceous deposits, constituting the base on which the whole oolitic series reposes. The upper portion of these deposits, including about two-thirds of their total depth, consists of beds of a deep blue marle containing only a few irregular and rubbly limestone beds. In the lower portion, the limestone beds increase in frequency, and assume the peculiar aspect which characterises the lias, presenting a series of thin stony beds separated by narrow argillaceous partings; so that quarries of this rock at a distance assume a striped and ribband-like appearance; in the lower beds of this limestone, the argillaceous partings often become very slight and almost disappear, as may be seen in the lias tract of South Wales: beds of blue marle with irregular calcareous masses, generally separate these strata from the red marle belonging to the subjacent new red sandstone formation.† The limestone

* Chiefly by the Rev. W. D. Conybeare.

† The subjoined sections will serve to convey an accurate idea of the detail of this formation as it exists in the neighbourhood of Bath; and afford a good type of its general arrangement, excepting that in the midland and north-eastern counties the thickness of these deposits must be at least double that here presented; and some of the stony beds, alternating in the marles which constitute the upper portion, assume occasionally a more decided character and importance. These local details will find their best place in pursuing the range and extent of this formation.

These sections are subdivided, as in the text, into the *Upper marles*, the *Stony or true lias beds*, and the *Lower marles* separating the lias from the new red sandstone formation.

The first section is from a fruitless trial for coal (abandoned in 1812) in the parish of Bath Easton; it begins in the very top of this formation, and extends through its lowest beds. The second also exhibits the whole formation, and is interesting as forming the upper part of the section afforded in the Paulton collieries, which include a greater geological depth (i. e. pass through a greater number of different formations) than has probably been ever actually verified in any other single point. The lower formations, ascertained in this section, will be given hereafter; the third section is given as exhibiting in greater detail the lowest members of the lias near their junction with the red marle.

beds, towards their centre, where most free from external

Section 1.

		feet.	in.
Upper marles.	Yellow clay	8	—
	Blue marle	27	—
	Stone	—	6
	Blue marle	18	—
	Stone	—	4
	Blue marle	9	—
	Blue rock	12	—
	Marle	1	—
	Stone	1	6
	Blue marle	24	—
	Rock	3	—
	Marle	6	—
	Stone	—	10
	Blue marle	41	—
	Stone	—	10
	Blue marle	7	—
	Stone	—	9
	Blue marle	7	—
	Stone	—	9
	Blue marle	6	—
	Stone	—	4
	Marle	3	—
	Marle and stone	11	6
	Stone	—	9
	Marle	2	6
	Stone	3	—
	Marle and stone	8	—
	Marle and stone	12	3
	Hard marle	4	—
True lias beds.	Hard rock	18	4
	Blue stone	5	1
	Hard blue stone	12	1
	Stone	2	—
	White lias rock	10	—
Lower marles.	Blue marle	6	—
	Stone	—	6
	Clay	—	9
	Rough blue marle	—	5
	Black marle	10	—
	Light blue marle	2	—
	Red ground of the new red sandstone formation.		

Section 2.

Upper beds sunk through to reach the coal measures, in a pit belonging to Mr. Simon Hill, on the brow of the down east of Paulton, Somersetshire.

		feet.	in.
	Inferior oolite	18	—
	Upper lias marles	120	—
True lias beds.	Grey and blue lias rock	6	—
	Sun bed or corngrit, divided into three thin beds slightly oolitic	1	6
	White lias	12	—

mixture, contain more than 90 per cent. of carbonate of lime ;† the residuum has never been distinctly analysed, but appears to consist of alumine and iron, and in some varieties traces of silex have been found: towards the edges of the beds, however, where they come in contact with the alternating strata of clay, the proportion of alumine is, as might be expected, more considerable. This limestone is particularly characterised by its dull earthy aspect, and large conchoidal fracture; in colour it varies in different beds from light slate blue, or smoke grey, to white: the former varieties usually constituting the upper; the latter, the lower portions of the formation. The blue lias, which contains much iron, affords a strong lime, distinguished by its property of setting under water; the white lias takes a high polish, and may readily be employed for the purposes of lithography. It must however be distinguished from the

		feet.	in.
Lower marles.	Blue marle	6	—
	Clay stone forming concretional and rubbly masses	3	—
	Black marle (excellent for manure)	6	—

Red ground of the new red sandstone formation.

The rest of this section will be given under the new red sandstone and Coal measures.

Section 3.

Westbury cliff on the west bank of the Severn, Gloucestershire, illustrating the lower beds of the lias formation.

	feet.	in.
White lias	10	—
Blue shale passing into marlestone	10	—
Black shale with iron-shot fissures	12	—
Green siliceous grit, highly micaceous, and containing abundant bones, well known here and at Aust by the name of the bone bed	1	—
Black shale	2	—
Green grit	—	6
Black shale.......	2	—
Greenish marlestone decomposing into balls	18	—

Red marle of the new red sandstone formation.

† The late Mr. Smeaton took the several undermentioned varieties of lias marlestone, and having dissolved 40 grains of each in aquafortis, obtained a residuum from each, which he weighed after drying them in the sun. (G. Notes.)

From the yellow lias of Axminster....	5¾ grains
Ditto with shining spangles	5½
Yellow such-stone of Glastonbury	5
Blue lias of Watchet	4½
Aberthaw	4½
Bath	4½
Axminster	3½

stone generally so applied on the continent, which is brought from the quarries of Solenhofen, and is of much more recent formation.

The slate-clay with which the lias alternates, is grey, brown, or black, is frequently bituminous, and readily divides into laminæ as thin as common pasteboard.

To the above general description of this formation we have added in the note below some further particulars extracted from memoranda kindly lent to the Editor by Mr. Greenough;*

* *Synonymes.* (Rudge's Glocestershire), Alum shale, Doggers, Scar of Whitby. The etymology of this word is unknown to me: it may perhaps be connected with the acknowledged excellence of some of the beds of this series as a cement. The Liais of the French is a very different substance.

Some of the beds of the lias are used as building stones, others as slabs, hearth-stones, grave-stones, &c.

At Kenton Mandeville the common slabs vary from 10 to 30 feet in length, and from 12 to 15 in width: but one has been raised containing 500 superficial feet.

Slabs of the Cotham stone are sometimes 2½ feet long, and seven or eight inches thick.

The marle-stone is used for walls, slabs, and flooring; that of Binton and Grafton in Warwickshire, which is waved like the Cottam, is used as a marble for chimney-pieces, also for paving, for stone seats, &c. At Puckeridge hill, south-east of Taunton in Somersetshire, it is burnt for manure. At Wingfoot, Red hill, and Bidford, about four miles from Stratford upon Avon, the lias assumes the character of a marble.

The lias is never variegated in colour like common marble, nor brecciated, nor does it admit of brilliancy or depth of tint, but it occasionally exhibits, especially in specimens cut and polished longitudinally, dendritical appearances (Cottam stone or marble), which may be supposed to be the consequence of the enlargement of the concretions in which this stone is found, since it occurs in detached masses beneath the surface: the upper surface of the stone presents branches and prominences which sometimes represent the interlacings of ivy. They are commonly used in the rough state for the rustic work of gateways.

The irregular beds consist of fibrous limestone and *cement* stones (septaria) so called because used in making Parker's cement. Where the fibres are not parallel to each other, they often form that irregular substance so common in the Coal-measures, to which an organic structure has often erroneously been attributed, and termed the cone-in-cone coral. The cement-stones are of different sizes; they are generally solid, and seem sometimes to have had a cornu ammonis or other shell, or wood, as a nucleus to form upon. Some of them have septa, which are occupied by calcareous spar or bitumen; the quantity of iron they contain is variable; some are coated with pyrites, or have lumps of it adhering to them; at Watchet, sulphate of strontian finely crystallized occurs in these concretions. When large and flât, these cement-stones are termed girdles. Along the Whitby coast, these girdles have given a partial protection to the shale, and thus occasioned a number of insulated and grotesque masses: they often turn red on exposure to air.

Some of the beds in the lias form a rich argillaceous iron-stone.

Near Axminster in Dorsetshire, the lias clay is so bituminous in some places, that it has been sunk through in search of coal.

and in a subsequent part of this section (under the head range and extent) some additional details concerning the characters assumed by it in its course through Yorkshire, will be found in the notes.

(*b*) *Mineral contents.* The lias is nearly destitute of metallic or earthy minerals. It does not appear commonly to contain any metallic substance except iron; which, though it occurs in layers under the form of clay-iron-stone, or disseminated through the mass in the form of pyrites, never constitutes a mineral vein. It is said, but I know not with what truth, that small pieces of galena have been found in the quarries near Bath, and galena and blende are said to occur also near Whitby. Sulphate of barytes has been found in small quantity by Mr. Gilding, jun. in the canal near Gloucester, and I have a specimen of it in wood from Lyme in Dorsetshire. Sulphate of strontian is also said to be found at Watchet. Though chert is abundant in the limestone above and below it, siliceous matter is so rare in this formation, that I never met with it except at Aberthaw and Dunraven in South Wales, where its fossil, the gryphus, is coated with chalcedony; but chert occurs in the lias of Watchet, and the lias contains veins of chert also near Cowbridge in South Wales. (G. Notes.)

The iron pyrites, which is very abundant, by its decomposition and action on the argillaceous strata, produces an efflorescence of the aluminous sulphate so extensively worked at Whitby: to the same cause must be ascribed the spontaneous inflammation often observed in the cliffs near Charmouth, Dorsetshire. (See Maton's West. Counties, p. 76. v. 1.)

(*c*) *Organic remains.* The organic remains contained in the lias are peculiarly interesting, as affording a greater number of animals of an higher order (that is to say of the vertebral class) than are exhibited in the list of any other formation, if we except the Stonesfield beds of calcareous-slate in the great oolitic series before described.

In this class we have first to notice the remains of two very remarkable extinct genera of oviparous quadrupeds, evidently belonging to the same class with the great natural order *Lacerta*, but yet differing very essentially in structure from all the genera at present known to exist, and in such particulars as evidently must have fitted them to live entirely in the sea. They appear therefore to hold the same place with regard to

At Thickerby seven miles east of Gainsborough, Mr. Hornby of the latter place, sunk for coal in a black hard slaty or shaly substance.

In the lordship of Thrussington in Leicestershire, seven miles west of Sexhill, coal has been tried for at a place called Coal-pit Lees, and £400 expended, but in vain.

recent lacertæ, that the cetacea do to other mammalia, and will form a division of the order lacerta to which the name *Enalio-Sauri* (marine lacertæ) may be conveniently applied. The investigation of their comparative anatomy, or rather osteology, is highly important, as laying open various new and interesting links in the chain of animated nature.

Two genera have been ascertained.

1. *Ichthyosaurus.** Animal with an head resembling the lacerta tribe, but extended anteriorly into a long pointed muzzle, armed with numerous conical teeth. Vertebræ resembling those of fish in being double concave (that is deeply cupped at each end) and as thin as those of the shark, in order to facilitate progression by a vibratory motion of the tail. From these double analogies, the name (the fish-like-lacerta) is derived. The extremities terminate in four paddles *sui generis*, composed of a series of flat polygonal bones greatly exceeding in number not only the phalanges of quadrupeds, but also the phalangic cartilages of the fins of fish. There are two or three species chiefly distinguished by the form of the teeth.

2. *Plesiosaurus.* The head of this animal is not yet perfectly ascertained: the vertebræ and extremities hold an intermediate place between the former genus and the recent lacertæ, and supply beautiful links in the series of organic structure.

* Although the Ichthyosaurus occurs, as will have been seen from preceding lists, in many beds of the oolitic series, we have reserved its description for this place, because the most numerous and perfect specimens have been found in this formation.

The remains of the Ichthyosaurus have been figured in the several plates published in the Philosophical Transactions, from 1814 to 1820 inclusive, with descriptions by Sir Everard Home (who proposes the name of Proteosaurus from supposed analogies to the Proteus; but this has not been generally received). A more detailed account of the osteology of this genus, and the only published description of the Plesiosaurus, will be found in the fifth volume of the Geological Transactions.

The Crocodile is said to have been discovered in lias, but the fact remains doubtful. The skeleton described by Dr. Stukely in the Philosophical Transactions, and supposed from the imperfect representation there given to have been a Crocodile by Cuvier, really belongs to the Plesiosaurus. The specimen from Whitby, figured also in the Philosophical Transactions, may possibly be a Crocodile, but is too incorrectly drawn to afford any certainty. The Ichthyosaurus undoubtedly occurs at Whitby, and is described and represented (but very imperfectly) in the third volume of the Wernerian Transactions. No true Crocodile bones have yet been discovered in the lias of the south-western counties, which constitute, however the most thoroughly examined district occupied by this formation; still as true species of the Crocodile certainly occur in other beds associated with the oolitic series, there is no improbability in their occurring here also.

Bones and palates of the *Turtle* have been found in this formation.

Fish of several species occur also in its strata. Barrow on Soar in Leicestershire, and Lyme in Dorsetshire, have afforded many fine specimens.

Figures of two or three different varieties may be referred to in Nicholls' History of Leicestershire, vol. 3. part. 1. plates 8 and 9, where they are conjectured to belong to a species either of Sparus or Chætodon; they are, however, in all probability *sui generis*, and unknown in a recent state. The lias fish are also figured in Townsend, plate 20.

The radius of a species of *Balista* (erroneously figured by Townsend as the jaw of some animal, character of Moses, plate 18.) is of common occurrence.

The leech-like palatal tritores of some species of fish are frequently found, and teeth resembling in form and arrangement those of the shark more rarely so. *(See Townsend, pl. 18. fig. 4.)

The order *Crustacea* affords one or two species of Cancri, apparently Crabs, also a species of Monoculus or Limulus of Lamarck.

At the head of the *Molluscæ* we may perhaps (although very doubtfully) enumerate the remains of the Sepia as occurring in the lias, since the collection of Mr. Miller of Bristol contains a specimen resembling the beak of this animal.

The following list contains the principal Testaceous Molluscæ found in the lias beds, with references as usual to the figures in Mr. Sowerby's Conchology,

CHAMBERED UNIVALVES.

Ammonites ellipticus. *T.* 92, fig. 4.
A. armatus. *T.* 96.
A. planicosta. *T.* 73.
**A,* Stellaris. *T.* 93.
**A.* Walcotii. *T.* 106.
**A.* Brookii. *T.* 190.
**A.* Bucklandi. *T.* 130.
**A.* Conybeari. *T.* 131.
A. finbriatus. *T.* 164.
A. Greenoughi. *T.* 132.
A. Henleyi. *T.* 172.

* There are some beds near the bottom of the lias series particularly distinguished by the number of vertebral remains; these are distinctly seen in the cliffs of Westbury and Aust on the banks of the Severn in Gloucestershire, and are well known to the collectors of that neighbourhood under the name of the Bone beds. See the section in the beginning of this article.

A. Loscombi. *T.* 183.
A. obtusus. *T.* 167.
A. annulatus. *T.* 222.
A. Heterophyllus. *T.* 266.
A. Birchii. *T.* 267.
A. Bechei. *T.* 280.
A. giganteus. *T.* 126.
A. angulatus. *T.* 107. fig. 1.
A. communis. *T.* 107, fig. 2. 3.

The Ammonites having the siphuncle in an elevated ridge between two furrows are characteristic of this formation; the species thus distinguished form a well marked natural division of this order; they are marked with an asterisk in this list.

Nautilus intermedius. *T.* 125.
N. striatus. *T.* 182.
N. truncatus. *T.* 123.
Scaphites æqualis. *T.* 18. fig. 1. 2. 3.
Belemnites; many varieties both fusiform and common. The alveolus of a large variety is figured by Mr. Sowerby as Orthocera conica. *T.* 60. fig. 1 & 2.

Univalves not chambered.
Helicina compressa. *T.* 10.
H. expansa. *T.* 273. fig. 1. 2. 3.
H. solarioides. *T.* 273. fig. 4.
Trochus imbricatus. *T.* 273. fig. 3. 4.
T. anglicus. *T.* 242.
T. similis. *T.* 142, lower figure.
Tornatilla. Fig. 18 and 19 of the shells represented in the introduction to Nicholls' History of Leicestershire.
Melania striata. *T.* 47.

Irregular Univalves. *Annelidæ* of Lamarck.
Dentalium cylindricum. *T.* 79.
Patella lævis. *T.* 139, fig. 4.

Bivalves.
Modiola lævis. *T.* 8. le. lo.
M. depressa. *T.* 8. *m.*
M. minima. *T.* 210, fig. 5. 6. 7.
M. hillana. *T.* 212, fig. 2.
Unio crasissima. *T.* 153.
U.... (a less oblong and more clumsy species, figured in Nicholls' Leicestershire, plate 9. vol. 3. fig. 6.)

Cardita lirata. *T.* 197. fig. 3.
Astarte
Arca
Cucullæa
Nucula } small species.
Terebratula (nonplicated varieties).
T. ornithecephala? *T.* 101. fig. 4.
T. acuta. *T.* 15. fig. 1.
Terebratula (plicated varieties).
T. crumena?
Spirifer or *Pentamerus*, a variety not yet figured; agreeing in general form with the former, but in the internal divisions of its valves with the latter.
Gryphæa incurva. *T.* 112. fig. 1.
G. obliquata. *T.* 112. fig. 3.
Ostrea.
Pecten.
Plagiostoma gigantea. *T.* 77.
P. punctata. *T.* 113, fig. 1.
Lima antiqua. *T.* 214, fig. 2.
Plicatula spinosa. *T.* 245.
Hippopodium. *T.* 250; and another smaller species not figured.
Perna?

The most characteristic shells of this formation are the Ammonites Bucklandi, the Gryphæa incurva, and the Plagiostoma gigantea.

At Watchet on the Somersetshire coast, beautiful specimens of compressed Ammonites presenting thin coats of iridescent nacre are found.

A large proportion of the Whitby fossils are in nodules of clay iron-stone; in the cement stone, the Ammonites are occasionally coated with a small quantity of blende and galena. (G. Notes.)

Echinus.
A variety of Cidaris papillata, with long slender smooth acicular spines.
Encrinies or *Crinoidea.*

The following species of the genus *Pentacrinite* occur in the upper beds of the lias formation, and will be found figured in Mr. Miller's monograph of this family.

P. Caput Medusæ.
P. Briareus.
P. subangularis.
P. basaltiformis.

P. tuberculatus. The last named species appears to characterise the upper beds of this formation, and is that found so abundantly about Shuckborough in Warwickshire.

CORALS.

A species of *Turbinolia* of Lamarck (*Madrepora turbinata* of former writers) occurs in the upper beds of the lias formation, especially at Fenny Compton tunnel on the Oxford canal.

VEGETABLE REMAINS.

These consist of fossil wood occasionally silicified,* and several species of ferns, flags, &c.

(*d*) *Range and extent.* This formation stretches across from the coasts of the German ocean in Yorkshire to those of the channel in Dorsetshire. To commence with its northern limit, it is seen lining the coast, and underlying the mountains of the eastern moorlands (composed of sand and sand-stone strata probably belonging to the inferior oolite,) from the Peak alum works on the south of Whitby nearly to the Tees mouth.†

* The wood is sometimes charred; sometimes impregnated with quartz, hornstone, agate, or pyrites. (G. Notes.)

Mr. Hornsey of Scarborough shewed me a specimen of jet from Whitby completely silicified, and encrusted with agate, in which are brilliant specks of pyrites. This beautiful specimen was presented to him by the Duchess of Leeds He has others from the same coast in which the fossil wood of the lias is completely silicified. (G. Notes.)

† The following particulars, connected with the phænomena of this formation in Yorkshire, are here inserted as belonging to its local rather than general history.

Mr. Smith has represented indeed the alum-shale of Yorkshire as belonging to another formation; but all other geologists who have examined the district, among whom Mr. Greenough and Professors Buckland and Sedgewick may be mentioned, are unanimous in assigning it to the lias. In the article dedicated to the express consideration of the differences of opinion which have arisen concerning the identification of the strata in this part of the island, the arguments on this subject will be stated; but this article is necessarily postponed to the end of the part of this work now published, in order to await the result of some further enquiries on the subject.

Over the alum-slate lies a bed of hard compact stone, six to twelve feet thick. The workmen call it *dogger*, a name by which they also designate the septaria or cement-stone, and the component parts appear to resemble each other. The colour of the recent fracture of the dogger is bluish grey, but on exposure it changes to a deep purple brown; and it appears to be divided into nearly cubical masses by transverse fissures filled with a soft ferruginous earth, containing thin ochreous plates having the earth between them. (P. M. vol. 51. p. 20.)

The whole of the upper part of the alum-shale resembles indurated clay, when first wrought; but by exposure to the atmosphere it suffers decomposition, and crumbles into thin layers. (N. J. vol. 25. p. 241.)

Turning southwards from the mouth of the Tees, the lias ranges beneath the west escarpment of the eastern moorlands,

The colour of the shale is a bluish grey. It varies in hardness. The upper part of the bed near Whitby may be crumbled in pieces between the fingers, but at a considerable depth it is as hard as roofing slate. (Ibid.)

The upper part feels soft and unctuous like indurated clay; the lamellar fracture is smooth and shining, the transverse dull and earthy. (P. M. v. 51. p. 207.)

It is traversed by fissures dividing it into rhombic portions. (Ibid.)

At the depth of about two hundred and fifty feet from the top of Boulby cliffs, which rise four hundred and fifty feet above the sea, the shale loses its unctuous feel, and becomes mixed with a large portion of sand, and mica in shining scales. It becomes of a light grey colour, and encloses seams of iron-stone; but below this part, the rock resumes its softness and smoothness. (P. M. v. 51. p. 207.)

It abounds in pyrites. (Ibid.)

The upper part is most abundant in sulphur, which decreases in going down, but the bituminous substance increases, and the rock becomes hard and slaty; so that a cubic yard of rock taken from the top of the stratum, is as valuable as five cubic yards taken at the depth of one hundred feet. (N. J. v. 25. p. 241.)

When a quantity of the schistus is laid in a heap, and moistened with sea water, it takes fire spontaneously, and will continue to burn till the whole of the combustible part was exhausted. A part of the cliff which fell some years ago was exposed to the tide; it took fire and continued to burn for two or three years. (Ibid.)

That part of the alum-shale of Whitby which is earthy rather than slaty, yields the greatest quantity of alum. A layer of brush-wood is made in the first instance, and shale is thrown down upon it, until a considerable mound is raised. The brush-wood is then lighted, and a slow combustion ensues; another layer of brush-wood is then placed beside the first, and is in like manner covered by a mound of shale: still others are added, and these mounds, with fires beneath, are extended on all sides; when the shale has effectually caught fire, it continues to burn without any addition of fuel. It is afterwards thrown into vats with water and boiled twenty-four hours; it is then conveyed into other vats where an alkali being added, it crystallises: it is then melted again, and purified by a second crystallisation. When so prepared it is shipped off for London and thence to Sweden and Russia. (G. Notes.)

The History of Whitby by Lionel Charlton, in one volume 4to. contains the best account of the alum works. (Ibid.)

Some parts of the alum shale at Whitby are useful as marle. (Ibid.)

It contains masses of iron ore (septaria), which, when of a globular form sometimes enclose naphtha (bitumen?); also sometimes wood more or less petrified and occasionally passing into jet (N. J. v. 25. p. 254); they are sometimes coated with pyrites.

It contains immense quantities of red iron ore at the depth of about two hundred feet from the top of the aluminous strata, in seams varying from a few inches to two feet. The iron ore yielded by analysis thirty to sixty per cent. of oxide of iron, phosphoric acid, lime, alumine and silex. It is smelted at Newcastle. (Ibid.)

It contains sulphate of lime in crystals, and carbonate of lime in veins. (Ibid.)

Under the aluminous shale formerly worked at Gisborough in Yorkshire, is a shale abounding in fossils, among which the pecten is the most frequent;

passes York on the east, and crosses the Humber a little eastward of the junction of the Trent and Ouse, stretching on-

it is in general much decomposed; the unweathered parts are besprinkled with specks of mica, and resemble the Die-earth of Shropshire; it is however evidently a member of the lias formation. (G. Notes.)

At Clay hill and Long steep hill, at the foot of Greenhow hill, which is composed entirely of lias, this substance appears in a very different character from that which it commonly assumes. It is true that it is occasionally limestone and occasionally shale, in both of which states it contains pectines, belemnites, &c. although neither ammonites nor nautili were perceived; but it is more commonly in the state of a thick irregular argillaceous flag-stone, like the dun-stone of Hereford, and this, becoming more sandy in the upper beds, at length assumes the aspect of Coleyweston slate, but the fossils are the same as those of the lias. (G. Notes.) Indeed all the most characteristic organic remains are found as abundantly in Yorkshire as in Dorsetshire, and the identity of the formation completely established by them; we may notice particularly, with this view, the remains of the Icthyosaurus and Plesiosaurus. The several species of the Ammonites belonging to the division, which have the siphuncle in an elevated ridge placed between two furrows traced along the back of each volution; the Ammonites armatus; the Gryphœa incurva; and the Plagiostoma gigantea.

Mr. Winch also gives the following list. (G. T. v. 5. p. 555.)

Echinus vulgaris. Parkinson, vol. iii. tab. 2. fig. 3.
Ammonites serratus. Sowerby, tab. 24.
armatus. .... 65.
heterophillus .... 168.
Nautilus.. lineatus................ 41.
imperialis .... 1.
discus .... 12.
Modiola.. depressa .... 8. middle fig.
Orthocera conica .... 6. fig. 1 & 2.
Belemnites.
Mya, two species.
Chama digitata. Sowerby, tab. 174.
Helix.
Trigonia.
Pentacrinite.
Pectenite resembling the common scallop.

These are all mineralised by clay-iron-stone, iron pyrites, and calcareous spar.

Gigantic reeds resembling arundo donax are found in the sea cliffs opposite High Whitby. They appear to have been rooted in a bed of shale or slate-clay, and their remains protrude into a stratum of sand-stone five feet thick. Those which stand erect retain their shape, but those which do not are compressed. Their tops seem to have been broken off; the woody matter has disappeared, leaving sand-stone casts. Casts of Euphorbiæ are also found in the sand-stone strata above the alum rocks. Wood mineralised by iron is frequently found at Kettleness and Stowbrow. Trunks and branches of fossil trees, the bark and softer parts of which have been changed into jet, are frequently met with in the alum shale: and leaves and impressions like those of the palm, are found in the sand-stone and iron-stone. (G. T. vol. 5. p. 556.)

wards beneath the scarp of the low oolite range of Lincolnshire, to the Wold hills on the borders of Nottingham and Leicester, and the celebrated quarries of Barton upon Soar, whence it continues still regularly accompanying the scarp of the hills formed by the inferior and great oolite through the counties of Nottingham, Warwick, and Gloucester. Its whole course throughout this extensive line to a few miles south of Gloucester* is remarkably regular, presenting an average breadth of about six miles, bounded on the south-west by the oolites, and on the north-west by the red marle; but beyond that point, its course becomes much more intricate; for while its eastern limit still continues to accompany the oolitic ranges through Somersetshire to the coast in Dorsetshire, being its line of junction with the superior formations, its western limit becomes very irregular, feathering in and out among the coal-fields which occur towards the æstuary of the Severn, and the upper part of the Bristol channel, in Gloucestershire, Somersetshire, Monmouthsire, and Glamorganshire, and attended by

The encroachments made on the cliffs by the sea and by the weather is so great, that the Abbey, built near Whitby in 656, was then nearly one mile from it, but in 1810 the sea had approached to within 200 yards of it. (N. J. v. 25. p. 241.)

* The western, or exterior and lower, limit of this tract from the Humber to Gloucester, may be thus stated more particularly; it follows a line about two miles (on the average) east of the river Trent, as far as its junction with the Soar, and then pursues in a similar manner the right bank of the latter river; forming occasionally a low escarpment, which is most conspicuous near the junction of the two streams. From the head of the Soar, it crosses by the west of Rugby towards the Warwickshire Avon, and keeps at a variable distance (never exceeding four miles) from the left or east bank of that river, approaching closely to it at Evesham: there are also several outlying hills of this formation on the right bank of the river between Warwick and Alcester; the same is the case between the confluence of the Avon and Severn, where the lias stretches between the fork of those rivers opposite Tewkesbury. This ground is correctly given in Mr. Smith's Map, but inadvertently coloured as lias in Mr. Greenough's; in the latter also, the lias does not extend far enough on the west of Gloucester, by about two miles; it ought to include Lassington and Woodbridge and the hill marked *g* on the west of the Severn, which flows in a denudation of the red marle to about three miles north of Gloucester.

In the midland counties, from the vale of Belvoir to the north of Oxfordshire, some peculiar beds of rubbly lias, often occurring as concretions, characterise the upper part of the lias marles, being separated however by about 50 feet of marle from the marly sandstones described in the preceding section: they contain a greater variety of shells than the regular lias beds; most of those figured in the Natural History of the vale of Belvoir in the first volume of Nichols' Leicestershire are from this part of the series; a thick Unio, and the Pentacrinites tuberculatus, are most characteristic. Shuckbrough hill in the south of Warwickshire, so well known for the occurrence of the star-like columnar joints of this fossil, is based on this bed.

numerous outlying masses. To render intelligible the course and position of the lias in this quarter, it is necessary to state that this district is principally occupied by three great basins of the coal formation, encircled by the subjacent rocks of mountain limestone and old red sandstone; the edges of these basins consist of strata thrown up at a high angle and often nearly vertical, forming bold and precipitous ranges of hills, among the vallies of which the horizontal strata of lias with the subjacent beds of red marle and brecciated or magnesian lime (hereafter to be described), form what the Wernerians would call up-fillings, being deposited *unconformably over* the extremities of the highly inclined strata of the coal measures, &c. which they cover very generally in the lower grounds; seldom exceeding a trifling elevation,* and cut through in various directions by the higher ranges of mountain lime and old red sandstone above mentioned, which of course gives to the whole a very irregular and perplexed outline. These three principal coal basins, for there are other small ones which it would only embarrass so general a description to specify, are, 1st. That of South Gloucester and North Somerset, bounded on the south by the range of the Mendip hills, on the west by the range which passes by Clifton forming the defile of the Avon, and on the north-east by a continuation of the same chain, trending round in that direction; all these chains exhibiting inclined strata of mountain lime and old red sandstone. The lias and subjacent horizontal beds are seen filling up the interior of this basin in the neighbourhood of Bristol, betweon that city and Bath, at Pucklechurch, and throughout the Somersetshire collieries. On the north-west of the ridge forming the edge of the coal basin, they are to be seen at Pyrton and Aust Passages, and numerous other points along the æstuary of the Severn, by which this coal-field is separated from that of the forest of Dean. On the south they stretch beyond the foot of the Mendips, through the marshes of the rivers

* This is to be understood as a general expression of the fact; in some parts, however, of the South Gloucester and Somerset basin, towards the Mendip hills, where the vallies are deeply excavated through the subjacent sandstone to the coal measures, the lias sometimes caps the brow of escarpment between 300 and 400 feet in height.

Above Crosscombe, on the south of the Mendips, between Shepton Mallet and Wells, the lias which here abuts abruptly against the elevated strata of mountain limestone, forms platforms about half way up the hills; and the groupe of Broadfield Down, which forms a portion of the western boundary of this basin, is skirted on the south-east with lias hills almost rivalling it in height. The details of this district, which it would be scarcely consistent with the limits of a general work like the present to trace at greater length, are represented with great fidelity in Mr. Greenough's Map.

Brue and Axe, from which they occasionally rise in low ridges, such as the Polden hills, and in one instance swell into a more lofty summit at Brent Knoll, an insulated hill rising from the marshes ou the southern coast of the Bristol channel—for although this hill has a thin cap of inferior oolite, yet its great mass is lias; thence they continue to skirt the coast to a few miles west of Watchett, near which place they exhibit cliffs occasionally rising to the height of one hundred feet, and extending to the greywacké chain of Quantock. The second coal basin to be mentioned as affecting the lias, is that of the forest of Dean, likewise encircled by an elevated border of mountain limestone and old red sandstone, prolonged chains of which proceeding from its western boundary, cross the Wye, producing the beautiful defile of that river, and forming the range of Pen cae Mawr, in Monmouthshire, between that river and the Uske. The interior of this coal basin is throughout too elevated to admit any up-filling of lias; but that rock is to be found abutting in horizontal strata against the southern edge of its exterior ridges in Gloucestershire, near the mouth of the Wye, and in Monmouthshire on the south-east of Newport; also at Godcliffe on the Bristol channel.

The third coal-field connected with the position of the lias, is the south-east portion of the great coal basin of South Wales, in Glamorganshire; where it is skirted on the south by two chains, or rather a double chain of mountain limestone, separated by the vale of the river Ely, in which the subjacent old red sandstone appears. This valley exhibits several lower ranges formed by up-fillings of lias, &c. commencing about five miles west of Landaff; whence with some interruptions they accompany the Ely, first on its left, and then on its right bank, to its junction with the channel near Pennarth Point; continuing thence to Lavernock point, where the lias is interrupted, but the inferior strata of red marle, &c. continue; and bending round the extremity of the southern chain of mountain lime, advance westwards along the coast. The lias again rises at Barry Island, and continues to skirt the coast in a western direction for the distance of fifteen miles nearly to the mouth of the Ogmore river, forming a range of bold cliffs, among which is the little harbour of Aberthaw, celebrated for the lime it exports. These cliffs stand directly opposite those of Watchett, on the Somerset coast.

In the valley of the Ogmore, near Bridge end, is another up-filling of lias, which advances a little within the edge of the coal-field.

Leaving these intricate districts, and resuming the progress of the lias to the south, it advances regularly beneath the oolitic ranges through the south-east of Somersetshire, into Dorsetshire; where the overlying strata of green sand, (which in this quarter extend successively over the edges of all the inferior strata as far as the red marle) cover and conceal it, forming the high ranges of the Blackdown hills. It is however laid open to view by denudation in all the eastern vallies of that chain, and may thus be traced to the coast of the channel; where, in the neighbourhood of Lyme, it is displayed in a range of cliffs, extending about four miles, and sinking at length beneath a covering of the inferior oolite and its sand.

The best places for studying this stratum are the cliffs of Whitby in Yorkshire, those of Fretherne and Westbury, on the æstuary of the Severn in Gloucestershire, of Watchett in Somersetshire, of Aberthaw in Glamorganshire, and of Lyme in Dorsetshire.

(*e*) *Height of hills.* The lias generally forms broad and level plains at the foot of the oolitic chain of hills. It may be generally observed, that all the argillaceous formations in most instances occur as constituting low tracts, in the present configuration of the Earth's surface;—a circumstance which is rather to be attributed to their having offered less resistance to the denuding causes which modified the inequalities of that surface, than to any thing connected with their original formation. These plains are sometimes diversified with low ridges, and a slight escarpment may often be traced following the lower limit of the formation. This escarpment is most conspicuous on the borders of Nottinghamshire and Leicestershire, where it forms a well marked range distinguished by the name of the Wold hills. Near the Mendips, as we have already noticed, the lias sometimes occurs on the brow of tolerably steep escarpments, but its maximum of elevation probably falls short of five hundred feet above the level of the sea.

(*f*) *Thickness.* The joint consideration of the dip of the beds constituting this formation and their horizontal extent, together with the relative levels of its superior and inferior limit, gives a result of between four and five hundred feet for their thickness in the midland counties.

(*g*) *Inclination of the strata.* This is usually very small not exceeding forty feet in the mile, which conformably with all the strata ranging through the island, from north-east to south-west, is in a south-easterly direction; but where it comes into contact with the ridges of mountain lime and older rocks in Glamorganshire and Somersetshire, it is occasionally much disturbed and affected by some of the faults which traverse them; such disturbances are however always partial and limited;

its general character of horizontal stratification, always prevailing. Near Watchett, these partial disturbances produce an appearance of alternation in the strata of lias and red rock marle.

(*h*) *Agricultural character.* The lias soil is generally cold and tenacious, better adapted to pasture than tillage: in more sheltered situations it is favorable to the growth of wood. (G. Notes.)

In Glamorganshire it produces very fine wheat; and the marle of the rag, or grey lias, is esteemed the richest in the country. (Mr. G. Williams. G. Notes.)

Samphire grows more plentifully and luxuriantly in the grey lias cliffs of Glamorganshire than on any other. (G. Notes.)

(*i*) *Water.* The springs are generally thrown out by the marle above the lias, near its junction with the lower beds of the sand underlying the inferior oolites, and it is therefore doubtful to which of the formations they should with most propriety be referred. The mineral waters of Ilmington (Warwickshire), Cheltenham, Bath, Glastonbury, and Alford near Castle Cary, appear to be thus situated: the former is a chalybeate, the latter contain various proportions of iron, carbonate of lime, sulphate of lime, sulphate of magnesia, sulphate of soda, muriate of soda, which last is often very abundant, &c. but the analyses of those of Bath and Cheltenham appear to have been alone carefully performed.

When completely within the district of the lias marles, water (excepting the ground springs) is only to be procured by sinking to the bottom beds.

CHAPTER IV.

RED MARLE, OR NEW RED SANDSTONE.

This Chapter will conclude our view of the formations above the great and important deposits of coal, and comprise the beds between the lias and those deposits: these are entirely referable to two formations very intimately connected together, viz. 1st. a series of marly and sandy beds intermixed with conglomerates derived from older rocks, containing gypsum and rock salt, and in one instance amygdaloidal trap: and 2ndly, a calcareous formation often brecciated and characterised by containing a considerable portion of magnesia: this lies beneath, or at least in the lower portion of the above series. The former deposits are commonly known by the name of Red marle or New red sandstone; the latter as the Magnesian limestone.

The first section of this chapter will therefore treat of the former, and the second of the latter of these formations; a third section will be added, dedicated to a comparison of these formations in this island and other parts of the globe, which in this instance it will be found more convenient to subjoin than prefix. (C.)

Section I.

RED MARLE.

(*a*) *Chemical and external characters.* This formation is by some also termed the *Red Rock*, or *Red Ground.* It is a very extensive deposit, stretching with little interruption from the northern bank of the Tees in Durham to the southern coast of Devonshire. Its texture is very various. It appears sometimes as a reddish marle or clay, sometimes as a sandstone; sometimes the clay and sandstone are interstratified or pass the one into the other; and it will farther appear that it is associated with, or contains beds of, a conglomerate consisting of masses of different rocks cemented by marle or by sand. When this deposit appears as a sandstone, its characters differ greatly in different places; it is occasionally calcareous, and sometimes of a slaty texture. Above all, this extensive deposite is remarkable for containing masses or beds of gypsum; and the great rock-salt formation of England occurs within it, or is subordinate to it; in some places the strata of coal dip beneath it.

Although it would perhaps be generalising too hastily to assert that these beds invariably follow in a constant order, yet it may be safely stated as a general truth, that the red marle containing gypsum usually occupies the higher, the sandstone the central, and the conglomerate the lower portions of this deposite; but the most remarkable of the subordinate beds connected with this formation are those of amygdaloidal trap, which occur in it in Devonshire: as these are confined to that county, they will be described in treating of its local characters in that quarter. (C.)

The general composition of these beds is argillaceous, argillo-siliceous, with a variable proportion of calcareous matter, sometimes effervescing weakly with acids, sometimes not at all. (C.)

The marle and sandstone are often red, but vary in their hue from chocolate to salmon colour; they are not unfrequently variegated, exhibiting streaks of light blue or verdigris, buff, or cream colour (G. Notes); this forms so prominent a character, that Werner denominated the formation "bunter sandstein," variegated sandstone. (C.)

At Westbury on the Severn in Gloucestershire, it is for the most part greyish blue and chocolate red, alternating at intervals of about a foot, but sometimes crossing each other, and sometimes intermingled. (G. Notes.)

The soil of Malborough in Devonshire is so red that the butchers at Dodbrooke market know the sheep by the colour of the fleece that come from thence. (Polwhele, G. Notes.)

From the prevalence of this striking colour, the soil of the formation has given rise to many local names; as in Exeter to Rougemont castle, now a prison; in Somersetshire to Radford, Red hill, and Redcliff; in Gloucestershire to Redbrook; in Worcestershire on the borders of Glamorganshire to Red marly; in Warwickshire to Radford; in Nottinghamshire to Retford; Radford, Ratcliffe and to Redhill at the junction of the Trent and Toar; in Derbyshire to Retford; and in Yorkshire, to Rotherham, Red mire, Red ho, and Red bar rocks, which appear on the sea coast between Gisborough and Hartlepool. (T. 155.)

The red marle is remarkable for its fissures, which are transverse to the strata, and separating the rock into rhombic forms. A striking example of this may be seen at the new cut, Bristol, near Bridgnorth in Shropshire, at Kidderminster in Worcestershire, and near Sutton and Ellaston in Staffordshire. G. Notes.)

At Westbury on Severn and other places, it is found in concentric lamellar concretions, which often crack into irregular hexahedral and pentagonal masses. (G. Notes.)

Fullers' earth is raised from the marle beds at Raddle pits near Braithweel, north-east of Rotherham, and at Renton in Yorkshire; and at Taschbrook, one mile from Warwick, a substance probably of the same nature, as it was intended as a substitute for soap, was raised by the Earl of Warwick. (G. Notes.)

The sandstone of this formation consists of rather fine grains of quartz, with a few specks of mica, cemented by clay and oxide of iron; it contains galls of clay, is friable, and affords large tracts of sand. The beds generally differ in colour, and though it rarely happens that any great variety of colour is seen in the same bed, yet between Exeter and Exminster in Devonshire, the white and red layers of sand are strangely mixed. Although it appears to consist principally of siliceous particles worn down by attrition, yet it exhibits unequivocal proofs of chemical action. It sometimes contains small white opake crystals of felspar, and in Cumberland fragments of flinty slate. It is sometimes amygdaloidal, and sometimes passes by insensible gradations into trap. (G. Notes.)

The slaty character of the sandstone is mostly derived from particles of mica which are generally grey, and lie in detached spots, not in regular layers. It occasionally passes into slaty marle. (G. Notes.)

This rock is generally unfit for the purposes of architecture, but in some places it has sufficient cohesion to afford some of the finest freestone in the kingdom. The tower of Kenton in Devonshire, is a proof that sometimes it is durable. (Polwhele. G. Notes.)

It affords an excellent white freestone at Runcorn and Manley in Cheshire. In the parish of Addingham (Cumberland?) the Druidical temple called Long Meg and her daughters, consists of red grit-stone; some of the stones are 18 feet in height and 14 in girth. (Pennant, G. Notes.)

From the softness of this sandstone it has been frequently excavated into extensive artificial caverns; such as those near Nottingham, which as they gave rise to the name of the place, Snodengaham, "the home of caverns," must have been of great antiquity, and probably may have formed the dwellings of the rude aborigines: there are similar but smaller excavations at Knaresborough (Yorks) and Guy's cliff, Warwick. (C.)

Some of the sandstone beds of this formation bear so near a resemblance to some of the grits associated with the coal formation, and to the softer strata of the old red sandstone underlying the mountain limestone, that a cursory observation of them would often lead to fallacious conclusions. It may however be generally recognised without much difficulty by the following distinctive characters; 1st, its containing gypsum; 2ndly, by the inferior consolidation of its stony beds; 3dly, by the regularity of its stratification, and the general parallelism of its beds to the horizon. (C.)

Further notices of the local characters of this formation will be incorporated in the account of its range and extent.

(*b*) *Mineral contents.* Besides the extensive deposit of rock salt and gypsum * noticed above, sulphate of strontian and ba-

* We have extracted the following notices from the notes obligingly given to the editor by Mr. Greenough with regard to the gysum and salt.

Alabaster. Provincially Plaister stone, and Hall Plaister. In Devonshire, Spear, i. e. Spar.

This substance is a considerable article of trade. The larger masses are worked into pillars, as at Kedleston in Derbyshire, or vases and other ornaments. The finer varieties of the fibrous gypsum are made into earrings and necklaces: the coarser kinds are used as moulds by the potters of Staffordshire, or is used for stucco, plaister, flooring, &c. for which it is prepared by burning, and threshing, or pulverizing with flails, after which it is passed through a riddle.

In America the virtues of this substance as a manure are highly extolled, but in this country our expectations in this respect have been disappointed.

No organic remains or metallic minerals have hitherto been found in the gypsum of this formation.

At Newbiggin in Cumberland it lies in red argillaceous marle, between two strata of sandstone; the upper solid, hard and fine-grained, the under loose and coarse-grained: in some places it rests on decayed wood-like umber. (Hist. & Antiq. of Cumb.)

One mile south of Whitehaven in Cumberland, the subterranean workings for alabaster extend 30 yards in a direct line; the passages are low, and of a size just sufficient to allow one man to enter them. From the main passage are two or three lateral ones, each extending perhaps 10 yards; at their extremities are large spaces in which the alabaster is blasted by gunpowder.

The alabaster is generally compact, forming a regular and conformable bed, but on blasting it, crystals of selenite often appear in druses. After exposure, it often exhibits parallel lines, the effect of stratification, which are not perceptible in the fresh fracture.

At the commencement of the last war with France, from 200 to 300 tons of the Newbiggin gypsum was sold per annum. (Hutch. Hist. Cumb.)

It is remarkable that the names of many places near which salt is found terminate with wich or wych, as Droitwich, or Nantwich, &c.; and the houses in which salt is manufactured Wych-houses. Wich, according to Skinner, is an Anglo-Saxon word for district or habitation.

The Aster trifolium, or farewell to summer, a kind of Michaelmas daisy, is considered an indication that brine is in the neighbourhood: its proper habitat is the sea-shore.

The making of salt from the brine-springs adjacent to Nantwich in Cheshire, formed a very important business in the time of Elizabeth, when there were 216 salt works of six leads walling each; in 1774 there were

rytes occur in this formation, and perhaps also yellow and other ores of copper, grey oxide of cobalt, and black oxide of manganese: the reason for speaking doubtfully on these latter articles will appear in the following statements. (C.)

At Seacome Ferry on the shores of the Mersey opposite to Liverpool, and for several miles along the coast, magnetic iron-sand, mixed with much iserine, ooses out of a cohering sand lying below a deep bank of clay, and which is considered as entering largely into the geological composition of that part of Cheshire. (G. T. vol. 4. p. 447.)

(*c*) *Organic remains.* None whatever have yet been found in any of the rocks connected with this formation, unless the magnesian limestone (which contains some marine animals) be considered as included in it. *

(*d*) *Range and extent.* The eastern or superior limit of this formation (being its junction with the lias) has already been traced through the island as forming the inferior limit of the latter rock, but we no longer find in this, as in the preceding formations, the western or inferior limit describing a line nearly parallel to the former, and thus together with it including a band or zone of nearly uniform breadth. The reason of this circumstance is, that the Red Marle, and its associated magnesian limestone, form the last of the nearly horizontal and conformable strata occupying the eastern and southern counties; and the strata of the succeeding formations are unconformably placed with regard to these, rising from beneath them at various

only two works of five large pans of wrought iron. The duty produced from them amounts annually to near £5000; from the whole district, including the works at Lawton and a small one at Droitwich, from £18,000, to £20,000.

Salt was an object of taxation at a very early period in this country. Ancus Martius, 640 years before our era, 'Salinarum vectigal instituit.' This tribute was continued on the Britons when our isle was possessed by the Romans, who worked the Droitwich mines, and who made salt a part of the pay of their soldiers' salarium or salary. Hence the custom at the Eton montem of asking for salt.

The ancient mode of making salt, and which even now I believe is practised in Germany, was to fling the brine on burning wood, by which means the water was evaporated, and the salt was left adhering to the ashes.

The Saxons, according to their ideas of liberty, divided the salina between the king, the nobles, and the freemen. Of the salt works at Nantwich, eight were the joint property of the king and Earl Edwin. The king had two-thirds of the profits, the earl one-third: Edwin had also a work near his manor of Aghton, which yielded sufficient salt for the consumption of his household. If the salt of this work was sold, the king was to have a tax of two pence upon it, and the earl one penny.

* There is a very remarkable variety of chert containing shells overlying the calcareo-magnesian conglomerates of the Mendip hills (Somersets.) in some places, but this should most probably be included under the deposite which these represent—the magnesian limestone formation.

and often very considerable angles, and towering into lofty groups and chains of mountains, around which the Red Marle skirts, occupying the extended plains at their base: so that the appearance of the whole may be described by the figure of a sea composed of horizontal beds of red marle, &c. surrounding elevated islands consisting of rocks of the coal-formation, or carboniferous mountain limestone, old red sandstone, transition slate, and greenstone, all variously and irregularly stratified. (C)

To trace this formation, as before, in its course from the north-east to the south-west; the first mountain chain thus skirted by it is the long mountain range which pervades the counties of Northumberland, Durham, York, and Derby, and joins on the west to the Cumberland mountains. (C.) It first occurs on the east side of this chain at the mouth of the Tees, where it appears as a fine-grained sandstone of a brick-red colour, which effervesces with acids; its limit on the north-east is a little above the northern bank of the Tees. The strata are numerous, and consist (as far as one can judge from the miner's language) of white, grey, or red sandstone, with occasional interposed strata of a more compact nature, red or blue shale (slate clay), coaly matter in thin layers, and gypsum in nodules and in beds of from one to three feet in thickness. The lowest bed in two of the deepest workings, was a white rock of a calcareous nature. (G. T. vol. 4. p. 2.)

It thence proceeds due south by the city of York to Nottingham, following the vales of the Ouse and the Trent. Through this part of its course, it has a pretty uniform breadth of from eight to twelve miles, and is regularly accompanied on the west by its attendant magnesian limestone. Gypsum occurs plentifully in the isle of Axholme, and various other places in Nottinghamshire. A considerable district occupied entirely by quartzose gravel occurs in the latter county, between the marle and magnesian limestone. Deposits of this kind extend to the depth of 200 or 300 yards, and are often consolidated into a soft pudding-stone, an example of which may be seen in the Castle hill at Nottingham. Although this gravel has been referred by some writers to a recent alluvial origin, it seems more probable that it is a form of the conglomerate rocks so generally attendant upon this formation. It constitutes the prevailing stratum throughout Sherwood forest. (C.)

To the south-west of Nottingham the district covered by the red marle expands into that vast tract of nearly level country which occupies the greater part of the midland counties. But before proceeding to the general description of this central plain, it will be more convenient first to trace its northern border, where the red marle skirts round the southern base of the Derbyshire mountains, thence pursuing the course of that formation to the north along the western base of the same chain, and afterwards returning to the central plain in our progress to the south. (C.)

Pursuing this order, we shall first have to describe the northern border of the great central plain which may be conveniently taken as the tract included between the Trent and the mountains on the north of Ashborne in *Derbyshire.* This tract is uniformly occupied with strata of the red marle formation, with the exception of some very inconsiderable and insulated patches of the older rocks, which, in one or two places, emerge from beneath the covering: of this we have examples at Wild park near Brailsford, eight miles north-west from Derby, and Birchwood park near Roston, four miles south-south-east from Ashborne, where the *older* variety of magnesian limestone associated with the carboniferous or mountain limestone, are thus thrown up; and in a line between the two points, coal-measures have been proved at Spring hall near Edmaston and Darley moor near Yeavely in Derbyshire. To the north-west of the latter place, a white gravel rock has also been found not far from the banks of the Dove near Snelston, &c. which, however, very possibly belongs to the conglomerates of this formation; and, with these exceptions, the whole of the district above stated is occupied by strata of the present formation (C), among which are visible some fine-grained micaceous *gritstone* beds, of considerable thickness, which occasionally are sandy, as at Normaton, south of Derby, &c. From the more clayey parts of this stratum, bricks and tiles are made. Occasionally it encloses streaks, or thin beds, of light blue or greenish blue earth, or marle. These are common in the red marle of *Nottinghamshire.*

In Derbyshire, some considerable deposits of gypsum have been found in the red marle, and some of them are worked for that mineral. Chellaston is about five or six miles south-east of Derby, and it appears that on the south-east of that town there are several quarries: part of Chellaston hill would present a naked and water-worn rock of gypsum, were it not for the alluvial clay that covers it. Near Alston, also, which is somewhat further on the south-east, and formerly at Ballington hill near Ambaston, there were other quarries: gypsum has also been seen in other places. It should seem that it occurs only in particular patches or nodules, where it occasions a rise or isolated hill, by the additional thickness which it gives to the stratum of marle in those places; but it occasionally forms very thin beds or layers, which sometimes are finely striated transversely to the strata. The principal demand for the pure white gypsum, or that slightly streaked with red, is by the potters in Staffordshire, who form their molds of the plaster of Paris which it yields; but some particular blocks are selected for the use of the turner, and the maker of alabaster ornaments. When mixed with blue or green earth, it is called flooring-stone; that which is found in very thin beds, is used for the building of walls and other common purposes. (F. 148, & seq.)

Advancing from Derbyshire into *Staffordshire*, we still find the same formation occupying the low district between the vallies of the rivers Dove and Trent: near the confluence of these streams, gypsum is dug in many places, and salt springs abound near the Trent, particularly at Weston near Stafford, where salt works have been established. (C.)

Near the head of the western branches of the Trent, the great central plain of the red marle unites with that occupying nearly the whole of *Cheshire*, the southern part of *Lancashire*, and the northern part of *Shropshire*, and watered by the Dee, the Weaver, and the Mersey. The vallies of these three rivers are occupied by the red marle formation, and the central valley, that of the Weaver, presents throughout its course, abundance of salt-springs containing above twenty-five per cent. of salt; gypsum is also plentiful. At Northwich in this vale, an extensive deposit of solid rock salt has also been found, consisting of two beds, together not less than sixty feet in thickness. These beds are supposed to form large insulated masses of this mineral, extending in length about a mile and a half, and in breadth about 1300 yards. There are other deposits of this mineral in the same vallies, but of less importance than this. (C. from Dr. Holland's paper in G. T. vol. 1.) The section presented by the workings at Northwich is as follows.

Section of the Strata sunk through to the second Bed of Rock Salt, at Witton near Northwich. (G. T. V. 1, p. 62.)

	feet.	inch.
1. Calcareous marl	15	—
2. Indurated Red Clay	4	6
3. Indurated blue Clay with sand	7	—
4. Argillaceous marl	1	—
5. Indurated blue Clay	1	—
6. Red Clay, with sulphate of lime irregularly intersecting it	4	—
7. Indurated brown Clay, with grains of sulphate of lime interspersed	4	—

	feet.	inch.
8. Indurated brown Clay with sulphate of lime, crystallized in irregular masses, and in large proportion	12	—
9. Indurated blue Clay, laminated with sulphate of lime	4	6
10. Argillaceous marl	4	—
11. Indurated brown Clay, laminated with sulphate of lime	3	—
12. Indurated blue Clay, with laminæ of sulphate of lime	3	—
13. Indurated red and blue Clay	12	—
14. Indurated brown Clay, with sand and sulphate of lime, irregularly interspersed through it. The fresh water (360 gallons per minute) finds its way through holes in this stratum, and has its level at 16 yards from the surface	13	—
15. Argillaceous marl	5	—
16. Indurated blue Clay with sand, and grains of sulphate of lime	3	9
17. Indurated brown clay, with a little sulphate of lime	15	—
18. Indurated blue Clay, with grains of sulphate of lime	1	6
19. Indurated brown Clay, with sulphate of lime	7	—
20. The first bed of Rock Salt	75	—
21. Layers of indurated Clay, with veins of rock salt running through them	31	6
	230	9

22. The second bed of Rock salt, which has been sunk into 105 to 108 feet.

The valley of the Weaver is divided from that of the Dee, by a low ridge of hills, including Delamere Forest, the crag on which Bierston Castle stands, and the Peckforton hills; and from that of the Mersey by a similar ridge connected with Alderley edge. In the Peckforton hills a small quantity of copper has been found, and at Alderley edge,* ores of lead,

* At Alderley edge in Cheshire, the sandstone of this deposit contains grey oxide of cobalt, galena and yellow copper ore, together with sulphate of barytes. In different parts of Shropshire, as at Hawkestone, Pym hill, &c. it is tinged by copper. Though iron abounds in this formation as a colouring matter, it contains no argillaceous iron stone, nor balls of iron stone, nor any variety of workable iron. (G. Notes)

It is sometimes slaty and micaceous, sometimes contains pebbles of quartz and agate. (G. Notes.)

See an account of Alderley Edge, in Monthly Mag. Vol. 1811. (G. Notes.)

copper and cobalt, and masses of sulphate of barytes; but it is probable that these ridges may not belong to the same formation as the red marle of the vallies, but may consist of elevated portions of some of the older sandstones. This is a point which at least deserves inquiry: if the rocks composing them are found to be thrown up at a considerable elevation, and to possess a considerable degree of consolidation, this suspicion would receive confirmation; but if on the other hand they are horizontally stratified, and of a loose texture, it would be negatived: the same remark also applies to Hawkestone and Pym hill in Shropshire, where traces of copper have likewise been found. Besides the extensive deposite in the valley of the Weaver, Salt springs are likewise found in that of the Mersey, and to the south near Wellington in Shropshire. (C.) The sandstone of the latter country is thus described. It consists for the most part of rather fine grains of quartz, with a few spangles of mica, cemented by clay and oxide of iron. Its colour is generally brownish red, and it has but little cohesion; on which account large tracts of loose deep sand are found in many parts of it. Sometimes it occurs nearly of a cream colour, and is then sufficiently hard to form an excellent building stone: it does not effervesce with acids, and no shells or other organic remains have been found in it. In some places, the loose sand on its surface contains rolled stones of quartz, granite, and porphyry, which also are dispersed over its surface, though they are rarely if ever observed at any considerable depth in the solid rock. (G. T. vi. p. 192.)

The red marle district continues to range in a northerly direction from the north-west corner of the Cheshire plain, skirting the western coast of *Lancashire*, where a red sandstone occurs in many places; and thus continues to line the western base of the great mountain chain so often mentioned: this district is erroneously coloured by Smith as old red sandstone? In the peninsula of Furness, the red marle is seen thus reposing against the foot of the Cumberland branch of those mountains; and to the north of this, it may be observed in a similar situation near Whitehaven in Cumberland, where a very satisfactory section is presented; the magnesian limestone subordinate the red marle formation being there seen reposing on the coal formation, and covered by the marle beds containing gypsum. (C.)

Further north, the same formation may be seen occupying the coasts of the Solway Firth, and the great plain through which the river Eden flows, and which here forms a kind of bay of low country, separating the Cumberland group of mountains from the prolongation of the central chain by Cross

Fell, &c. Gypsum is abundant in this tract; hence the red marle extends into the south of Dumfrieshire. (C.)

Returning to the great plain in the centre of England, we here find a tract of about eighty miles in length and sixty in breadth, principally occupied by this formation; several islands of the older rocks, however, rising in various places through it, as described in the beginning of this article. These are 1st. the sienite, greenstone, and slate district of Charnwood forest in Leicestershire; 2dly, the coal district surrounding Ashby de la Zouch in the same county, connected with which are several patches of the older or carboniferous variety of magnesian limestone, and a patch of millstone grit at Stanton bridge on the Trent, (see pl. 2. fig. 2); 3dly, the coal-field of Warwickshire; 4thly, the Coal-field in the south of Staffordshire, with the transition limestone on which it reposes; 5thly, the lower and northern range of the Lickey hill near Bromsgrove in *Worcestershire*, which exhibits elevated strata of quartzose sandstone, probably transition quartz rock; and has near it some small patches of transition limestone and rock.

The salt works of Droitwich in the latter county are situated in the south-west portion of this great marly plain. (C.)

The prevailing rock around Droitwitch is a fine grained calcareo-agillaceous sandstone, of a brownish red colour, with occasional patches and spots which are greenish blue. At Doder hill, where a vertical section of it is exposed, it contains beds of a greenish grey colour, and of a more indurated texture, but which do not appear to differ materially in composition from the red sandstone. These contain slender veins of crystallized gypsum, the forms of which are very distinct where the widening of the vein has produced small cavities. (G. T. v. ii. p. 95.) But around Worcester it sometimes resembles mere marle, with occasional patches of a blue or of a green colour; sometimes a fine sandstone and sometimes a breccia. (K. 110.) In various parts of Worcestershire, there are alternations of this marle with small and large grained gritstones. (K. 109.) In one of the pits, the strata sunk through, were, mould 3 feet, marle 35 feet, gypsum 40 feet, a river of brine 22 inches; gypsum 75 feet, a rock of salt bored into only five feet, but probably extending much deeper. (G. T.)

Hence, the red marle extends southwards down the valley of the Severn, abutting against the elevated escarpment of the sienitic chain of Malvern. (C.)

The course of the red marle, near the æstuary of the Severn in Gloucestershire, Somersetshire, Monmouthshire, and Glamorganshire, has already been sufficiently indicated in tracing that of the lias, which it here regularly accompanies, seldom extending to any considerable distance beyond it, and as it will be requisite to revert to the same district in describing the magnesian limestone, it is only necessary here to add the following particulars. (C.)

In some places near *Bristol*, the lower part of the sandstone of this formation, which there overlies the *Gloucestershire* and *Somersetshire* coal-field, alternates with layers of a blue or greenish colour.

The red clay belonging to this formation, contains, in the neighbourhood of Bristol, a small quantity of sulphate of barytes, and abounds with sulphate of strontian in the form of veins and even large beds, and as usual containing gypsum. (G. T. vol. iv. p. 214.)

On the top of the limestone strata forming the cliffs on each side of the Avon, lies a yellowish sandstone, which has sometimes the appearance of a breccia. In its fissure are found crystals of carbonate of lime and of sulphate of strontian, the latter often in a radiated form, and sometimes in balls weighing many pounds. At Redland this stone is covered by the lias limestone in horizontal strata, containing ammonites, gryphites and anomiæ in abundance. (G. T. vol. iv. p. 196.)

On the west of the exterior ridges of this coal-field, the red marle is beautifully displayed at Aust cliff on the Severn, where it contains large interstratified masses of gypsum, and is also traversed by veins of sulphate of strontian: this cliff is capped by lias. (C.)

On the Welsh side of the Bristol channel, gypsum occurs in this formation in the cliffs of the *Glamorganshire* from Pennarth to Lavernock, and geodes filled with crystals of sulphate of strontian occur in the island of Barry; the conglomerates associated with this rock prevail near Landaff and in the vale of the Ely. (C.)

In the south of Somersetshire this formation is seen resting on one side against the Mendip hills, and on the other against the greywacké chains of Quantock and Brandon (being covered by lias in the intermediate tract); gypsum occurs plentifully in it near Somerton, as also in the cliffs of Watchett, which are traversed by it in every direction. In this neighbourhood extensive beds of conglomerate, apparently associated with this formation, occur, surrounding the hills of greywacké. (C.) These conglomerates and sandstones assume very various appearances, but under every form of aggregation the same materials may be traced. Where the component parts are large, as in the conglomerates, the nodules consist of some varieties of the rocks that compose the greywacké formation; in many places there are nodules of a limestone very similar to that of the beds enclosed in the greywacké, and which therefore are considered as *subordinate* to that formation. The conglomerates and sandstones are found in all the intervening vallies, and the great valley on the western side of the Quantock hills is wholly composed of them. They are not however confined to the vallies, but are sometimes found on the sides of the hills, at a very considerable elevation; (G. T. vol. iii. p. 356) and even so high as near the summit of the Quantock range, where rounded fragments of greywacké, cemented by a deep red clay, form a mass of extreme hardness.

By this description of these deposites it appears that they have both, in this neighbourhood, resulted from the ruin of the greywacké formation, since they not only enclose fragments of it, but also of the limestone imbedded in it, and the fragments are sometimes rounded. In some places the quantity of calcareous nodules in the conglomerate is so great that it is quarried for the purpose of obtaining lime from it, as near the village of Alcombe, at the foot of one of the lateral branches of Grabbist hill, and at Torr, Tor Weston, and Vellow, where by the quarriers it is termed *Popple* (pebble ?) rock. In other places, it is almost entirely siliceous. In another, it contained the green carbonate and sulphuret of copper. The sandstone, at least, may be considered as a member of the numerous series of rocks constituting what is commonly called the red marle, sometimes the red rock. Near Timberscombe the conglomerate is covered by a friable and marly sandstone. The rich vale of Taunton Dean is believed to consist of rocks of this description. (G. T. vol. iii. p. 359.)

From the vale of Taunton in the south of Somersetshire, the principal mass of the red marle, sandstone and conglomerate, (after detaching that long narrow tongue of these formations, which, as has been observed, runs in between the chains of Quantock and Exmoor, and advances westwards among the grauwacké hills almost as far as Porlock) proceeds south into Devonshire, its line of junction with the transition chains ranging a little west of the river Tone, and afterwards nearly following the Exe and ranging between it and the Culm as far as Silverton (about five miles north of Exeter). There is, however, an insulated and outlying groupe of transition hills within this tract, surrounding Collumpton both on the north and south. From Silverton, a long tongue of this formation runs in westwards among the transition hills along the valley of Crediton, whence the junction trends round by Upton Pine and Pinhoe to Exeter, proceeding thence with a nearly uniform southern course by Chudleigh and the river Teign to the channel at the south-west angle of Torbay, where it ends; that portion of it which ranges between the river Teign and Torbay being indeed only a tongue of this formation lying among the transition formations, since at Babicombe and the northern cape of Torbay, insulated groupes of transition limestone skirt it on the east, as does the great transition district on the west. Mr. Greenough's map has however given rather too great an extension to these eastern groupes, and made them cut off the red marle of the middle of Torbay from the main tract of the formation, with which we have however been assured that it is continuous. (C.)

The formations of red marle, sandstone and conglomerate, occasionally associated with amygdaloidal trap, occupy a great part of East Devon between the line of junction just described and the overlying platform of green-sand which crowns the ridges proceeding from Blackdown hill, and forms their eastern boundary. The red marle has also a corresponding cap of overlying green-sand along Haldon hill on the west of Exeter, which advances within two miles of the granitic ranges of Dartmoor, so that some of the most ancient and modern formations are here brought into close vicinity. As this district has been more fully and carefully examined than any other occupied by this formation, and presents more important and interesting varieties than are usually to be found in it, we shall subjoing the following particulars collected from various sources. (C.)

Near Honiton in *Devonshire* is a red sandstone having an argillo-ferruginous cement; it is in the state of a coarse-grained gravel almost entirely disintegrated. It contains rounded pebbles, which are two or three inches in diameter: it then approaches to a conglomerate pudding-stone, but near Exeter it assumes the character of an arenaceous sandstone, and becomes more compact and uniform in its texture and composition. This conglomerate is in nearly horizontal strata, which probably extend eastward below the chalk, while in other directions they lie upon greywacké; as for instance, on a hill consisting of that rock, north of Exeter, and overlooking the town, and at other places north and north-east of Exeter. On the south and south-west of that place, these sandstones and conglomerates form the surface of the country for several miles towards Ivy bridge, near which also is found slaty and compact greywacké. It is however clear that the cliffs at Budleigh Salterton near Teignmouth in Devonshire, which are of considerable height, and those also of Teignmouth itself, though consisting in part of conglomerate, belong to the red marle or newer red sandstone, since they contain gypsum, which forms a principal characteristic of that rock. (G. T. vol. i. p. 98 & seq.)

The nature of the conglomerate belonging to this formation around Exeter, is best seen by the openings or quarries in its neighbourhood. That of Heavitree is situated about a mile and a half from Exeter on the road to Honiton. It is worked to the extent of a quarter of a mile in length, and to the depth of about 100 feet, in a plane intersecting that of the strata. The rock of this quarry, is a conglomerate evidently stratified; the strata are from six to eight feet in thickness, and dip south-east at an angle of about 15 degrees. As long as this rock preserves the character of conglomerate, it is compact and tenacious; and, according to the workmen employed in the quarry, it hardens more and more by exposure to the air. But as soon as the nature of the stratum changes to an arenaceous sandstone, it is tender and friable. It is very common to see blocks of it in this last state, and sometimes of a great size, included in the middle of the conglomerate. The cement of this rock is argillo-ferruginous; and by itself does not effervesce with acids, as it is easy to prove by making use of pieces of the pure sandstone for that purpose; but it produces so brisk an effervescence from the intimate mixture of calcareous particles, that it might be very easily mistaken for limestone. The substances which enter into the composition of this conglomerate are numerous; and it may first be remarked, that these pieces are of very different sizes and forms, sometimes rolled and rounded, sometimes pointed with sharp angles, from very minute grains to the size of several inches in diameter. There are found in it rhomboidal crystals of calcareous spar and crystals of felspar, most frequently of an opake white and decomposed; pieces of chert; greywacké; yellowish limestone; rolled masses of a species of porphyry somewhat resembling the antique, the base of which is a reddish brown colour, not effervescing with acids, and containing numerous, small, and well defined crystals of felspar imbedded in it; pieces of a rock which is itself compounded, having the appearance of a porphyry, the base earthy and including small grains of quartz, crystals of felspar, and pieces of bluish carbonate of lime; and a whitish tender steatite in small angular fragments. (G .T. vol. 1.)

While speaking of the neighbourhood of Exeter, we ought, if adhering closely to the principle of pursuing the geographical course of these formations, to notice the masses of amygdaloidal trap which occur in that vicinity; but wishing to reserve the consideration of this very interesting circumstance for the close of this article, we proceed at once to the coast, a mode of arrangement which will enable us to keep together the consideration of the conglomerates of this district. (C.)

In the cliffs about half way between Sidmouth and Seaton in Devonshire, the red marle contains gypsum, very much resembling the mineral called mountain leather; nor was it difficult to trace on the spot, the gradual transition of the transparent crystalline gypsum into this amianthiform state. (K. 106.) The red cliffs of Budleigh Salterton near Teignmouth, which are of considerable height, and those also of Teignmouth itself, consist of alternations of argillaceous beds of sandstone and of breccia (conglomerate) and the red strata on the opposite side, near Powderham, are alternately soft and stony, but always intermixed more or less with strata of breccia; and they are inclined in various directions. (K. 109.) *

* The Rev. J. J. Conybeare has given in the Annals of Philosophy for April 1821, a fuller description of the range of strata from Dawlish to Teignmouth; as it contains a more precise examination of the rolled masses of various rocks included in the breccia of this formation than any account hitherto published, we have subjoined it nearly entire.

In these strata the rock exhibits itself under the several characters of a sandstone, either loosely compacted, or altogether pulverulent; a marle, more or less indurated: and a breccia composed of fragments of various sizes. Near to Dawlish, the sandy form is more frequent, towards Teignmouth the breccia, the base of which is usually marle, of an unctuous and argillaceous character. The marle has frequently those patches of white and purple, which have been often noticed as characteristic. The mineral contents of the rock seem to be few. *Calcareous spar* occurs in small patches a little south of Dawlish. *Gypsum* I could not detect either here or under the elevated plains of Haldon. On Blackdown, however, I have found it in small nodules. A sand sufficiently charged with, and indurated by iron, to be termed *ironstone*, traverses in all directions, the cliff to the north of Dawlish (see Deluc, vol. ii. p. 85), and the *earthy brown oxide of manganese* is found in numerous and small cavities nearly through the whole range of the coast. In one instance I detected a small portion of the black oxide of cobalt, precisely resembling that found at Alderley edge (Cheshire) in the same strata.* But the most remarkable feature in the rock appears to be the variety of substances contained in its brecciated form. Deluc has mentioned two only of these, the limestone and greywacké, though he insinuates that there are others, and appears (if I rightly understand the paragraph) to apprehend that of these the limestone only can be traced to any neighbouring rock, an opinion which, in its full extent, seems untenable. The following catalogue presents a tolerably faithful description of the fragments occurring in the breccia of Dawlish and Teignmouth, with the addition of some few from other quarters.

Granitic and Porphyritic Rocks.—These form a very considerable portion of the imbedded fragments. A 1. A minute aggregation of pale reddish-

* I am indebted to the Rev. the Dean of Bristol for an addition to this meagre catalogue. That gentleman has had the kindness to forward me some specimens of chalcedony, which he discovered in this rock not far from Torquay. It is coarse, and appears under the form of spherical nodules, either hollow or investing portions of the marle itself.

Amygdaloidal trap of the vicinity of Exeter. Along a line of from five miles north to five miles south of Exeter, many

brown crystalline felspar, quartz, and common schorl. A 2. Same with felspar, somewhat darker, and more crystalline. A 3. Same mixture, with lighter-coloured felspar, and the schorl distributed in small contemporaneous veins, as well as intermixed in the mass. B. Felspar same as A nodules of quartz, and minute portions, apparently crystalline, of *chlorit?* Structure semiporphyritic. C 1. Minute aggregate of earthy felspar, of a pale dirty-red, quartz, and chlorit. C 2. Same with the felspar, less earthy, and of a deeper red. D. Porphyritic base, of a purplish-white apparently a minute aggregation of earthy felspar and quartz, imbedded quartz in small nodules, and some crystalline felspar (semivitreous) D 2. The same with the quartz so much predominant in its base as to give it, at first sight, the aspect of a sandstone, or greywacke.† D 3. Same with imbedded semivitreous felspar, and common felspar in various stages of decomposition (from the bed of the river Exe). D 4. Base more felspathic, and of a deeper purple, much disintegrated, with the exception of the semivitreous felspar. Many cavities filled with earthy felspar. D 5. Porphyry, base compact felspar, of a greyish-white, having imbedded small nodules of quartz, and penetrated by numerous cavities, apparently left by the disintegration of crystals of felspar, and the loss of the powdery matter so produced. (This remarkable character I have observed in more than one porphyry from Cornwall.) E. Base, a minute aggregation of earthy felspar, quartz, and chlorit, coloured green by the latter. imbedded minute crystals of flesh-coloured felspar, and small nodules of quartz. (Descent of Haldon on the eastern side.) All these occur in various stages of disintegration. Those porphyries approaching the nearest in colour to the marle which surrounds them are, when far advanced in this state, not readily distinguishable from that substance, the outline of the original fragment being so broken down and lost, that it appears to pass insensibly into the imbedding mass. Hence, perhaps, some geologists, of no inconsiderable authority, have been induced to suspect that these, and, by consequence, the other fragments imbedded in the red marle, were of a formation contemporaneous with itself.‡ A minute and accurate inspection of the coast between Dawlish and Teignmouth must, however, I think, convince us of the truth of the commoner, or at least earlier, opinion which regards them as derived from the breaking up of the inferior strata. Other fragments imbedded in the marle are *greywacké, or compact sandstone.* F 1. *Compact greywacke*, of a dirty-white, much ironshot, having the aspect of a sandstone with a very small intermixture of argillaceous matter. F. 2. *Compact greywacké*, quartz more predominant, and closely aggregated ironshot throughout of a reddish-grey. F 3. *Same*, of a greyish-black, with contemporaneous veins of white quartz. G 1. *Black compact siliceous rock*, of a very close texture, resembling *lydian stone.* G 2. *The same* intersected in all directions by small veins of ragged quartz, so as nearly to resemble a breccia, occasional cavities filled with brown manganese ochre. H. Small fragments, apparently of the reddish *greywacké slate*, provincially termed *shillat.* This list might be increased by the enumeration of some more trifling varieties in the compact greywacké. *Calcareous rock.* I 1. *Semi-*

---

† It is distinguishable, however, by its fusing readily before the blowpipe into a vitreous globule. On breaking too the larger masses, the interior is found to be somewhat more felspathic.

‡ See Dr. Kidd's Essay, p. 109. I have every reason to believe that in this opinion my respected friend was by no means singular.

points occur in which masses of this rock are found interposed between the beds of this sandstone. As these points nearly

crystalline limestone, of a dirty-white, without organic remains. (Near the headland termed the parson and clerk) 12. *Dark grey limestone*, without organic remains, near Teignmouth. 13. *Same*, full of coralloids. Same spot.

It may be added, that insulated fragments, and occasionally crystals of semivitreous felspar are met with in the marly beds. A pit near Exeter afforded me an interesting specimen of three macles united in a single groupe. Generally this variety of felspar seems to have resisted the action of those causes which have produced the disintegration of its parent rock.

The fragments which I have attempted to describe are, for the most part, mixed promiscuously in the same strata. Occasionally particular substances predominate, but rarely, or never (as far as my observation went) to the total exclusion of all others. The porphyritic and quartzose fragments have usually their angles but slightly rounded; in some cases, not even perceptibly so. The calcareous portions have generally more the appearance of being worn (as would from their softness be the case) by attrition. These facts, added to the consideration that the porphyritic or felspathic portions bear no resemblance to the one solid rock which appears to be subordinate to this formation (namely, the amygdaloid of Thorverton) will, perhaps, be sufficient to establish the mechanical origin of the breccia in question. We shall then have to inquire whence its contents might be derived.

The *calcareous fragments* bear a resemblance sufficiently close to the limestones of Chudleigh and Babicomb. A limestone more abundant in coralloids is found yet nearer in the neighbourhood of Lindridge. The limestone also of Bickington, near Ashburton, contains many of these fossils. The fragments of the *greywacké* class may readily be traced to the rocks of that species which lie in most places immediately beneath the marle, and with which indeed the transition limestones of the country are interstratified. Of the *granitic and porphyritic* fragments, those marked A, 1, 2, 3, have all the characters of a rock frequently occurring on the confines of the Dartmoor granite, and not unfrequently intermixed either as veins or irregular masses, both with that rock and with the neighbouring schistus. It will be found thus distributed a little beyond Bovey Tracey. I have met with aggregates nearly similar at the junction of granite and schist at Ivy Bridge, and at Buckland in the Moor. The remaining felspathic fragments I have little hesitation in referring to that class of rocks which are known by the name of Elvans, and found in numberless instances traversing the metalliferous slate of Devon and Cornwall. In the latter country, they have been more frequently observed, both from the greater extent of those sections of the Killas which are offered by its coasts, and the frequency and magnitude of the excavations made by the miner. In Devon I have noticed them near Tavistock, near Buckland Monachorum, and in the course as the West Okement, and have no doubt that they might be detected in various other quarters, especially near the junctions of the granite and slate. The only instance of dissimilarity which I have observed is the occurence in some cases of large crystalline masses of the felspar, which I have termed semivitreous, and stated to form a part of the rocks marked D 1, 2, 3, 4. My limited collection of Elvans does not afford any analogous specimen, but when we remember that nearly every mine in Cornwall presents one or more varieties of this rock, and how endless are the minute shades of difference which characterize them, it will, I think, be allowed that there is nothing improbable in the supposition that the

follow the junction of the red sandstone and transition districts, the trap must occur towards the lower part of the sandstone series: that it is, however, associated with the sandstone, and not, as might have been supposed from this circumstance, with the transition series, is said to be distinctly proved by clear instances of its alternation with the former. The points along which it occurs are, proceeding from north to south, 1. Near the mouth of the tongue of red marle which, as has been observed, penetrates among the greywacké chains towards Crediton, and close to its north edge, in a groupe of quarries lying near Killerton, Silverton, and Thorverton. 2. On the south edge of the same tongue of red marle at Upton Pyne, Poltimore and Poucham. 3. A little north-east of Exeter, and again south-west of it in going to St. Ides. 4. Near Dunchidiock. (C.)

Dr. Berger gives the following particulars of this rock.

At Upton Pyne, a village five miles north of Exeter, the same conglomerate as that found at Heavitree, occurs beneath

whole contents of this breccia have been furnished by the inferior rocks of its immediate neighbourhood, by those, perhaps, whose edges are yet covered by it at a depth to which our labours and investigations have but lsttle chance of penetrating.

You will scarcely need to be reminded that Mr. Leonard Horner arrived at a like conclusion from his examination of the rock marle and adjacent strata in Somersetshire.§ It struck me as singular that among the fragments which fell under my inspection I observed no traces of hornblende rock, or greenstone, although the latter especially, and in some instances small portions of the former are to be found on the borders of Dartmoor. The cliffs of Henoch present so large and striking a specimen of greenstone as long ago to have attracted notice, and the town of Bovey Tracey stands on a rock of the same nature.

I forbear to speculate on the probability that the whole extent of the red marle was produced by the degradation of the rocks which have left their fragments still imbedded in its mass. The total absence of those organic remains, which occur so abundantly in the strata immediately below as well as above, and the general want of consolidation in its various and heterogeneous beds, certainly argue that its formation took place under different circumstances, and by a different process from that of the subjacent slate and limestone, or the superincumbent lias. The strata at Dawlish are not everywhere of uniform thickness; they dip at an angle hardly exceeding 15° to south-east by south. On this coast they are usually capped by the debris of the green-sand formation which covers the neighbouring heights of Haldon. At Dawlish these debris are much more plentiful than at Teignmouth. It may be remarked, that while they cover so large a space towards the coast, they are of much scarcer occurrence on the plain of Bovey, which lies under the opposite declivity of Haldon. Some, however, apparently water-worn, are found on that tract. I cannot conclude without expressing a wish that the whole extent of this formation were carefully examined by some abler and more instructed observer.

§ See Geological Transactions, vol. iii.

a red argillaceous sandstone several feet in thickness. The conglomerate lies on felspar in mass, containing a few laminæ of calcareous spar and some crystals of quartz, forming the roof of a mine of black oxide of manganese which is worked to open day. The red argillaceous sandstone occupies the surface of the country from Upton Pyne to Thorverton, where there are several quarries, all of which are described as being in a *calcareous amygdaloid*, varying in nature considerably in different places. In some parts, the nodules are small, and very closely united in clusters in the base, forming nearly a homogeneous mass, with here and there nodules of a much larger size than the rest, imbedded in it. In other places the nodules are about the bigness of a pea, all of the same size, and consist of rhomboidal sparry laminæ. There are other places where the base of the amygdaloid has the appearance of a sandstone, in which a small number of calcareous nodules are imbedded, which are externally coloured green by the steatite, and exactly resemble those entering into the composition of some of the amygdaloids of Derbyshire, and of the Pentland hills near Edinburgh. (G. T. vol. i. p. 99 & seq.)

The Rev. J. J. Conybeare has given the following more precise mineralogical description. (Annals, Sept. 1821.)

Its general aspect is that of a granular mass, somewhat loosely compacted, of a purplish-brown colour, more or less intense (given most probably by the oxide of manganese in which it abounds.) In this paste are imbedded, or rather intermixed, in such quantities as to form a very considerable part of the whole mass, minute portions of calcareous spar, mica, or chlorite, in a state of semi-disintegration, and indurated clay (lithomarge?), sometimes tinged by copper, and sometimes by manganese. This latter substance, as well as the calc spar, frequently traverses the rock in small veins. The cells of the amygdaloidal portions are filled or lined with brown oxide of manganese, with calc spar and a coarse jasper. The nodules of the latter are not remarkable either for their size or beauty. The character of the rock is so obscured by this abundant admixture of substances apparently adventitious, as to render it very difficult to pronounce with any certainty as to its essential constituents. These we should, I apprehend, in the present state of our knowledge, assume to be granular or earthy felspar, and one or more of the following: hornblende, augite, bronzite, or hyperstene, probably the second of these. My specimens do not afford distinct indications of any of them. The more compact portions fuse before the blowpipe, sometimes into black glass more or less slaggy, sometimes into a dirty-white enamel more or less mixed with black patches. The criterion, therefore, proposed by Cordier, here fails us.* A portion of the rock broken into small fragments, and exposed for an hour to the heat of a Black's furnace, gave a black glass much resembling that produced by various forms of the dolomite under the same circumstances.

The same obscurity which is attached to the mineralogical character of this rock seems to extend in some measure to its relations with the conglomerate in which it occurs. In some places it covers, and in others is covered by sandstone. On the road from Killerton to Silverton, near a house occupied (in the year 1812) by Mrs. Brown, we saw it resting on the large-grained conglomerate; and at one of the Radden quarries, near Thorverton, covered by a sandstone bed of from three to ten feet in thickness. Its line of separation from the sandstone is sometimes tolerably distinct. In one quarry at Thorverton, a line of sandy clay, not quite a foot thick, prevents their actual contact. At other places, especially at the Radden quarries, the two substances appear to pass so insensibly into each other as to induce for the moment a conjecture that both were the result of a common deposition, modified in its characters by the partial intrusion of some extraneous matter. This phænomenon has already been notied by Mr. Greenough. "What mineralogist," he asks, "can draw a line of demarcation between the red marle and the toadstone at Heavitree." (Essay, p. 215.) Your geological readers have probably already anticipated that a vulcanist would at once decide that the whole of the amygdaloidal beds was a series of *whin-dykes*; while others will be disposed to regard them as concretions or depositions more nearly connected and contemporaneous with the strata which envelope them. The difficulty would probably vanish before a more accurate investigation of their character and position, which I beg to recommend to such mineralogists as may travel westward.

It may be added that at the Radden quarries we noticed the occasional tendency of this rock to split into basaltiform balls; and in one spot observed it traversed by nearly horizontal veins of its own substance, differing slightly from the mass by their greater compactness, and the largeness of the nodules which they contained. The veins of extraneous matter were mostly vertical, or at a very high angle.

(*e*) *Height of Hills, &c.* Although the great central plain of red marle gives rise to several tributary streams of the Avon, flowing towards the Bristol Channel, and of the Trent flowing

* I am acquainted with the experiments of M. Cordier only through the notice given of them in M. Bouet's Geologie de l'Ecosse.

towards the German Ocean, yet it is ascertained by the level of several canals traversing this district, that the line along which the waters thus divide is only between 300 & 400 feet above the level of the sea. A similar separation of water-courses takes place in the Cheshire and Shropshire plain, between the Weaver flowing to the Irish Channel, tributaries of the Severn and tributaries of the Trent; but the elevation of this line is yet less considerable, being only between 200 & 300 feet; the high ground of Ashley heath, situated in this quarter, however, rises to the height of 803 feet: but the general character of the districts occupied by this formation, is to be destitute of any considerable eminences. (C.)

(*f*) *Thickness.* The thickness of this formation appears to be very variable. In Durham, pits have been sunk in it in the fruitless search for coal, to the depth of 708 feet without penetrating below the beds of this rock; while, on the other hand, at Pucklechurch in Gloucestershire, shafts have been sunk from the lias beds to the coal-measures, and passing entirely through this formation, which was thus found to be only 153 feet in thickness.

The red marle between Darlington and Yarm, which is about 10 miles distant from it a little on the south of east, and therefore near its junction with the magnesian lime, is at least 120 fathoms thick. (G. Map.)

Near Evesham in Worcestershire the red marle was sunk through 600 feet in fruitless search for coal. (G. Notes.)

(*g*) *Inclination.* The dip of the beds of this formation is generally very trifling, and in a direction conformable to that of the lias, and other superior strata. (C.)

(*h*) *Agricultural character.* Throughout its range we find some of the richest land in England, consisting of red marle, which is every where most fertile for wheat, barley, beans, peas, and is equally distinguished for the goodness of its cider. (T. 155.)

Wheat produced in Exminster parish in Devonshire, is said to be thinner hulled, and to afford a larger proportion of white flour than any other, and sells higher. Polwhele says that in that district, and around Collumpton and Bradnich, it produces strong crops of corn, but is more profitable when kept in grass for bullocks. The trefoil springs up every where spontaneously. Throughout these districts are hills or downs on which sheep are bred. (G. Notes.)

It is a remark which I have heard from several experienced land-surveyors, that the best tracts of land which they have any where met with in the course of their practice, have been upon this stratum: and I think I shall not be much wide of the mark in saying, that the best land which I saw in or near Derbyshire, was on the red marle about Barton Blount and Ash, Rolleston Park in Staffordshire, &c. In general, however, the Derbyshire red marle is inclined to be too tenacious and cold, and in some parts would pay amply for draining. Marling was very extensively practised upon it at a former period, as the large ancient pits testify; but at present it is practised in very few places. (Farey's Derbyshire Survey.)

It forms a good manure for poorer land. (G. Notes.)

Most of the Rye grown in England is grown on the more sandy districts of the sandstone beds of this stratum, which are not strong enough to bear wheat. (Rev. W. Buckland. G. Notes.)

According to Robinson, the soil produced by the out-crop of the sandstone beds is very sterile and barren, producing only fern, heath, bent, and a lean hungry grass, except by the sides of the rivers which traverse it, or where its nature is changed by burning, liming and manuring it. This applies probably to the sandstone, as distinguished from the marly beds of this formation. (G. Notes.)

(*i*) *Water*. It is necessary to sink through the marles of this formation into its sandstone and conglomerate beds to procure water. There are, besides the salt springs which characterize it, many mineral springs in its course, which contain besides sea salt other purging salts, such are Hartlepool (Durham), Thirsk and Croft, and Knaresborough and Rippon (Yorks), Gainsborough (Lincoln), Moreton (Shrops), Orston and Thoroton (Nottingham), Leamington and Newnham Regis (Warwick), Tewkesbury (Gloucester), &c.

Section II.

NEWER MAGNESIAN, OR CONGLOMERATE LIMESTONE.

Synonyme. First Flœtz Limestone of *Werner*.

Much confusion has arisen from neglecting to distinguish between the magnesian limestone associated with the red marle, and the older rocks of similar composition associated with the mountain limestone. Since the geological relations and geographical position of these rocks are altogether different, it is absolutely necessary to treat of them separately: they are likewise distinguished by their organic remains, and by the frequent occurrence of extensive beds of calcareous conglomerate connected with the newer magnesian limestone. (C.)*

(*a*) *Chemical and external Characters.* This limestone contains about 20 per cent of magnesia, and prevails considerably in England.

Analysis by Smithson Tennant, Esq. of the stone of York Minster.

Carbonic acid	47.00
Lime	33.24
Magnesia..........	19.36
Iron and clay	0.40
	100.

That of Westminster hall contains about 2 per cent. less of magnesia. That of Denton near the Tees, consists, according to the analysis of the Rev. J. Holme, about the same proportion of the latter, with a very small proportion of bitumen, and less than 1 per cent. of water. (G. Notes.)

A magnesian lime from Eldon, analysed by Sir H. Davy, yielded Carbonate of lime 52, Carbonate of magnesia 45.2, Iron 1.1, Residuum 1.7. Another from Denton, not far from the Tees in Durham, analysed by the Rev. J. Holme, yielded 11 per cent. more carbonate of lime, and 11 per cent. less of carbonate of magnesia. (G. T. vol. 4. p. 7.)

It differs from common limestone in external character, in having generally a granular sandy structure, a glimmering lustre, and a yellow colour; and in the course of the range from Nottingham northwards, its surface in many places is covered by a poor herbage, uncommon to limestone, and attributable to the magnesia it contains, which is known to be unfavourable to vegetation. It is associated with a conglomerate limestone. This conglomerate often exhibits very distinct fragments of the oldest mountain limestone, passing gradually from an aggregate compound of very large pebbles of this rock, to one in which the grains are so small as scarcely to be distinguishable.

In a quarry at Hartlepool on the coast of Northumberland, is a stratum of hard white oolite, the grains composing it being about the size of a mustard seed.

* Though the quantity of magnesia contained has been considered as in some degree characterising this formation, yet it is a character which, taken singly, cannot be relied upon, since magnesia in considerable quantity has likewise been found in some of the chalk of France, in some of our own oolites, in the mountain or carboniferous limestone, in that associated with transition rocks, and, as need not be stated, in primitive dolomite: we have not however ventured to change a name generally received. (C.)

This limestone often forms in large concretional and botryoidal masses; the concretions are often as large as a cannon ball, and sometimes grouped together like chain shot. These concretions are dispersed through an arenaceous form of similar materials.

Mr. Smith in his geological map of Yorkshire subdivides the magnesian limestone, or, as he calls it, red-land limestone, thus:

1. A hard bluish white thin bedded stone which at Kinnersley, Knottingly and Brotherston, makes the lime celebrated for agricultural purposes.
2. Red and blue clay and gypsum
3. A soft yellowish calcareous freestone or magnesian limestone.

These beds are separated from the superior red marle, by a thick conglomerate.

Mr. Buckland has observed in Yorkshire beds closely resembling the Rauchwacke or cellular limestone of the continent, associated with magnesian limestone. See farther parliculars in the local account.

The following more detailed particulars relating to the external characters and œconomical application of the beds constituting this formation, together with their disposition and order in which they occur, are extracted from notes kindly lent to the Editor by Mr. Greenough.

Its general colour is buff, but it also occurs white, of various shades of yellow, also of fawn and salmon colour; occasionally of a brown or reddish hue, which prevails chiefly near the partings (Farey) where it is often dendritic.

Its texture is frequently arenaceous, but is sometimes composed of minute rhombic crystals, and is occasionally oolitic, (Mr. Wynch.) It is often carious, the cavities being sometimes lined with calcareous spar. Between Shields and Hartlepool it is crystalline and cellular, from which cause it resists the stroke of the hammer: it is slaty at Baldon hill, Marsdon Rocks, &c. near Newcastle.

The limestone of Sunderland is flexible, near Ravenstone, it contains beds as compact as the Portland, and of the same colour.

At Ferrybridge in Yorkshire it is fetid; its odour does not however arise from the presence of bitumen, but according to Mr. Aikin from sulphuretted hydrogen.

Large masses of it are detached by the agency of fire. In this process that portion of the stone which is impregnated with iron becomes brick-red. Considerable quantities of fuel are required to burn it into lime, and it is apt occasionally to vitrify. Near Sunderland, the brown is generally quarried;

it partakes of the nature of swinestone, and from containing some inflammable substance, requires only a small quantity of coal to reduce it into lime. That worked at Denton, was found by Mr. Holme to be bituminous.

It dissolves slowly and leaves a milky appearance in the liquid.

The lime from the stone of Leigh and Ardwick has the property of setting under water, and was used for the tarras cistern of the late Drury Lane Theatre

(Uses.) It is employed as a building stone, and has a pleasing tint: at Broadworth it is worked into cisterns, and at Langwith in Derbyshire into flooring and stair-cases. It is rarely uniform enough in its grain, or of sufficient hardness to deserve the name of marble, but slabs of it are polished at Knaresborough, Sunderland, &c. It does not burn to a white lime, but furnishes a strong mortar; that of Leigh is celebrated as a cement, and is said to be used, when mixed with bullocks blood, as a cement for mending boilers worn by use. It is said that if a damp room be white-washed with it, the dampness will be cured. At Ferrybridge, Mansfield, &c. it is burnt for manure, but is generally considered to be injurious to the land, unless used in very moderate quantities, in which case it is esteemed to be highly beneficial.

At Pallion near Sunderland, beds of a dirty light brown colour are quarried and sold as marbles; they take a tolerably good polish.

Over the magnesian limestone is a hard stone abounding in shells which are very imperfect: it is red on the surface, blue internally. There is a large quarry of it on the road to Pleasley close to the turnpike about two miles and a half from Bolsover in Derbyshire.

In Yorkshire, as we have already observed, it occurs in two beds separated by variegated marle containing gypsum.

Near Nottingham, its upper beds are separated by partings of red and verdigris coloured clay, and it contains quartz pebbles.

At Bolsover, the beds are separated by pipe-clay, which is used in the arts.

The lowest beds are blue limestone containing galena and yielding excellent lime, and are quarried at Bolsover, Excroft, Barlborough, Palterton and Houghton.

Towards the bottom of this series several beds of compact blue limestone occur imbedded in blue clay and abounding in anomia and other shells, have long been known at Stoney Houghton near Pleasley, at Palterton, &c.

At Whitley, near Allercoats in Northumberland, the lower beds alternate with shale or slate clay. One part of the bed rests upon this substance, another upon one of the sandstones of the coal series; it is again seen covering sandstone on the side of the river under Knaresborough Castle.

The conglomerate form of this limestone is very cavernous. Wokey hole in Somersetshire, which yields only to the caverns of the Derbyshire peak in extent, is entirely situated in it.*

(*b*) *Mineral Contents.* In the conglomerate beds associated with this formation, calamine, blende, and some galena have occasionally been found, particularly near the Mendip hills; but it seems probable that these minerals may in some instances have been derived from the detritus of the older metalliferous strata of the carboniferous or mountain limestone of that neighbourhood. Many of the principal mines of that district are however entirely seated in this rock.

Galena is also mentioned as occurring in strings in the magnesian limestone of Nottingham and Durham.

Nodules of hæmatites iron-ore, used for burnishing, are imbedded in and found scattered over the surface of the under beds. At Bolsover it contains pyrites. (White Watson, G. Notes.)

* A very singular formation of shelly chert occurs resting upon, and perhaps associated with, the calcareo-magnesian breccias covering the side and even top of the Mendip hills, in the parishes of East and West Harptree: the most abundant shells are a Modiola, a variety of Ostrea gregarea, a peculiar Pecten or Lima, and a longitudinally striated Telliniform bivalve; a Plagiostoma and Ammonite also occur; but the whole series are found as casts only, and frequently very obscure. The chert contains sulphate of barytes, often in great quantity, and sometimes assumes a conglomerate form, including fragments of the limestone and old red sandstone of the adjacent hills; it more rarely passes, by the intermixture of calcareous matter, into a siliceous limestone; it is associated with beds of ochreous sand.

From its first appearance it might be supposed to be an insulated and outlying mass of chert belonging to the green sand formation, but a comparison of the organic remains does not confirm this suspicion, and there is no other deposit of that formation nearer than fifteen miles. Not far from this vicinity also, a very similar chert (excepting that shells have not yet been discovered in it) is found near the Pitcot collieries (in the parish of Stratton on the fosse) interposed between the magnesian conglomerate and the incumbent red marle; and it is very probable that a more careful examination may detect the shells here also, since in the places where they certainly exist they are often very obscure. Were our knowledge of the organic remains of the magnesian limestone more perfect and full, we should be better able to determine the probable relations of this chert to it; but upon the whole the evidence seems in favour of its connection with it, although we have not felt sufficiently confident to introduce it otherwise than in a note.

The Druidical circles at Stanton Drew are built of this chert. (C.)

It contains veins of sulphate of barytes at the Huddleston quarry, near Sherburn, between Ferrybridge and York. (G. Notes.)

It is traversed by veins of sulphate of barytes, near Nottingham, at Bramham Moor, between Aberford and Wetherby, and between Ferrybridge and York in Yorkshire. (G. Notes.)

It encloses arragonite near Whitehaven. (G. Map.)

(*c*) *Organic Remains.* These are rarely met with in this limestone. The most remarkable one was found in a quarry at Low Pallion in Durham. It is the impression of a fish which appears to belong to the genus Chætodon (G. T. V. 4, Plate 2.) In length it is about 8½ inches and 4½ in breadth. The dorsal fin reaches from the middle of the back to the tail. In Humbleton quarry, situated one mile from Bishop Wearmouth, on the road to Durham, were found the following remains, imbedded in hard buff-coloured crystalline limestone. viz. Casts of the internal part of the vertebral column of the cap encrinite. (See Parkinson, v. 2, tab. 10, fig. 4.) A species of Donax with hair-like spines. Casts of reticulated alcyonite. (Parkinson, v. 2, tab. 10. fig. 1, 2, 3.) Smooth-shelled bivalves, from the size of a pea to that of the cockle, resembling those of the genus donax. Small round bodies, delineated by Parkinson, vol. 2, tab. 8. fig. 10. Casts of bivalves resembling muscles. Casts of Arcæ and Anomiæ (Sowerby, Brit. Min. tab. 55.) Impressions of a reticulated marine production resembling the genus Flustra. (G. T. V. 4. p. 10.) These organic remains may serve to distinguish this from the older formations. (G. T. vol. iv. p. 12.)

The lower beds at Whitley near Allercoats in Northumberland contain bivalves and entrochi. (G. Notes.)

(*d*) *Range and extent.* The principal range of hills consisting of it, extends from *Sunderland* on the north-east coast, to *Nottingham.* This range is not very elevated; near its northern and southern terminations it attains the height of about six hundred feet above the sea, but the middle part is not so high, and the hills of which the chain is constituted are round topped. Magnesian limestone, exactly at the northern extremity of its western boundary, overlies the coal strata in the cliffs at Cullercoats in *Northumberland.* (G. T. vol. iv. p. 3.) It overlies the ninety fathom dyke, which appears between the strata of coal in a nearly vertical position; and it is concluded that the magnesian limestone is of a newer formation than the coal and the dyke, since the dyke has traversed the coal, but has not affected the limestone.* This rock, along the coast of Durham, is covered by a species of breccia, the cement of which is a marle-like substance consisting chiefly of magnesian carbonate of lime; and with this breccia the wide chasms or interruptions in the cliff are filled. The upper strata of the limestone are thin and slaty, and inclining to a buff-colour. Below, the stratification is less distinct, and the colour is hair brown, the texture crystalline and cellular: a variety that is flexible has lately been found. It partakes of the nature of swinestone; and from containing some inflammable matter, it requires only a small quantity of coal to reduce it to lime. (G. T. vol. iv. p. 6.)

It is well ascertained that the magnesian limestone of Durham and Northumberland rests upon the coal-measures; for at Pallion, a little to the west of Sunderland, and but a short distance from the western limit of the magnesian limestone, it was only 12 fathoms thick, and was found to overlie the coal strata, which were bored through to the depth of 140 fathoms, without finding a seam of coal worth working. No coal-mine has yet been won by sinking a shaft through the limestone, although the workings of some collieries, situated on its western boundary, have been carried underneath it. At Hart, near Hartlepool, which is almost the southern extremity of the limestone on that coast, it was bored to the depth of 52 fathoms, without penetrating through it. (G. T. vol. iv. p. 8.)

The dip of the coal strata south of Newcastle is not in conformity with that of the magnesian limestone. It is a circumstance, however, too well ascertained to admit of a doubt, though difficult to be accounted for, that the coal is deteriorated in quality where it is covered by the limestone. (G. T. vol. iv. p. 9.) Botryoidal massess of fetid limestone devoid of magnesia, in balls varying from the size of a pea to two feet in diameter, imbedded in a soft, marly, magnesian limestone, are found at Hartlepool, in a quarry at Building hill near Sunderland, and on the sea coast a mile or two north of

* The subjoined sketch of the position of the magnesian lime-coal measures, and dyke is copied from that given in the volume of the Geological Transactions above referred to.

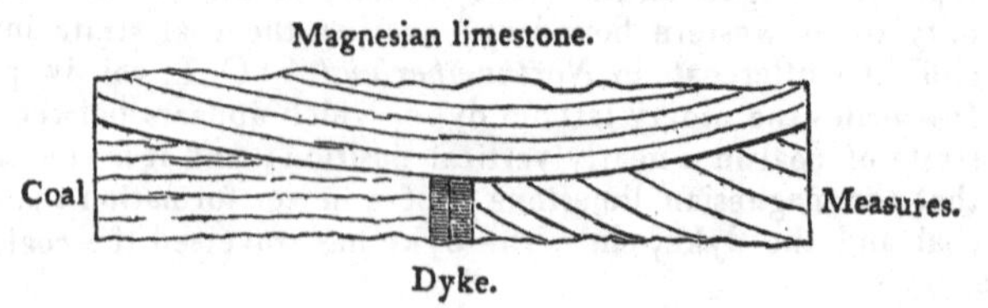

Monk Wearmouth. These balls are radiated from the centre, their colour hair-brown, fracture shining, cross fracture splendent approaching to vitreous; white calcareous spar is often observed in them. The magnesian limestone of this district contains few organic remains. A representation of a fish found in it is given in the 4th volume of the Geological Transactions. It contains but few shells. (G. T. vol. iv. p. 10.)

The beds constituting this range, are described as being about three hundred feet in thickness, on the east of the coal-field in *Derbyshire*, which is near its southern extremity. Their general colour is yellow, which is often as bright as gamboge, with almost all intermediate shades, to a very light straw, and white. Many of the beds have a granular texture, and a brown or reddish hue, particularly near their joints. Those beds in which the magnesian earth abounds most, generally pass for a grit-stone. A considerable number of the upper beds are incapable of calcination. (F.)

The coal-beds of the Derbyshire field extend beneath the magnesian limestone, and have been worked under it at Bilborough and Nuthall, which are situated on it, a few miles north-west of Nottingham; (F. 166.) and all the coal strata on the east of the great zig-gag *fault*, from Trowel in Nottinghamshire, to the north of Aston in Yorkshire, have an easy dip to the east, similar to that of the magnesian limestone, (F.168,) and being worked in several places beneath that limestone, probably pass beneath it with the same gentle dip.

The existence of the newer magnesian limestone does not yet appear to have been ascertained within the limits of the central plain of red marle; for the limestone rocks on the borders of the Ashby de la Zouch coal-field, which have by some writers been referred to this formation, are determined by their organic remains and geological position, to belong to the carboniferous or mountain limestone underlying the coal, and those of the Dudley coal-field appear of still earlier origin, and referable to the transition series. (C.)

The limestone described by Mr. Bakewell as overlying the coal at Bradford near Manchester, is probably the younger magnesian limestone, as also some other similar patches on the south of the coal-field between Manchester and Preston; for the localities of which see Smith. (C.)

In the Shropshire plain however conglomerates of this formation are said to occur at Caerdeston and Loton, and some other places. (C.)

On the western side of the Cumberland mountains, magnesian limestone occurs overlying the coal-measures near Whitehaven; and as connected with the red marle deposits of the vale of

Carlisle, it may be mentioned that beds of conglomerate limestone are found in that quarter overlying the coal on the south of Dumfrieshire.

In the south-western counties, especially in Somersetshire, conglomerates are not uncommon towards the lower parts of the red marle series; the shaft sunk at Bath Easton, (see p. 261) in search of coal, passed a bed of this description; near Wick Rocks, on the south of the defile, they occur in horizontal strata, reposing against the elevated beds of the coal-measures and mountain limestone, on the eastern side of the Somerset and Gloucester basin; they form hangings on its exterior chains of mountain lime and old red sandstone along its northern beds near Tortworth (Gloucestershire); and at its western side, near Redland, similar beds occur; beyond the elevated ridge forming the western border of this basin, the calcareo-magnesian conglomerates are seen stretching from Hung road on the Avon to Portishead point, and on the north side of the Avon, running up the valley below King's Weston hill, extending to the east as far as Sneed park. (C.)

The whole of the calcareous chain of Leigh Down, extending on the south of the Avon, is also surrounded by these conglomerates, as is also Broadfield Down; which, being interposed between Leigh Down and the Mendips, completes the south-west angle of the line of mountain or carboniferous limestone bounding the coal-field. (C.)

The Mendip hills on the south of this coal-field, are invested in like manner by the same conglomerate, which occurs near Wokey hole, on the south of Cheddar Cliffs, and in the valley of Roborough, as also all along the northern skirts of the chain, in beds abutting against or covering the elevated strata of mountain or carboniferous lime, &c. and frequently crowning even the very summits of the hills. (C.)

In Glamorganshire, this conglomerate is seen in a similar position, alternating with and passing into compact beds of magnesian limestone throughout the parish of Sully, near Cowbridge, and in several points along the valley of the river Ely: considerable deposits of it also occur, intruding within the limits of the coal-basin, between Bridge-end and Lantrissent, and in this district the coal is worked beneath a conglomerate of this formation at Torygelly colliery near Lanharran. (C.)

There can be little doubt but that traces of these beds will be found among the conglomerate rocks before described in the south of Somerset and Devon: this is certainly the case at Saurpford Peveril in the latter country. (C.)

(e) Height, &c. According to Mr. Winch, Brandon Mount, which rises 875 feet above the sea, is in this formation; it more generally constitutes only low hills. (C.)

(f) Thickness. The thickness of these beds appears to be very variable. In Glamorganshire, they may be seen varying from thirty feet to as many inches in nearly contiguous cliffs. In Derbyshire they are said to be about three hundred feet. (C)

(g) The inclination of these beds is trifling, and conformable to that of the superior strata. (C.)

The stratification of this rock is very distinct, the individual courses of stone not generally exceeding the thickness of a common brick. According to Mr. Winch it varies from three or four inches to as many feet, along the coast of Durham from Shields to Hartlepool; the uppermost beds are thin, but lower down the stratification becomes more distinct. (G. Notes.)

Near Mansfield in Nottinghamshire, the beds are arched and contorted, but four and a half miles from that place they are so horizontal as to serve for a foot way. (G. Notes.)

(h) *Agricultural Character.* The soil made by the yellow limestone strata, is generally of a medium quality and degree of tenacity; it is much improved, either by the peak lime, where the canals admit of its being applied, or by the lime from the blue beds near the bottom of this series. It seems best adapted for arable land, on account of its proneness to shar-grass, pry-grass, or spiked fescue *(festuca pinnata)*, a light green sharp grass, which scarcely any thing will eat, which soon prevails when the yellow limestone lands are laid down to grass. (F.)

It crumbles into a dust, which after rain becomes slippery and tenacious, but in general it affords no considerable thickness of soil, which is of a chocolate brown colour. The luxuriance of the bromus pinnatus (spiked heath brome grass) is characteristic of the soil, and the sweet scented violet is very plentiful. Yellow rose trees with single and double flowers grow luxuriantly in this soil, as at Shire Oaks near Whitwell in Derbyshire. In some situations, as at Hardwicke, it is favorable to the growth of oak; in others as at Pleasley it flourishes a few years and then ceases; the elm grows well, but the broad-leaved is subject to crack: the walnut grows to a large size. (White Watson. G. Notes.)

(i) Water. The cavernous structure of the conglomerate varieties of this rock in the south-western counties forms vast reservoirs of water; but unlike the springs percolating through uniformly porous strata, they are not calculated to afford a constant supply, but when once tapped may be soon exhausted; this was experienced in a branch of the Somerset coal canal

near Radstock, which was carried through the conglomerate in order that it might, as was hoped, be fed by these natural reservoirs; their whole contents however soon ran off; and they defeated instead of answering the intended purpose, by draining off the water of the canal; which was consequently obliged to be puddled along the whole line. (C.)

Are not the mineral springs of Rippon and Knaresborough in Yorkshire on this formation?

Section III.

FOREIGN LOCALITIES. *

In quitting England, before we pass beyond the British islands, we may observe that the new red sandstone, accompanied by its calcareous conglomerates, stretches from both the northern angles of this country into Scotland, extending on the north-east up the valley of the Tweed, and on the north-west along the southern plains of Dumfries; in both cases resting against the great southern chain of transition mountains which forms the natural barrier of Scotland in that direction. We are not yet informed whether the red sandstone which occasionally covers the great central coal-district of Scotland belongs to this deposit, or is only a variety of the regular coal grits; but there can be little doubt that much of the sandstone of the Western Isles belongs to it.

The formations which we have described in this article occur abundantly on the continent. As far, indeed, as our knowledge of the geological structure of the whole face of the earth extends, there is reason to suppose that a much greater portion of its surface is occupied by these, than by any other single class of rocks.

Before we attempt to collect together from the notices as yet obtained, a general view of the local distribution of this series, it will be necessary to premise a few words concerning the arrangement of its constituent members in those places where they have been most accurately examined, especially in the north of Germany, where the researches of Karsten, Voight, Raumer, Von Buch, and more particularly of Freisleben, have filled up the slight, but generally accurate descriptions of Lehman, and elucidated the systematic, but occasionally confused views of Werner. And we have here the advantage of being able to refer the reader for further information to an extremely able abstract of their statements published by Mr. Weaver in the Annals of Philosophy for September 1821.

* By the Rev. W. D. Conybeare.

The members then of this series may be enumerated in the following order beginning with the highest and most recent.

1st. Beds of marle and variegated sandstone containing rock salt and gypsum exactly corresponding with the English red marle. It contains also occasionally some subordinate calcareous beds coarsely oolitic, or rather formed of middle-sized globular concretions sometimes dispersed through a sandy matrix, and thus passing into sandstone. This appears to agree with some of the upper beds of magnesian limestone in the English series; but the great mass of that formation is probably rather to be referred to the next member of the German suite: argillaceous iron-stone and thin traces of coal are likewise found incidently in this part of the series.

2. A calcareous formation containing fetid limestone, (Stinkstein), cellular limestone (hehlen kalstein or rauchwacke,) and compact marly limestone (zechstein) alternating with marly beds, follows. Towards the bottom of this part of the series is a bed of slaty marle-stone richly impregnated with copper pyrites, (kupferschiefer flœtz) for which it is extensively worked. This bed is considered as characterising this part of the series. Much of the alpine limestone must be referred to this formation, and also that of Carniola, which is associated with the bituminous marle-slate containing the mercurial mines of Idria. This constitutes the first flœtz limestone of Werner and his disciples. Near the Hartz, these calcareous beds rest on a marly sandstone of very variable character, occasionally passing into a calcareous conglomerate called the Weissliègende.

The magnesian limestone of England answers in position to these deposites, but no metalliferous beds have yet been observed in this formation in this country: well characterised rauchwacke may however be observed in Yorkshire.

The organic remains contained in these formations are principally skeletons of Saurian animals, and fish in the copper slate, together with small gryphites (perhaps Chamæ) ammonites and belemnites, &c. a ferriferous limestone which replaces the zechstein in the Thuringerwald, is replete with a species of gryphites (or perhaps Chama,) called by Von Schlottheim gryphites aculeatus; vegetable impressions of ferns, seeds, &c. occur in the coal-shale.

3. Interposed between the last series and the coal which it always covers, is a great deposit of red sandstone and conglomerates associated with various masses of porphyry, basaltic trap and amygdaloid. It is locally called the *red dead lyer (rothe todte liegende)* because the metals worked in the former beds here cease; this is termed by Freisleben the older sandstone in distinction to the sandstone beds of No. 1, and is the first

flœtz sandstone of the Wernerians; it has been confounded with our old red sandstone from this resemblance of names, and from the difficulty of discriminating between quartzose conglomerates nearly allied in external character; but since the rock thus named in England is uniformly beneath the principal deposits of coal, and on the other hand the rothe todte liegende of Germany as constantly above them; and since its character and composition closely agree with the conglomerates of Devonshire, which like it are associated with amygdaloidal and porphyritic trap, there is no reason for hesitating to refer it to the same epoch: we therefore consider it as included in the present series of formations.*

* Mr. Weaver in the very useful compilation before referred to, endeavours however to establish the claims of the rothe todte liegende to the greater antiquity of the English old red sandstone. His view of the subject is that the three formations associated in the great carboniferous series, namely, the old red sandstone, the carboniferous or mountain limestone, and the regular coal measures, may be intermixed without any determined or constant order of superposition; and he cites the division of the mountain limestone series in Northumberland, where its beds alternate with sandstone and shale, and present even near the bottom of the series, in two or three instances, thin seams or rather traces of coal. He therefore thinks it probable that the coal, which in England generally forms the upper member of this series, may in Germany as generally form the lower; and he appears to be led to adopt this explanation principally from the occurrence of porphyritic and trapean rocks, in the rothe todte which correspond with those in the old red sandstone of Scotland.

It may be objected to this view, however, that it supposes a deviation from the general geological order of formations as deduced from a very wide induction; and secondly, that it supposes it without necessity: for not only in our own islands, where it is a constant fact, but in the Netherlands also (the coal fields of which the author of this notice has personally examined,) the same order of superposition prevails; the great coal deposit is always the upper member of the series; the limestone deposit the central, and the old red sandstone the lowest. By most continental writers these latter rocks are however classed with the transition series, the old red sandstone being considered as a variety of greywacke, which has hitherto prevented our own writers from recognising this exact identity of arrangement; and although it is indeed true that in Northumberland, traces of coal occur near the bottom of the limestone deposit and perhaps beneath it, still the general rule holds good there also: for the upper part of the series contained in that country, contains frequent and thick seams of coal, and in this part no limestone is found: then, after an interval of an intermediate character corresponding to the millstone grit and shale formation of Derbyshire, containing sandstone and shale with two or three thin beds both of limestone and coal, follows the great limestone deposit containing eighteen beds, many of them of considerable thickness, separated by shale and sandstone; and lastly, the old red sandstone. This part of the series contains only two thin and unworkable seams of coal; for the coal-beds which are found associated with limestone in the north of Northumberland, belong to the intermediate formation between the limestone and regular coal measures; so that here there is in fact no exception to the general rule; for all the workable coal is above the great central

To trace these beds hastily through the continent, we may begin with France. The new red sandstone which crosses the

calcareous deposit, and below it are only imperfect traces of coal, and those of the rarest occurrence.

That on the other hand the rothe todte liegende as uniformly occupies a place above the great coal deposit, the evidence of Freisleben is decisive. The accounts of Lehman, Karsten, Voight and Von Buch, confirm this position: the description, map, and sections of Von Raumer, also prove that it is similarly placed in Upper Silesia and Bohemia; for the crop of the coal beds is distinctly represented as occupying a line between the red sandstone and transition rocks. Freisleben considers it as a formation distinct from and reposing upon that of coal; Von Raumer as so closely allied to the coal-measures, as to constitute with them but a single formation, the lowest beds being in his opinion associated with the coal; the instances he cites are however by no means satisfactory, and probably indicate nothing more than the incidental occurrence of traces of coal in the red sandstone, which is not an uncommon circumstance even in its youngest members which contain gypsum and salt.

In one of the sections given by Lehman, he indeed represents a bed of red conglomerate (either our millstone-grit or old red sandstone) as basing the coal, but he carefully distinguishes this from the true rothe todte which occurs in the same section above the coal.

All these writers, therefore, with one voice confirm the position of the coal beneath the rothe todte liegende, either as a distinct formation, or at any rate the *lowest* member of the same. It is therefore clear that the relations of position of this formation to the great coal deposit are directly contradictory to those of the old red sandstone of this country.

Let us enquire, then, whether there are any circumstances connected with this rock which render it necessary to resort to the supposition of this great inversion of a geological arrangement which has been found to hold constantly good through the British Islands, and also in the most extensive of the continental coal-districts—that of the Netherlands: for though it would certainly be rash to pronounce against the possibility of this supposition, it must yet be acknowledged that it ought not to be admitted without cogent arguments in its favour. The descriptions however of this formation will be found closely to agree with the conglomerates of our new red sandstone in Devonshire, and even the occurrence of porphyritic and trapean rocks in association with it, is there paralleled; for it will have been seen in the preceding article that there is no ground to doubt that these latter rocks are in that district truly associated with the new red sandstone, and not, as has been suggested, with the subjacent transition strata. The writer of this article particularly examined the rothe todte liegende beneath the copper marle-slate of the Thuringerwald, in company with Professor Buckland, in the summer of 1816, and both were struck with its identity of character with the rocks of Heavitree near Exeter; and the latter has since enjoyed repeated opportunities of studying most of the principal deposites ascribed to this formation on the continent, and has seen nothing which did not strongly confirm him in this opinion. It should be added, however, that three great conglomerates of nearly similar appearance, occur in the neighbourhood of the English coal districts. 1st. The lowest beneath the great carboniferous limestone, associated with the old red sandstone; in this coal has never yet been found (at least in any workable quantity) in these islands; secondly above the carboniferous limestone, and between it and the great coal deposit, associated with the sandstone called in Derbyshire, &c. the millstone grit; this does occasionally, though not very commonly, alternate with, and cover regular beds of coal, and

channel from Devonsire, is seen, though not extensively, skirting the transition rocks of Brittany; but the lias and oolite advance so near to them, as almost to overlie and conceal it; as also seems to be the case in the centre of that country, against the northern edge of that great group of primitive ridges branching off from the Cevennes; but we want information on this district.

If we pass from the centre of France into Spain (as this seems the most convenient place to include that nearly detached country), we find, after crossing the Pyrenees, the rock salt of this formation associated as usual with gypsum at Cardona (a description of which will be found in the 4th volume of the Geological Transactions). The celebrated conglomerate mountain of Montserrat in the same quarter, is perhaps referable to the same æra; and we find gypsum and rock salt abundantly distributed along the course of the Ebro from below Saragossa to above Tudela. It is difficult to speak of a country whose geology has yet never received a strictly scientific examination; but enough is known to teach us that the central and western

also alternates with the upper part of the limestone series: and thirdly, the conglomerates forming the lowest member of that great series of sandstone deposites which covers the whole of the coal-measures, and which it has been the object of the present article to describe. Identity of geological position, and resemblance in character, seem to combine in persuading us to refer the rothe todte to the last of the three, rather than to either of the former. Should it however prove that they are really associated with the coal formation, as Von Raumer believes them to be, they may perhaps be classed with the second; but there is every reason to withhold our assent from their proposed identification with the first and oldest of them; for in a case where external characters are nearly similar (since one quartzose breccia cannot materially differ from another) our surest guide must be the position in the geological series, and this *rule will hardly permit us to class a formation uniformly below, with one uniformly above, the principal deposit of coal.* It may be asked, however, if the rothe todte be not our old red sandstone, what rock in the continental series does correspond to it; to this it may be answered generally, that the entire absence of a formation is a less anomaly in geology than its false position, and that the old red sandstone is thus absent in many of our own coal-fields, as for instance in those of Staffordshire, where the coal-measures repose immediately on transition limestone, and even in the western extremity of the great South Welsh coal-field, which exhibits this formation in its greatest thickness; in every other part it vanishes in the same manner. In some instances, however, the old red sandstone is decidedly to be seen on the continent, as at Hug, on the Meuse, placed exactly as with us; but it has always been spoken of by continental geologists as a variety of grauwacke, and the carboniferous limestone reposing on it as transition lime; and the same nomenclature has probably been adopted in other localities, for Von Raumer speaks of a conglomerate and limestone associated with grauwacke in Siberia; which, from the description he gives, and its relations to the adjoining coal-formation, may very probably answer to our old red sandstone and carboniferous limestone.

districts are principally occupied by primitive chains, while the east and south-east exhibits little but calcareous mountains, among which gypsum is plentifully interspersed. As we shall hereafter see that the limestone formation, answering in age to our magnesian limestone, swells into great importance on the continent, and constitutes large mountain zones encircling the Alps, &c., which are in like manner characterised by the intermixture of gypsum, it is no improbable conjecture that these deposits belong to the same period.

To return to the coast of France. Further west the red marle appears in great thickness, underlying the Jura chain near Lons le Saulnier, where it exhibits as usual salt-springs and gypsum.

At no great distance we find it encircling the Vosges, where vast masses of its conglomerates invest the primitive chains; and in like manner surrounding the opposite German chains of the Black forest and Bergstrasse, as may particularly be seen near Heidelberg: hence it spreads northwards as far as the transition slate district which stretches from the north-east of France across the Rhine, forming the chains of the forest of the Ardennes, the Rhingau, the Westerwald, &c. and skirting those on the south and east. Omalius d'Halloy, in his Geologie du Nord de la France, has described it in part of this line, where he remarks that, as in England, this is the oldest of the horizontal formations, the coal formations which rest on the opposite side of this slate district being highly inclined; and gives a description which strictly applies to our own red marle, &c. Although this tract is more naturally connected perhaps with those of similar formation in the north of Germany, yet as these may be traced in a line probably continuous through Poland to the very extremities of Russia, it is more convenient, before departing so widely from the central countries of Europe, to review in the first place the course of the new red sandstone through the south of Germany and the adjacent countries.

In this quarter we find these formations forming a zone on either side of the Alps; on the north interposed between the older rocks and great Nagelflue of Switzerland, which was once itself considered as belonging to them, but has been proved by subsequent researches to be of much more recent date, and contemporaneous with the sandstones of the basin of Paris. The new red sandstone is here intimately associated with alpine limestone, which corresponds with the calcareous formations already described as coeval with our magnesian limestone; and gypsum and salt may be found interspersed through the whole series. A similar character applies to the zone on the south side of the Alps; here the new red sandstone may be seen to the greatest advantage in the valley of the Adige ten miles north of Trent, and in the valley of Avisio which descends from the Val di Tassa into that of the Adige. In the same neighbourhood a porphyry occurs associated with these formations on the south of the Alps only.

The reader is referred for further particulars to the excellent memoir of Professor Buckland; Annals of Philosophy, June 1821.

It is probably the limestone of this formation belonging to the Southern Alpine zone, which extends into Carinthia, Istria, Dalmatia, &c : the limestone of the Apennines, and much of that in Greece, may also perhaps be referred to the same æra.

Returning to resume our survey of the cours of this formation from the point where we left it in the north of Germany, we may first observe that there is a tolerably extensive sandstone district in the centre of Germany, between Wurtzburg and Bamberg, which may belong to this formation, and appears in some places to contain gypsum, but cannot yet be considered as satisfactorily identified. Having passed, however the calcareous district containing the caverns celebrated for the fossil remains of bears, &c. in proceeding westwards, between Bayreuth and the Fichtelgebirge, a zone of red marle is passed which appears to skirt the transition and primitive ridges of the Bohemer Wald towards the south, and which continues on the north-west round the projecting chain of the Thuringer Wald. Here we may observe all the formations noticed at the head of this article, viz. the red marle and gypsum; the calcareous beds associated with the cupriferous marle-slate, and at the bottom the rothe todte: a shell limestone answering to our lias rests on these beds, and separates this from a similar district encircling the detached ancient groupe of the Hartz mountains: here, and in the continuation of this district towards Halle, the rothe todte is to be observed in many places in contact with the coal formation and always above it. Rock Salt is found in numerous points in this quarter, along the line between Osnaburg and Magdeburg throughout the south of Hanover. (See the works of Freisleben.)

The zone of these rocks surrounding the Thuringer wald, continues to skirt the prolongation of the same great mountain band through Saxony, where it assumes the name of the Erzegebirge, through Silesia, where it changes its appellation for that of the Riesengebirge. It occurs on both sides this chain, extending on the south into the great basin of Bohemia, and covering the coal formation of that country and the adjoining parts of Silesia. This district has been fully described by Von Raumer,* and in part also by Von Buch in his account of Glatz.

These formations appear to skirt in like manner both sides of the Carparthian chain, which is still only a continuation of this same great primitive band traversing central Europe. The most extensive salt mines which have ever been worked are to be found in the northern sandstone zone at Wielictzka on the south of Cracow, and salt is also worked along the inner zone in several vallies descending to the west from the chain where it trends round Transylvania.† The primitive ridge of the Carpathians, turning eastwards on the south of the Danube near its mouth, assumes the name of Mount Balkan, and proceeds to the coast of the Euxine, which cuts it off; but the transition rocks on the south of the peninsula of the Crimea, appear to form a portion of its northern exterior chain, and the Caucasus to form its prolongation; both these ranges are skirted by conglomerates, probably of this formation.

The sand of this formation, containing gypsum, appears to be very abundant in the north and east of European Russia. Mr. Strangways has recently laid much important information on the mineralogical relations of this vast empire before the Geological Society, in which all the particulars hitherto collected are given. It will here suffice to observe, that if a line be drawn from Riga north of Moscow to the banks of the river Oural, this formation will be found plentifully distributed on the north and east of it, especially along the Volga and its branches on the north-east of Moscow: it appears indeed to extend to, and invest the Oural mountains with the intermixture of a cupriferous sand, probably allied to the cupriferous beds associated in this formation in Germany and the Tyrol, &c.

On the south of the Oural chain, it appears to stretch to the Caspian, and to spread very extensively in the adjoining regions of Asia. Dr. Kidd observes in his Geological Essays, "that when it is known that rock salt is used as building-stone at

* The reader will find an admirable digest of the statements of Freisleben, Voight, and Raumer, on the secondary formations surrounding the Hartz, Thuringerwald, and Reisengebirge, by Mr. Weaver, in the Annals of Philosophy for October, November, and December, 1821: we have already stated our hesitation in adopting his proposed identification of the Rothe todte with our older red sandstone, but in every other particular our statements will be found in exact accordance.

† Mr. Fichtel says, that on the north this zone extends from Wielictzka into Moldavia, in which interval he enumerates 58 places where salt is worked or salt-springs found, and on the south from Eperics, 4 or 500 miles eastwards through Transylvania, affording 159 localities of salt.

Ormuz*; and that the sand of the great desert of Persia is of a brick red colour; and that salt abounds throughout that desert; there can be very little doubt in the mind of a geologist that the rock marle formation abounds in that part of the world. The same formation probably abounds also about the streams of the Indus and the Ganges, after they have quitted the mountains; for it is observed by Major Rennell, that in the tract between the Indus and the Chelum are wonderfully productive salt mines, affording masses of salt hard enough to be formed into vessels, &c." †

Does not the occurrence of a salt lake in the centre of Asia Minor lead us likewise to expect this formation in that quarter? We know that rock salt is used in Caramania as a building stone, in consequence of its hardness and the dryness of the air. (Chardin.) Similar indications are said to exist in Thibet and Tartary.

With respect to Africa, we may again quote Dr. Kidd. "I need not insist on the existence of rock salt in Africa; and though we have little satisfactory evidence of the formation to which it belongs, yet as syenitic rocks occur in Upper Egypt, and magnesian limestone in Lower Egypt, and Vitruvius mentions a spring of tar near Carthage §; we have good geological reasons for expecting that in those parts of Africa the rock marle may be found."

Much of the sandstone of southern Africa appears to belong to this formation.

Rock salt is said to exist in more than one place in New Holland.

In North America these formations appear, from the travels of Messrs. Lewis and Clarke, to be very abundant in the vast plain of the Mississippi, between the Alleghany mountains and the great range of stony mountains bordering the opposite sides of the northern half of this continent.‡

* Kinneir, p. 13. and Voyage of Nearchus, p. 322.
† Rennell's Hindostan, p. 69.
§ Vitruvius, lib. viii. cap. 3.
‡ In America the localities of this mineral have been but little explored. It appears, however, to exist in many places under one or both of its usual forms. In Peru are numerous mines, situated at a very great elevation above the sea; some are near Potosi. The salt is very hard, and usually of a violet color – It has also been found in several parts of Chili, &c.

In California it is found in very solid masses; and in St. Domingo, near lake Xaraguay, it exists in a mountain.

In the United States, salt springs are numerous in several districts. These springs sometimes flow naturally, but are more frequently formed by sinking wells in those places, where this salt is known to exist, as in certain marshes, and in *salt licks*, so called, having formerly been the resort of wild animals to *lick* the clay, impregnated with this muriate. These

They extend into Louisiana and Mexico; in the latter, however, few regular beds of salt are found, but it occurs principally disseminated in argillaceous soils near the top of the Cordilleras. There is a salt lake in the valley of Mexico.

Humboldt mentions among the secondary rocks skirting the islands of Orinoco some calcareous beds mingled with gypsum and rock salt, and associated with beds of clay and sand containing the same minerals which appear clearly to belong to this formation.

springs are found on the banks of the Hockhocking, Scioto, Wabash, Tennessee, Kanhaway, Great Sandy, and various other rivers, all west of the Alleghany mountains, and emptying their waters into the Ohio. They occur also in the state of New York near the Onondago and Cayuga lakes; those of Onondago rise in a marsh on the border of the lake, at some distance from hard ground; they are richly impregnated, one gallon of the water sometimes containing a quarter to half a pound of the salt. Some springs, however, on the eastern waters of the Ohio are considerably richer than these.

The whole quantity of salt, annually extracted from saline springs in the United States, undoubtedly exceeds 600,000 bushels. Of this the springs of Onondago and Cayuga alone furnish about 300,000 bushels, and the Wabash *saline*, which belongs to the United States, yields 130,000 bushels. (Cleaveland's Mineralogy.)

SYNOPTICAL TABLE.

Book III.

MEDIAL or CARBONIFEROUS ORDER.

Independent Coal-formation of Werner.

Chapter I. *General view.*

Section I. Introductory.
Section II. Of the Coal-measures; with a preliminary view of analagous deposits in other geological situations.
Section III. Of the Millstone-grit and Shale.
Section IV. Of the Carboniferous or Mountain Limestone.
Section V. Of the Old Red Sandstone.

The characters of these formations are treated generally under the same heads as in the former Books: but the description of their range and extent, height, and other local phœnomena, are necessarily referred to the following Chapters, which treat of the several English Coal-districts in their geographical order.

Chapter II. *Coal-district north of Trent, or Grand Penine chain.*

Section I. Coal formations in the fields of
- *a.* Northumberland and Durham.
- *b.* North of Yorkshire.
- *c.* South York, Nottingham and Derby.
- *d.* South of Derby.
- *e.* North Stafford.
- *f.* South Lancashire.
- *g.* North Lancashire.
- *h.* Cumberland or Whitehaven.
- *i.* Foot of Cross Fell.

Section II. Millstone-grit & Shale throughout the Penine chain.
Section III. Mountain Limestone ditto.
Section IV. Old Red Sandstone, on north-west of the chain.

Chapter III. *Central Coal-district.*

Coal-formation, &c. in the fields of
- Section I. Ashby de la Zouch.
- Section II. Warwickshire.
- Section III. South Stafford or Dudley
- Section IV. Indications near the Lickey hill, &c.

Chapter IV. *Western Coal-district; divided into*

Section I. North Western or North Welsh — Coal & adjoining formations in the Coal-fields of
- 1. Isle of Anglesea.
- 2. Flintshire.

Section II. Middle Western or Shropshire. — Coal, &c. in the fields of
- *a*. Plain of Shrewsbury.
- *b*. Coal-brook-dale.
- *c*. The Clee hills and South Shropshire.
- *d*. Near the Abberley hill

Section III. South Western.
- *a*. Coal-formation in the basins of
  - 1. South Wales.
  - 2. Forest of Dean.
  - 3. South Gloucester and Somerset.
- *b*. Millstone-grit in ditto.
- *c*. Mountain Limestone in ditto.
- *d*. Old Red Sandstone in ditto.

Chapter V. *Particular account of the Trap rocks occurring in the several Coal-districts above specified.*

Chapter VI. *Comparative view of the distribution of the Carboniferous formations in other countries.*

BOOK III.*

MEDIAL OR CARBONIFEROUS ORDER.

CHAPTER I.

General view of the formations comprised in this order.

Section I. *Introductory notice.*

It is intended to comprise, in the present book, an account of the rocks associated together in the districts which afford the principal deposit of fossil-coal, and indeed the only one capable of being applied to purposes of extensive utility which appears to exist in the whole geological series.

The class of rocks thus constituted will contain not only the great coal-deposit itself, but those of the limestone and sandstone also on which it reposes; which, though entitled to the character of distinct formations, are yet so intimately connected with the above, both geographically and geologically, that it is impossible to separate their consideration.

This series of rocks is by some geologists referred to the flœtz, by others to the transition class of the Wernerians: we have preferred instituting a particular order for its reception, a proceeding justified by its proportional importance in the geological scale, its peculiar characters, and the many inconveniences arising from following either of the above conflicting examples. For this order we have proposed the name of *Medial*, wishing to adopt an appellation entirely free from theory, and indicating only the central place of this groupe in the five-fold division of the geological series which results from assigning to it a separate class. The epithet *carboniferous* is of obvious application to this series.

Had we been obliged to refer it either to the flœtz or transition class of the Wernerians, we should not have hesitated in preferring the latter branch of the alternative; since at least ten characters will be found in common between the carboniferous and transition † class, for one which would lead to an opposite

* Chiefly by the Rev. W. D. Conybeare.

† We only use the term 'transition rocks' as indicating a part of the general series without any theoretical views.

arrangement: for instance,—1st, in those countries (as England and the Netherlands) where all the beds above this series really deserve, from their generally horizontal position, the name of flœtz rocks, the carboniferous strata are most frequently much inclined, and exhibit every variety of contortion and disturbance,—agreeing in these respects with the transition rocks, but being entirely unconformable with the more recent: 2dly, the limestones of this series exhibit, in chemical and external characters, in mineral contents, and in organic remains, a very near alliance with those of transition, but differ in all these respects most strikingly from the more recent calcareous beds: it often requires much consideration to distinguish them from the former, but the merest tyro in Geology would at once discriminate them from the latter: 3dly, the sandstones in the lower part of this series approach closely in character to the more obviously mechanical varieties of greywacké, and indeed so completely pass into that rock, that in many instances the limits between this series and that of transition rocks can only be arbitrarily assigned.

In entering on our description of this series, we are at once struck with its far greater importance in a statistical and œconomical point of view above the formations we have previously described. We have hitherto had to notice little more than a few varieties of stone fit for architectural purposes; and of those which are more usually considered as the sources of mineral wealth, the trifling metalliferous deposits occasionally occuring in some of the rocks associated with the new red sandstone have afforded the only trace; but here we enter on a new scene in this respect.

The manufacturing industry of this island, colossal as is the fabric which it has raised, rests principally on no other base than our fortunate position with regard to the rocks of this series. Should our coal mines ever be exhausted, it would melt away at once, and it need not be said that the effect produced on private and domestic comfort would be equally fatal with the diminution of public wealth; we should lose many of the advantages of our high civilization, and much of our cultivated grounds must be again shaded with forests, to afford fuel to a remnant of our present population. That there is a progressive tendency to approach this limit, is certain; but ages * may yet pass before it is felt very sensibly; and when it does approach, the increasing difficulty and expense of working the mines of coal will operate by successive and gradual checks

* See some remarks on this subject in the section on the Northumberland and Durham coal-field.

against its consumption, through a long period, so that the transition may not be very violent: our manufactures would first feel the shock; the excess of population supported by them would cease to be called into existence, as the demand for their labour ceased; the cultivation of poor lands would become less profitable, and their conversion into forests more so.

The iron ores associated with the carboniferous strata being thus conveniently situated in the immediate vicinity of coal and limestone—the fuel and the flux requisite to work them, have led to the establishment of all our great Iron-Works within the limits of the coal-fields, and consequently produced a great condensation of manufacturing population.

The principal supply of lead used in this country is also derived from this series; with regard to zinc and copper, it contributes a smaller proportion than the transition and primitive series.

To trace the position of the horizontal strata above the coal formation, disposed as they are in uniform and parallel bands stretching across the island in similar lines of bearing, is easy, and requires nothing more than a sufficiently general knowledge of the districts in which they occur; but to reduce our description of the coal-fields, scattered as they are in unconnected basins and exhibiting every possible mode of disorder and derangement, into a regular systematic form, is a task of much more difficult accomplishment; not however that we are to imagine the non-existence in this series of an arrangement, as certain and constant as in those described in the two preceding books, but that the more violent convulsions to which it has been subjected, the elevated and deranged position of its strata, and the unconformable disposition of the more recent and overlying formations with regard to it, necessarily produce appearances of much greater intricacy, and require a more patient examination for their development; and even when that is accomplished, render the description of their local distribution much more embarrassed, than when we had only to pursue the simple lines indicating the extent of the younger deposits: the order of superposition, however, is as clearly ascertained in the one instance as the other.

The following general principles may be laid down, to guide us in our order of treating this subject. In the first place, the series of rock formations which ought to be considered together with the coal-measures, should be taken as including the four following subordinate series. I. *Coal-measures.* II. *Millstone grit and shale.* III. *Carboniferous* or *Mountain limestone.* IV. *Old red sandstone.* These are so much associated together in the same districts, (entering as component parts into the same chain of hills, &c.) that in describing any extensive tract of country, they must be kept together under the view, or an inextricable confusion will result.

Secondly. The coal-fields of England will, from geographical position, naturally fall under the following arrangement. 1. *The great Northern district*; including all the coal-fields north of Trent. 2. *The Central district*; including Leicester, Warwick, Stafford, and Shropshire. 3. *The Western district*; which may be subdivided into *North-western*, including North Wales, and the *South-western*, including South Wales, Gloucester, and Somersetshire. Physical circumstances also confirm this geographical arrangement.

Pursuing these principles, first, a general account of the characters of the several subordinate formations may be given; and secondly, a particular description of each district as far as is consistent with the limits of the work; tracing through each, the position and relations of the several rock formations above described.

To the first of these objects the four remaining sections of the present chapter will be dedicated; these will treat of the above formations under the same general heads which have been adopted in the former books, excluding however that of "range and extent", which together with the local phœnomena of the several districts specified, will form the subject of the ensuing chapters. Two chapters in the nature of an Appendix will close the book; the former devoted to the consideration of the Trap rocks associated in various parts of this series; the latter to a comparative view of the distribution of this series in foreign countries.

Section II. *Of the Coal-Measures, or Great Coal-Formation.*

Preliminary remarks on the limitation of this term, and the relations of this and other carbonaceous deposits.

In speaking of the coal-formation, we must be understood as applying that term emphatically to the great and principal deposit of that mineral, interposed between the newer red or saliferous sandstone, and the great carboniferous limestone and older sandstone formations; or, where these are absent, resting on transition rocks. This is the deposit distinguished by the Wernerians as the *Independent* coal-formation.

It may be useful here to observe that, besides this great deposit, thin seams of carbonaceous matter may be traced in several other geological positions, and that such are sometimes, though very rarely, sufficiently productive to be worked for fuel; yielding however generally a coal of very inferior quality.

In taking the following brief view of the several geological localities in which substances either identical with coal, or in an high degree analogous to it, passing into it by a series of insensible gradations, occur, we shall find that these range through the whole suite of formations, beginning in the most recent, and terminating only amidst the oldest with which we are acquainted; and we shall have to remark that the more recent deposits are unequivocally of vegetable origin, and that there is great reason to ascribe those of the middle periods to the same source in every instance where bitumen is present; yet it seems scarcely possible to ascribe the non-bituminous varieties of carbonaceous beds which occur in the rocks usually esteemed primitive (namely anthracite and plumbago) to similar causes; and in this case therefore we seem obliged to admit carbon, in its simple state as well as in its well known compounds, as an original mineral substance. It might appear therefore that a line of distinction might be drawn between those carbonaceous formations which are of derivative origin, and only introduced as extraneous materials from the vegetable into the mineral kingdom, and those which have belonged primarily to the latter; but in fact to assign such a line is impossible, since the clearly marked and unequivocal extreme forms are found blended together, and passing into each other by a series of middle terms. All that can be done, therefore, in the present state of science is, to state the difficulty and leave it for solution to that more advanced period towards which we are now only securing the approaches, by preparing a firm ground-work of induction from facts. What may be considered as ascertained concerning the conversion of vegetable matter into bituminous and coaly matter will be found shortly summed up from the able statements of Hatchett and Mac Culloch, under the head 'chemical and external characters' of coal.

We now proceed with the proposed enumeration, beginning with the most recent deposits which admit of comparison with the coal-formation.

A. *Alluvial. Peat.* This substance, arising sometimes from the subversion of forests covered by sphagnum palustre, and other mosses, and sometimes from the growth of various maritime and semi-maritime plants on the marshes bordering the coasts, is found among the most modern alluvia, generally covering them; often containing works of human art imbedded, and in many instances still in the act of progressive increase. It belongs therefore entirely to an order of causes still in action; the upper parts of its mass present the fibres of the vegetables whence it originates, and which still cover its surface (principally sphagnum palustre) in an almost unchanged state; in the middle part the texture is gradually obliterated, and the mass passes into a compact peat; in the lowest portion this change is carried still farther, and substances very analogous to jet are found; in some instances beds of peat alternate with beds of mud or sand deposited in lakes, or of silt and sand formed in the æstuaries of rivers; in these cases they appear exactly to represent an imperfect and unmatured coal-formation. See Dr. Mac Culloch's excellent memoirs on this subject.

The present work will contain a further account of this phœnomenon in that part of the last book in which alluvial formations are treated.

B. *Diluvial. (Associated with accumulations of gravel apparently resulting from the last great catastrophe that has affected the earth's surface, but unconnected with the order of causes still acting.)* The carbonaceous deposits of this period consist of beds of fossil wood (Lignite) in some places retaining its texture in the most distinct manner, and passing by a series of gradations from this state to that of jet. The mean terms of this series appear on chemical examination to consist of woody fibre in a state of semi-carbonizatiou, impregnated with bitumen, and a small portion of resin; so that of its original proximate principles as a vegetable, the extract has disappeared, the woody fibre and much of the resin, &c. is apparently modified into bitumen.

Bovey Heathfield, in Devonshire, affords an excellent example of this deposit; on the continent similar instances may be cited on the banks of the Rhine between Cologne and Bonn.

C. *Associated with the overlying basaltic formation known as the newest flœtz trap.* This is also a lignite, nearly agreeing with the former. England affords no example of it, but it abounds in the basaltic area of north eastern Ireland, and in almost every tract of this formation on the continent, especially in Hesse and Bohemia.

D. *Carbonaceous strata associated with the plastic clay and other formations above the chalk.*

The Isle of Wight furnishes an example. See Book I. of this work.

Similar instances occur in the basin of Paris, &c.

E. *In the sands immediately under the chalk;* interposed between it and the oolitic series. See in this work the description of the iron sand, especially near Hastings and at Lulworth cove.

F. *In the oolitic series*, the Kimmeridge clay, interposed between the upper and middle division, contains beds of bituminous shale affording an imperfect fuel. Lower down, the sands resting on the lias of Yorkshire contain regular strata of workable coal, though of inferior quality, accompanied by vegetable impressions. The account of this has, from the uncertainty of its exact place in the oolitic series, been necessary postponed to an appendix.

G. *The newer red or saliferous sandstone* is said on the continent to contain occasionally thin seams of coal, though such do not appear to have been yet noticed in this country.

H. The descending series next conducts us to the great coal-formation, the subject of the present book.

I. *The slates, &c. of transition,* (as they are called) sometimes contain beds of Anthracite; examples of this we believe occur in Devonshire.

K. *Beds of anthracite and plumbago,* (which is the purest known form of carbon) occur in mica slate and other rocks esteemed primitive; no example of this position is as far as we are informed known in this country, but it is frequent on the continent. These transition and primitive carbonaceous beds appear to be destitute of bitumen: bitumen however has been found in the veins of transition rocks; e. g. accompanying yellow copper in Carharack mine Cornwall.

In thus stating the occasional occurrence of carbonaceous beds in other formations, it is necessary carefully to guard against the error of supposing that any supplies of this mineral, capable of being profitably worked, are to be found any where without the limits of the coal-district of which we are now treating; an error that has led to much waste of capital in fruitless speculation. The local deposits above mentioned are objects of Geological curiosity, not of statistical interest.

We now proceed to consider generally the great coal-formation so called, as distinguished from the above partial deposits.

(*a*) *Chemical and external characters.* The coal-measures consist of a series of alternating beds of coal, slate clay, and sandstone; the alternations being frequently and indefinitely repeated. The coal-beds are principally distinguished by the proportion of bitumen which they yield: three species may be ascertained. The first yielding about forty per cent. of bitumen, is known by the names of *slaty coal, binding or caking coal, crozzling coal, cherry coal, &c.* In ignition it swells, agglutinates, and emits much smoke, which inflames at a sufficient temperature: the second variety yields about 20 per cent. of bitumen, and has been termed *cannel coal and caking coal;* it inflames readily, but does not agglutinate: the third variety contains little or no bitumen; it has been called *culm, coarse coal, stone coal, Kilkenny coal, &c.;* in ignition it exhibits little or no flame, and does not agglutinate; this last variety forms the anthracite of mineralogists, when these characters are carried to their furthest point. It must be understood however, that much confusion has prevailed in the application of the above names; particularly in those of slaty and of stone coal.

The variable proportions of bitumen in all these species of coal were considered by Mr. Kirwan as united with charcoal, which constitutes the predominating ingredient; the residuum, besides these two substances, consisting only of between two or three per cent. of earthy ashes. Mr. Kirwan supposes that the bitumen exists in different states in these varieties, and when in the state of Asphaltum, communicates to some of them the property of caking. These coals are often much mixed with clay, and pass into bituminous shale.

The above are the views of the chemical composition of coal exhibited by the analyses of Kirwan.

Dr. Mac Culloch, however, has been led by subsequent experiments (see G. T. vol. 2.) to consider coal rather as itself forming a series of links connected with the bitumens, varying by the diminishing proportions which the hydrogen bears to the carbon in their composition; than as a mixture of bitumen and charcoal, as distinct principles. We cannot better explain his opinions than by subjoining the condensed statement of them which he has lately published in his Classification of Rocks.

All the bitumens, from naphtha to asphaltum, consist of compounds, apparently indefinite, of carbon and hydrogen principally; the small quantities of oxygen and azote which they contain, appearing to have little or no effect in modifying their mineral characters. In the most fluid, the hydrogen predominates, diminishing progressively according to the order of their relative tenacity or solidity. Where asphaltum ends this series, cannel coal, with some interruption in composition, and a considerable one in texture, commences that of the coals. From this variety, down to the most perfect anthracite, there is a similarly indefinite range of composition; the hydrogen gradually diminishing as the coal becomes less inflammable, as it is less capable of being separated into bitumen and charcoal by distillation, and as it yields a smaller comparative proportion of the former. Thus the composition of the bitumens illustrates that of the several varieties of coal. The most perfect anthracite appears to yield no bitumen, yet it still contains hydrogen, perhaps in every case; as that element is present even in common charcoal, which is itself a compound substance. Where anthracite passes to plumbago, which may in fact be considered as the true end of this series, the hydrogen seems to have disappeared; and, if this substance be not mere carbon, as it probably is not, from the apparent combustion which it undergoes on exposure to air, when its base has been extricated from iron under water, it undoubtedly approaches nearer to that element than any of the preceding substances.

The paper of Dr. Mac Culloch above referred to (G. T. vol. 2), contains some observations on the experiments which have been made with a view to illustrate the origin of coal, and its formation from vegetable matter, modified by the action of fire or water, or both, which are highly valuable, and strongly marked by the masterly precision which characterises all his researches.

1. He shews that the substances produced from wood by the action of fire, by Sir James Hall and others, and supposed to be true bitumens and coal, were forms of a peculiar compound, resembling indeed the bitumens in colour and inflammability, but essentially different in many of its properties, being insoluble in naphtha, &c.; containing oxygen and azote in proportions different from those in which the same substances exist in the bitumens, and in particular a considerable quantity of the former. He proposes that this peculiar substance should be distinguished by the name of *bistre*, which is already applied to it in the arts; hence these experiments afford no countenance to the idea that fire has been employed to convert vegetable matters into bitumen.

2. By a series of experiments on peat and various lignites, the gradual progress of bituminization in these was ascertained; and it was found that in jet, the extreme term of the lignite series, no chemical difference from coal existed, excepting that a greater quantity of acid was given over in the process of its distillation. The general result of these experiments is thus stated.

'Examining therefore the alteration produced by water on common turf or submerged wood, we have all the evidence of demonstration, that its action is sufficient to convert them into substances capable of yielding bitumen on distillation.

'That the same action, having operated through a longer period, has produced the change on the brown coal of Bovey, is rendered extremely probable by the geognostic relations of that coal. From this to the harder lignites, surturbrand and jet, the transition is so gradual, that there seems no reason to limit the power of water to produce the effect of bituminization in all these varieties, nor is there aught in this change so dissonant from other chemical actions, as to make us hesitate in adopting this cause.' (G. T. vol. 2. p. 19.)

3. Although it thus appears that water, rather than fire, has been the agent employed in the first bituminization of vegetable matter, yet there is a wide interval between the external characters of these bituminous lignites and true coal, although their chemical composition is nearly the same. As, therefore, many philosophers of high reputation have supposed that fire has been the probable agent in this part of the conversion, Dr. Mac Culloch felt it incumbent to examine what analogies in favour of this hypothesis result from experiment; and he represents those results as so far according with this theory, that by the application of heat under compression to jet, it was fused into a substance possessing the true characters of coal. It is possible, therefore, that the agency of fire applied to beds of lignite and peat, may convert, not wood, but vegetable matter previously bituminized, into coal.

Dr. Mac Culloch does not however pronounce in favour of this theory, but merely points out its possibility; fully admitting, on the other hand, that we cannot presume to state the period which nature has used in her operations, nor during how long a space the causes have continued to act, before the vegetable matter has undergone its ultimate change into coal; nor, therefore, whether the long continued agency of water and pressure may not have produced the required changes.

We ourselves, on the assumption that coal really is derived from altered vegetable matter, should infinitely prefer that view which refers the whole of this change to water; thinking it greatly more probable, that the same agents which have converted wood into jet should also have accomplished the last and least important steps of the same process, by converting it into coal. We cannot consider this as a sufficiently 'dignus vindice nodus' to evoke the god of fire for its solution; we are certain, from the nature and contents of the strata associated with coal, that water was present; but had fire been the agent employed in consolidating the coal, it is difficult to understand why it has not also consolidated the shales and sandstone. The coal districts no where present any appearances which obviously suggest the idea of igneous agency, except where they are traversed by whin-dykes.

The question concerning the vegetable origin of coal will be farther considered in treating of the vegetable remains which so peculiarly characterise it. The great chemical argument appears to be, that the coal exhibits exactly the same results as the most decided lignites, the process being however still farther advanced.

The slate-clay of the coal-measures differs from clay-slate by its less solid and indurated state; it is known in different collieries by the names of *black* or *blue metal*, *shale*, *clunch*, *cleft*, *bind*, &c.

The sandstones of the coal measures are usually gritty, micaceous, and tender; they afford freestones for buildings, whetstones, grindstones, &c.; some varieties of a large schistose structure are raised as flag-stones for paving; others, more finely laminated, as roofing slates: *Plate*, *Post*, *Pennant*, are names locally applied to these sandstones.

(*b*) *Mineral contents.* Besides these strata, *Clay-iron-stone*, either in the form of continuous beds, or courses of nodules, is of common occurrence in the coal-fields. These yield on an average about 30 per cent. of metal; they are provincially termed *Mine*, and *Pins*. The occurrence of this most useful of metals in immediate connection with the fuel requisite for its reduction, and the limestone which facilitates that reduction, is an instance of arrangement so happily suited to the purposes of human industry, that it can hardly be considered as recurring unnecessarily to final causes, if we conceive that this distribution of the rude materials of the earth, was determined with a view to the convenience of its inhabitants.

Iron pyrites is often abundantly disseminated among the coal.

Associated with the ironstone, small quantities of galena, and more rarely of blende, are sometimes observed.

Petroleum sometimes occurs, as might naturally be expected, among the coal.

(*c*) *Organic remains.** From the greater abundance and importance of the remains which the vegetable kingdom has

* The references in this article are to
Parkinson's organic remains, vol. I.
Martin s Petrificata Derbiensia.
A Memoir by the Rev. Mr. Steinhauer in the American Philosophical Transactions
Flora zur Vorwelt and Petrifactenkunde by Schlotheim.
Flora der Vorwelt by Sternberg.
The Edinburgh Philosophical Transactions.

yielded to the coal strata, we shall consider these before we proceed to the scanty animal remains they afford. The remains of the coal-field exhibit the trunks, leaves, and more rarely pericarps of various vegetables, all greatly distinguished from known recent species, but apparently agreeing more nearly with the vegetation of a hot than temperate climate, and with moist rather than dry situations. In considering their original habits, we must, as Count Sternberg observes, transport our thoughts to an epoch when the vast tracts now occupied by more recent marine deposits were still beneath their parent ocean, from which scattered groups of primitive islands alone emerged, covered by the vegetation of which these relics are still preserved. The rivers, which in such a condition of things could have existed only as torrents, would frequently tear up this vegetation, and deposit it along the bottom of the adjacent basins.

The vegetation of a country covered with lakes and marshes consists, along the margin of the waters, in gramina, and particularly arundinaceous and other aquatic plants; and on the hills rising above the level of those waters, but still in their neighbourhood, in various trees, shrubs, ferns, &c. And such is in effect the ancient vegetation which we are now considering. It offers but few genera, and on the largest allowance not more than 400 species. The trunks or stems thus discovered belong principally to arundinaceous plants, approximating to those now known, and to a very peculiar order, distinguished by the cortical part being entirely covered by regular impressions resulting from the petioles of fallen leaves, ranging round them in spiral lines; these have been supposed to belong partly to the Palmaceous order, and partly to anomalous forms, constituting a transition between these and the coniferous plants; such a link has been already established in Professor Sprengel's Natural System. We must observe, however, that so much of their organization is, in their present state, not capable of being ascertained; and so much of what can be ascertained entirely *sui generis*, that any attempts even at assigning the natural class to which they belong must be received with great hesitation.

Other trunks have a fluted character.

Of the leaves which occur in the coal strata, many undoubtedly belong to the order Filices; others have been but perhaps on insufficient ground, referred to Hippuris Equisetum, &c.; these we shall specify in the following detail.

To consider these remains more particularly, we shall begin with those of trunks or stems; and here we shall be materially assisted by the accurate and precise figures of Count Sternberg,* and by the investigations of Mr. Steinhauer, who, in a very valuable but perhaps occasionally rather fanciful memoir, printed in the Transactions of the American Philosophical Society, has described many of the remains of this class which he had collected during his residence in England from our coal-fields.

I. We shall begin with the *Arundinaceous plants*, to which the generic name *Calamites* has been applied by Sternberg and Schlotheim.

Parkinson, O. R. vol. 1. pl. 3. fig. 3. Lluid tab. 5. fig. 184—6.

Steinhauer, Pl. 5. fig. 1. 2. Martin's Pet. Derb. Plates 8. 25. 26.

Schlottheim, Pl. 20. Sternberg, Pl. 13. fig. 3.

Generic character. A jointed stem, longitudinally striated. In the most perfect specimens (which have been found many feet long) the termination is in a conical point; and the base of each joint in the upper part appears to have been surrounded with a whorl of leaves. Some specimens are finely striated, others widely. The joints are sometimes short, sometimes long; so that either there must have been great variety in different individuals of this species, or, which is most probable, it forms an extensive genus comprehending many species. Some specimens much resemble the young shoots of the Surinam bamboo (Steinhauer); but Sternberg observes, "elles se distinguent des bambousiers en ce que les divisions n'en sont point marquées par des nœuds saillans, mais par des coutures; elles sont en outre rayées plus distinctement."

II. Stems surrounded with impressions of the attachment of leaves spirally or quincuncially disposed.

A. Variolate or Verrucate. (Variolariæ, Sternberg.)

Generic character. Stem surrounded by verrucate impressions or depressed areolæ, with a rising in the middle having a central speck. These areolæ are very like the papillæ of the Echinus papillatus; they support long cylindrical and seemingly tubular acini or leaves? Mr. Steinhauer imagined that he had traced these acini, as he calls them, extending in rays to the distance of 20 feet, round a stem of this kind lying horizontally in a soft bed of clay; and infers, from their position, that the stem *had grown horizontally in its present position*,

* In referring to the foreign works of Sternberg, Schlottheim, &c. we have been careful to cite only such species as had fallen under our own observation from the English coal-pits; we believe, however, that the vegetation of the continental coal stata and our own is similar throughout, with very few exceptions.

vegetating amidst the mud: but we must surely attribute this improbable idea to an error of observation.

The stem, in its present state, generally exhibits a solid but often compressed cylinder of sandstone or clay, invested with a carbonaceous coat; it has usually a deep groove running along one side, and in its interior an included cylinder may be observed, traversing the sandstone in a direction parallel to, but not coincident with, the axis of the stem, approaching more nearly to its grooved side.

Mr. Steinhauer considers this as representing the pith, which resisted decomposition longer than the ligneous portion; he accounts for its position by supposing that it subsided towards the side of the decomposing plant which lay lowest (in which point he says it is always found), and attributes the groove to a rise in the middle of the inferior surface of the matrix in which it lay, squeezed upwards by the unequal pressure of the incumbent mass while unequally consolidated. These views, however, we cannot entirely adopt, but rather regard these remains as having proceeded from a succulent plant which never had a much larger portion of ligneous fibre than that which has shrunk into the carbonaceous investment still surrounding it. It is probable, however, that the included cylinder may represent the pith, but it is said that other smaller longitudinal tubuli, parallel to the above, may be traced through its substance. A termination has been in one instance observed, drawing from a thickness of three inches to an obtuse point; the trunk has also been seen subdividing into two nearly equal branches.

The diameter is from two or four (the most usual size) to even twelve inches.

Sternberg believes that the family to which this plant belonged must have had many analogies with the arborescent Euphorbiæ and the Cacti. On the whole it seems certain that these also must have been succulent plants; only one species is certainly known.

Steinhauer, Pl. 4. fig. 1 to 6. Martin, Pet. Derb. Pl. 11. 12. 13.

Parkinson, O. R. vol. 1. Pl. 3. fig. 1. Sternberg, Pl. 12, fig. 1. 2.

The leaves or acini are beautifully shewn in this last figure.

B. The leaf-bearing impressions, forming scales spirally surrounding the stem.

Lepido dendra (Sternberg, from λεπις a scale, and δενδρον wood.

Phytolithi cancellati (Steinhauer and Martin.)

Count Sternberg, taking for the basis of his arrangement the varieties in the configuration of the circumambient scales, has divided this family thus:

Tribus 1. (*Lepidotæ*) squamis convexis.

a. scutatæ. Species 7. figured in his first part, and many additional in the second.

b. escutatæ. Species 1.

Tribus 2. (*Alveolariæ*) squamis subconcavis. Species 3.

These general divisions may probably stand, but the subdivision into species appears to require reconsideration; for the character assumed as the basis of the classification—the figure of the scales—appears to vary in different parts of the same individual, as will be obvious in comparing the scales on the basis of the stem, and the branches and extreme shoots of Sternberg's first species; and, if Steinhauer's observations are correct, a still more formidable difficulty will oppose itself to the use of this character without extreme caution; for, according to him, the impressions of the different integuments of these stems, (or as he calls them from the parts of the original plant whence he conceives them to have resulted) the epidermal, cortical, and ligneous impressions of the same species, present varieties of aspect which at first sight appear quite irreconcilable: we proceed to such of his species as we are acquainted with among the English series, and as are clearly distinct.

1. *Lepidodendron* dichotomum. S. C. Stem dichotomous and branching; lower scales obovate: superior scales rhomboidal, scutate in the middle, the scuta marked horizontally with three glandules at the insertion of the leaves; narrow linear leaves 12 to 18 inches long.

Sternberg figures a magnificent specimen 12 feet long, several of the extreme shoots surrounded by their long acicular leaves, and in one instance assuming an appearance like infrutescence; thus much more is known of this than of any other among the coal plants. Detached parts are frequent in our own coal. Sternberg observes, "la stature elancée de ces arbres, la dichotomie des branches, les feuilles extremement longues et etroites, qui en environnent les tiges sur lesquelles elles sont rangées en spirales continues, sont, si je ne me trompe, des indices characteristiques d'une espèce d'arbre inconnue."

Martin Pet. Derb. T. 50. represents the middle of the stem at T. 14. upper, and perhaps also lower figure, its extreme shoots.

Strernberg's 2nd, 4th, and 7th species we are not certain of as English fossils.

His 3rd (*Lepidodendron* aculeatum), 5th (*L.* rimosum), and 6th (*L.* undulatum), at first sight appear greatly to resemble those represented by Steinhauer, as the epidermal, cortical, and ligneous impressions of a single species, and figured by him Plate 6. fig. 2 to 6. If his ideas are correct, he well describes it as a protean fossil; he assures us that he has traced frequently two of these impressions in the same specimen: "it frequently happens that the cast and impression are from different integuments, the space separating them being occupied by carbonised or bituminised vegetable matter; sometimes the impression was epidermal and the cast cortical; at others, both impression and cast were from the same integument. The manner of accounting for these varieties," he adds, "is obvious; it only requires us to suppose the cast and the impression or matrix to have been formed while part of these integuments were still in their natural state, which being thus inclosed was afterwards changed into bitumen or carbon."

The above paragraph was written before the publication of the second part of Count Sternberg's work, and the less perfect state of the epidermal impression figured by Mr. Steinhauer rendered it difficult to institute an exact comparison. In the second part of the Count's work, however, he figures another species of L Lycopodiodes, T. 16. fig. 1. 2. 4. which appears much more exactly coincident with Steinhauer's figure, and in these figures the difference of the external and internal impression will be seen distinctly exhibited. We still, however, feel inclined to consider the 5th and 6th species of Sternberg as corresponding with Steinhauer's cortical and ligneous impressions. Sternberg gives the following specific characters.

> L. *Lycopodiodes.* Candice arboreo, dichotome ramoso, squamis rhomboidalibus utrinque acuminatis scuto sublente tantum distinguendo nec definiendo.

Steinhauer's memoir adds that "The appearance of some specimens seems strongly to suggest the idea that the bark was furnished with, or composed of, strong longitudinal fibres, and almost all betray a tendency to be striated in a vertical direction."

"The few specimens which afford traces of a pith inform us that it also was very finely striated in a longitudinal direction, but afford us no further information respecting the internal organisation of the orignal."

The quantity of coaly matter investing these specimens is more considerable than in the former cases.

Martin's Pet. Derb. Plate 13. appears referable to one of the appearances above described.

The second division *(Esculatæ)* of Sternberg's first tribe of this family contains but one species, which has not yet been described from our coal-pits.

Of his second tribe, *Alveolacæ*, he only figures one, though he refers to three species. The one species figured by him has not been described as English, but Steinhauer gives two species under the names Phytolithus tesselatus, Plate 7. fig. 2. and Phytolithus notatus, Pl. 7. fig. 6., which clearly belong to this tribe, and might stand in Sternberg's arrangement as Lepidodendron tessellatum and L. notatum.

The species of this tribe are, however, always fluted longitudinally, and thus approach to the fluted specimens constituting the family Syringodendron of Sternberg; and as the scales only appear on the outer or epidermal impressions, the internal or ligneous resembling the appearances of the Syringodendron: this tribe appears to require further examination.

The very singular Phytolithus parmatus described by Steinhauer, undoubtedly belongs to the Lepidodendra; Steinhauer, Plate 6. fig. 1., Plate 7. fig. 1.; but we feel uncertain whether a particular subdivision of that family ought not to be created for its reception. The surface is reticulated by the circumambient spiral rows of scales as before, destitute however of the marks so interesting in the Lepidodendron aculeatum, being simple depressed spaces: "but the most remarkable and inexplicable part of the organisation of this fossil consists in a series of circular or oval larger scutillæ or shields, placed close to each other in a right line across the surface" (or rather extending in a longitudinal row down it); "in the epidermal appearance each of these is surrounded by a raised margin, the included disk swells towards the central umbo or boss in curiously disposed rugæ arranged in a manner slightly resembling the curves on the back of an engine-turned watch case, and the boss is generally more or less excavated in the centre. These shields are often two inches or more in diameter; their series is bounded on each side by a rather indistinct ridge, beyond which the usual reticulated or squamous appearance of the surface may be observed."

"The total appearance of the fossil has a curious resemblance to that of some of the Iungermaniæ preparing for fructification, when highly magnified."

Mr. Allan has described and figured in the Edinburgh Philosophical Transactions for 1821, a much more complete specimen of this interesting species.

III. We now proceed to a third great class of the stems preserved in the coal-strata;—those remarkable for a fluted surface, but distinguished from the Calamitæ by having no joints, and by the wider intervals between the flutings.

Syringodendra (Sternberg.)

We have before, however, remarked that some of the figures referred to this division resemble the ligneous impression represented by Mr. Steinhauer as belonging to the Phytolithus notatus, which the epidermal impressions would refer to the second tribe of Count Sternberg's Lepidodendra.

1. *Syringodendron* organum of Sternberg. T. 13. fig. 1. resembles the ligneous impression of Steinhauer (Pl. 7. fig. 3.) above-mentioned.

2. *Syringodendron* pes capreoli (Sternberg, Pl. 13. fig. 2.) closely agrees with Phytolithus Dawsoni (Steinhauer, Plate 4. fig. 2.)

3. A *Syringodendron* agreeing with Tab. 16. fig. 1. Schlotheim Petrifactenkunde, by him called Palmacites sulcatus, is common in our coal-fields.

IV. A variety not reducible to any of the above heads, is figured by Steinhauer, Pl. 5. fig. 3, under the name *Phytolithus* transversus; a cylindrical trunk, transversely closely striated, without any traces of leaves or fibres; the general appearance like that of a large earth worm; perhaps a creeping root.

In closing this branch of our inquiry into the coal-fossils, we may remark that in all of them there is but very little appearance of ligneous matter; the carbonised or bituminous coat surrounding them, and a few traces of what has been called the Pith, forming the whole representatives of any substance of that nature; the main body consisting of an inorganic mass of clay or sandstone, most unequivocally mechanical in its origin. If these trunks had ever presented solid woody masses, it is almost impossible to conceive how such a substitution of materials could have taken place; and what is more important, it is quite contrary to the analogy of the changes which wood transferred into a fossil state is in every other instance found to undergo; for if we examine the numerous fragments of wood found in the London clay, the oolites, and lias, we shall find that it has been either simply carbonised, or else charged by infiltration through its pores with calcareous spar or pyrites; the woody structure and fibre being in every case distinctly preserved; and even in that more remarkable change in which wood passes by a siliceous infiltration into the state of wood-opal, its original structure is still preserved, often in its minutest parts, and the woody fibre seems rather masked by its siliceous investment than destroyed.* But if we suppose the originals of the coal-plants to have been succulent vegetables, this difficulty will vanish, and many of them, as we have seen, resemble this class rather than any other. It is more easy too to conceive how plants of a marshy seat should have been washed down into the basins which now constitute our coal-fields, and there covered with deposits of sand and clay, &c , for the fact will then be analogous to what is often observed in the present day on the borders of our lakes.

These plants are sometimes parallel to the coal-strata in which they lie, sometimes shoot as it were through them perpendicularly, and have been observed in this latter position rising 15 feet or more. Mr. Trevelyan has drawn some striking examples of this exhibited in the cliffs of the Durham and Northumberland coast: these positions probably arose only from the distribution of gravity in the trunks as they subsided amidst the thin, loose, and unconsolidated materials of the strata in which they are buried.

Mr. Brongniart has published in the Annales des Mines for 1821, a plate exhibiting many of these erect stems, from the open coal mines of St. Etienne, Department de la Loire.

The above statements will shew that the accounts which have been given of trunks closely resembling peeled oaks, &c. discovered in our coal-fields, must have originated in the hasty judgment of an eye unpractised in examining these remains.

The roots of these stems have in few instances been ascertained. Mr. Brewster, however, in the Edinburgh Phil. Trans. for 1821, figures a stem with branching roots found at Niteshill, but the whole of the external carbonaceous coat being stripped off, its species cannot be identified. Count Sternberg has figured in his second part, T. 14. a magnificent specimen of the trunk of Lepidodendrun aculeatum, in which the base spreads out in a sudden concave swell.

Sternberg, in the second part of his great work, has instituted the new genera Schlotheimia, Annularia, and Rotularia, from some of the verticillate leaves above alluded to: his generic characters are

> *Schlotheimia.*—Candex articulatus ad articulos contractus verticillo foliosus. Both the species he describes occur in the English mines.
>
> *Annularia.*—Folia in verticillum disposita annulo proprio inserta. We may cite *A.* reflexa, T. 19. fig. 5. as occurring in our mines.

* The greatest quantity of woody matter in the coal strata we have ever seen was an irregular compressed stem about two inches thick, which more resembled a tuberose root than any thing else.

Rotularia.—Folia Verticillata in parvæ rotæ formam expansæ.

R. cuneifolia, T. 26. fig. 4. is very common in our coal mines. See Parkinson, org. rem. vol. 1. pl. 5. fig. 3.

Leafy impressions.

We have already mentioned the leaves belonging to the various stems above-described, when they could be clearly ascertained; those we have now to consider are probably in the great majority of instances quite distinct from them, and are generally indeed found in distinct strata: they appear to belong to low shrubby vegetables, such as ferns, rather than to arborescent plants.

In this branch of the inquiry we have to regret the want of good and accurate figures of English specimens to which we might refer. We may mention another difficulty which affects the subject generally: a few detached leaves, or single pinnæ, afford a very imperfect view of the general infrondescence of a plant, and the appearances of the imperfectly-developed leaves of the youngest shoots may often be mistaken for the marks of distinct species. The subject requires rather to be studied from the large slabs in the coal-pits themselves, than from the small and mutilated fragments which find their way into cabinets. Count Sternberg unites these requisite and local advantages with all the precision of science, and when his work shall have been continued so as to include these remains, much new light will probably break in: at present Schlotheim's figures in his Flora der Vorwelt appear to be the best extant.

A. The first class of leaves we shall consider, are those which are *Verticillate.*

Martin. Pet. Derb. T. 20. Schlotheim, 1. 2. 3. 4.

There are several varieties; the leaves in some more, in others less, numerous; sometimes pointed, at others truncated and serrated at the end; sometimes there appears to be only a single stem bearing these verticillæ; this is often striated, and appear to be smaller varieties of those which we have already distinguished as Calamitæ; perhaps young individuals of them: at other times the verticillæ appear to issue from jointed branches at right angles to the main stem.

These impressions have been compared by different authors to Hippuris Equisetum, Asperula, Galium, Rubia, Molluga, and Casuarina: a list quite sufficient to shew the entire uncertainty of the subject.

B. Filiciform leaves. The following may be mentioned as the principal varieties.

1. Martin, Tab. 10. Park. o. r. pl. 4. fig. 1. 2. Schlotheim, fig. 22; Pinnæ decurrent and alternate; by some considered as a Pteris, by Schlotheim an unknown species of Polypodium or Lonchitis.

2. ——— Schlotheim, fig. 7. 8. Pinnæ decurrent and alternate, supposed Pteris aquilina; fig. 8 is closely related to Parkinson's O. R. pl. 4. fig. 1. 2. there considered as an osmunda.

3. ——— Schlotheim, fig. 11. Pinnœ sessile, alternate; supposed Polypodium oreoptieris; the other supposed Polypodia figured by Schlotheim fig. 9. 13. 14. 15. are likewise found in our coal-fields; fig. 9 agrees with Parkinson's O. R. pl. 4. fig. 7. and fig. 14 with pl. 5. fig. 9.

4. Osmunda gigantea. Sternberg, Tab. 22.

5. Parkinson figures a plant with five-lobed petiolate leaves, plate 4. fig. 5. and compares it with Dicksonia.

6. In plate 5. fig. 2. of the same work is a plant with three-lobed petiolate leaves compared to Adianthum.

7. Schlotheim, fig. 18. a. also compared to Adianthum, but widely different from the last.

C. Kidney-shaped leaves, figured by Martin, Pet. Der. T. 34. fig. 1. 2. are also common in the coal-fields; as they somewhat resemble the expanded wings of the papilionaceous tribe, they have given rise to a fanciful account of the discovery of fossil butterflies in Fletcher of Madeley's curious description of the great landslip in that neighbourhood.

D. Impressions resembling Confervæ, not figured.

Pericarpial remains.

Specimens referable to this class appear to be figured in Martin's Pet. Derb. T. 21. 51. 53; but we will not hazard any observations on them.

The greater abundance of these vegetable remains, and their importance in illustrating the probable history of the coal-formation, have induced us to depart from our usual arrangement, in giving the precedence to these before entering on the consideration of those derived from the animal kingdom, which are very rare and entirely confined to a few species of Testacea; excepting that, in one instance, a fragment supposed to have been part of the radius of a Balistes has been found.

About the middle of the coal-series in Derbyshire, however, and in the ninth bed of shale (reckoning in the ascending order) a bed of ironstone occurs, which is so full of different species of

Mytili, &c. as to be distinguished by the name of the *muscle-band.* It will be seen by the following list, extracted from Mr. Farey's Index to Sowerby's Mineral Conchology, that Testacea are also found in some of the other beds.

CHAMBERED UNIVALVES.		*Stratum in which found.*
Ammonites Listeri.	(p. 132. Sowerby's Brit. Min. *T.* 435)	3d. Shale.
Ammonites Walcotii.	*T.* 106.	Argillaceous iron-stone, Coalbrook D.
Orthocera Steinhaueri.	*T.* 80. fig. 4.	3d. Shale.
BIVALVES.		
Terebratula crumena.	*T.* 83. fig. 2. 3.	1st. Shale.
Lingula Mytiloides.	*T.* 19. fig. 1. 2.	9th. Shale.
Mytilus crassus.	Brit. Min. *T.* 386.	Ibid.
Unio acutus.	*T.* 33. fig. 5. 6. 7.	Ibid.
U. uniformis.	*T.* 33. fig. 4.	Ibid.
U. subconstrictus.	*T.* 33. fig. 1. 2. 3.	12th. Shale.

The localities given are all in Derbyshire, and the beds are numbered in an ascending order.

A most important question with regard to the inferences to be deduced from these shells, as affecting our conclusions concerning the circumstances under which the coal-deposits were formed, here presents itself;—do they belong to the marine or fluviatile class?

There can be no doubt that the Ammonites, Orthocera, and Terebratula are marine; but on the other hand the Unios have been considered as fluviatile, and the Mytilus crassus as belonging perhaps to the fluviatile genus Anodonta. We are greatly inclined, however, to hesitate in admitting these representations. It is indeed well known to conchologists that the genus Unio was instituted in order to separate the fluviatile from the marine muscles; but it must be equally known to the students of fossil conchology that the form of hinge, assumed as its distinguishing character, belongs to several species found in a fossil state, under circumstances that preclude the suspicion of their being other than marine: thus shells called, on account of their hinge, Unio, are found in many of the oolites, the lias, &c. (see the lists of those formations); these occur with shells undoubtedly marine, and are too numerous, and too constantly attendant on the strata, to be considered as fluviatile shells accidentally introduced. Now some of the so-called Unios which occur in the oolitic series, are closely allied to those of the coal-measures, and as we are nearly certain they are marine in the former, we can have no good reason to pronounce them certainly fluviatile in the latter: it is nevertheless very remarkable, that these problematical shells are the only ones at all common in the coal-measures, and that remains of undoubtedly marine origin should be so very rare; we shall however find in the next section, that in the millstone-grit and shale formation the alternation of coal-strata with calcareous beds of marine origin is clearly ascertained.

We proceed briefly to notice the theoretical deductions concerning the origin of the coal-formation, and the circumstances under which this singular collection of strata were accumulated, which may be built on the nature of the remains they include.

It has been argued from the abundance of vegetable remains in the coal-field, as well as from the composition of the coal itself. that this mineral owes its origin to the vegetable kingdom; but it has been contended, on the other hand, that if we ascribe this origin to the strata of coal, we are bound by parity of reasoning to maintain, that all calcareous beds have been formed from the detritus of shells, because in many such beds, abundance of testaceous remains are preserved, and the same analogy of composition also exists in this instance; and it is further urged that as such an origin cannot with any plausibility be ascribed to the anthracite and plumbago which occur in primitive rocks, so neither ought we to ascribe it to substances so similar, when found in this part of the series.

The force of these objections must be duly allowed; yet when we observe the remains above described so peculiarly characterising, by their abundance, the carboniferous strata; when we observe the cortical part of these vegetables generally converted into a coal identical with the contiguous strata of that mineral; and when we observe in the lignites of Bovey Heathfield, the most decided wood pass into a substance nowise differing from common coal in chemical characters, the impression left on the mind seems clearly in favor of the hypothesis which derives these fossil combustibles from the vegetable kingdom, and this opinion is strongly sanctioned by the experiments and inferences of Hatchett and Mac Culloch.

The nature of the remains found in the coal-strata gives no countenance to the idea that the bituminization of animal matter can in any sensible degree have contributed to form them.

How then, on the above supposition, are we to account for such a surprising accumulation of vegetable matter arranged in repeated strata (sometimes to the number of sixty and even more in a single district,) separated from each other by intervening deposits of clay and sand?

It seems certain that the coal-strata were deposited within, and perhaps along, the borders of great accumulations of water, whether fresh or salt; the testacea occurring in them sufficiently prove this; and, as we have seen, it is also certain that in some periods of the coal-formation (and more especially with regard to those beds of coal which are occasionally associated with the millstone-grit and limestone shales to be described in the next section) the water was salt, and that the evidence of its ever having been otherwise is far from convincing. It hardly seems necessary therefore to have recourse to a series of reciprocating inundations of the sea and fresh-water lakes; but we may more naturally suppose these deposits to have been entirely formed within the former, and their disposition in limited basins seems farther to indicate that they were accumulated in friths or æstuaries.

Now the partial filling up of lakes and æstuaries offers us the only analogies in the actual order of things with which we can compare the deposits of coal; for in such situations we often find a series of strata of peat, and sometimes submerged wood, alternating with others of sand, clay, and gravel, and presenting therefore the model of a coal-field on a small scale, and in an immature state.

The lignites of Bovey Tracey, which seem evidently to have been accumulated (but at an earlier period) in the filling up of an æstuary, are disposed in regular strata (see the section given in the first volume of Parkinson's Organic Remains) alternating with clay and gravel evidently derived from the detritus of the neighbouring granitic chains. The thickness of the beds in this instance, and the structure of the whole deposit, give it a still nearer resemblance to a regular coal-field. We must here suppose the wintry torrents to have swept away a part of the vegetation of the neighbouring hills and buried them in the æstuary, together with the alluvial detritus collected in its course; the latter would, from its gravity, have sunk first and formed the floor; the wood would have floated till, having lost its more volatile parts by decomposition and become saturated with moisture, it likewise subsided upon these, being perhaps also loaded by fresh alluvia drifted down upon its surface; the reiterated devastations of successive seasons must have produced the repetition and alternation of the beds.

In this instance, then, it is evident that these or similar causes must have acted; and if we suppose a like order of causes to have operated more extensively and for a longer period during the formation of the coal-strata, we shall find such an hypothesis sufficiently in accordance with their general phœnomena.

It seems probable that the coal-strata were originally much more nearly horizontal than at present, and that they owe their present contorted disposition to subsequent convulsions. In the last book of this work we shall have to consider evidence nearly demonstrative of this.

It may be objected that, if the coal had really been deposited in æstuaries, we ought to find fuci and algæ among its vegetables; but their absence is only a circumstance common to this and every other formation, though the great majority of them are undoubtedly of sub-marine origin, and many that are clearly such, as for instance the Stonesfield slate, contain other vegetable remains.

(*d*) *Range and extent.* This article is necessarily deferred, from the arrangement adopted in the present book, to the chapters which treat of the local distribution and phœnomena of the coal-fields.

(*e*) *Elevation.* The particulars referable to this head will likewise be found incorporated in the following chapters; but it may be stated generally, that the coal-measures often form hills exceeding one thousand feet in height, but generally inferior to those formed by the subjacent grit, &c.

(*f*) *Thickness.* The greatest ascertained thickness is probably that of the Northumberland series, exceeding 180 fathoms, but this circumstance varies so greatly that no general remark can be made.

(*g*) *Inclination.* The strata are generally inclined, and frequently at a very high angle, being entirely unconformable to those more horizontal beds which we have hitherto described, and which overlie them; they frequently also exhibit contortions as rapid and singular as those which have been so often described and figured in the transition slate rocks: appearances of this kind are displayed in a manner particularly striking on the coasts of Bridesbay Pembrokeshire, near Littlehaven. It may be observed that where the associated solid masses of limestone and sandstone are elevated in high angles, but still disposed in nearly regular planes, the more tender argillaceous beds are generally twisted, and as it were crumpled together. The Mendip hills and adjacent collieries in Somersetshire afford an excellent illustration of this fact, which strongly suggests the idea of a mechanical force which has elevated the more solid rocks en masse; while the more yielding materials, giving way to its lateral pressure, have become irregularly contorted. These phœnomena cannot be attributed to any internal power like crystallisation; for they appear to be common to all rocks, even those most decidedly mechanical in their structure; we have already seen, p. 183, that they were equally to be noticed in the most recent members of the oolitic series in the Isle of Purbeck. The *faults*, or as they may be most appropriately termed, *dislocations*, of the coal-fields, are other and still more irresistible evidences of their having been affected by violent mechanical convulsions subsequently to their original formation. These consist of fissures traversing the strata, extending often for several miles, and penetrating to a depth in very few instances ascertained; they are accompanied by a subsidence of the strata on one side of their line, or (which amounts to the same thing) an elevation of them on the other; so that it appears that the same force which has rent the rocks thus asunder, has caused one side of the fractured mass to rise, or the other to sink; it being difficult, if not impossible to say (since in either case the relative motions of the disjoined masses would be the same) in which direction the absolute motion has taken place. Thus, the same strata are found at different levels on opposite sides of these faults, which appear to derive their name from their baffling for a time the pursuit of the miner; they are also called *traps*, and the elevation or subsidence of the strata described as their *trap up* or *down*, probably from a northern word signifying a step. The change of level occasioned by these dislocations sometimes exceeds 500 feet; whence we may infer the immence violence of the convulsion which had power to produce motions of such vast masses to such an extent. The fissures are usually filled by clay, which has subsequently filtered in, and often includes fragments disrupted from the contiguous strata; their direction usually approaches to vertical.

(*h*) *Agricultural character.* "The soil of the coal-districts," says Mr. Farey, "inclines much to clay, and is generally of an inferior quality." "Draining and liming seem to be essential to the proper occupancy of a farm in these districts; when laid down in pasture, small daisys and other insignificant weeds are more disposed to prevail than grass, on their strong soils." This character appears generally applicable to the argillaceous soils on the coal-measures: we have however known them to prove favourable to the cultivation of apple orchards. Sometimes the grits appear in particular portions of the coal-fields in such abundance as to constitute a sandy soil, but it is usually poor, hungry, and heathy.

(*i*) *Phænomena of Water.* The alternation of porous strata of grit, and retentive ones of clay, being generally frequent in the coal-measures, water is usually to be procured at inconsiderable depths. The clay filling the fissures of the faults also holds up the water, and throws out springs; as these faults, traverse the strata in all directions, it often happens that limited tracts are bounded on all sides by faults running into one another, so that such tracts are as it were insulated by them, and being surrounded on all sides by water-tight walls, their waters may be consequently drained without affecting those of the contiguous district; a circumstance of great advantage to the miner, who has thus on his hands the drainage of this limited tract only, instead of that of the whole adjacent country; hence it is necessary for him to use caution in piercing these faults, neglecting which he may at once overflow his own works, and drain all the neighbouring wells. The waters percolating among the coal-measures are very often chalybeate.

Salt springs have been sometimes found, but generally near the red marle, from which they may possibly have been derived, as at Measham in the Ashby de la Zouch field, and in that of South Gloucester; those however near Newcastle are too far from the red marle to be assigned to this cause: purging salts are found in the waters of Mosshouse lane and Tarleton on the north-east of Ormskirk, Lancashire, &c.

Thermal springs sometimes occur in the coal-measures; an example of this may be seen in the valley of the Taafe in South Wales, about six miles above Cardiff, under singular circumstances, for a thermal and strongly chalybeate spring there rises within the shingly bed which forms the wintry channel of the river.

Section III.

MILLSTONE-GRIT, AND SHALE.

(*a*) *Chemical and external characters.* The coal-measures generally repose on a series of beds which are usually designated by the name of Millstone-grit and Shale. The millstone-grit is most commonly seen under the form of a coarse-grained sandstone consisting of quartzose particles of various sizes (often sufficiently large to give the rock the character of a pudding-stone) agglutinated by an argillaceous cement. This sandstone differs from those which accompany the coal-measures, principally by its greater induration. It has every appearance of a rock mechanically formed from the detritus of pre-existent materials; and rounded particles of felspar may occasionally be traced in it. It sometimes, however, (though comparatively seldom) assumes a finer texture, in which the mechanical structure becomes less evident, and even passes into a hard and solid cherty rock.

The shale beds of this series present scarcely any characters which may serve to distinguish them from the slate-clay beds of the coal-measures.

Where the series is most completely developed, the mill-stone-grit beds predominate in its upper region, the shale beds in the lower: alternations, however, of each kind of rock may be observed throughout.

This series presents occasional beds both of limestone and of coal, particularly in those parts where the shale predominates. The coal-beds are few in number, usually very thin, and of a very indifferent quality. The limestone-beds resemble in many points the mountain limestone, which will be next described: they generally exhibit, however, a blacker colour, and contain more bitumen.

Considered in a general point of view, this series is intermediate in character and composition, as it is in position, between the main coal-measures which it supports and the mountain lime which it covers, forming the natural link between them.

(*b*) *Mineral contents.* Nodules of clay-iron-stone occur in beds similar to those in the coal series, in the shales of this series, sometimes assuming the forms of septaria. Iron pyrites also is abundant; and occasionally the metalliferous veins, whose principal seat is in the subjacent mountain limestone, extend upwards into this series of strata. In Derbyshire there are several instances of lead veins worked in the shale of this series, but they are usually thin. The celebrated copper mine of Ecton in Staffordshire, is represented by Farey as being situated in the limestone associated with this shale, and in Northumberland the lead veins are worked even in the mill-stone-grit; but since these veins are always most productive in the subjacent limestones, it seems most proper to enumerate their contents under that head. The variety of calcareous spar called Satin spar, appears to belong to the shales of this series. Bitumen, under most of the forms which it is capable of assuming, has been found in the rocks of this series, particularly in the shale. In driving a level through the shale in Derbyshire, springs of naphtha issued forth so plentifully as to cover the surface of the water in the level; and as inflammation took place on the approach of a candle, the spot attracted the notice of tourists under the title of a burning spring. Petroleum, elastic bitumen, and asphaltum are likewise found. It is probable that these substances are derived from the partial decomposition of the coal of the superior and accompanying strata, which may be affected perhaps by the heat arising from the decomposition of the associated pyrites.

(c) *Organic remains.* Vegetable impressions, analagous to those of the superior strata of coal, are also found in the shales in this series; and nodules of iron-stone containing muscles, probably the same species with those before described as occurring in the coal-strata. In the limestone beds of this series, again, the organic remains appear to be decidedly of marine origin, and analagous to those of the subjacent limestone. Thus, in a blue limestone of this class at Newton hall near Corbridge, Mr. Winch specifies pectines, and large ostreæ as occurring; and in the limestone covering the thin coal of Newton near Felton (both subordinate members of this series) a bed contains impressions of bivalve shells, as mentioned by the same writer, but the species is unfortunately not distinguished.

(d) *Range and extent.* See the local details in the ensuing chapter.

(e) *Elevation.* The mountains formed by these beds are often between 1 and 2000 feet in height, and sometimes between 2 and 3000, above the level of the sea; for particulars see the ensuing chapters.

(f) *Thickness.* Sometimes exceeds 120 fathoms, for particulars see the ensuing chapters.

(g) *Inclination.* The phœnomena in this respect are the same with those of the coal-measures, with which these subjacent beds are perfectly conformable. The solid strata of millstone-grit, however elevated, usually form rectilinear planes, while the more yielding shales are often singularly contorted. The same faults traverse these beds and the coal-measures.

(h) *Agricultural character.* The millstone-grit usually forms the surface of barren and elevated moorlands covered with mountain peat, the moisture of these elevated regions being favorable to the growth of the mosses, &c. which form them; and, as it appears, the soil afforded by the gritstone also concurring to facilitate their production, for they are not common in calcareous mountains of equal height. The shales associated in this part of the series are very variable even in different parts of the course of the same beds, sometimes affording good, and sometimes poor soils; perhaps the different proportions of calcareous matter dispersed through them may be the cause of this.

(i) *Phœnomena of water.* This part of the series agrees with the coal-measures in this respect. The well-known waters of Harrowgate are believed to be in this part of the series.

Section IV.

CARBONIFEROUS OR MOUNTAIN LIMESTONE.

(*a*) *Chemical and external characters.* The series just described reposes on that important assemblage of calcareous strata (occasionally alternating with beds of a different composition, e. g. shale-grit and amygdaloid) which has been often described by the name of mountain limestone, from its usually forming considerable hills; of metalliferous limestone, from its mineral riches; and of entrochal or encrinal limestone, from its organic remains.* We prefer to all these that of *carboniferous limestone*, derived from its association in the coal districts, as expressing a character more constant and more peculiar than any of the former. The texture of this rock is generally imperfectly crystalline, and sufficiently close and hard to afford marbles susceptible of a durable polish. Its prevailing colour is grey; passing, on the one hand, into greyish-white and yellow, and on the other, into greyish-blue and black; occasionally also a red shade of colour may be observed. Its purest beds appear to contain about 96 per cent. of calcareous matter; but by the admixture of other ingredients, it often passes into magnesian limestone, ferruginous limestone, bituminous limestone, and fetid limestone. It usually presents beds of very considerable thickness; a continuous series of which often extends many hundred feet in

* It has more than once been proposed to designate this by the Wernerian name " first flœtz limestone." We have already assigned our reasons for objecting generally to all attempts to refer any part of the carboniferous series to the so-called flœtz formations, and against the application of the principle in this instance we must enter our most decided protest, for no canons of nomenclature can be more positive than, 1st, that names should be applied in the sense in which they were imposed by their original authors, unless that sense has been modified by common subsequent usage; and 2dly, that where a name includes a description, it should not be applied in cases where the implied description is inapplicable. Now with regard to the first of these rules, Werner constantly applied his own term, ' first flœtz limestone,' not to our carboniferous limestone, but to that associated with the bituminous and cupriferous marle-slate, which is now generally admitted to be coeval with our newer magnesian limestone. The continental writers who employ this term, still generally retain this application; and when they have occasion to describe the carboniferous limestone, appear to refer it to the transition series. With reference to the second note, we must observe that the carboniferous limestone seldom occurs in horizontal or flœtz stratification, being more often highly inclined. In different countries, however, the same rocks vary so much in this particular, that it is absurd to assume it as a basis of nomenclature, otherwise we must bring ourselves to talk of the " highly inclined flœtz limestone of the Jura chain, &c."

depth without the intervention of any other rock, the strata being divided only by thin partings of clay; but sometimes this series exhibits alternations of various heterogeneous rocks, particularly toadstone, grit, and shale; so that the proportion of the limestone beds in constituting the series of strata here denominated from them, as its characteristic feature, varies; they being sometimes exclusively prevalent, and sometimes forming little more than one-third of the whole series: occasionally also the limestone beds themselves become more thinly laminated.

These limestone beds contain nodules of chert disposed in layers in a manner precisely analogous to those of flint in the chalk strata.

It is a prevailing character of this limestone to be full of caverns and fissures. All the caverns in this island (with some very trifling exceptions presented by the transition limestone of Devonshire, and the new magnesian limestone in Somersetshire,) occur in this rock. Rivers which flow across it are often suddenly engulphed, and pursue to considerable distances a subterranean course; and the hills constituted by it usually exhibit rocky dales and mural precipices. Hence it presents much of the most picturesque and romantic scenery of which England can boast.*

* The following information respecting the caverns and subterranean rivers of the mountain lime has been extracted from the Notes obligingly lent to the Editors by Mr. Greenough.

Caverns are extremely common in the limestone district. The following are amongst the most remarkable:—

In Westmoreland, there is a great cavern in the main limestone, at Dunall, five miles from Dufton.

In Durham, Hetherburn cave, near Stanhope, in the 70 fathom limestone, runs above a mile underground.

In Yorkshire. Gigglewick scar; Kingsdale; Wethercat cave near Ingleton; Tiernham's mine, and Old Cam Rake, Coniston moor; Barefoot-wive's hole; Hardrawkin. In West's guide to the Lakes, caverns are mentioned by the names of Hurtlepot, Sandpot, Donk cave on the base of Ingleborough, Gate Kirk cave on the south-east of Whernside, Greenside cave, Cathnot hole, Hardraw scar near Hawes in Wensleydale, Alan or Alumn pot near the village of Sebside, Long Churn, Dickenpot, Halpit hole and Huntpit hole. In the Gentleman's Magazine for 1761 are mentioned others by the names of Blackside cove, Sir William's cove, Atkinson's chamber, Johnson's Jacket hole, and Gaper Gill.

In Lancashire. Dunald mill-hole near Kellet, eight miles from Lancaster, near the road to Kirby Lonsdale. Yardhouse cave and Gingling cave in Kingsdale; and smaller caverns in Yealand.

In Derbyshire. Bagshaw's cavern, south-west of Bradwell. Bamford-hole near Eyam. Bondog hole near Wirksworth. Charleswark cavern near Eyam. Chelmerton cavern. Cresslow mine cavern. Cumberland or Rutland cavern near Matlock. Devil's hall in Foreside mine, Castleton. Dove-hole cavern, Dove-dale. Drake mine cavern. Elden hole (see Cat-

(*b*) *Mineral contents.* This series forms the principal depositary of the British Lead mines; those of Northumberland, Durham, York, Derbyshire, and Somersetshire, are all situated in it; † it also affords ores of some other metals; all

cott. Phil. Trans. & Rees). Golconda near Hopten. Knowles mine cavern. Merlin's cave. Orchard mine cavern. Peaks hole, Castleton (Catcott, p. 234.) Placket mine cavern. Pool's hole, Buxton (Catcott, p. 236.) Ranter mine. Reynard's hall and cave in Dove-dale. Speedwell or Navigation mine cavern, Castleton.

In Staffordshire. Thor's house, or Thyrsis cave in Wotton dale. Ribden and Ruden caves near Calden here receive the waters of the Manifold, &c. Ludchurch, between Swithamley and Wharnford, is 208 yards long and 40 or 50 deep, (Plott.) Hobchurch cave near Welton mill. And Plott mentions caves at Wharnford, north-east of Leek; a large one at Yelpersley Tor; one at Kinsare; under Kinfare edge; at Holloway near Stourbridge; and under Peakstones in Olverton parish under Long Hursh hill.

In Gloucestershire. In Catcott on the Deluge mention is made of caverns in St. Vincent's Rock near Clifton. Penpark hole on Durdham Down is by some suspected to be the workings of an old mine: and there are huge caverns near Colford, which serve to ventilate the mines.

In Somersetshire. Lockston cavern and Banwell cavern, 12 miles north-west of Wokey, (see Ph. Trans. by Lowthorp, vol. 2. p. 368.) There are also caverns in Charterhouse liberty, Mendip, and Green-ore farm, 4 miles from Wokey; and a cave called the Lamb cavern near East Harptree.

Subterranean rivers are frequent all over the limestone district. Such are the Manifold, which passes beneath the limestone hills about three miles south-west of Ecton mine in Staffordshire, and after traversing a cavern through the base of the limestone hills four miles in length, re-appears near Ilam. The Hamps river also breaks out near the same place (Farey). The Ribble is also an underground river; as also is the Greta (Goldsmith, Nat. Hist.) Such also are Horton Beck, and Bransil Beck, at the foot of Pennigant in Yorkshire. The stream which forms the cataract in Speedwell mine near Castleton, is, according to the inhabitants of that place, proved to be the same as the stream which traverses the Peak cavern at that place.

The streams forming these subterranean rivers are absorbed by funnel-shaped hollows on the surface of the limestone, called provincially *Swallows*, and *Swallet holes*.

† The following notice, communicated by Mr. John Taylor, of the produce of lead in England, may be here appropriately introduced. Where two asterisks (**) are prefixed, the mines belong entirely to the carboniferous series: where a single asterisk (*) they are partly in these and partly in other formations: where no asterisk, entirely in other rocks.

In round numbers the annual produce is as follows,—

ENGLAND.—**Durham, Cumberland & Yorkshire	Tons	19,000	
**Derbyshire, probably	„	1,000	
*Shropshire	„	800	
Devon and Cornwall (transition and primitive rocks)	„	800	
			21,600
*WALES. .... (Flintshire and Denbighshire)			7,500
*SCOTLAND			2,800
			Tons 31,900

the following minerals have been found in it more or less abundantly; sulphuret, carbonate, and phosphate of lead; antimoniated lead ore; sulphuret, and carbonate of copper; sulphuret, carbonate and oxide of zinc; iron pyrites, hæmatitic iron ore, and also exide of iron, disseminated in beds of ferruginous limestone associated with pearl spar and brown spar. Many varieties of common calcareous spar, arragonite, many varieties of fluor, selenite, carbonate and sulphate of barytes, sulphate of strontian, quartz crystals (Derbyshire and Bristol diamonds). Bitumen is also found filling cavities in this lime and associated with the vein-stones, particularly the elastic variety, though not so commonly as in the shale of the preceding series. If we include the minerals contained in the trap rocks, sometimes alternating in this series, we must add zeolite, onyx, and perhaps alabaster to this list.

(*c*) *Organic remains.** The animal remains preserved in the carboniferous limestone are strongly distinguished from those of the oolitic series and lias, belonging in a majority of instances to entirely distinct genera. These genera are very

The ores in Durham, Cumberland, and Yorkshire, are generally smelted by blast furnaces on the mines, as at Alston moor, Wear dale, Arkendale, Greenhoe hill, Grassington, &c.: in general these ores contain very little silver, and refining them is not much practised. Some lead is however refined for silver by the London Lead Company, and perhaps by some others, but I do not know to what extent.

In Derbyshire the lead ores are smelted by persons who own smelting-houses with reverberatory furnaces, and who buy the ores of the miners. There are not more than three or four of such now at work, and I do not know of any silver refinery.

Shropshire I know nothing of.

Devon. Huel Betsey lead mines ores, smelted on the spot, and silver extracted; produce 12 ounces per ton. Beeralston ores, smelted at Bristol, and silver extracted; 80 to 100 ounces per ton.

Cornwall. Sir C. Hawkins mine near St. Michael's, ores smelted and refined on the spot, produce of silver said to be 30 to 40 ounces per ton.—Huel St. Vincent silver mine in Calstock, it is said has produced 1000 lbs of silver in a very few months. Ore, Horn silver (Qu.) and native, refined on the spot.

Wales.—There are several lead smelting houses in Flint and Denbighshire (reverberatory furnaces). Lord Grovesnor's, M. Roskill's, and others near Holywell. Mr. Jones near Mold, Messrs. Hunt & Co. at Minera. Some smelt their own ores, and others purchase of the miner. Very little silver extracted.

Scotland—I do not know much about, but believe the ores are smelted at Lead Hills.

* We have to acknowledge the kind assistance of Mr. Miller, who has furnished several important additions and corrections to this list. The references, when not otherwise specified, are as usual to Sowerby's Min. Conchology. Martin's Derbyshire Petrifactions has been cited for some species not yet figured in the former work.

generally common to this and the transition limestone, and many species are also common. In this, as in every other respect, its alliance is much more close with the older, than with the more recent deposits; and were it not preferable to constitute a distinct class for the reception of the carboniferous series, it ought undoubtedly rather to be referred to the transition suite than any more recent order.

Vertebral remains are very rare. Of *Testacea* though the species are many, the genera are comparatively few; while the *Zoophytal* families, particularly *Encrinites* and *Corallites* are in the greatest profusion.

To proceed to particulars.

Vertebræ of fish, *Sharks teeth*, and many singular *palatal tritores*, and the radius of a *Balistes*, exhibit proofs of the existence of vertebral animals in this formation.†

Of the *Crustaceous tribe*, *Trilobites* are found in this formation, as in the transition limestone; but the species are apparently distinct: one of these, Oniscites Derbiensis, is figured in Martin's Derbyshire Petrifactions, P. 45.* 1 & 2. P. 45. 1 & 2.; and another species, as yet undescribed, occurs in the carboniferous limestone near Bristol, whose transverse folds are tuberculated.

CHAMBERED UNIVALVES.

Ammonites sphœricus *T.* 53, fig. 2.
A. striatus. *T.* 53, fig. 1.
A. Luidii. Martin's Pet. Derb. *T.* 36, fig. 1.
Nautilus discus. *T.* 13.
N. pentagonus. *T.* 249, fig. 1.
N. bilobatus. *T.* 261.
N. complanatus. *T.* 60, fig. 5.
Orthocera Breynii. *T.* 60, fig. 5.
O. undulata. *T.* 59.
O. gigantea. *T.* 246.
O. cordiformis. *T.* 247.

Some of the Orthoceratites nearly resemble the alveoli of common belemnites; but the occurrence of true belemnites in this rock appears doubtful.

Conularia (a conical shell divided by imperfect septa) *C.* quadrisulcata. *T.* 260, fig. 3 to 6.
C. Teres. *T.* 260, fig. 1. 2.

* Mr. Whitehurst (Theory of the Earth) mentions the discovery of a crocodile in this formation in Derbyshire; having made enquiries in that county respecting the specimen alluded to, we have been informed that a mutilated orthoceratite had been mistaken for the scales of this animal.

UNIVALVES NOT CHAMBERED.

Euomphalus catillus. *T.* 45, fig. 3. 4.
E. nodosus. *T.* 46.
E. pentangulatus. *T.* 45.
Cirrus acutus. *T.* 141, fig. 1.
Nerita; of this genus four species occur, as yet undescribed.
Helix carinatus. *T.* 10, *n.*
cirriformis. *T.* 171, fig. 2.
striatus. *T.* 171, fig. 1.
Nautilites? hiulcus. Mart. P. 40, fig. 1 & 2. It has no chambers, and must according to Mr. Miller form the type of a new genus.
Melania constricta. *T.* 218, fig. 2.
Turbo.
Planorbis? equalis. *T.* 140, fig. 1. Mr. Miller considers this shell to be rather a depressed variety of *Cirrus* acutus.

BIVALVES EQUIVALVED.

Modiola.
Mya.
Cardium elongatum. *T.* 82, fig. 2.

BIVALVES UNEQUIVALVED.

Terebratula crumena. *T.* 83, fig. 2. 3.
T. lateralis. *T.* 83, fig. 1.
T. biplicata. *T.* 90, fig. 1.
T. Wilsoni. *T.* 118, fig. 3.
T. Mantiæ. *T.* 277, fig. 1.
Spirifer cuspidatus. *T.* 120.
S. trigonalis. *T.* 265.
S. oblatus. *T.* 268.
S. glaber. *T.* 269, fig. 1. 2.
S. obtusus. *T.* 269, fig. 2. 4.
S. striatus. *T.* 270.
S. pinguis. *T.* 271.
S. triangularis. Mart. *T.* 36, fig. 2.
Producta aculeata. *T.* 68, fig. 4.
P. scabricula. *T.* 69, fig. 1.
P. Flemingii. *T.* 68, fig. 2.
P. longispina. *T.* 68, fig. 1.
P. Scotica. *T.* 68, fig. 3.
P. spinosa. *T.* 69, fig. 2.
P. spinulosa. *T.* 68, fig. 3.
P. striata. Martin's Derb. fossils. *T.* 22, f. 1.
P. gigantea. Mart. Pt. 15.
P. crassa. Mart. Pt. 4. 16.

Remains of the Echini are occasionally found, but rarely: among these the plates and spines of a peculiar species of Cidaris may be distinguished. Mr. Miller has observed a new genus nearly allied to the Echinus, but differing from it in the greater number of plates in each area: to this he has given the name of *Hystricites.* The fossil from Kentucky figured in Parkinson's Organic Remains, vol. 2. plate 13. fig. 36. among the Encrinites also appears to be allied to the Echini, and occurs in this formation.

It appears from Mr. Miller's work on Encrinites that the following genera and species of that interesting family occur in this formation.

Poteriocrinites crassus
tenuis
Platycrinites.. lævis
rugosus
tuberculatus
granulatus
striatus
pentangularis
Cyathocrinites lævis
tuberculatus
quinquangularis
Actinocrinites.. Triacontadactylus
Polydactylus
Rhedocrinites.. verus
quinquangularis

besides fragments of some other varieties not sufficiently complete to be ascertained.

All these species are distinguished from those occurring in the lias and more recent beds, by the thinness of the ossicula forming the cup containing the viscera, &c. In the more recent varieties (if the term may be allowed) these pieces form thick wedge-like joints, adhering by broad articulating surfaces; but in those above enumerated they consist of thin plates adhering by sutures only: of course, in proportion as the shell surrounding the abdominal cavity is thus reduced in thickness, the interior cavity becomes more enlarged.

Many of the same genera, and some of the same species, are found in transition limestone; so that in this class, as in the Testacea, a greater alliance exists (as far as regards the organic remains) between the carboniferous limestone and the transition formations, than with those of more recent origin.

Figures exhibiting the whole anatomical detail of all these species will be found in Mr. Miller's work above referred to, which from its accurate, precise, and scientific views, yields only to the researches of M. Cuvier on the fossil remains of Vertebral animals.

From the abundance and variety of these remains, the name *Encrinal limestone* has often been given to this formation.

Coralloid remains are equally plentiful; but here we have to regret the want of any work which can afford any thing like a complete enumeration of the fossil species, and our list must be therefore very imperfect.

Caryophyllæa. Knorr. vol. 2. G. 1. fig. 1.
Park. Or. Rem. vol. 2. *T.* 6. fig. 8.
Turbinolia. Park. Or. Rem. vol. 2. *T.* 4. fig. 1. 2. 3.
Astrea basaltiformis. Lithostrotion. Luid. *T.* 23.
and 3 undescribed species.
Favosites ramosus. Luid. *T.* 3. fig. 95.
Tubiporus? Park. Or. Rem. vol. 2. *T.* 1. fig. 1.
T. 2. fig. 1.
T. 3. fig. 1.
Retepora. Martin's Derbyshire, *T.* 43. fig. 1 & 2.
Milleporites flustriformis.
frustulata? Lamk. flabelliformis. Miller's Manusc.

An elegant flexible Coral, belonging to Lamouroux 5th Division, occurs, though rarely, in the mountain limestone near Bristol, which serves as the type of a new genus named *Fenestrella* by Mr. Miller.

Vegetable remains corresponding to those of the coal-measures are occasionally, though rarely, found in the upper beds of this formation.

When the organic remains of this formation occur in the cherty beds associated in it, the testaceous matter of the original has almost always disappeared, bearing only a siliceous cast.

(*d*) *Range and extent.* See the local details in the ensuing chapters.

(*e*) *Height.* We also refer to the ensuing chapters for the particulars relating to this head, observing generally that this limestone commonly constitutes hilly and even mountainous tracts, many eminences formed by it exceeding 1000 feet above the sea level; hence it has been often designated as the *Mountain Limestone.* It is generally, however, overtopped by the ranges of the superjacent millstone-grit.

(*f*) *Thickness.* The thickness of this series is very considerable, certainly exceeding in some instances 900 feet; but, like every other formation, it is very variable in this respect.

(*g*) *Inclination.* The strata of this formation, like those of the superjacent grit and coal-measures, are often highly inclined; this is especially the case in the south-western counties, where they sometimes become perfectly vertical. They sometimes present inarched or saddle-shaped curvatures, the point of the arch being unbroken, which proves that the masses must have been to a certain degree soft and pliant when they assumed this form. Contortious are, as we have before observed, less usual in these more solid strata, than in the tender and yielding beds of the coal-measures; yet such sometimes occur, especially where the limestone beds are thin and alternate with shale. Instances may be seen on the southern coast of Pembrokeshire.

Extensive faults traverse this formation.*

When the beds of this limestone run thick, and are not separated from each other by argillaceous partings, it is often extremely difficult to ascertain its stratification, a rock of it under these circumstances often appearing at first sight as a vast unstratified mass: a careful examination will however generally detect the direction of the planes of stratification in some part of it; and when this clue is once obtained, and the eye is kept constantly directed to the face of the rock, in lines parallel to those planes, the stratified structure of the whole seldom fails to develop itself.

(*h*) *Agricultural character*. The elevated and exposed situation so generally occupied by this formation, has occasioned the greater part of the districts in which it prevails to remain in an uninclosed and unimproved state; and the surface, being generally covered with large rocky fragments, requires much labour to clear it. Mr. Farey observes that the rubble and debris on the limestone tract produce good pasture land when limed, without which great part of these lands would be covered with unproductive heath.

(*i*) *Phænomena of springs*. Where the beds of this limestone alternate with argillaceous strata, the latter throw out the waters in the usual manner; but where, as is more usually the case, the series consists exclusively of calcareous beds extending to a great depth, water is of very uncertain and irregu-

* Some of these derangements appear to be of a magnitude which almost startles the imagination. At the foot of Ingleborough is a subsided mass of coal-measures at the base of the whole limestone series; a fault therefore seems to have taken place equal to the whole thickness of this series. Mr. Farey describes the western edge of the Derbyshire limestone tract as abutting against a similar fault which must exceed 900 feet. In Somersetshire, near Clevedon, a large tract of limestone covered with coal-measures appears to have subsided to such an extent, that at one end the strata of this tract crop out three miles to the west of the main chain to which they must have originally appertained. Undulations are probably combined with, and extend the effects of, the faults in some of these instances.

lar occurrence. The rains which fall on it (and frequently also the springs which flow from other strata over it), sink through its numerous fissures, and form subterranean streams flowing through the many caverns with which, as we have already noticed, it abounds. A bed of shale, which separates the limestone from the old red sandstone, throws out its springs abundantly. The following particulars are extracted from Mr. Greenough's Notes.

Springs rarely appear on the sides or summits of limestone hills, but break out in great numbers, and often with extraordinary impetuosity round their bases. Of this nature is the celebrated spring of St. Winifred at Holywell in Flintshire, situate in all probability at the point where the limestone first comes in contact with the coal-measures. The quantity of water thrown up by it is 84 hogsheads per minute, and though this stream has little more than a mile to run before it arrives at the sea, yet eleven mills are put in motion by it, of which three are placed abreast. There is another very powerful spring in the same neighbourhood dedicated to St. Osward; and still another called St. Mary's Well rises at the temperature of 51°. The recollection of every one who has passed along the confines of the limestone district will supply him with many analagous instances. About Denton in Yorkshire, the roaring of the waters is incessant.

Some of these springs vary very little in quantity, either in droughts or after the heaviest rains; others again are intermitting. The most celebrated of these occurs at the foot of Giggleswick Scar on the road from Settle to Kirby Lonsdale. The ebbing and flowing well in Derbyshire is supposed by Mr. Farey to be artificial.

The hot springs at Buxton, Matlock, and Clifton are upon these beds.

There is no water more apparently pure and pellucid than that which is furnished by this limestone, which however holds a considerable quantity of calcareous matter in solution. The iron manufacturers always use it in preference for grinding their tools, conceiving that it is less liable to produce rust than any other. If obliged to use other water, they put a piece of lime into it before they venture to dip their steel. Where the current is slow, the calcareous matter subsides; hence the stalactites so common in limestone caverns; hence the property which several streams possess of incrusting substances over which they flow; such as the streams at Smedley's mine near Matlock (Phil. Trans. vol. 8. p. 406.), at Alport near Youlgrave, Little Longsden, Crossbrook dale, Slatey and Stoney Middleton in Derbyshire.

Section V.

OLD RED SANDSTONE.

The Carboniferous limestone is sometimes separated from this rock by a thick shale much resembling that associated with the millstone-grit; this may be distinguished as the *lower limestone shale.*

(*a*) *Chemical and external characters.* The old red sandstone is a coarse-grained, micaceous sandstone, evidently of mechanical origin, constituted apparently of abraded quartz, felspar and mica, and containing fragments of quartz, clay-slate, flinty-slate, &c.; sometimes passing into the state of a quartzose conglomerate, sometimes possessing a structure coarsely schistose (and thus affording slates for paving), and sometimes, particularly towards its lower regions, becoming finely schistose, and passing into a fine-grained micaceous sandstone slate. It alternates with argillaceous beds, sometimes soft but more usually indurated and often slaty; the colour is usually dirty iron-red or dark brown, but sometimes passing into grey. It approaches in its lowest beds very nearly to the characters of the greywacké upon which it reposes, and indeed graduates insensibly into that rock; so that the line of separation between them is frequently only an imaginary and arbitrary demarcation. This rock contains in several places calcareous concretions, which produce a rock of a pseudo-brecciated appearance, known by the name of *Corn-stone*; and has also some unimportant beds of limestone subordinate to it.

The superior consolidation of many of the beds of this rock will generally serve to distinguish it without much difficulty from the newer red sandstone, when a tract of any extent is examined; for although doubt may often remain from the examination of a single quarry, more extensive observation of the general features of a district will seldom leave any. Rock formations usually bear external marks of their relative antiquity, which the eye of the experienced geologist readily perceives. It is more difficult to distinguish this rock from the sandstones of the millstone-grit series, and those alternating with mountain limestone; and in fact it can only be considered as a lower link in the great chain of beds to which those belong; its prevalent and characteristic colour forms its best distinction.

(*b*) *Mineral contents.* No important minerals yet appear to have been procured from this series; pyrites, calcareous spar, common and fibrous, and sulphate of strontian sometimes occur.

(*c*) *Organic remains.* It is generally destitute of organic remains; but towards its lower regions, where it approaches the limestone of the transition series, some beds of micaceous sandstone-slate occur, containing anomiæ and encrinites similar to those in the transition limestones, which will be described hereafter. Vegetables similar to those of the coal are said in some instances to occur.

(*d*) *Range and extent.* The particulars falling under this head will be given, together with the local details, in the ensuing chapters. It may, however, be here stated generally, that this rock is most abundant in the vicinity of the south-western coal-fields, especially that of South Wales, adjoining to which it forms an immense tract in Brecon, Monmouth, and Herefordshires, occupying the whole of the latter county; in the other coal districts it is only of partial and limited occurence.

(*e*) *Height.* The old red sandstone frequently forms mountains between two and three thousand feet above the sea level; in this respect it yields only to the transition and primitive chains of this island, surpassing those of every other formation.

(*f*) *Thickness.* In the borders of the forest of Dean, this formation, there interposed between the carboniferous and transition limestone, exceeds 2000 feet. In Herefordshire and Brecon its thickness must be considerably greater than even this, while in some parts of Gloucestershire, near Tortworth, it cannot exceed two or three hundred feet.

(*g*) *Inclination.* The remarks already made with regard to the mountain limestone may be considered as equally applicable to this also, both being conformable inter se.

(*h*) *Agricultural character.* Where argillaceous beds alternate with this rock, it often affords a very fertile soil, as the rich fields and luxuriant orchards of Herefordshire abundantly testify; but where the sandstone exclusively prevails, sterile heaths are the result: the summits of mountains of this formation are usually covered with mosses.

(*i*) *Phænomena of water.* Springs, usually descend from the morasses on the hills of this formation, and the argillaceous beds alternating in the series, are in general sufficiently frequent to occasion such a distribution of the rain waters percolating through the sandstones, as brings them within the reach of wells of no great depth.

Such is the general outline of the four series of rocks which are usually associated in the districts affording coal; and it will be seen in the following survey of our coal-fields, that with very few exceptions, each of these series is of constant occurrence in all such districts, (that is, limiting the observation to England); and that wherever they occur, they uniformly succeed each other in the order above described; but although a general identity of structure is thus apparent in all our coal-fields, yet in the detail there exists much local variety. This, indeed, is the general complexion of all geological analogies: beds which occupy the same geological position, usually present a near resemblance, if considered on the great scale; subject to much difference, if examined more minutely. The aggregate series of strata is marked by permanent features, but the individual strata composing that aggregate undergo frequent changes; for instance, the first series of rocks above described (that of the coal-measures) viewed as an whole, is an uniform assemblage of alternating strata of coal, shale, grit, and ironstone. But any attempt to trace any individual stratum of the above substances to a considerable distance, even in the same coal-field, is usually vain; and much less can the same individual strata be recognised in different coal-fields. The strata of the mountain limestone series are far more uniform than those of coal; and yet, as we shall presently see, these appear at one extremity of the same range of mountains, under the form of a thick series of limestone beds, divided only by three alternations of toad-stone; while at the opposite extremity of the chain they are divided into numerous beds of much less thickness, separated by alternations of grit and shale. This kind of general resemblance and partial difference is indeed exactly what we should be led to expect, whatever hypothesis of the formation of these strata we may adopt; for it seems impossible that any general causes which could be supposed to act in the formation of strata, should not, while prevailing over extensive tracts of country, have had their effects modified by many local circumstances.

CHAPTER II.

Coal district north of Trent, or Grand Penine Chain.

Introductory view of the general features of this district.

Having in the preceding chapter sketched the general structure and arrangement common to all our coal-fields, we may now proceed to consider the disposition of the several component series of rocks in the various particular districts; following the division before suggested, which will lead us to begin with the great northern coal district, including all the coal-fields north of the rivers Trent and Mersey, as far as the Scotch border.

This district is traversed longitudinally by a grand central chain of mountains, extending north and south from the borders of Scotland to the centre of Derbyshire; the different portions of this chain are at present known under various local names, but the Roman Colonists of Britain, whose attention so prominent a feature in the physical geography of the island could not escape, denominated them the PENINE ALPS, as appears from the following passage in Richard of Cirencester's description of the Roman state of Britain; he is treating of the Roman province MAXIMA, which includes all the northern counties of present England. "Totam in æquales fere partes provinciam dividunt montes Alpes Penini dicti. Hi ad fluviam Trivonam (*The* TRENT) surgentes, continuâ serie per 150 milliaria septentrionem versus decurrunt." This appellation was probably derived from the British term PEN, head or summit, which enters so commonly into the composition of the names of the hills amidst which that language is still spoken, and which may be traced in some names even in this chain, e.g. Penygent, Pendle hill, &c. though usually superseded by more recent Saxon terms, as Fell, &c. This appellation was also familiar to the Romans, as the crest of the Alps near Mount St. Bernard was anciently designated Alpes Peninæ; however this may be, it will be useful to distinguish this ridge of mountains by some collective appellation; and as they have clearly a title to this, as their earliest known, if not their original designation, we shall therefore henceforth call them the PENINE CHAIN. The particular names of the principal summits of this chain will be enumerated in the list of heights; its general extent and range has been already indicated. It may be traced branching off from the transition chain of mountains which traverses the south of Scotland, but does not attain any considerable elevation until it reaches Gelts-dale Forest in Northumberland, near which is Cross Fell, the highest summit of the chain, rising 2901 feet above the sea; it proceeds thence to the south by Stainmoor Forest, to form the western moorlands of Yorkshire, where it exhibits the summits of Whernside, Ingleborough, and Penygent; between Stainmore and Ingleborough it abuts against, and joins the group of transition mountains comprising the English Lake district (which we shall call the Cumbrian group;) these, in fact, form a sort of excrescence swelling out on the western side from the Penine chain, and surpassing it in height; but as they belong to older formations, and deviate from the general line of the Penine chain, they must be treated of in a subsequent article. After throwing off this Cumbrian branch, the Penine chain proceeds still to the south, following the boundary between the counties of Lancaster and York. Pendle hill and the range of Blackstone Edge are in this tract. Further south it enters Derbyshire and Staffordshire, forming the "High Peak" of the one (a Saxon translation of the British name Pen) and the moorlands of the other; and here the chain finally expires on the banks of the Trent.

The whole of the Penine chain is composed of the four series of rocks above described as associated in our coal districts. In the central ranges, the second series, or millstone-grit and shale, predominates, composing all the most elevated summits. The third series, that of the carboniferous limestone, also occurs extensively towards the northern parts of the chain; occupying the middle region of the hills, especially on their western escarpment. A zone of this rock is here detached from the Penine chain, and completely encircles the transition mountains of the Cumbrian group, wrapping round them in what the Wernerians call mantle-shaped stratification. Hence, through the middle regions of the chain, the carboniferous limestone is entirely concealed by the super-strata of millstone-grit; it re-appears however at the southern extremity, where it forms an extensive and elevated district in Derbyshire. The old red sandstone has been only ascertained to exist in the northern parts of this chain, where it forms the fundamental rock beneath the western escarpment of the Cross Fell range, as is also the case near Ingleborough.

The three lower formations then of the carboniferous suite occupy the central mass of the chain; and may all be studied together in the single mountain of Cross Fell. The upper series, or great coal-formation, constitutes the exterior and lower ranges surrounding the chain on the east, south, and west: in the last direction, however, they make a considerable deflection, receding from the Penine chain and again approaching it. This is occasioned by the intervention of the Cumbrian group, round which they bend like the subjacent carboniferous limestone.

In the middle regions of this chain, the strata form a regular saddle, dipping east and west; but this regularity is disturbed, by what must be considered as faults of very astonishing magnitude, at the two extremities; for on the north, the strata continue to crop out to the very edge of the chain towards the west; the transition rocks, on which the whole rests, appearing at the foot of the escarpment, beyond which (instead of a regular repetition of the same series dipping in a contrary direction as from the axis of the chain) shattered traces of the coal-formation only are found, immediately succeeded and overlaid by horizontal deposits of new red sandstone: this part of the chain will be the object of more particular description hereafter. Towards the south, according to Farey, a derangement somewhat similar, though less enormous, has affected the strata of Derbyshire and Staffordshire, where the limestone tract exhibits on the west, by successive croppings out, its lowest strata; against which the highest strata of the millstone-grit directly abut. The structure, however, of this part of the chain, requires further examination. The general inclination of the strata composing this chain, is not very considerable, excepting where affected by local derangements, and does not usually exceed three or four degrees; presenting in this respect an exception to the highly elevated position more usually characterising the beds of this formation.

The features of this chain are often very wild and picturesque: it exhibits all the scenery and accompaniments of a considerable mountain range; precipices, torrents, and cataracts. The caverns, cliffs, and rocky dales of Ingleborough and the Peak, are too well known to need description. Two facts observed in the moorlands of Staffordshire will serve to illustrate the depth of the ravines and abrupt escarpment of the mountains in that part of the chain. The Sun, when nearest the tropic of Capricorn, never rises to the inhabitants of Narrowdale for nearly a quarter of a year; and during the season when it is visible, never rises till one o'clock p. m.: on the other hand, at Leek, the Sun is, at a certain time of the year, seen to set twice in the same evening, in consequence of the intervention of a precipitous mountain at a considerable distance from the town, for after it sets behind the top of the mountain, it breaks out again on the northern side, which is steep, before it reaches the horizon in its fall: so that, within a very few miles, the inhabitants have the rising Sun, when he has, in fact, passed his meridian, and the setting Sun twice on the same evening. There is a beautiful description of one of the rocky dales which characterise this chain in Rokeby, Canto II. stanzas 7 & 8.

Having thus given a general and collective view of the structure of the Penine chain, we shall next proceed to consider more in the detail the disposition and phœnomena of the several rock series composing it; beginning with the exterior ridges of coal, and proceeding regularly to the old red sandstone.

Section I.

Particular view of the Coal District north of Trent, or Grand Penine Chain.

We now proceed to consider more in detail the disposition of each of the four great assemblages of strata entering into the composition of the coal districts, throughout the tracts connected with the Penine chain of mountains already generally described, and thus forming the first great division of our coal districts; that, namely, north of the Trent.

In pursuance of our plan, we must begin with the first of these series of strata, the great coal-formation. This, as has already been stated, encircles the whole Penine chain on the east, south, and north; not, however, in one continuous line, but in a series of detached coal-fields, often of very considerable extent. These it will now be our business to trace around the chain, beginning on the north-east, proceeding thence down its eastern side to the south, and thence re-ascending by the western side again to the north. The coal-fields we shall have thus to notice will occur in the following order. On the eastern side, (*a*) *The great coal-field of Northumberland and Durham.* (*b*) Some small detached coal-fields in the *north of Yorkshire.* (*c*) *The great coal-field of South Yorkshire, Nottingham, and Derby.* On the south we find only, (*d*) some successful trials for coal in the *neighbourhood of Ashborne*, which have, however, never been further prosecuted. On the west, (*e*) *the coal-field of North Stafford.* (*f*) *The great Manchester, or South Lancashire coal-field.* (*g*) *The North Lancashire coal-field.* (*h*) *The Whitehaven coal-field.* (*i*) The indications of coal at the foot of the *western escarpment of Cross Fell.* Of these we now proceed to treat in order.

(*a*) THE GREAT COAL-FIELD OF NORTHUMBERLAND AND DURHAM.

The coal-field of the north-eastern extremity of England extends over a great part of the counties of *Northumberland and Durham.* Of this district we have two accounts; one by N. J. Winch, Esq. published in the 4th volume of the Transactions of the Geological Society; the other by Dr. Thomson, in the Annals of Philosophy for November and December, 1814. We proceed to select from both, recommending them, and particularly the former, to the perusal of the reader afterwards. This part of England is highly interesting both in a geological and economical point of view. Geological, since the facts it discloses are extremely curious; and economical since it supplies London, and the east and south coasts of England with coal, besides what is consumed by the inhabitants themselves.

The *coal-measures* of this field commence near the river Coquet on the north, and extend nearly to the Tees on the south: the length of this tract is about 58 miles, and its greatest breadth about 24. The strata of the millstone-grit and shale series pass under those of the coal-measures; which latter pass beneath the magnesian limestone; the northernmost point of which is near the mouth of the river Tyne.

The strata of this, as of many other coal-fields, appear to dip from the surface, and rise again to it after attaining a certain depth; so that a section of them gives the idea of a form of a boat. A place called Jarrow, which is about five miles from the mouth of the Tyne, and on its southern bank, is the spot beneath which the beds of *coal in the coal-measures*, are found at their greatest depth. One of the thickest beds, called the High Main, is 960 feet below the grass at Jarrow, and rises on all sides; but as the dip of the strata (which averages one inch in 20) is not uniform in every part in the surrounding district, that bed does not rise to the surface at equal distances around that place.

We may assume that Jarrow is the centre* of the coal-measures, since it is the spot beneath which the *High Main* coal is found at the greatest depth beneath the surface. There is a considerable though not a perfect uniformity in the distance of the several coal-beds from each other. Hence, as the High Main coal rises to the surface of the alluvial soil, around Jarrow,

* In forming this conclusion, it is necessarily inferred that the coal-measures once extended far to the east of the present coast of Durham, and *above* the level of the sea: that they exist *beneath* it, is proved by the workings under it.

we may conclude that the beds of coal above and below the High Main, arise also, at a distance from it, proportionate to their depth beneath it.

The inequality of the surface does not affect the dip or inclination of the strata constituting the *coal-measures*; so that when they are interrupted or cut off by the intervention of a valley, they will be found on the sides of the opposite hills at the same levels, as if the beds had once been continuous. The conclusion is obvious, that the present irregularities of hill and dale have been occasioned by the partial destruction or dispersion of the uppermost strata constituting the coal-formation.

The beds of coal and of the other strata composing the coal-measures, are not every where of uniform thickness. They occasionally enlarge or contract so greatly, that it is only by an extensive comparison of the whole series, that any certainty is arrived at of that general uniformity of stratification which is known actually to exist. From the best information, they are calculated at 1620 feet in thickness.

The beds of coal, &c. basset out at the surface one after the other; each on the *east* of that which immediately precedes it in point of age. These beds are sometimes visible; but are more commonly covered by alluvial soil.*

The whole surface of the coal-measures is calculated at 180 square miles. The greatest number of the numerous mines are situated on both sides the river Tyne, but are not far distant from its banks. There are several in the northern part of the district, and many about five miles south of the Tyne, about mid-way between Newcastle and Durham.

In the coal-measures forty beds of coal have been seen; but a considerable number of these are insignificant in point of thickness. The two most important beds are those distinguished by the names of *High Main* and *Low Main*. The thickness of the first is six feet, of the second six feet six inches. The High Main is about 60 fathoms above the Low Main coal, which latter is at St. Anthon's colliery, not far from Newcastle, 135 fathoms from the surface. Between them occur eight beds of coal, one of which is four feet thick, another is three feet thick. Seven beds have been seen under the Low Main, but the quality

* The alluvium contains masses of different rocks composing the whole district; and amongst them, portions of hard black *basalt* are found every where in abundance. From this stone, the ancient inhabitants of Britain formed the heads of their battle-axes, which are commonly called celts. They resemble in shape the tomahawks found in the South Sea islands. Barbed arrow-heads, neatly finished, and made of pale-coloured flint, are frequently picked up on the moors, and are called *elf-bolts*. It also contains portions of the trap rocks of the Cheviot range, and masses of fine-grained granite appear on the surface of the whole country.

is inferior, and more nearly approaches that of the coal in the millstone-grit and shale series.

The superior excellence of the coal of this formation in quality, over every other, is sufficiently known. The quantity of coal raised annually in this district, and sent to London, and the whole east and south coasts of Great Britain, is quite enormous. Shields and Sunderland are the two places from which they are exported; and a curious distribution of the trade has taken place, depending upon the size of the two rivers. The Tyne vessels are large, and are therefore chiefly destined for the London market. The Wear vessels, on the contrary, are so small, that they can make their way into the small rivers and harbours all over the kingdom, and therefore they supply the whole east and south coasts as far west as Plymouth.

To form an idea, says Dr. Thomson, of the quantity of coal contained in the formation called the coal-measures, let us suppose it to extend in length from north to south 23 miles, and that its average breadth is eight miles. This makes a surface amounting to rather more than 180 square miles, or 557,568,000 square yards. The utmost thickness of all the beds of coal put together does not exceed 44 feet; but there are eleven beds not workable, the thickness of each amounting only to a few inches. If they be deducted, the amount of the rest will be 36 feet, or 12 yards. Perhaps five of the other beds likewise should be struck off, as they amount altogether only to six feet, and therefore at present are not considered as worth working. The remainder will be ten yards; so that the whole coal in this formation amounts to 5,575,680 cubic yards. How much of this is already removed by mining I do not know, but the Newcastle collieries have been wrought for so many years to an enormous extent, that the quantity already mined must be considerable. I conceive the quantity of coal exported yearly from this formation exceeds two millions of chaldrons; for the county of Durham alone exports 1½ millions. A chaldron weighs 1.4 ton; so that 2.8 millions of tons of coals are annually raised in these counties out of this formation. Now a ton of coal is very nearly one cubic yard; so that the yearly loss from mining amounts to 2.8 millions, or (adding a third for waste) to 3.7 millions of yards. According to this statement, the Newcastle coals may be mined to the present extent for 1500 years before they be exhausted. But from this number we must deduct the amount of the years during which they have been already wrought. We need not be afraid then, of any sudden injury to Great Britain from the exhaustion of the coal mines. It is necessary to keep in mind, likewise, that I have taken the greatest thickness of the coal-beds. Now as this thickness is far from uniform, a considerable deduction (I should conceive one-third of the whole) must be made in order to obtain the medium thickness; so that we may state in round numbers that this formation, at the present rate of waste, will supply coal for 1000 years, but its price will be continually on the increase, on account of the continually increasing expense of mining. It appears that in the above estimate of Dr. Thomson's all the beds of coal are calculated upon as co-extensive throughout the whole field; whereas allowance ought to have been made for the smaller extent of the upper beds which first crop out. It is also probable that the consumption of coal now materially exceeds that taken into the account: for both these reasons we must deduct a century or two from the calculation.

Besides the coal exported to different parts of England, a large quantity is consumed in the two counties, which cannot easily be calculated. About thirty years ago a practice was adopted at the pits, where the coal was of a fragile nature, of erecting screens to separate the small from the sounder coal. This system is now become universal, and immense heaps of coal are thus raised at the mouths of the pits. These soon take fire from the heat of the decomposing pyrites, and continue to burn for several years.* Not less than 100,000 chaldrons are thus annually destroyed on the Tyne, and nearly an equal quantity on the Wear. Two such heaps in combustion may be seen at present (Nov. 1814) on the north side of Newcastle. If you travel from Berwick to Newcastle, and enter this last town in the dark, about three miles from the town you see two immense fires; one on the left hand, about three miles from the road, which has been burning these eight years. The heap of coal is said to cover 12 acres. The other on the right hand is nearer the road, and therefore appears more bright; it has been burning these three or four years. These fires are not visible during the day, but only during the night. It has often occurred to me, says Dr. Thomson, that this small coal might be converted into coke with profit, and certainly in all cases where coal gas is wanted it would answer as well as any coal whatever. And surely such an important waste, amounting to 20 per cent. on the whole consumption of an article so essential to our commercial greatness, and of which we may already calculate the limited supply, calls loudly for timely legislative interference.

* Beneath the heaps that have taken fire, a bed of blackish brown scoria is formed, which greatly resembles basalt, and is used for mending the roads. (G. T. vol. 4. p. 54.)

Wallis, in his history of Northumberland, gives an account of a fire happening in the High Main coal-bed about 140 years ago, on the Town Moor and Fenham estates, which continued to burn for 30 years. It began at Benwell, about a quarter of a mile north of the Tyne, and at last extended itself northward into the grounds of Fenham, nearly a mile from where it first appeared. There were eruptions at Fenham in nearly twenty places; sulphur and sal ammoniac were sublimed from the apertures; but no stones of magnitude were ejected. Red ashes and burnt clay, the relics of this pseudo-volcano, are still to be seen on the western declivity of Benwell hill; and it is credibly reported that the soil in some part of the Fenham estate has been rendered unproductive by the action of the fire.

The *choak-damp*, the *fire-damp*, and *after-damp* or *stythe*, are the miners terms for the gasses with which the coal mines are affected, and of these the second, both from its immediate violence and as occasioning the other kinds of damps, is the most to be dreaded. The accidents arising from it have become more common of late years; but it should not for a moment be supposed that they arise from any want of skill or attention in the professional surveyors of the mines. The following seem to be the causes in which the gas originates.

1st. The coal appears to part with a portion of carburetted hydrogen, when newly exposed to the atmosphere; a fact rendered probable by the well known circumstance of the coal being more inflammable when fresh from the pit, than after long exposure to the air. 2nd. The pyritous shales that form the floors of the coal seams, decompose the water that lodges in them, and this process is constantly operating on a great scale in the extensive waste of old mines. In whatever mode we suppose the gas to be generated, it is disengaged abundantly from the High Main, but more particularly from the Low Main coal-seam, and in a quantity and with a rapidity that are truly surprising. It is well known that the gas frequently fires in a shaft, long before the coal-seam is reached by the sinkers: and that the pit-men occasionally open with their picks, crevices in the coal or shale, which emit 700 hogsheads of fire-damp in a minute. These blowers (as they are termed) continue in a state of activity for many months together, and seem to derive their energy from communicating with immense reservoirs of air. All these causes unfortunately unite in the deep and valuable collieries situated between the great north road and the sea. Their air-courses are 30 or 40 miles in length, and here, as might be expected, the most tremendous explosions ensue.

The after-damp or stythe, which follows these blasts, is a

mixture of the carbonic acid and azotic gases resulting from the combustion of the carburetted hydrogen in atmospheric air, and more lives are destroyed by this than by the violence of the fire-damp.

To guard against these accidents, every precaution is taken that prudence can devise, in conducting and in ventilating the mines. Before the pitmen descend, wastemen, whose business is to examine those places where danger is suspected to lurk, traverse with flint-mills * the most distant and neglected parts of the workings, in order to ascertain whether atmospheric air circulates through them. Large furnaces are kept burning at the up-cast shafts, in aid of which, at Wall's-end colliery, a powerful air-pump, worked by a steam-engine, is employed to quicken the draft: this alone draws out of the mine 1000 hogsheads of air in a minute. A kind of trap-door, invented by Mr. Buddle, has also been introduced into the workings of this colliery. This is suspended from the roof by hinges, wherever a door is found necessary to prevent the escape of air. It is propped up close to the roof in a horizontal position; but in case of an explosion, the blast removes the prop, whereby the door falls down and closes the aperture.

Sandstone is termed *post* by the miners of the *Coal-measures*; but when the bed is very hard it is termed *whin*; which therefore is not applied to basalt only, though most frequently. A bed 66 feet thick crops out at the hill called Gateshead Fell, on the south of Newcastle, and is quarried for grindstones which are of good quality. Great Britain and even the Continent are supplied chiefly from this place. The softer parts of the bed are used as filtering stones. There are about 25 beds of sandstone in the Coal-measures; the greater part of them are thin.

The beds of *shale in the Coal-measures* amount to about 32 in number. Shale is called *metal* or *metal-stone* by the miners; thus they have *grey*, *blue*, or *black* metal, according to the colour of the shale; when very indurated it is called *whin* by the Newcastle colliers. The beds of shale are usually thinner than those of the sandstone with which they alternate. Both the sandstone and shale form the roof and floor of the coal-beds; but the latter much more frequently than the former. Each is to be seen in immediate contact with the coal, without the smallest sensible alteration in its properties. But a hard bituminous shale often forms the floor of the coal-beds, which is used by the manufacturers of fire-bricks.

* An apparatus for producing light, without the danger of inflammation, by the friction of flint, now generally superseded by the safety-lamp.

In the coal-measures, *potters' clay* occurs immediately below the vegetable soil. Its colour is bluish or smoke-grey, and sometimes yellow approaching to orange, in consequence of a mixture of iron ochre. It is used in the manufacture of coarse earthen-ware, bricks, and tiles.

The trap rocks occurring in connexion with the coal-measures, whether as dykes, overlying masses, or beds, will be noticed in a separate article in the Appendix.

The minerals that accompany the coal strata are the following; clay-ironstone forming thin beds or nodules in the strata of shale; in the nodules of clay-ironstone are found galena and iron pyrites; and the latter is found in great abundance crystallized and disseminated in the beds both of coal and of shale. Calcareous spar is common, either blended with the coal, or in the form of stalagmites.

The organic remains found in the coal strata are, according to Messrs. Winch & Thomson,—In the shale, the impressions of several plants, amongst which is a variety of fern. Another fern or two is found in the nodules of clay-ironstone, as well as impressions of cones. Impressions of the bark of a plant resembling euphorbia, in iron pyrites, are found in several collieries; of another plant in coal: vegetable impressions in sandstone: the cast of a cane-like vegetable: an aggregate of black quartz crystals diverging from the centre, having the interstices filled with yellow ochre; it is supposed to be a mineralized tree, and is found at Bigge's Main colliery, and often in large masses on the sea-shore. Bivalve shells, like those of the freshwater muscle, in dark grey ironstone, and in black shale and ironstone: the same in black shale in Hebburn colliery, at the depth of 780 feet.

In one of the sandstone strata, termed *fire-stone*, a tree has lately been discovered, 28 or 30 feet in length. The trunk and larger branches are siliceous, while the bark, the small branches, and the leaves, are converted into coal; and it is believed, that the small veins of coal, called by the miners *coal-pipes*, owe their origin universally to small branches of trees. It is stated by Mr. Winch, as a remarkable and interesting fact, that, while the trunks of trees in the Whitby alum-shale are mineralized by calcareous spar, clay-ironstone, and iron-pyrites, and their bark is converted into *jet*; those buried in the Newcastle sandstones are always mineralized by silex, and their bark is changed into common coal. (Ann. Dec. 1817. p. 68.) In the introductory chapter a more precise account of these remains will be found.

———

Having thus examined the contents and disposition of the various beds entering as constituent members into what are termed the coal-measures of this field, it remains to complete our survey of it, that we should notice the difficulties which perplex and impede the operations of the miner; and which arise from the frequent derangement and dislocation of the strata; the results and proofs of ancient convulsions of the globe.

The principal class of these comprises what are generally termed faults; the phenomena of which are these. The strata are rent to an immense depth by fissures usually approaching to a perpendicular direction; which not only separate them, but are also accompanied with the elevation and depression of the portions of strata occurring on their opposite sides, in such a manner that the same stratum is found on the different sides of the fissure, at very different levels; the difference sometimes amounting to several hundred fathoms. These fissures do not remain empty, but are filled with various substances in the coal-field of which we are now treating.

These, if large, are locally called *dykes;* but if inconsiderable, *troubles*, *slips*, or *hitches*.

The most celebrated of these is called the *Main* or *Great dyke*, or 90 *fathom dyke*. The latter name has been given to it, because the beds on the northern side are 90 fathoms lower than those on the southern side of it; its underlie is inconsiderable. In some places, its width is not great, but in Montagu colliery, it is 22 yards wide, and it is *filled with hard and soft sandstone*. This dyke is visible in the cliff at Whitley quarry, a short distance north of the mouth of the Tyne, from which place it traverses the coal strata in the general direction of north-north-east and south-south-west, but not in a straight line; and it is considered to be probable that it passes into the formations underlying the coal-measures: a small string of galena has been observed in it at Whitby. From the southern side of this dyke, two others branch off, one to the south-east and the other to the south-west. The latter is very remarkable: it is called from its breadth, the 70 *yard dyke*, and is filled by a body of *hard and soft sandstone*. This intersects the upper or Beaumont seam of coal, which is not thrown out of its level by the interruption. The seam however decreases in thickness from the distance of 15 or 16 yards, and the *coal first becomes sooty*, and at length assumes the appearance of *coak*. *This phenomenon is unknown elsewhere, except in the vicinity of basaltic dykes.* The south-eastern branch is only 20 yards in breadth.

There are several other dykes of the same kind, which, following the same law as the cross veins in the lead mine district, *elevate the strata on the side to which they dip.*

The dykes are an endless source of difficulty and expense to the coal-owner, throwing the seams out of their levels, and filling the mines with water and fire-damp. At the same time they are not without their use; when veins are filled, as is often the case, with stiff clay, numerous springs are dammed up, and brought to the surface; and by means of those dykes which throw down the strata, valuable beds of coal are preserved within the field, which would otherwise have cropped out, and been lost altogether. Several valuable beds of coal would not now have existed in the country to the north of the main dyke, but for the general depression of the beds occasioned by that chasm.

In many instances dykes occur, filled by rocks of the trap formation; but these will be treated of in a separate article in the appendix.

Having treated so copiously of this important coal-field, the others which we have to notice may be dismissed more briefly, since in general circumstances they all agree.

(*b*) DETACHED COAL-FIELDS IN THE NORTH OF YORKSHIRE.

Proceeding into the north of Yorkshire, the superincumbent beds of magnesian limestone, extending themselves far to the west, overlie and conceal the coal formation, coming immediately into contact with the inferior strata of the millstone grit and carboniferous limestone formations. Near Middleham, and at Scrafton, Leyburn, Thorp, Fell near Burnsell, and as far west as Kettlewell, on a hill called centre lights, there are several small detached coal basins provincially termed swilleys, lying in hollows in the gritstone. They are of limited extent, and the seam is seldom more than twenty inches thick. At Hudswell Moor, the lowest and thickest part of the coal is one yard, but the stratum diminishes and vanishes at the edges. The extent of this coal is about one mile in each direction (Bakewell, p. 370.) It may be doubted however, whether these unimportant beds should not rather be referred to the thin coal seams subordinate to the millstone grit series, than to the principal coal measures.

(*c*) GREAT COAL-FIELD OF SOUTH YORKSHIRE, NOTTINGHAM, AND DERBYSHIRE.

Advancing to the southern parts of Yorkshire, we arrive at the great Yorkshire and Derbyshire coal-field which rivals, or even surpasses in importance, that of Northumberland, with which it so closely agrees in the direction, inclination, and character of its strata, that it may not be improperly considered as a re-emergence of the same beds from beneath the covering of magnesian limestone which has concealed them through so long an interval.

This coal-field occupies an area nearly triangular, but with a truncated apex. The base, or broadest part being at the northern extremity; and the apex, or narrowest, at the southern; its greatest length, which is from north to south between Leeds in Yorkshire and Nottingham, is above 60 miles. Its greatest breadth, from east to west, which is in the Yorkshire portion, is about 22 miles.

Like those of the Northumberland Coal-field its strata range from north to south; dip to the east, where they sink beneath the super strata of magnesian lime, and rise to the west and north-west, in which directions the lowest measures at length crop out against the rocks of the millstone-grit series, which constitute the higher ridges of the Penine Chain.

Mr. Farey has inserted in his agricultural report on Derbyshire, a list of all the principal coal-pits in this field: which might be more easily consulted, were it not confused by the insertion of the pits of several other coal-fields unconnected with this, and disposed in an alphabetical order; the very worst arrangement that can possibly be adopted with a view to geological reference, for which a disposition according to geographical situation is almost indispensable. He has likewise furnished several other particulars from which the following are extracted.

It is ascertained that the strata of which the whole formation consists, are numerous. There are 20 gritstone beds, numerous strata of shale, bind, and clunch, alternating with several beds of coal of different thickness and value, the lowest of these is termed the millstone-grit, beneath which no workable coal is found Some of the gritstone beds are of great thickness, and are described as consisting of grains of semi-transparent silex, united by an argillaceous cement; in some of the beds there are subordinate ones, in which the cement is very small in quantity, and from which are quarried grindstones for cutlers, &c. The beds of *shale* consist of a slaty argillaceous substance of a black or brown colour, rarely of a light yellow; their joints are ochreous; and the springs of water issuing from them are tinged by iron. Some of them contain roundish or ovate masses of *argillaceous ironstone*, and even thin strata of it, in which are coaly impressions of vegetables. One of these beds of ironstone which occurs towards the middle of the coal series in a line traversing the field in the parallel of Tupton is remarkable for abundant impressions of muscle shells; whence it is known by the name of the muscle band. It is worked as an ornamental marble. The thickness of this bed is 8 or 10 inches. These ironstones dip, of course, with the strata, beneath the grass; and the workings of them, which are numerous, are begun at the surface, and pursued until it becomes dangerous, from the loose nature of the stratum in which they lie, to follow them deeper. In some places, where the texture of the bed is favourable, the iron-ore has been followed down 35 or 40 yards. This ore also is found in the beds which are by the miners called *Binds*, which appear to be beds of indurated loam, or of sand and clay mixed and indurated, and which are enclosed in the shale just noticed; the bind falls to pieces on exposure, however hard it may be in its natural state and position. When the sand abounds, and the bed is very hard, it is called stone-bind, and it then contains scales of mica. Binds are black, or blue, yellow, grey, &c. Some of the very hard black binds are used as black chalk by the stone mason. Others when decomposed become good brick clay, as well as the *Clunches*, which we have yet to notice. Clunch is indurated clay, and yields those infusible kinds which are adapted for fire-bricks; it varies in hardness, and is black, grey, yellow, white, &c. Clunch is generally found immediately beneath each bed of coal, and at the places where it bassets or crops out on the surface, becomes soft clay.

A hard argillaceous rock called *Crowstone* forms in some places the floor of the coal-beds. This may perhaps be considered as a variety of the clunch, still more highly indurated.

Potters clay of various hues and qualities occurs in this coal-field.

In consequence of the disturbance created by the faults presently to be described, Mr. Farey had not, at the period of publishing his report, ascertained the exact number or order of the coal-seams in this field: but according to the manager of the Alfreton coal works (see Bakewell, 384), in the whole of this range there are thirty different beds of coal, varying from six inches to 11 feet; and the total thickness of coal is 26 yards. This, however, is only offered as an approximation.

There is an account of some of the coal-seams near the south-eastern extremity of the field in Townsend's Vind. Mos. p. 163. Every variety of coal appears to occur in this field. Hard stone coals, which neither flame nor run together in the burning; soft or crozzling coals, which do both; Cannel coal; and irridescent or peacock coal are mentioned; and we are told that the same bed of coal in different parts of its course varies from one to the other quality, i. e. from an hard to a crozzling coal.

According to Mr. Farey, the strata of this field are traversed and dislocated by an immense *fault* proceeding from near the termination of the magnesian limestone range on the south, northwards, in a zig-zag direction, on the western side of that limestone, quite into Yorkshire. Of this fault, nothing has, we believe, been said in regard to its size or contents: but it is believed to be owing to it that the coal strata of Derbyshire and Yorkshire, through which it passes, are on the west of it, so dislocated, that it would be extremely difficult so to connect the beds of coal, and the interposed substances, as to form a reasonable conclusion as to their number and nature. The beds of coal east of the fault, are known to pass beneath the magnesian limestone, since they are worked beneath it.

The rise of the strata is said to be much more rapid on the western than on the eastern side of this fault. It must be added, however, that the existence of this fault rests entirely on the authority of Mr. Farey, and is disputed by many miners. Many other faults, however, and some of considerable magnitude, are ascertained to traverse the field in various directions. The vegetable remains of these coal-measures agree with those discovered in the Northumberland field: most of them are accurately figured in Martin's "Petrifactions of Derbyshire."

This coal-field terminates abruptly on the south near Nottingham; horizontal strata of the newer red sandstone and red marle prevailing on the south of a parallel of latitude passing through that town, and abutting against the inclined strata of the coal-formation, carboniferous lime, &c. Mr. Farey was at first inclined to attribute this relative position of these formations to the effects of an enormous dislocation or fault: but the more probable explanation appears to be that in this, as in many other instances, the rocks of the coal series had assumed their inclined position previously to the formation of the newer red sandstone: which, being deposited in horizontal beds, by a necessary consequence, was brought successively in contact, at the same level with the various beds of older formation as their inclined position caused them to rise in succession to the surface, in the same manner that the waters of the sea washing the foot of a cliff composed of inclined strata, would successively bathe all its strata. And this explanation, as appears from some later papers in Tilloch's Journal, Mr. Farey is now himself inclined to adopt.

(*d*) BETWEEN ASHBORNE AND DERBY.

We have next to speak of the indications of coal on the south of the Penine chain,—between it and the Trent. The horizontal deposits of the newer red sandstone, which, as we have seen, sweep in what is called an unconformable and overlying position, round this extremity of the chain, generally conceal all older rocks from observation. Coal has, however, been proved to exist within this tract at two points; both so nearly contiguous, as probably to form parts of one coal-field, situated about half way between Ashborne and Derby. These points are Sprinx-hall in Edlaston parish, and Darley moor. On the east and west of the coal-field thus assumed, are patches of mountain limestone at Wild park, in Magginton parish, and Birchwood park in Boston. Millstone-grit also occurs on the Trent at Stanton bridge.

(*e*) COAL-FIELDS OF NORTH STAFFORD.

Turning from the south along the western side of the Penine chain, we meet in the north of Staffordshire with two detached coal-fields; namely that of Cheadle, and that of Newcastle-under-Line, called the Pottery coal-field.

The Cheadle coal-field appears to consist of an insulated basin of the lower members of the coal series surrounded by, and reposing upon, millstone-grit; which latter rock, according to Mr. Farey "from Dilhorn northward round by Ipstone edge, and southwards to near Oak Moor Mills, declines towards Cheadle as a centre; being covered by the lower part of the coal series in Ipstone, Foxley and Kingsley. Near Cheadle, and to the south of it the coals are thicker and better in quality; but the great thickness of quartz gravel which occurs southward and round that town prevents the tracing these very satisfactorily." p. 173.

Of the Pottery coal-field which next claims our attention, a short description is inserted in Pitt's Topopgraphical History of Staffordshire. This coal-field appears to occupy a triangular area of which the apex is situated near Mole Copt hill; hence the sides diverge to the south-south-east and south-south-west; in each direction about ten miles. Newcastle-under-Line is situated about the middle of the base (which may be estimated at about seven miles) but falls rather within the area. From the two sides the strata dip towards the centre of the area. On the north-eastern side, the inclination westward is estimated at one foot in four; on the north-western, where the strata crop out against Mole Copt and the Harecastle hills, the inclination eastward is still more rapid. We are not informed in what manner the strata lie along the base line by Newcastle, whether they dip north and so render this coal-field a complete and insulated basin, or whether the strata continue to range in their former planes, and thus the coal-field in this direction terminates either by a fault or by the overlaying of the superior formations. As the red marle occurs near Newcastle on the south, the latter is perhaps the more probable supposition. Between Burslem, which is nearly in the centre of the coal-field, and its eastern limit at the range of hills east of Norton church, it has been clearly ascertained that there are 32 beds of coal, of various thicknesses, generally from about three to ten feet each.

On the north-east and north-west, the coal-field appears to be bounded by the cropping out of the millstone-grit on which the coal strata rest. Mole Copt and the Harecastle hills exhibit this rock, while beneath these on the further side, the carboniferous limestone shews itself towards Congleton.

(*f*) THE MANCHESTER, OR SOUTH LANCASHIRE COAL-FIELD.

We have next to notice a far more extensive and important coal-field, that of Manchester, or South Lancashire; or, as it is called by Mr. Farey, the great Derbyshire and Lancashire coal-field. This commences in the north-western parts of Derbyshire, and ranges thence to the south-western parts of Lancashire: forming a crescent-like figure, with Manchester nearly in the centre: the chord or span between the opposite horns of which, is about 40 miles. Speaking generally, it may be said that the strata rise towards the exterior edge of this crescent, along which the inferior strata of millstone-grit crop out from beneath them, and dip towards its inner edge; along which they are covered by the superior strata of the newer sandstone formation, containing occasionally beds of the calcareo-magnesian conglomerate. Great disturbances however interrupt the regularity of the disposition here sketched out. In the first place, what may be called the south-eastern horn of the crescent, forming the portion of the coal-field which lies within Derbyshire and Cheshire, bifurcates at the village of Disley in the latter county, being divided into two branches by an intermediate ridge or "saddle of millstone-grit; the eastern branch forming a trough of which the strata crop out on both sides against the mill-stone-grit." To this part of the field, Mr. Farey has applied the appellation of the Goyte Trough, from a small river of that name which runs through it. It extends about 15 miles from Disley southwards, to near Mearbrooke in Staffordshire.

Of the western branch of this bifurcation in the coal-field, Mr. Farey gives the following account. "From the ridge at Disley, the strata decline or dip again to the west, but not so rapidly as they rose. This occasions the coal-measures in this field again to cover the millstone-grit for some miles south-south-west of Disley, when a fault commences, which proceeds south; and between this fault and Macclesfield the measures again basset west."

Mr. Farey adds that he was unable from the limits of his observations to state how much further to the west this undulating of the strata might continue. But that, from some excursions he had made to other parts of this field in Cheshire and Lancashire, he was "induced to think that enormous faults occur in these districts, which will render the elucidation of their highly valuable strata, containing more than 50 seams of coal in a few hundred yards of sinking, a work of some labour and difficulty."

It is much to be regretted that we have no precise information with regard to the remaining and far more important part of this great coal-field, which lies within Lancashire. But it is to be hoped that the scientific spirit which has always prevailed at Manchester, will not much longer suffer this deficiency to exist on a point so peculiarly connected with the local interests of that great and opulent town. At present, however, we find nothing to add to the very general view already given, with the exception of the short memoir of Mr. Bakewell's, published in the second volume of the Geological Society's Transactions, which relates only to a small and insulated portion of this field (by him entitled, the coal-field of Bradford), extending little more than two miles in length, by 2000 yards in breadth.

This tract is situated on the river Medlock, a short distance east-south-east of Manchester, and the phœnomena presented by it, are shortly as follows. It is surrounded on every side except the east by the prevailing red sandstone of the environs of Manchester, of which the relation to the coal-measures appears in this neighbourhood not to have been ascerteined;

but which, most probably is the newer red sandstone; a conjecture strengthened from the circumstance, that beds of limestone are interposed between it and the highest coal strata, a position which exactly agrees with that of the magnesian lime. Beneath this limestone several beds of coal basset out, rising to the north under an angle of 30°. One of these, near the middle of the field, is four feet in thickness. To the north of these inclined beds an interval occurs in which the direction of the beds becomes suddenly vertical; and one of the vertical beds so exactly resembles, both in itself and its concomitant strata, the four-foot coal above mentioned, that little doubt can exist of its being a continuation of that bed, broken off and thrown into its present position: with these vertical beds the coal-measures terminate for some distance on the north, an interval of the red sandstone succeeding; beyond which, however, at the distance of 1400 yards, they again emerge in the collieries of Droylsden, rising as at first towards the north. These circumstances appear only to indicate that the coal-measures have here been dislocated by considerable faults and subsidences; and that the newer red sandstone, in consequence of its having been deposited after those subsidences had taken place, has insinuated itself into the vacuities which they occasioned. Such a position must be familiar to those who have attentively examined the analogous formations in the south-west of England.

The north-west horn of the crescent formed by the great Manchester coal-field appears to be about Prescott, not far from Liverpool.

(g) THE NORTH LANCASHIRE COAL-FIELD.

Towards the opposite, or northern extremity of Lancashire, another coal-field occurs, half way between Lancaster and Ingleton. It is of small extent, and has never been thoroughly examined. It probably forms a small insulated basin, surrounded by millstone-grit: but its northern extremity approaches so nearly to the transition slate, on which the mountain limestone of Ingleborough rests, that it is difficult to explain its position, without having recourse to the theory of a considerable subsidence having depressed the coal-measures, and thus brought them down nearly to the level of the slate in that quarter. It is believed in confirmation of this idea, that a second ridge of limestone exists on the south of Ingleton, dipping rapidly in a direction contrary to that which forms the base of the adjacent mountains, namely to the south-east, and so sinking beneath this coal-field. But the observations are very imperfect.

(*h*) THE WHITEHAVEN COAL-FIELD.

In our progress round the Penine Chain, we have now arrived at that point where the transition group of the Cumbrian Mountains bursts from its side like an immense excrescence; around which we may still trace the formations which constitute the whole mass of the Penine Chain, ranging in nearly a complete circle; being forced, as it were, to make a long detour, as the layers of a piece of wood are seen to do in surrounding a knot. Our present concern is only with the highest of these formations, the coal-measures. These are indeed not to be met with on the south of the circle above described; for the æstuary of Morecambe bay cuts off the space in which they might be expected to exist. But on the western coast we find them near Egremont, south of Whitehaven,* whence they range in a large arc of a circle without interruption to beyond Hesket, a distance of about forty miles; and then turning towards the south, continue, though with several interruptions, to extend towards Orton, thus completing the whole of the northern semicircle round the Cumbrian mountain group, and returning to the point where this inosculates (if the expression may be allowed) with the Penine chain. The newer red sandstone of the plain of Carlisle covers it on the north throughout the whole of this range. A good

* The following particulars are extracted from Townshend's Vind. Mos. Cumberland abounds with coal near to many of its eastern mountains, and in various districts between Sebergham and Whitehaven, from whence collieries proceed along the coast by Cockermouth to Maryport, forming a district of about one hundred square miles, in which three coal-fields are particularly noticed.

1. Howgill, west of Whitehaven, two miles and a half wide, from the rivulet called Pow, on the eastern side, to more than one thousand yards under the sea. In this seven beds have been worked.

2. Whingill, north-east of Whitehaven, extends 3000 yards in length, by 2800 in width. The beds are from four to ten feet in thickness, and dip one yard in ten. In the depth of 165 fathoms they work seven large beds, and have noticed eighteen thin ones. The faults are about 120 feet, up or down, and run from east to west.

3. Beside these, a more extensive coal-field has been discovered to the south and south-west of Whitehaven, which is yet unexplored.†

At Preston How, south-west of Whitehaven, on the Croft Pit, they cut fourteen beds of coal before they met with any one which is considerable. But to reward their perseverance, the fifteenth bed proved more than five feet in thickness; and the seventeenth, separated from the former by twenty-four beds of slate, ironstone, sandstone, and one small bed of coal, is nearly eight feet in thickness.

† Dixon's Life of Dr. Brownrigg.

section is exhibited in the cliff of St. Bees near Whitehaven, where the strata being inclined to the south, the coal is seen sinking beneath superstrata of magnesian limestone, and these in their turn beneath beds of red marle, containing gypsum.

(*i*) INDICATIONS OF COAL AT FOOT OF THE WESTERN ESCARPMENT OF CROSS FELL.

Returning to the Penine chain at the great escarpment over which Cross Fell dominates, we find the horizontal strata of the newer red sandstone extending closely in many points to the very foot of that escarpment, and thus brought into contact with the older sandstone on which the whole escarpment is based. Between Melmerby and Merton Pike, however, a long and narrow stripe of transition rocks (greenstone and slate) intervenes, extending almost 12 miles: and on the north-west of this tract, beds of carboniferous limestone and coal occur in their regular order of succession, dipping west beneath the newer sandstone of the plain under an angle so rapid as to be nearly vertical. The beds are, however, thin, and greatly shattered and deranged. There are pits at Melmerby Lane Head, Hay Gate, Gale Hall, and Ourby Town Head.

We have thus brought to a conclusion our survey of the coal-districts connected with the Penine chain. Had it been our object to extend our enquiries into the adjacent portion of Scotland, we should have found that in Dumfries similar relations prevail. The coal-fields of that county occurring in small basins surrounded by mountain limestone, which finally rest against the transition rocks of the Lead Hill mountains: and the whole being partially covered by the newer red sandstone containing beds of a calcareous conglomerate, probably magnesian.

Section II.

Formation of Millstone-Grit and Shale throughout the Penine Chain.

Having thus completed our survey of the several Coal-fields connected with the Penine chain, the next object that demands our attention, is the tracing through the same tract, the disposition of the second series of beds into which we have divided the rocks associated in the carboniferous districts, namely, that in which the millstone-grit forms the prevailing feature; and this will be found to constitute the most elevated and extensive portion of the district under consideration.

We begin as before with the northern extremity of the chain. In this quarter Mr. Winch has classed the beds of which we are now about to treat together with the subjacent strata of carboniferous limestone, under one order of rocks, to which he bestows the common name of Lead measures. But the distinction which is here proposed, is imperatively required by the structure of the southern parts of the chain in Derbyshire: and even in the northern, the following considerations will evince its propriety. The sections of the workings at Hely, Aldstone Moor, and Dufton, as given by that gentleman (which taken in succession afford nearly a complete view of the series of beds occurring within this district,) exhibit a total depth of about 450 fathoms. Of these, the beds constituting the upper portion to the depth of 150 fathoms (in which the millstone-grit occurs) *contain only two thin beds of limestone, each but one fathom in thickness, but in the lower portion* 19 *beds of limestone occur, many of very considerable thickness, amounting together to one-third of the total depth of this portion.* So that the lower portion may be considered as distinctly characterised by the abundance, and the upper bed by the almost total absence of limestone. Workable seams of coal also occasionally occur in the upper portion, but none in the lower, although some faint traces of the same mineral may even there be traced.

Guided by these principles, we select for our present observation only the upper portion of the beds comprised in the sections published by Mr. Winch; those namely, which occur in the workings at Hely field on the Derwent, and Aldstone Moor, above the thick calcareous bed called the Tumblers, and great Limestone.*

* *Section of the Lead-mine strata at Hely Field on the river Derwent.*

	Fs.	Y.	Ft.	In.
Slate sill	2	1	—	—
Plate	3	1	—	—
Different Girdle beds	2	—	—	—
Plate	2	1	—	—
Freestone (fine-grained sandstone)	7	—	—	—
Coarse hazle	1	1	—	—
Plate and Blue whin	1	—	—	—
Plate and Grey beds	—	—	2	—
Hard stone and Whin	1	—	2	—
Plate and Whin	1	1	2	—
Plate	2	1	—	—
Millstone grit	5	—	—	—
Plate	4	1	—	—
Hard hazle	2	1	—	—
Grey beds. (Thin layers of slate-clay and sandstone alternating)	1	—	—	—
Freestone	7	—	—	—

As has been already stated, these sections assign a thickness of at least 150 fathoms to the formation we are now describing;

	Fs.	Y.	Ft.	In.
Plate	1	—	—	—
Hazle or Slate	2	1	—	—
Plate or Famp	2	—	—	—
Hazle and Plate	2	1	—	—
Plate	2	—	—	—
Hazle or Slate	1	1	—	—
Plate and Grey beds	1	1	—	—
Thin stratum of Grey beds	15	—	—	—
Fathoms	74	—	—	—

Section of the Lead-mine Strata on Aldstone Moor, Cumberland.

	Fs.	Y.	Ft.	In.
Grindstone sill	4	—	—	—
Plate	6	—	—	—
Hazle	1	—	—	—
Plate	2	—	—	—
LIMESTONE	1	—	—	—
Crow Coal occasionally.				
Hazle or Upper Coal sill	1	—	—	—
Plate	8	—	—	—
Hazle	1	1	1	—
Plate	2	1	—	—
Hazle	2	—	2	—
Plate	1	—	—	—
Upper Slate sill	4	—	—	—
Plate	1	1	—	—
Lower Slate sill	4	—	—	—
Plate	5	—	2	—
Whetstone sill. (Fine grained Micaceous sandstone)	1	1	—	—
Plate. (Ferruginous sandstone)	2	—	—	—
Iron-stone with Coal	1	—	1	—
Freestone with Iron pyrites	5	1	—	—
Plate	6	—	1	—
Girdle beds	—	1	2	—
Plate	3	1	—	—
Pattison's sill or hazle. (Very hard grey sandstone with specks of mica)	1	1	—	—
Plate	3	—	—	—
LITTLE LIMESTONE	1	1	1	—
Little hazle	—	1	2	—
Plate	2	—	—	—
Coal occasionally.				
High Coal sill. (Hard grey sandstone with specks of mica)	1	1	1	—
Plate	1	—	—	—
Coal occasionally.				
Low Coal sill	2	—	—	—
Plate	3	—	2	—
Fathoms	77	4	—	—
Preceding section	74	—	—	—
	151	4	—	—

and this is probably short of the truth; since we have no evidence that the highest beds at Aldstone immediately succeed to the lowest at Hely. But the interval is probably not considerable, so that the above estimate may be admitted as tolerably correct. The prevailing rock of this series is shale, (known by the provincial name of *Plate*), with which various beds of sandstone, differing in hardness and texture, and according to these differences distinguished, as freestones, hazles, whetstones, grindstone, and millstone, occur: of the latter only one bed is worked; the thickness of which is about 30 feet. This is one of the uppermost strata on the Derwent, where it crops out, and does not occur further west. A similar rock is found in the north-east of Northumberland at Scramerstone, four miles south of Berwick, and at Craster near Howick: it entirely agrees with the character given of this rock in the general account of the formations. The freestones of this formation frequently contain vegetable impressions.

Towards the lower part of this formation, two thin beds of limestone, each about one fathom in thickness, occur; and alternating with them, some occasional seams of coal. In the mountainous tract dependant on Cross Fell, these seams are so thin as to be of little importance; but in the flat country in the north, they appear to dilate considerably; for several valuable coal-pits are worked on the north of the Coquet, in beds which must be referred to this formation. The feature which distinguishes these coal-measures from those of the principal coal-formation before described, is their alternating with strata of limestone. In one instance this limestone is said to contain bivalve shells, but the species is not mentioned: particular attention should be given to this portion of our strata, since, according to the views of some, they might be expected to present similar phœnomena with regard to the alternation of beds containing fluviatile and marine reliquia with those so strikingly exhibited by the most recent formations in the Isle of Wight,—it being generally asserted that the shells accompanying the coal-strata are fluviatile, while the limestone beds, which in this part of the series alternate with the coal, seem closely to agree with the inferior or carboniferous limestone, which is undoubtedly of marine formation. We have however already suggested our doubts whether the shells in the coal-measures are really fluviatile.

The collieries to which these remarks apply, extend over the whole of the flat country between the mouths of the Tweed and Coquet. Their stratification is less regular than that of the great coal-field, and undulates with the surface of the country.

To return to the rocks of this formation as exhibited in the mountainous tract, we find that they constitute the superstrata throughout it, the subjacent beds of limestone appearing only where exposed towards the middle or lower regions of the valleys, or by the slope of the great western escarpment; while the lower beds of this series, as detailed in the section before given, form the summit of the lofty Cross Fell and all the neighbouring eminences.

Here the strata appear to dip 2° 15′ to the north-east, so that, on crossing the range from east to west, they will be seen cropping out, one after the other, and forming parallel ridges extending from south-east to north-west.* The principal disturbance which interferes with this regularity of position, is occasioned by a thick metalliferous vein called Burtreeford Dyke, which crosses the strata near the head of Weardale, elevating them in some places to above 80 fathoms on the eastern side, and in others greatly depressing them. Contiguous to this dyke the strata rise at an angle of 45°.

The same constitution continues to prevail through the mountains of the north of Yorkshire. The beds of the formation at present described, composing their higher regions, and the subjacent limestones appearing in their vallies and towards their base, as may be seen to the greatest advantage in the neighbourhood of Ingleborough, and as far south as Clitheroe; beyond which, the shale and millstone-grit constitute the entire mass of the mountains, (the lower formations being entirely concealed through an interval of near 50 miles), until they again emerge in Derbyshire. Concerning this interval, we have less precise information than with respect to any other part of the chain; but it is also from its uniformity, far less interesting, presenting the rocks of this single formation exclusively, in strata which generally appear to dip from the central ridge towards the east on the one side, and the west on the other. About Pendle hill, however, which rests upon the limestone of Clitheroe, it is said that the stratification is more disturbed, and near this point the mountains extend more to the westward than usual, forming the heights of Bolland forest, which appear to consist entirely of the beds of the present formation. The copper mine of Anglezark near Chorley, long but falsely celebrated as the only one in England producing carbonate of barytes, is situated in this district. And here we have again to lament that the details of the geology of Lancashire are less

* Mr. Winch states the dip to be to the south-east and the line of bearing from south-west to north-east. But as such a direction is altogether inconsistent with the ascertained position of the places where the same strata are worked, I have ventured to correct it as above, supposing an error of the press.

known than those of almost any other county, and to express our hope that the naturalists of Liverpool and Manchester will shortly enable us to speak with greater precision of their immediate environs.

Concerning the central regions of this chain, Mr. Bakewell presents us with the following notices. "Millstone-grit forms the summit of Blackstone Edge, Pule Moss,* East and West Nab, and all the higher hills in that part of Yorkshire. A perforation of three miles was lately made through Pule Moss, 750 feet below the summit of the hill, to form a tunnel for a canal from Huddersfield to Manchester. The tunnel appears to have been principally carried through the shale which lies immediately upon the limestone; the strata are elevated, and inclined in an opposite direction, and are intersected by a large dyke, containing a vein described by Mr. Outram, the engineer, to be limestone, (Phil. Trans. for 1796). At a considerable distance from the entrance, a number of balls were found, composed of argillaceous ironstone." Mr. Bakewell also states that the millstone-grit extends from a little east of Halifax to Blackstone Edge, and near Todmorden, where its beds are bent in an opposite direction, and then generally follow the curvature of the hills.

We have thus traced this formation from Northumberland into Derbyshire, where our information again becomes precise in consequence of the researches of Mr. Whitehurst and Mr. Farey. The millstone-grit and shale together constitute in Derbyshire a series of strata very closely agreeing in their aggregate thickness, with that presented by the same rocks in Northumberland, amounting to 145 fathoms. In this series, as here exhibited, the millstone-grit (exhibiting the same characters which have already been sufficiently described) constitutes the upper portion, extending to the depth of 120 yards, the lower 170 yards being occupied principally by the shale, containing however, some alternating beds of fine-grained siliceous grit beds, and nodules of ironstone, and some subordinate, and apparently only local beds of limestone. In Derbyshire therefore, it appears that the separation between that part of this series in which the millstone-grit prevails, and that characterised by the shale, is so well marked, that they may be subdivided, as they have been by Mr. Farey, into distinct formations. This distinction does not appear to be equally applicable to other districts.

The millstone-grit in this part of the country, ranges at some

* The whole of Romalds moor also consists of millstone-grit, which some of our topographical writers have mistaken for granite.

distance round, three sides of the great central tract of carboniferous or mountain limestone, in a figure resembling an horse-shoe. Tracing it from south-east to north, and thence round to south-west, it may be said to begin near the junction of the Derwent and Ecclesburn, about four miles north of Derby; it there occupies both banks of the Derwent; then keeps the west side of that river as far as Cromford; where having crossed, it ranges along the eastern bank for many miles, as far as its source in the high peak; all the way occupying a narrow band of country, between the superstrata of the coal-formation of Derbyshire and Yorkshire and the substrata of shale. From the northern angle of Derbyshire it bends again to the south by west; passes near Buxton, and thence pursues its course through Staffordshire in a line somewhat broken and irregular, which may be most clearly described as forming a narrow band round the several coal-fields in that district noticed in the preceding article. The hills formed by this rock usually present a bold escarpment, and are often crowned by rude piles of crags, exhibiting some of the wildest rock scenery of the district. The interval between this circling range of mountains and the central calcareous group is usually a lower district occupied by the shale; in which however, several insulated mountains appear, each bearing a cap of millstone-grit, while the shale may be traced all round their base. Mr. Farey has given a list of 20 instances of this circumstance, which, as he strongly observes, proved in the most convincing manner, that the surrounding vallies owe their origin to denudation. The principal of these insulated mountains is Kinder-Scout, the loftiest eminence in Derbyshire.

The beds of limestone subordinate in the shale, constitute the feature perhaps most worthy of attention in this tract. Some of these beds afford a beautiful black marble. The most considerable tracts of this limestone are situated near the extremity of the carboniferous limestone tract, on the south-west, where the shale limestone abuts against both sides of a vast promontory formed by the hills of the latter variety, extending on one side to Mixon-hay in Stafford, and including the copper mines of Ecton hill, and on the other side to Atlow in Derby. As these tracts are placed over the same line of bearing, Mr. Farey thinks they may have once been united, the strata having been continuous over the intervening limestone; which may, as he thinks, have been elevated into its present position, and been subsequently stripped of the superstrata in question by denudation. There is another considerable tract of this limestone near Bakewell.

The strata of this limestone often present very singular contortions. The most remarkable example of which may be seen two miles and a half east-north-east of Ashbourne.

Section III.

Carboniferous Limestone of the Penine Chain.

The chain we have been so long considering exhibits the rocks of this formation wherever its lowest beds are brought to view, either along the line of its western escarpment, or by the deep excavation of the vallies traversing it, for nearly 100 miles from its northern extremity as far south as Clitheroe; and two calcareous branches detached from this part of its course embrace and encircle the transition mountains of the Cumbrian group: hence, after remaining concealed through an interval of about 40 or 50 miles, it again emerges in Derbyshire, where it forms a mountain plain about 20 miles in length.

The section published by Mr. Forster, and those inserted in Mr. Winch's paper, afford the best materials for the description of this formation, in the most northerly part of its course; as in this district the limestone beds repeatedly alternate with others of siliceous grit and slate-clay, and bear to these the proportion only of 1 to $2\frac{1}{2}$, it is here a matter of much greater difficulty than in Derbyshire, to draw an exact line of demarcation between this and the preceding formation. The reasons which have induced us, however, to draw such a line, have been sufficiently explained in the foregoing article. According to this division the strata assigned to the present formation, beginning with the first important bed of limestone (that distinguished locally as the Tumblers and great Limestone, which exceed 10 fathom in thickness,) and extending thence to the lowest limestone yet discovered in the chain, will constitute an aggregate of about 254 fathoms in thickness; of which about 93 fathoms is formed by 19 beds of limestone, the remainder being clay-slate and grit. We subjoin an abridgement of Mr. Winch's section, which will give a clear and precise view of the internal structure of this portion of our chain.

		Fs.	Y.	F.	In.
1.	Tumblers and great lime	10	1	—	—
	Slate slay and sandstone	14	1	—	—
2.	Limestone	4	—	—	—
	Slate-clay and sandstone	9	—	—	—
3.	Limestone	2	—	—	—
	Slate-clay and sandstone	6	—	2	—

		Fs.	Y.	F.	In.
4.	Limestone	—	1	2	—
	Slate-clay and sandstone	3	1	—	—
5.	Limestone	9	—	—	—
	Slate-clay, &c. with thin coal occasionally	4	—	—	—
6.	Cockleshell limestone	—	—	1	6
	Slate-clay, &c.	3	—	1	—
7.	Limestone	—	1	I	6
	Slate-clay, &c.	4	1	—	—
8.	Tyne Bottom limestone	3	1	1	—
	Whin sill basalt *	10	—	—	—
	Slate-clay, &c.	11	1	—	—
9.	Jew limestone	3	—	—	—
	Slate-clay, &c.	7	1	—	—
10.	Little limestone	1	—	—	—
	Slate-clay, &c.	15	1	—	—
11.	Smithy lime	4	1	—	—
	Sandstone	1	—	—	—
12.	Limestone	4	1	—	—
	Sandstone	1	—	—	—
13.	Limestone	3	—	—	—
	Slate-clay	1	1	—	—
14.	Limestone	1	1	—	—
	Slate-clay	1	1	—	—
15.	Robinsons's great lime	14	—	—	—
	Slate-clay and sandstone	3	—	—	—
16.	Great Randal or limestone ..	21	—	—	—
	Slate-clay, &c.	4	—	—	—
17.	Limestone	4	—	—	—
	Slate-clay, &c.	28	—	—	—
18.	Limestone	2	—	—	—
	Slate-clay, &c. containing a 7-inch seam of coal	10	—	—	—
	Sandstone	28	1	—	—
19.	Limestone	3	1	—	—
	Total	254	—	—	—

Beneath this are 40 fathoms of sandstone and slate-clay, and then the old red sandstone.

* This basaltic stratum offers one of the most interesting phœnomena of this district. But in order to collect together under one point of view all the circumstances which bear on the occurrence of rocks of the trap formation among those associated with the coal districts, the particular description of this and all similar facts will be postponed to an appendix on this subject, with which the account of the coal districts will conclude.

All these limestones appear to contain the Encrinus. Most of them also bivalve shells; and that called the Cockleshell limestone, oysters ? of the diameter of four or five inches. They seem to agree together in every essential character, as well as in their extraneous and native fossils.

This formation is the great repository of the metallic veins of this district.

The fissures which contain lead ore in the mining district, are exactly similar to those described by Williams in his mineral kingdom. Such as range from north to south are called *cross veins*, or (occasionally) *dykes*; they are generally of great magnitude, and seldom carry ore; the most valuable mineral depositories are fissures from three to six feet wide, running for the most part from north-east to south-west, and cutting the cross veins; the cross veins being frequently rendered productive to some distance from the points of intersection.

The same vein is productive in different degrees at different depths, according to the bed which it traverses. Generally speaking, veins are most productive between the grindstone sill and the four-fathom limestone; none have been worked in Aldstone moor below the level of the Tyne bottom limestone; but the Dufton mines are situated in the lower beds, though none are worked in the Melmerby scar limestone.

The limestones are the chief depositories of ore, particularly that called the *great* limestone, which is considered to have produced as much lead as all the other sills together. Next to the limestones, the strata of sandstone called *hazles* are the most productive of ore; but the lead-bearing veins appear compressed between these hard sills. In Arkendale the sills of chert yield considerable quantities of galena, but this rock does not occur in the mining field further north. In shale the veins are comparatively barren, and in traversing these soft strata weak veins 'hade' considerably.

The hade of the veins is variable in degree, and in direction. When the veins in Weardale point east and west, they hade towards the south; but in Allendale and in the Aldstone moor country they generally hade towards the north: the strata are universally elevated on the side towards which the veins dip.

Veins, that are otherwise favourably circumstanced for producing ore, are more particularly so if the throw or alteration in the level of the beds of limestone, occasioned by the vein, does not exceed one or two fathoms: for then both cheeks of the veins correspond in their nature, and limestone does not become opposed to shale or any other barren stratum.

The beds above described forms the whole of the middle and lower regions of the escarpment of Cross Fell, extending about two-thirds up that mountain. They are seen near this point to rest on the old red sandstone and greywacké slate, as described in Mr. Buckland's paper. Greywacké-slate also occurs near the head of Swale-dale, perhaps connected with this tract. The limestone beds, having continued to Kirby Stephen,* a branch of mountain limestone is thrown off from them to the north-west; which, ranging by Orton, Hesket, Ireby, Cockermouth, Egremont, and Ravenglass, skirts the slate mountains of Cumberland on the north-east, north, and north-west. Farther to the south near Ingleborough, another similar branch is detached, which skirts the southern portion of the same mountain, occupying all the lower portion of the valley of the Kent, as far as Kendall; thence crossing the æssuary south of Ulverstone, near which place the hæmatitic iron ore is procured abundantly from this rock;† and proceeding over the mouth of the Daddon, till it almost joins the former branch, and thus completes a calcareous ring encircling the transition district of the lakes. This calcareous ring is attended, as has been before stated, by an exterior zone of the coal-formation. These branches have never been accurately described or minutely examined. It should seem, however, that the beds of limestone are in them less interrupted by heterogeneous strata, and of greater thickness than farther north, composing almost exclusively the substance of entire mountains. This description particularly applies to the southern branch; and may be extended to the base of Ingleborough. The vast base of Ingleborough, near 30 miles in circuit, consists of limestone; which extends in a similar manner beneath the neighbouring mountains of Whernside, Pennegent, Greg roof, Colm hill, &c. The summits of these mountains consist of the millstone-grit formation. A thin seam of coal also occurs near the top of Whernside and Colm hill. At the foot of Ingleborough, a contact of the mountain limestone and transition slate, was observed by Lord Webb Seymour and Mr. Playfair, going the Askrigg road from Ingleton. About a mile and a half from the latter, an opening appeared in the side of the hill, on the

* Mr. Greenough's Map represents the junction of the limestone and slate incorrectly in the neighbourhood of Dent dale and Houghill fells. The hills marked *m. n. o. p.* ought to be have been coloured slate, not limestone, as also the Riggs, and the southern point of Houghill fells; the southern branch of limestone is also incorrectly given. Whin fell is not limestone, and, e contra, the hills south of the road from Kendal to Newby bridge, as far as the sands of Morecombe bay, are limestone, though in many cases coloured in the map as slate.

† One perpendicular vein of iron ore traversing the limestone is 30 yards wide. Large reniform nodules of hæmatites, some even weighing 4 cwt. are found in the loose ore.

right hand, about 100 yards from the road, formed by a large stone which lay horizontally, and supported by two others standing upright. On going up to the spot, they found it to be the mouth of a small cave; the stone lying horizontally being part of a limestone bed, and the two upright stones vertical plates of argillaceous schist. The limestone bed which formed the roof of the cave, was nearly horizontal, declining south-east; the slate nearly vertical, stretching from west-north-west to east-south-east. The higher regions of the mountain are described by these observers as consisting of strata of limestone and grit, nearly horizontal and alternating. A junction almost similar may be seen at a cascade on the river Greata, called Thornton Ford: but here, on the south side of the river, a breccia, containing fragments of the slate imbedded in a calcareous cement, is seen interposed between the limestone and slate. The latter rock here occurs at the height of 7 or 800 feet above the level of the sea. The same slate extends lower down the valley, and is quarried nearer Ingleton.

The numerous caverns in the district surrounding Ingleborough are well known; and there is not one of the many rivulets which run from the base of the mountain, that has not a subterraneous passage of some extent: all the springs rise about the summit, among the strata of grit, and sink or fall into some hole as soon as they descend to the limestone rocks;* where, passing under ground for some way, they burst out again toward the base. Similar caverns occur in all the adjacent hills. Of these the Yordas Cave on the side of the mountain of Greg roof is the principal.

Perpendicular precipices of limestone, provincially termed scars, exceeding 300 feet in height, are common. The romantic and bold scenery presented by the calcareous mountains of Yorkshire has before been noticed.

Calcareous rocks continue to prevail down the course of the Ribble as far as Clitheroe, and that of the Air as far as Skipton.

These notices have been confined to the formation we have been describing as exhibited by the out-crop of its strata on the western side of this grand mountain chain. The same beds are also laid open by the deep excavation of most, if not of all

* The source of the river Air affords a good example of this. It issues from Malham Tarn, a circular lake about a mile in diameter, on the summit of a lofty moor. Proceeding hence, it soon loses itself, and descends through a subterraneous passage; whence it again issues at the foot of Malham Cove, a perpendicular limestone rock 288 feet high. During heavy rains the subterraneous passage is not sufficient to carry off all the water; the remainder of which makes its way over the surface, till it reaches the top of the rock, and precipitates itself thence in a magnificent cascade.

the vallies traversing the eastern slope of the same chain. They are thus seen in the vallies of the North and South Tyne for some miles above their confluence; in Weardale, Teesdale, Gretadale, Swaledale as low down as Richmond, Yoredale as far as Middleham, Coverdale, Netherdale and Wharfedale.

Between Clitheroe and the north of Derbyshire, the rocks of this formation are every where concealed by the formations of millstone-grit and shale.

The mountain limestone tract of *Derbyshire* extends from Castleton, which is its northernmost point, about 25 miles south of that place; its breadth does not appear any where to exceed about fifteen miles. Its form on the surface is very irregular. Buxton is situated on the north-western edge of this tract, Castleton on its north-eastern, and Matlock is on its south-eastern extremity. The surface of this district is occupied by the *out-crop of four strata of limestone, and of three beds of toadstone which lie between the strata of limestone.*

This tract is encircled (as has been already stated in treating of these formations) by superstrata of shale and millstone-grit: from beneath these, it rises on the eastern side, the strata ascending, though under a small angle, towards the west: but along its northern, western, and south-western edge, there ranges (as we are informed by Mr. Farey) a great fault, which by elevating the limestone tract, or depressing the district beyond it, has produced the effect of bringing the lowest bed of the limestone into immediate contact at the same level, with the strata of the shale formation; from which, were there no such disturbance, it would be separated by the intervention of the three upper beds of limestone and all those of toadstone. This is described by Mr. Farey as the great limestone fault. It is observable, that from the line joining the two eastern extremities of this fault, as from a hinge, all the strata rise more rapidly to the west.

The *lowest stratum of limestone*, being that, on the outgoing or outcrop of which are situated the Peak forest, Buxton and many towns on the south of it, passes across Dove dale and Wetton dale: the Weaver hills consist of it. In it are many caverns, as the immense one called Elden hole, north of Peak Forest town; the Devil's hall, connected by a tunnel with the Speedwell mine; Pool's hole near Buxton, and several of less note.

The thickness of the lower limestone is not known; it certainly exceeds 250 feet. We are consequently ignorant of the rock on which it rests. It is regularly stratified, consisting of very many beds, several of which are of considerable thickness; some thin ones are described as being a *freestone* (being of a more compact grain than usual): its colour varies from white to a yellowish stone colour: it rarely includes dark coloured beds. Small entrochi, numerous anomia and other shells and organic remains, occur throughout the whole of this stratum: in some mines, a thin bed of clay had been found in it. The lime yielded by this stratum is preferred to that of the strata above it. A bed of toadstone lies on it, but we propose to notice together the three beds of this substance, and therefore proceed to the

Second stratum of limestone. This is about 210 feet in thickness, and consists also of many beds; the superior ones are often of a dark colour, and contain nodules of black chert, shells of the genus anomia, madrepores, &c.; some of the beds are quite black. It contains layers of clay, and towards the lower part of it, some dark beds of limestone contain white madrepores. Imbedded masses of toadstone occur in it. On this lies another bed of toadstone, to which succeeds a

Third stratum of limestone. This, like the two preceding strata, consists of many beds whose average thickness is about 150 feet: and it is worthy of note that several of them are of magnesian limestone. In some places the upper beds partake so greatly of the nature of chert, as to be unfit for the purposes of the lime-burner; these cherty masses are usually called in Derbyshire, *dunstone*, or bastard limestone. Here and there are masses of white chert or china-stone. Some few beds contain entrochi; and towards the lower part are beds of a very black limestone, which, as it takes a very brilliant polish, is termed black marble. It contains thin beds of clay. On this stratum lies the third bed of toadstone, on which reposes the

Upper limestone. This, like the preceding, is about 150 feet in thickness. In it, as in the three lower strata, some thin beds of clay are found, and it contains imbedded masses of toadstone, though rarely. The upper beds are of that variety of limestone called swine-stone, and are often dark-coloured or black: near the top are found layers of nodules of black chert, similar in their arrangement to the flint nodules in chalk; in the upper beds also the shells called anomia, and others, are common. The middle beds contain vast assemblages of entrochi, and are occasionally quarried as marbles; and it is remarkable that in some places, where these middle beds basset out on the surface, masses are ploughed up from beneath the alluvial soil, exhibiting the casts of the inside of entrochi in chert; these are commonly called *screw-stones.* Blocks of these were heretofore used in the forming of mill-stones, which were employed instead of the French buhr-stone. This stratum contains beds of what is termed white chert or china-stone, of which considerable quantities are used in the Staffordshire potteries.* (F. 271 & seq.)

The impression of a crocodile was said to have been found in this stratum by Mr. H. Watson at Ashford (see Whitehurst); but we have been informed on enquiry, that an orthoceratite was mistaken for some part of this animal.

The description of the intervening strata of toadstone might here with propriety be introduced; but for the reasons before assigned, when mentioning the whin sill of Northumberland, it appears more convenient to refer them to the appendix on the occurrence of rocks of the trap formation among those associated in the coal districts: we shall here only remark, therefore, that the lowest is 66 feet in thickness; the middle 138 feet; and the upper 48. We have thus a total ascertained thickness of 1010 feet for the rocks constituting the calcareous tract of Derbyshire, of which 760 feet is limestone; and 252 feet, toadstone. This will serve as a point of comparison with the account already inserted of the beds of this formation in Northumberland. The thicknesses, however, of these beds, and especially of the toadstone, are very variable. We now pass to the consideration of the mines of Derbyshire, which are entirely situated in this tract.

The out-going of the strata just described, forms the great *Lead district* of Derbyshire; very numerous veins have been worked in it principally for lead, but the ores of zinc, manganese, copper, and iron, also occur in them; but they are more plentiful and productive when in the limestone, than when in the other strata. It has been supposed that lead ore has not been found in the toadstone, but nineteen instances of its discovery in that situation, in strings and short branches, are mentioned. A vein, somewhat approaching the *perpendicular*, is in Derbyshire termed a *rake vein*. Rake veins are from two or three, to thirty or forty feet wide. A large cavity, often nearly *horizontal*, between beds of limestone, and containing spars and ore, is termed a *pipe vein*. Veins (or rather beds) of this description are sometimes of considerable height, and from two to 500 feet wide, and are commonly connected with the surface

* It is in a mountain composed of limestone, that the beautiful masses of various coloured fluor spar, termed Blue John, are found. The mountain has no appearance of regular stratification, and is full of fissures and caverns of immense depth: the fluor occurs in those nearly horizontal beds, or rather openings, which are termed *pipe veins*, and is found of a roundish form, in which it seems to have crystallized; but the centre is frequently hollow. It is from these masses that elegant vases, &c. are manufactured by Mawe & Co. (M. 69.)

by means of a rake vein; when without this kind of connexion, the nearly horizontal deposite is termed a *flat-work*, which is rare. The direction of the veins containing ore, appears to be nearly east and west, and their hade or underlie beneath the surface, towards the north or south; but in this respect, it is said that a vein will change two or three times from north to south: these veins are crossed by others whose direction on the surface is nearly north and south. The east and west veins in descending, are always *cut off by the strata of toadstone*,* which therefore pass through and divide them; and it is worthy of note, that when the vein is again found in the stratum of limestone beneath the toadstone, it is not *immediately on a line* with the upper part, nor exactly of the same nature; in this case a vein is said to have *squinted.* The toadstone is said sometimes to assume the consistence of clay. It has been before noticed that the limestone strata contain thin beds of clay, termed by the miner *way-boards*; these sometimes pass through and divide the veins of ore in the same manner as the toadstone does: and so complete is the separation of the veins of ore by the clay and the toadstone, that not even the water in the upper part of the vein penetrates through them into the part beneath. The sides or walls of a rake vein are commonly lined by fluor, or cawk, or calcareous spar, termed by the miner *vein-stuff*; between, or against these, lies the ore, which sometimes fills up the space between them, and is then termed a *rib of ore.* But it sometimes happens that the vein-stuff of each wall of the vein is nearly compact, both so completely occupying the vein, that they meet together in close contact in the middle; forming what might be termed from its appearance, a vertical crack down the vein. The two faces in contact, appear as though they had been polished, and are ribbed or somewhat fluted horizontally; and the face of each is sometimes covered by a remarkably thin coating of lead ore; these planes, when separated, are the *slickensides* of the mineralogist. This circumstance is altogether remarkable in itself, but an extraordinary effect ensues when one side of the vein-stuff is removed. The other side then cracks, especially if small holes be made in it, and fragments fly off with loud explosions, and continue so to do for some days; as is the case in the Gang mine in Cromford,

* We must, however, exercise some caution in adopting this opinion; for since the publication of the works whence the above information was extracted, indications have in some instances been found of the passage of the lead veins, into the toadstone. The subject is yet veiled in some obscurity, but it is certain that the ore generally stops where the vein descends into the toadstone, in which the lead has hitherto been found only in very small quantities, chiefly in strings.

where the miner, availing himself of this circumstance, makes with his pick small holes about six inches apart and four inches deep, in one surface, after the other is removed; and the consequence is, that on his return in a few hours, he finds every part so treated, ready broken to his hand. (M. passim. F. 243, and seq.)

Of those remarkable derangements in the strata of Derbyshire, termed *faults* by the miner, we have no very clear geological account. The direction of some of them on the surface is detailed in the 1st volume of Farey's General View of Derbyshire, &c. to which we refer the reader: occasionally they appear to be very extensive, and their consequences very extraordinary; but a knowledge of their width, dip, and contents is yet a geological desideratum. Some of them have intersected the veins of lead ore, and are said to have introduced rounded quartz pebbles or gravel, alluvial clay, and other extraneous mineral matters, into them.

This limestone tract is, as usual in this formation, distinguished by the abrupt and wild features of its narrow rocky dales, by numerous caverns; and by the frequent engulphment of its streams in subterraneous courses, called swallow-holes. Mr. Farey's report contains a very copious list of these objects.

Section IV.

OLD RED SANDSTONE.

On the North-West of the Penine Chain.

To complete our description of the rock formations entering into the composition of the Penine chain, it now only remains to mention the old red sandstone. This rock, so extensive in the coal districts of the south-west of England, has yet been observed only in one limited portion of this chain; namely, under the escarpment of Cross Fell; where it may be traced for 15 miles from near Melmerby to near Murton; occupying an intermediate position between the mountain limestone and the adjacent tract of greywacke slate, described by Mr. Buckland in the 4th volume of the Geological Transactions. It appears here in its common form of a coarse puddingstone.

CHAPTER III.

Central Coal District.

Under this division we shall include I. The coal-field surrounding Ashby-de-la-Zouch, on the borders of Leicestershire and Staffordshire. II. That of Warwickshire. III. That of south Staffordshire or Dudley. IV. Indications of coal near the Lickey Hill. As these do not appear to form constituent parts of one whole, in the same manner with those connected with the Penine chain, we shall here so far depart from our former method, as to treat of the various formations accompanying each of these three coal-fields, in describing that individual field.

Section I.

THE ASHBY COAL-FIELD.

This district occupies an area of irregular figure, nearly in the centre of which, the town of Ashby-de-la-Zouch is situated. The longer diameter of this area, from north-west to south-east is about 10 miles; the shorter, from south-west to north-east, about eight miles.

The eastern extremity of this area approaches almost closely to the transition district of Charnwood forest. The coal-formations occupying it are much broken, and may perhaps when fully examined be found to constitute two small detached basins, rather than to belong to one continuous field. Of these portions, one ranges by Ashby Wold, about three miles on the west of Ashby; the other by Cole Orton, about the same distance on the east.

The former, or Ashby Wold portion, ranges from Swepston four miles south of Ashby, to Bretby in Derbyshire, about the same distance on the north-west; the outcrop of the beds sweeps in a curved line between these two places, the inclination of the strata being inwards, that is, towards Ashby: but between this outcrop and that town, another crop has been traced near Brothorpe, dipping in a contrary direction; a circumstance which has induced some of those practically acquainted with this district, to consider it as forming a long elliptical basin, of which the axis ranges from north-north-west to south-south-east, between the villages before mentioned of Bretby and Swepston. More than 20 coal works have been opened on this line. The deepest of these is sunk 246 yards. One of the coal-beds attains the extraordinary thickness of from 17 to 21 feet. This is probably occasioned by the running together of two or more coal-seams, a circumstance of which other examples occur in the coal-fields of Warwickshire and Dudley.*

The eastern portion of the Ashby coal district may be said to commence in the pits belonging to Sir George Beaumont, about a mile and a half north-east from the above town. Two coal beds, each a yard and a half thick, are here worked: the strata dip to east-north-east 1 in 12. On Cole Orton Moor, several coal seams which have been proved to lie above these, have been worked to the depth of 116 feet. The dip here continues the same; that is directly towards the transition group of the Charnwood forest hills which rise at the distance of less than a mile from this point; this circumstance appears embarrassing, as it would naturally have been expected that all the strata would have been found to rise and crop out on their so near approximation to this chain: somewhat further however, and close to this chain, such is found to be the case; for at Thringston the coal-measures have been proved, and found to rise, although very confusedly, towards it.

At Stanton Harold and other places on the north of Cole Orton, as far as Ticknall, coal has been raised, probably from a continuation of the same series.

It should generally be observed, that the strata of the red ground and newer red sandstone are so confusedly intermixed on the surface with the coal rocks of this district, as to render the elucidation of the details of its structure, an almost hopeless task.

The millstone grit and shale have not yet been ascertained to exist in this district; but a line of detached carboniferous

* A section of one of Lord Moira's pits on Ashby Wold presents—

		Ft.	In.
Various strata of bind occasionally containing ironstone		53	7
Coal		3	—
Bind, &c.		128	5
Coal		3	4
Bind, &c.		177	7
Coal		4	10
Bind, &c.		46	9
Kennel Coal		2	9½
Bind, &c.		128	3
Main Coal	Rider coal	5	5
	Fire clay	3	—
	Coal	15	—
	Total	571	10½

limestone, of a very remarkable character, flanks it on the north-west. These require a more particular description. There are eight of these detached points, each occupying but a few acres in extent, surrounded, and as it were insulated, by overlying strata of the newer sandstone formation.

The limestone thus occurring, is strongly impregnated with magnesia, resembling in this several of the strata of the third bed of carboniferous limestone in Derbyshire. This circumstance has led some geologists to confound it with the newer magnesian limestone, from which however it is clearly distinguished by its extraneous fossils, which entirely agree with those of the mountain limestone. The quarries of Ticknall afford abundant proofs of this.

Some of these islets of limestone are nearly horizontal in their stratification, and in that case lie low: others are highly inclined and rise into hills, but of no great elevation.

As a guide to those who may be inclined to attempt the further elucidation of this district, a list of these points is subjoined, following the order in which they occur in proceeding from south to north.

1. The most southerly of these points is about five miles from Ashby on the Loughborough road, a quarter south of that road, at the point where it is crossed by the projected canal; here on each side of the little streamlet of Grace Dieu, are quarries of this limestone. The strata are viewed on the large scale, nearly horizontal, but have several partial undulations. The sienitic crags of Charnwood forest, rise within a quarter of a mile on the south, and the coal-measures are not far distant on the west; but the relations of the limestone with neither formation is distinctly displayed.

2. One mile north from this spot, close to the village of Osgathorpe, are quarries of the same rock, which here also appears nearly horizontal.

3. Half a mile north of this village, rises Barrow hill, entirely composed of this rock: the strata here are highly inclined, but the stratification is very indistinct.

4. Three quarters of a mile north-north-west, and in a line with this hill, is Cloud hill, altogether of the same nature: here the strata rise towards the east seventy degrees.

5. On the further side of a plain occupied by red ground, and at the distance of one mile and a half north-north-west from Cloud hill, rises Breedon hill; which although it cannot boast any considerable elevation, is rendered by its insulated position and the church tower on its summit, a very conspicuous object. Dr. Beddoes has described the strata of this hill as rising on all sides towards its centre; but on a very careful examination

the strata were observed to rise uniformly to the east. It is not however always easy to distinguish the lines separating them; and to this the Doctor's error, (if it can be one) must be attributed. The angle of their inclination is from 45° to 60°. This hill affords the best station for commanding a general view of the Charnwood Forest Chain, which is seen stretching on the south in a bold range of broken and conical summits.

6. In Stanton Park in the parish of Stanton-harold, are quarries of the same rock horizontally stratified.

7. Also in Calke Park, near the east of a place called Diminsdale.

8. Close to the north entrance of the village of Ticknall, are very extensive quarries of this rock horizontally stratified. The lower strata here are of a blue colour, and these abound in entrochi, terebratulites, &c. The upper are of a pale yellow or green colour and thicker; this latter description applies generally to the character of the stone in all the other quarries.

It does not appear on what substratum these calcareous masses rest; but it may be observed that the inclined masses all rise to the east; which is the direction they would naturally assume, if we suppose the transition chain of Charnwood forest to be prolonged towards the north, beneath this covering of the newer sandstone; and thus to form the basis of this system of hills.

Section II.

THE WARWICKSHIRE COAL-FIELD.

This coal-field extends in length about sixteen miles from Wyken and Sow, (villages about three miles east from Coventry,) on the south-east, to Polesworth and Wareston, (about five miles east from Tamworth,) on the north west. The average breadth of the coal tract may be about three miles. All the strata rise to east-north-east; the inclination becoming more rapid towards the eastern edge of the field, where it in many places exceeds an angle of 45 degrees with the horizon, and decreasing towards the west to 1 foot in three, and lastly in 5. The outgoing of the more inclined strata on the east, forms a well defined, although low escarpment, which presents in some places the strata of the coal-measures; in others, the subjacent strata of millstone grit. Beneath this escarpment is a level plain, in which all the inferior strata are covered up and concealed by horizontal and overlying strata of the newer red sandstone and marle; which also mantle completely round the coal-field, whose measures are carried beneath them on the western side by their dip.

The principal coal-works are at Griff and Bedworth, near the south of the field: at Griff four beds of coal are worked, the depth of the first being 117 yards, and the principal seam being three yards in thickness. The Bedworth works are upon the same beds, but here the first and second coal-seams of Griff run together, and constitute one five-yard seam. The interposed strata of shale which separate them at Griff, and which are there found to be in the eastern shafts thirty-three yards and in the western twenty-five, gradually thinning away in proceeding westwards, till at length they entirely vanish.

Between Griff and Nun Eaton, in the bottom of the valley called Griff hollow, and thence ascending the northern hill, large masses of a very compact greenstone may be observed traversing the coal-shale: they do not appear to be dykes, but portions of two beds regularly and conformably interstratified among those of shale. (For a further account, see the appendix to this article.)

The millstone-grit, which has been mentioned as the lowest formation exhibited in this coal-field, may be seen to the greatest advantage on the edge of the escarpment half way between Atherstone and Nun Eaton. The character here assumed by this rock, is that of a very compact and cherty sandstone: its strata rise to east-north-east in an angle of 45°.

The opposite dips of this coal-field and the nearest parts of that described in the preceding article, shew that they are detached; although the interval between them at one point does not exceed six or seven miles. The overlying beds of the newer red sandstone completely conceal the substrata, on which both these coal-fields rest in that direction.

Section III.

DUDLEY, OR SOUTH STAFFORDSHIRE COAL-FIELD.

This coal-field extends in length about twenty miles, from near Stourbridge on the south-west over Cannock Chase, to Beverton near Badgeley on the north-east. Its greatest breadth near Dudley, may be about four miles. Its superficial area has been found by actual survey to equal sixty square miles: of this the northern portion from Cannock Chase to near Darlaston and Bilston, affords many coal-seams of eight, six and four feet in thickness. The southern portion, extending thence to near

Stourbridge, is about seven or eight miles in length and four in breadth. It is advantageously distinguished from the former by the occurrence throughout, of a coal-bed of the enormous thickness of 30 feet. This is considered as lying above the Cannock Chase beds, and cropping out round Bilston and Darlaston.

No satisfactory account has yet been published of the northern portions of this district; but the southern has been fully illustrated by a very able memoir of Mr. Keirs, published in Shaw's History of Staffordshire, and by a paper of Dr. Thomson's in the Annals of Philosophy.

To the spectator viewing this district, it appears to be traversed from north-west to south-east by a line of hills, not absolutely continuous indeed, but yet having an uniform general direction in the interval, near the centre of which stands the town of Dudley. On examination, however, the hills on the north, and those on the south of that town, will be found to differ entirely in their constitution, although they agree in their line of bearing. The northern chain (composed of three insulated oblong hills) is entirely constituted of limestone, disposed in highly inclined strata, rising on all sides from the base to the ridge of the hills, and forming on the summit an acute arch; or as the Wernerians would say, exhibiting a saddle-shaped stratification. Against the sides of these hills all the coal-measures (as reposing on the limestone) crop out at a considerable angle; but become more flat in proportion as they recede from these hills; which constitute, if the expression may be allowed, the centre of elevation. The other chain of hills on the south of Dudley, is entirely composed of one mass of basalt and amygdaloid; the relations of which to the coal-measures have not been clearly ascertained, further than that they preserve their usual level in approaching the chain; and evidently do not crop out round it, as round that of limestone. Two alternations remain with regard to this chain, it may be either the protruding edge of a vast basaltic dyke traversing the coal-field, or an over-lying mass; which latter is the opinion generally entertained on the spot. The particular description of this chain is of course reserved for the appended article on the trap rocks of the coal districts.

On the west, near Wolverhampton, and south, near Stourbridge, the coal-measures appear to dip beneath the beds of the newer red sandstone formation; since therefore, we find the measures of this and the Warwickshire coal-field dipping in opposite directions beneath these superstrata, it seems probable that they may extend continuously below this.

On the eastern limit of the coal-field near Walsall, the same limestone with that of Dudley again rises: and the coal-measures may again be observed to crop out against it; thus lying as it were in a trough between these two towns.

The following information is extracted from 'A geological sketch of the country round Birmingham,' by Dr. Thomson, inserted in the Annals of Philosophy, vol. 8. p. 164—170.

Some years ago Lord Dudley cut an underground canal to his limestone quarries near Dudley, in the course of which undertaking all the coal-beds betweeu the limestone and the ten-yard-coal were cut through. All the coal-pits in this country go as low as the ten-yard coal, in which their great workings always exist. Hence all the different beds which constitute the coal-formation in this place have been cut through, and are known. The following table exhibits the name and thicknesses of these different beds as determined by Lord Dudley's canal, and by a coal-work at Tividale, in the parish of Rowley, wrought by Mr. Keir. This table I have taken from Mr. Keir's paper above-mentioned, making such alterations in it as will serve to render it more intelligible to the reader. I begin with the lowest bed, which lies immediately over the limestone, and terminate with that bed which constitutes the immediate surface of the earth:—

Names of the Beds.	Local Names of Ditto	Thickness.		
		Yds.	Ft.	In.
1. Slate-clay	Wild measures	30	0	0
2. Limestone	Limestone	10	0	0
3. Slate-clay	Wild measures	76	2	0
4. Coal	1. *Coal*	0	2	0
5. Slate-clay	Wild measures	40	0	0
6. Coal	2. *Coal*	5	0	0
7. Slate-clay	Black measures	2	2	0
8. Coal	3. *Good Coal*	3	1	0
9. Gravel ?	Rough spoil	2	0	0
10. Coal	4. *Good Coal*	3	0	0
11. Slate-clay	Wild measures	9	0	0
12. Slate-clay	Pot-clay	2	0	0
13. Coal	5. *Heathing Coal*	2	0	0
14. Slate-clay	Clunch and iron-stone	7	0	0
15. Coal	6. *Main Coal*	10	1	6
16. Bituminous shale	Black batt	0	0	7
17. Slate-clay	Catch earth	0	2	9
	Carried forward	204	2	10

Names of the Beds.	Local Names of Ditto.	Thickness.		
		Yds.	Ft.	In.
	Brought over	204	2	10
18. Coal.............	7. Chance *Coal*.........	0	0	10
19. Bituminous shale ..	Black batt	2	0	0
20. Slate-clay	Clunch and iron-stone .	0	2	9
21. Sandstone	Rock or rock binds...	2	2	10
22. Slate-clay	Clunch binds	1	1	0
23. Coal.............	8. Chance *Coal*	0	0	9
24. Slate-clay	Clunch parting.......	0	0	10
25. Sandstone	Strong rock	1	1	0
26. Sandstone	Rock with laminæ of coal	1	1	0
27. Sandstone	Strong rock..........	1	1	0
28. Sandstone	Rock binds..........	5	1	0
29. Slate-clay	Clunch with iron-stone	4	2	0
30. Sandstone	Rock binds..........	5	2	0
31. Slate-clay	Clunch with iron-stone	0	2	9
32. Slate-clay	Clunch binds	8	2	0
33. Slate-clay	Penny-earth with iron-stone	2	1	0
34. Coal	9. *Coal*	0	1	3
35. Slate-clay	Black clunch	2	1	0
36. Coal	10. Broach *Coal*	1	0	9
37. Slate-clay	Kind clunch	0	1	0
38. Sandstone	Rock binds	2	1	0
39. Shale ?...........	Parting emitting fire-damp	0	0	3
40. Sandstone	Rock binds	0	2	0
41. Sandstone	Rock...............	0	2	0
42. Slate-clay	Fine clunch	4	0	0
43. Slate-clay	Fire-clay	1	1	0
44. Coal	11. *Coal* called two-foot-coal	0	1	6
45. Clay	Soft clunch..........	2	2	9
46. Slate-clay	Clunch binds.........	4	0	0
47. Slate-clay	Kind clunch with iron-stone	3	2	2
48. Sandstone	Black rocky stuff.....	2	2	0
49. Clay with coal	Smutt	0	0	3
50. Sandstone	Rocky black stuff	0	1	0
51. Slate-clay	Wild stuff...........	5	1	8
	Carried over	277	2	2

Names of the Beds.	Local Names of Ditto.	Thickness.		
		Yds.	Ft.	In.
	Brought over	277	2	2
52. Slate-clay	Binds with balls of grey rock	3	2	0
53. Slate-clay	Red wild stuff	2	1	6
54. Sandstone	Greenish rock	1	1	0
55. Slate-clay	Red wild stuff	13	2	6
56. Slate-clay	Grey clunch	2	1	3
57. Slate-clay	White clunch........	1	0	3
58. Clay mixed with coal	Smutt	0	0	10
59. Slate-clay	Clunch with iron-stone in it..............	2	2	3
60. Sandstone	Rock with coal interspersed	1	2	0
61. Slate-clay	Red-coloured roach...	1	2	0
62. Clay	Blue clay	0	1	0
63. Slate-clay	Brown-coloured roach.	2	0	0
64. Red-clay.........	Brick clay	1	2	6
65. Soil	Soil................	0	1	0
	Total thickness	313	1	3

From this table we see that the beds distinguished by different names in this coal-formation amount to 65, and that its whole thickness is 313 yards, 1 foot, and 3 inches, or about 156 fathoms. The main coal, which is the great object of the colliers in that country, is about 60½ fathoms below the surface in the neighbourhood of Dudley. The beds of coal are 11 in number, five above and five below the main coal. The first bed occurs at the depth of 55 yards, or 27½ fathoms below the surface, but none of the beds above the main coal are considered as worth working. The beds below the main coal are of very considerable thickness. None of them are wrought in the neighbourhood of Dudley; but on the north side of Bilston, and in Cannock Chase, are the beds which supply the country with fuel. The main coal, or ten-yard-coal, consists, in fact, of 13 different beds, some of them lying close to each other, and others separated from each other by very thin beds of slate-clay, called *partings*. The following table exhibits the names and thickness of these different beds, as stated by Mr. Keir, in the Tividale Colliery. I have compared them with some other collieries, and found them nearly the same :—

	Yds.	Ft.	In.
1. Roof floor, or top floor..............	1	1	0
Parting of four inches.			
2. Top slipper, or spires.........	0	2	2
3. Jays	0	2	0
White-stone, called patchel, one inch.			
4. Lambs	0	1	0
5. Tow, or Tough, or Kitts, or Heath ...	0	1	6
6. Benches	0	1	6
7. Brassils, or Corns	0	1	6
Foot coal parting (sometimes only).			
8. Foot coal, or bottom slipper, or fire coal	0	1	8
John coal parting one inch.			
9. John coal, or slips, or veins..........	1	0	0
Hard-stone, 10 inches, sometimes less.			
10. Stone coal, or long coal	1	1	0
11. Sawyer, or springs..................	0	1	6
12. Slipper	0	2	6
Humphrey parting.			
13. Humphrey's, or Bottombench, or Kid..	0	2	3
Total thickness of coal....	9	1	7

About five yards of this main coal, namely, the lambs, the brassils, upper part of John coal, bottom part of stone-coal, and sawyer, consist of coal of the best quality, which is employed in private houses. The quality of the remainder is inferior. On that account it is used only in the iron furnaces, which abound in this part of the kingdom. The coal is of the species of slate-coal. It does not cake; and burns away more rapidly than Newcastle coal, leaving behing it a white ash. But it makes a more agreeable fire, and does not require to be stirred.

The coal-beds dip towards the south, and rise towards the north; so that at Bilston the main coal crops out, and disappears altogether. A very curious phœnomenon takes place at Bloomfield Colliery, to the south of Bilston. The two upper beds of the main coal, called the roof floor and top slipper, separate from the rest, and are distinguished by the name of the *flying reed*. This separation grows wider, and at Bradley Colliery amounts to 12 feet, four beds of shale, slate-clay, and ironstone, being interposed. These two upper beds crop out, while the rest of the main coal goes on to Bilston, and is only eight yards thick.

To give the reader some idea of the degree of regularity which the different beds exhibit in this district, I shall give a table of the different beds bored through at Bradley colliery, near Bilston, beginning, as before, with the lowest, and terminating with the surface bed :—

Names of the Beds.	Local Names of Ditto.	Thickness.		
		Yds.	Ft.	In.
1. Coal	*Heathing Coal*			
2. Slate-clay	Clunch	3	1	0
3. Shale	Table batt	0	2	0
4. Coal	*Coal*	0	0	6
5. Shale	Hard batt	0	1	0
6. Clay-iron-stone	Iron-stone	1	0	0
7. Slate-clay	White clay	0	2	0
8. Slate-clay	Blue clay	0	0	6
9. Clay	Short earth	0	1	6
10. Coal	*Main Coal*	8	1	3
11. Shale	Black batt	0	2	6
12. Clay-iron-stone	Iron-stone	0	0	8
13. Slate-clay	Blue binds	1	2	0
14. Shale	Batt	1	1	0
15. Coal	*Flying reed*	1	2	0
16. Shale	Batt	0	2	0
17. Slate-clay	Blue clunch	3	0	0
18. Slate-clay	Ditto containing four thin iron-stone beds	4	0	0
19. Sandstone	Grey rock	0	1	0
20. Slate-clay	Clunch	0	1	6
21. Sandstone ?	Peldon	0	2	0
22. Sandstone	Grey rock	1	0	0
23. Slate-clay	Blue clunch	6	0	0
24. Sandstone	Grey rock	1	0	0
25. Slate-clay	Blue clunch	8	0	0
26. Red sand	Sand	10	0	0
27. Soil	Soil	0	2	0
	Total	56	2	5

We see from this table that the greater number of the beds which cover the main coal at Tividale have cropped out and disappeared before the main coal got to Bradley. At Tividale the main coal is $60\frac{1}{2}$ fathoms below the surface; at Bradley it is only $20\frac{1}{2}$; making a difference of 40 fathoms. Thus we see that the dip south is pretty considerable, amounting probably to 1 foot in 90. Indeed, if we subtract the flying reed, and all the beds between it and the main coal, amounting to about three fathoms, we should increase the dip somewhat.

The curious phœnomenon of the flying reed seems to show very clearly that the different beds of which the main coal consists, were deposited at different times, and at considerable intervals from each other. During one of these intervals the beds separating the flying reed from the rest of the main coal seem to have been deposited towards the north of the field, while no deposit whatever took place towards the south of the field.

The substances which occur in this coal are the same as those found in the coal of other coal-fields; namely, 1. Iron pyrites, which occurs chiefly in that bed of the main coal called Brassils, and which furnishes a coal of the best quality. 2. Galena in very small plates and strings; it occurs likewise in the Newcastle coal. 3. Gypsum and calcareous spar: both of these (chiefly the former) may be seen occasionally in thin plates encrusting pieces of coal. When the coal is in small fragments it is called *mucks* by the colliers. These small fragments are left in the mine, and constitute nearly one-third of the whole coal in the bed. The pillars left standing probably amount to another third; so that the miners in this country extract only one-third of the coals, and leave two-thirds in the mine. This wasteful mode of working is to be ascribed to the low price of coals. As far as I have had an opportunity of judging, and I have been in most of the coal countries of Great Britain, the price of coals at Birmingham is less than any where else except Glasgow.* The consequence is, that the small coal will not bear the expense of removal. It is, therefore, left in the pits in prodigious quantities, where it is speedily destroyed by the weather. It is a pity that this enormous waste, which must hereafter be dreadfully felt in that country, could not be prevented. The consumption of coals in this part of England is prodigious. All the neighbouring counties, to a considerable distance, are supplied by means of the numerous canals of which Birmingham constitutes the centre. Besides this, an immense quantity of coal is required for the iron works, which are established in the neighbourhood of Dudley to the amount of 68. These smelted an immense quantity of iron; probably

* I consider the wonderful rapidity with which Glasgow has advanced in population, manufactures, and trade, as owing in a great measure to this circumstance. The inhabitants pay less for their coals than is paid in every other part of Great Britain.

more than the quantity manufactured in all the rest of Great Britain. But the low price to which iron has of late sunk (about £9 or £10 sterling per ton) has in a great measure destroyed this formerly lucrative manufactory. No less than 32 of the 68 furnaces have stopped, or *been blown out*, as the phrase is in Staffordshire. The Welsh iron manufacturers, it seems, produce a greater proportion of iron from their ore, and work with less coals than they can do in Staffordshire. They are able, in consequence, to undersell them. This opposition has been carried so far as to sink the price of iron much lower than it seems possible to manufacture it at. Before the late peace it sold at £18 per ton, which was almost double its present price.

Tracts of coal occur in this coal-field distinguished by a blacker colour, possessed of less lustre, and burning with less flame, than the common coal. Such tracts are called *blacks*. It contains less bitumen, and approaches nearer to coke than the rest of the coal. In cracks of the superincumbent beds there occur shining pieces of coal, like Kilkenny coal. According to Mr. Kier, it is imbedded in cubic cells, formed by thin planes of calcareous spar, intersecting each other at right angles.

This coal-field contains a less number of sand-stone beds than the coal-formations in Scotland and the north of England. The slate-clay, called in that country *clunch*, is harder, has more lustre, and is composed of finer particles, than slate-clay in the coal-formation generally is. The *butt* is a black slate or shale, which may be split into very thin fragments, and which in general contains much less bitumen than bituminous shale. It approaches more nearly to *drawing-slate* in its appearance, only its slaty fragments are much thinner. The clay-iron-stone occurs in various beds, but is only wrought in two; namely, in the bed that occurs under the broach coal in the neighbourhood of Wednesbury, and in that which occurs under the main coal. This last is the bed usually wrought for iron ore. This ore is what mineralogists term *clay-iron-stone*. It is, in fact, a carbonated hydrate of iron, usually mixed with clay. Probably the proportion of clay is greater here than in Wales. This would account for the greater produce of the Welch ones, and the smaller quantity of fuel which they consume. This iron ore, when taken out of the mine, is built up in heaps called *blooms*, four feet long, three feet wide, and 22 inches high. It is considered as weighing 35 cwt., each cwt. being 120 lb.: 1000 or 1200 such blooms are usually got from an acre of good mine.

In the Marquis of Anglesea's park, called Beaudesert, there is a mine of cannel coal, which is reserved for the exclusive use of the Marquis's family. This coal has a brownish-black colour, and much less lustre than common coal. The fracture is flat conchoidal, and quite smooth; but the cross fracture is more rough, and on that account has a blacker appearance. This coal is hard, and does not soil the fingers. Interspersed through it are numerous specks of a brown matter, very similar in appearance to Bovey coal.

Many *faults* occur in the coal-field. They are rents in the beds, which are usually filled with clay. Very frequently the height of the beds varies on the two sides of a fault. By a great fault which occurs near Bilston, the dip of the coal is reversed; that is to say, the coal-beds on the south side of the fault dip south, and those on the north side dip north. But this is an unusual occurrence.

Ironstone is found in several of the measures. But of these two only are worked for the ore. Vegetable impressions are common. The deepest pits are sunk about 140 yards.

Millstone-grit does not appear to have been noticed in this district; but the coal-shales repose immediately on the limestone.

Neither does the subjacent limestone itself belong to the same formation with that constituting the ordinary basis of the other coal-fields; but appears to be of much higher date, agreeing closely in its extraneous fossils and general character with the transition limestone which (as we shall have an opportunity of demonstrating hereafter) is separated from the mountain limestone by the important formation of the old red sandstone. The limestone of Dudley and Walsall will therefore be described in treating of the transition rocks.

This absence of millstone-grit, carboniferous limestone, and old red sandstone, and the immediate contiguity of the coal-measures and a transition rock, constitute a remarkable and important character in this coal-field: and in this it resembles some others in Shropshire, of which we shall hereafter have occasion to speak.

A little on the west of Dudley there is a considerable tract in which some of the collieries having taken fire from spontaneous ignition, many years back, the conflagration has spread to a great extent, and still continues in great activity: this produces many singular effects: smoke and steam may in some places be observed to rise; the vegetation on the surface is accelerated, and the ponds become thermal: where the ignited mass of coal has been nearer the surface, the argillaceous strata covering it have become converted by heat into a species of porcelain jasper, beautifully striped by the various degrees of oxidation induced in the iron it contains.

Section IV.

INDICATIONS OF COAL NEAR THE LICKEY HILL.

Mr. Buckland has observed some broken and confused traces of coal-measures and of transition lime near the north foot of the quartz rock of Lickey hill, a few miles south from this coal-field; and it has been reported that old coal-works exist in Ashby heath, about six miles south-east of the Lickey. But this statement rests on no good authority.

CHAPTER IV.

Western Coal Districts.

The coal-fields referred to this division, may be generally described as disposed around the transition district of North and South Wales. They may conveniently be subdivided into A. the north-western, including the coal-fields of Anglesey and Flintshire, B. the western, or those of Shropshire, and C. the south-western, or the three important coal-basins of South Wales, of Monmouthshire, and that of South Gloucester and Somerset.

Section I.

North-western, or North Welsh Coal-fields.

1. ISLE OF ANGLESEY.

The first then of these coal-fields which requires our attention is that contained in the Isle of Anglesey.

A remarkable valley traverses the whole of this island, running nearly parallel with the Menai Straits, and at the distance of about six miles from them: this opens on the south into the æstuary of Maltraeth, and on the north into Redwharf Bay. This valley is flanked on both sides by parallel belts of carboniferous limestone, in the depression between which, coal has been worked near the Maltraeth æstuary, and probably extends along the whole line. A few years since, shafts were sunk in the vicinity of Trefdraeth; and new trials have recently been made with success at Pen tre beren, about five miles north-east from the old pits. The veins are described as thick and extensive, and the coal of excellent quality.

2. FLINTSHIRE.

Carboniferous Limestone of North Wales, and Coal-field of Flintshire.

Before proceeding to the coal-field of Flintshire, it will be necessary to trace the line of carboniferous limestone skirting the edge of the North Welsh mountains, which, on arriving in that county (Flint), form the base of its coal-measures.

This commences near the port of Crickhaeth, on the south of Carnarvon, and hence proceeds northwards in an interrupted line along the shore, until it arrives at that town: here its strata dip in an angle of 10° or 12° towards the water, and rest on a conglomerate of quartz pebbles in a calcareous cement. From this place it proceeds along the Menai, forming its eastern bank as far as Bangor Ferry; beyond which point it is cut off, by the approach of mountains of older formation to the sea. The line recommences in the lofty promontory of Ormes head, a large detached hill, situated at the eastern extremity of the mouth of the æstuary of the Conway. This vast unconnected rock has no doubt been formerly an island, being at present joined to the main land only by a neck of low marshes. Some copper, though in no great quantity, is procured from two mines near the top of the mountain. The ore is malachite, found between the limestone in strata about two inches thick: the limestone resembles in appearance those varieties that contain magnesia. The limestone cliffs commence on the coast about four miles south-east of this promontory, and follow the line of the coast by Abergeley (where are some old lead mines) to the mouth of the Clwyd. In the vale of the Clwyd may be traced the old red sandstone, on which this formation appears to rest: it is perhaps co-extensive with this rock, but has never been particularly noticed. Beyond the Clwyd, the limestone is prolonged through Flintshire, where it forms the base of the coal-field we are about to describe; its direction continues to be westerly, as far as the neighbourhood of Holywell (where extensive lead-mines are worked in it); thence it turns southward, passes a little on the east of

Mold, skirts the little river Alain with a bold ridge of precipices, forms the Eglwysegg rocks in Langollen vale, which it crosses towards the south of that valley. It is broken and interrupted for a time by the slate mountain of Selattyn, (between Chirk and Oswestry) but again resumed in the hill of Lanymyneck, where its strata are nearly horizontal, and contain, as at Ormeshead, malachite; here the line expires.

The coal-formation which rests on this rock in Flintshire, commences with beds of shale and sandstone, answering in position and character to the shale and millstone-grit of Derbyshire. The coal is of different thickness, from three-quarters to five yards. The beds dip from one yard in four, to two in three. They immerge beneath the æstuary of the Dee; are discovered again on its opposite side, on the south of the peninsula of Wiral in Cheshire, where they finally sink beneath superstrata of the newer red sandstone, and are possibly prolonged beneath these, until they re-emerge in the great Lancashire coal-field.

The coal-measures extend from north to south somewhat more than 30 miles, from Llanassa, near the western cape of the æstuary of the Dee, to near Oswestry in Shropshire; forming an exterior belt, co-extensive with the range of the mountain lime from the north of the Clwyd; where the carboniferous lime is partially interrupted by the mountain of Selattyn; the coal-shales rest immediately on the transition slate of which that mountain is composed. Common, cannel, and peacock coal, are found.

Section II.

Middle Western Coal-fields.

(*a*) COAL-FIELDS OF THE PLAIN OF SHREWSBURY.

After quitting the coal-field just described, a district of much more confusion is entered upon: the general relations of which it will be necessary to sketch, before attempting any account of its coal-fields.

Near the point at which our survey of that coal-field terminated, the Severn quits the transition mountains, to flow through what may be termed, the plain of Shrewsbury, which is skirted at a variable distance on the south-west, by the prolonged chain of those mountains, whence frequent branches are detached, (under the names of the Long Mountain, Breddin Hills, the Stiperstone, and Longmont Forest) which advance far into the plains; and close to the last at Church Stretton, a still more important range of hills, of transition trap, slate, and quartz rock, which, under the names of Caer Caradoc, the Wrekin, &c. project many miles to the north-east, crossing the Severn, and extending almost to Newport. Thus the plain of Shrewsbury is bounded on the west, south, and east, by mountain chains of this age and class. The plain itself is generally overspread by strata of the newer red sandstone; but many broken patches of coal-strata, almost too limited in extent to be worthy of mention, are scattered over it, and in the intervals between the transition chains by which it is indented. No distinct account of these has yet been made public. Next to these narrow and broken coal-fields of the plain of Shrewsbury, we find on the eastern side of the Wrekin chain, one of much greater importance, that, namely, of Coalbrook Dale.

(*b*) THE COALBROOK DALE COAL-FIELD.

This coal-field, like that of Dudley, reposes on transition limestone, a long belt of which skirts the transition chain of the Wrekin, on the eastern side extending into Herefordshire: but sometimes between the limestone and the coal, a bed locally termed *die earth* (from the fact that beneath it the coal beds die, or cease) is interposed. This however in fact, is only a loose and impure form of the calcareous strata, adulterated by the mixture of particles of clay and sand. The nature of the fossils it contains clearly refers it to the same formation, which will be described in treating of the transition rocks. Trap rocks are also interposed in some places between the lime and coal; an account of which will be found under the proper head.

The coal-measures rise west-north-west at an angle of about 6°. On the eastern side, towards which they dip, they are succeeded by the strata of the great red sandstone tract, now generally referred to the newer sandstone; it has not however been absolutely ascertained whether they are prolonged beneath this sandstone, or cut off abruptly against it: the former opinion however is entertained by many of the most experienced miners. And as at the distance of 12 miles to the west, the coal-strata of the Dudley coal-field appear to emerge from beneath the same formation with an opposite inclination, they have been supposed to have a subterraneous connexion. However this may be, it does not seem possible for any one who has attentively studied the two sandstone and coal-formations in their character and relations, as exhibited in the clearest manner in the south-western counties, to entertain a doubt as to the class of formations which most properly claims the sandstone in question as one of its members. These remarks are rendered necessary, since it has been described as the old red sandstone by a most able and accurate geologist, from whom it is impossible to differ without hesitation and deference; but whose researches do not appear to have extended to those parts of the island which alone afford a completely satisfactory solution of this debated question.

This coal-field ranges from Wombridge in the parallel of Wellington, to Coal-port on the Severn, a length of about six miles; its greatest breadth is about two miles. It is composed of the usual members, namely, of *quartzose sandstone*, of *indurated clay*, of *clay porphyry*, of *slate clay*, and of *coal*, alternating with each other without much regularity, except that *each bed of coal is always immediately covered by indurated or slaty clay*, and not by sandstone. The series immediately belonging to the coal-formation is most complete in the deep of Madely colliery, where a pit has been sunk to the depth of 729 feet through all the beds, *eighty-six in number*, that constitute this formation.

The *sandstones*, which make part of the first 30 strata, are fine-grained, considerably micaceous, and often contain thin plates or minute fragments of coal. The 31st and 33d strata are coarse-grained sandstone entirely penetrated by petroleum; they are, both together, fifteen feet and a half thick, and furnish the supply of petroleum that issues from the *tar-spring* at Coal-port. At the depth of 430 feet occurs the first bed of very coarse sandstone or grit; its thickness is about 15 feet. The next bed of sandstone deserving notice, occurs at the depth of 576 feet, is about 18 feet thick, is fine-grained and very hard, and is often mixed with a little petroleum: the name given to it by the colliers is the *big flint*. The lowest sandstone, called the *little flint*, is the 85th in number, and is about 15 feet thick; the lower part is very coarse and full of pebbles of quartz; the upper is of a fine grain, and sometimes is rendered very dense and hard by an intimate mixture of iron ore; it occurs at the depth of 705 feet. Vegetable impressions are met with in most of the sandstone beds, but it is not known that they contain shells.

The *clay-porphyry* occurs only once in the whole series; it forms a bed nine inches thick, at the depth of 73 feet from the surface. It consists of a highly indurated clay of a liver-brown colour, in which are imbedded grains of quartz, of hornblend, and of felspar.

The *indurated clay* is in some beds compact, dull, and smooth; it is then termed *clod:* in others it is glossy, unctuous, and tending to a slaty texture, and is then called *clunch.* It encloses beds of *clay-ironstone* in the form of compressed balls, some vegetable impressions, and a few shells. The beds of iron ore are five or six in number; they all lie in the indurated clays, and all consist of balls or broad flat masses.

The *slaty clay*, called by the miners *basses*, is of a bluish black colour and slaty texture; it usually contains pyrites, and is always either mixed with coal, or combined with petroleum; in the former case it passes insensibly into slaty coal, and in the latter into cannel coal.

The first bed of *coal* occurs at the depth of 102 feet from the surface, is not more than four inches thick, and is very sulphureous; nine other beds of a similar nature, but somewhat thicker, lie between this and the depth of 396 feet. They are termed stinking coal, and are employed only in the burning of lime. The first bed of coal that is worked is five feet thick, and occurs in this colliery at the depth of 496 feet; between this bed and the big-flint sandstone, are two beds of coal, one ten inches, the other three feet thick. Between the big and little flints, which are about 100 feet apart, lie nine beds of coal of the aggregate thickness of about 16 feet. Beneath this, and the lowest bed of the whole formation, is a sulphureous eight-inch coal. The best coal of the above beds, usually presents a mixture of slate-coal and pitch-coal, rarely of cannel-coal; none of it possesses the quality of caking.

Of the numerous beds visible in this colliery, some are wanting in the neighbouring ones, and there exists also a considerable difference in the thicknesses of their respective beds.

(c) COAL-FIELDS OF CLEE HILLS AND BILLINGSLEY.

(1) *Coal-fields of the Clee Hills.*

A few miles south from the preceding coal-field, rise the Brown Clee Hill and the Titterstone Clee Hill, the former lying three or four miles to the north of the latter. Both of these, which rank among the most considerable mountains of Shropshire, exhibit coal-measures towards their central regions, the highest summits being formed of overlying masses of basalt. It is to be regretted that we are not yet in possession of any precise account of the relation of these rocks and the adjacent country. The following notices are chiefly extracted from Tracts in Natural History, by Robert Townson, LL.D.

These mountains belong to the flat topped hills, but are very irregular in their forms. They are about five or six miles in length, and about half as much in breadth. They resemble each other in their products; both contain coal and ironstone, which in both are in some parts covered by a thick bed of basalt; and this basalt in each forms two irregular ridges, higher than the other parts of the hill. They further agree in their strata dipping all round from their circumference to their centre like the sides of a bowl. But they differ greatly in the quantity of coal they yield. The coal in the Brown Clee Hill only lies in thin strata, whilst the principal stratum in the Titterstone is six feet thick. Three other beds, of less importance occur; cannel coal is likewise here found. On this hill, however, there are six different coal-fields. The most extensive and valuable is the Cornbrook; which is about a mile long, and half a mile broad. This is generally covered by basalt; and has four coal-beds. The Newberry coal-field, which is on the south end of the hill, is about half a mile long, by a quarter broad. This has the same number of beds with the preceding, but they are always about one-third thicker: the basalt does not cover the coal in the field, nor is it to be found in it. The other coal-fields, which, with one exception, are likewise never covered by basalt, are of small extent, and have only one stratum of coal, from 18 inches to two feet and a half thick, or the same divided in two by a thin bed of clay. The Hill Work coal-field, one of the six, lies upon, or is surrounded by, the Cornbrook coal-field; and where the coal in the latter field is cut off by a fault in the neighbourhood of the former, the miners in working in that direction, have always come to basalt.

All these little coal-fields, with their accompanying strata, dip all round from their circumference to their centres, and are to be considered, not as parts of one great bowl, but as so many small ones."

It may be observed, with regard to this account, that it contains some apparent inconsistencies; for, in the first place, each hill is described, as forming a single bowl or basin (i. e. arrangement or system of concave strata), and afterwards the Titterstone is said to contain no less than six independent systems of this kind. Perhaps the greater basin, constituted by the whole mass of the hill may be thus subdivided, by its strata pursuing an undulating line, instead of an uniform course. This description is, however, sufficiently clear, to indicate great derangement in the stratification of this hill. Mr. Bakewell asserts, that a vast basaltic dyke, more than one hundred yards wide, intersects the hill, cutting through the coal-measures, a part of which it forces to the surface, and rising from an unknown depth. These circumstances are well worthy of notice.

But it is to be regretted that the exact spot where the observation was made, is not distinctly recorded; for where facts have so immediate bearing on disputed points of theory, they cannot be too rigorously and scrupulously ascertained.

We have to lament also the absence of any precise account of the substrata on which the coal-formation of the Titterstone rests. A limestone, however, may be observed at Cainsham on the south-west, and Weton on the north-west foot of the hill, which agrees with the carboniferous rather than with the transition variety; it may be inferred perhaps that this extends completely under and round the base of the hill; still lower the mountain appears to be skirted by old red sandstone, which probably forms the base of this limestone, and separates it from the transition limestone which appears on the west near Ludlow; these points, however, must at present be spoken of with much diffidence.

Still less is known with regard to the structure and relations of the northern or brown Clee hill, but it seems probable that it exhibits a general agreement with the Titterstone. The whole of this country is most interesting, and the public may perhaps soon be favoured with a satisfactory description of it from the accurate pen of Mr. Aikin.

(2) *Coal-field of Billingsley.*

On the east of the Clee hills, and between them and the Severn, another coal-district ranges from Deuse hill and Billingsley on the north, to the borders of Shropshire and Worcestershire on the south, a length of about eight miles, coal being worked in several points along this line; but whether this tract consists of one continuous field, or several smaller ones, we have no precise information. Coal is also worked near Over Arley on the Severn, adjoining this tract on the west; but we are in possession of no particulars relating to any part of this district.

Amygdaloid with calcareous glands also occurs near Kinlet.

(*d*) COAL-FIELDS NEAR THE FOOT OF THE ABBERLEY HILLS.

At Pensex near the north-west foot of the Abberley hills (which consist principally of transition limestone capped by basaltic peaks), is a small patch (rather than field) of coal-measure; and another about three miles to the west; but the relations of these have not yet been clearly ascertained; they are situated in a line with the Billingsley coal-field above described, aud only a few miles distant.

Section III.

Great South-western Coal District.

This division is assumed as including the three following principal coal-fields, viz. the grand South Welsh basin, the forest of Dean basin, and that of South Gloucester aud Somerset, together withs ome smaller basins adjacent to the two last, which may be considered (if the figure may be allowed) as satellites attendant on them.

All these coal-basins are closely related and connected with each other by contiguity of position, by resting on a common base of old red sandstone, and by the general analogies of their structure throughout: their strata near the edges of the basins are often very highly inclined, and are partially covered and concealed on the south-east side of the great basin of South Wales, and throughout a great portion of that of South Gloucester and Somerset, by horizontal depositions of more recent formation, consisting of the calcareo-magnesian conglomerate, sandstone, and marle of the newer sandstone formation, and of lias; in many instances in Somersetshire the shafts are begun in lias and sunk completely through the newer sandstone to the coal-measures. Some of these extend to the enormous depth of 200 fathoms: within the Somersetshire coal-field even the lower members of the oolite series appear forming the summits of Dundry and other hills, the coal-measures being exposed and worked in vallies of denudation below; and at one point a shaft has been driven even from these rocks to the coal.

In the very important and extensive coal districts comprised under the present article, the several members which have been enumerated in the beginning of our observations on the coal-fields as associated in them, are displayed on the most striking and satisfactory scale; and it is here perhaps that their relations may be most advantageously studied, especially those of the older red sandstone, its distinction from the newer, aud the position occupied by both with relation to the coal: here also we see the utter impossibility of considering them in the light under which Mr. Jameson has regarded them as members of one great formation, with which beds of limestone and coal are accidently associated; since here, and indeed throughout England, these rocks are entirely unconformable in their position, and are marked, in a manner yet more strong perhaps than in any other instance which can be adduced, as the results of a distant and different order of causes.

In treating of this district we shall pursue the arrangement before adopted in our description of the Penine chain, first giving an outline of the principal facts relating to the coal-formation in the several basins, and then in succession of the millstone-grit, the mountain lime, and the old red sandstone.

I. COAL BASIN OF SOUTH WALES.

First then as most important we may enter on the coal basin of South Wales.

The *Great Coal-field of South Wales*, extending from Pontipool on the east to St. Bride's Bay, south of St. David's head, on the west, belongs to the independent Coal-formation; and is situated in a large limestone basin. The limestone crops out at the surface all round the coal, except where its continuity is interrupted by Swansea and Caermarthen bays. The deepest part of this basin is in the neighbourhood of Neath, which is near its centre: and below Neath, or a little to the west of it, the lowest strata of coal are nearly 700 fathoms lower than the outcrop of some of the superior strata in the more hilly parts of this district. The bed of coal which is nearest to the surface, lies, (near Neath) about 60 fathoms beneath it, and rises to it about a mile north and south, and also a few miles east and west of the deepest part of the basin. So that we are to imagine the inferior beds of coal rising to the surface all round the outcrop of the superior stratum, and between it and the basset edges of the limestone basin. If a line be drawn from Pontipool on the east to St. Bride's bay on the west, it may be said that all the beds of coal on the north of that line crop out on the north of it, at distances proportionate to their depth beneath the surface: so also those on the south of it, except near Pontipool, where they rise towards the east.

It appears however that, though the lowest bed of coal is so far beneath the greatest elevations of this district as near 700 fathoms, the miner finds it without any very considerable descent: for the whole country is intersected by deep vallies in a north and south direction, which consequently cut the strata of the coal. The miner therefore, taking advantage of this circumstance, drives levels into the hills, and there finds the beds of coal and ironstone; there are however many mines in vallies and low places.

There are (according to Mr. Martin) 12 beds of coal from

three to nine feet thick, making together 70½ feet; and 11 others from 18 inches to three feet, making 24¼ feet; together 95 feet of workable coal, besides numerous other beds from six to 18 inches thick. By taking the average length and breadth of the Coal-field, the amount is about 100 square miles, containing 95 feet of coal in 23 distinct strata, which will produce in the common way of working, 100,000 tons per *acre*, or 64,000,000 tons per square mile.

The coal on the north-eastern side of the basin is of a coaking quality, on the north-western it is what is termed stone-coal (the large of which is used for drying malt and hops, and the small, which is called *culm*, for the burning of limestone); on the south side, from Pontipool to Caermarthen bay, the coal is principally of a bituminous or binding quality.

Near the western termination of the basin, beyond Caermarthen bay, it is shallower, so that the beds of coal, found in the deeper parts, are not found there.

The lower part of the coal series, as worked at Merthyr Tydvil and the neighbourhood, is distinguished by the predominance of shale, the upper by the predominance of a coarse grit of loose texture abounding with specks of coaly matter, and agreeing with the rock termed *Pennant* in Somersetshire; these beds are often schistose, sometimes sufficiently so to be employed as tiles; a great thickness of them separates the lower from what may be called the upper coal series, and it is of this rock that the summits of all the principal mountains in the interior part of the basin consist; the lower series which (from the more perishable nature of its materials) forms a belt of rather lower ground interposed between these mountains and the edge of the basin, contains numerous strata of coal and 16 of ironstone; the ironstone occurs in continuous beds and in layers of detached nodules; it is principally found in the lower series, and some of its most valuable beds occur beneath the lowest coal. The strata alternating with the coal and ironstone in this lower series, consist almost exclusively of argillaceous slate: between it and the exterior limestone, millstone grit is often, but not universally interposed. The upper coal series, which forms an interior ellipsis, is said to contain several beds; but the accounts hitherto published do not sufficiently distinguish them from those beneath the Pennant.

The inclination of the strata is much more rapid on the south edge of the basin than on the north, being often at the angle 45° or upwards on the former, while that on the north is generally under 10°. On the western termination of the basin in Bride's Bay, the strata exhibit the most extraordinary marks of confusion and derangement, being vertical and twisted into every possible form of contortion.

This Coal-field is traversed by *dykes* or *faults*, generally in a north and south direction, which throw all the strata from 50 to 100 fathoms up or down. They are usually filled with clay, but Mr. Townsend mentions an enormous fault, many fathoms thick, filled with fragments of the disrupted strata, which traverses the collieries of Lansamlet near Swansea, effecting a rise of 240 feet in the strata.

(2) COAL BASIN OF THE FOREST OF DEAN.

This forms an irregular elliptical basin occupying the whole of the forest tract. The interior portion which contains the coal measures ranges round Colford as a centre; the longest diameter from north-north-east to south-south-west, being about ten miles, the shorter about six. It is said to possess seventeen coal-beds, which together contain 37 feet in thickness of clear coal. All the strata dip uniformly towards the centre of the basin; the exterior ridges of mountain lime and old red sandstone inclosing the coal-measures, are prolonged across the Wye towards the west and form a mountain tract between that river and the Usk in Monmouthshire, the whole of this *coal tract* constitutes very high ground.

On the north of the forest of Dean basin, and at the distance of a few miles, is another small coal-field at Newent; it is surrounded and concealed by overlying strata of the newer red sandstone, and its relations have not as yet been distinctly ascertained.

(3) COAL BASIN OF SOMERSETSHIRE AND SOUTH GLOUCESTERSHIRE.

(a) *Coal-measures.*

These occupy an irregular area of which the longest diameter, from near Iron Acton on the north to Coalford at the foot of the Mendip hills on the south, is near 25 miles; the shorter, from the Newton Collieries near Bath on the east to those of Bedminster near Bristol on the west, about 11 miles; the course of the river Avon nearly coincides with this diameter, bisecting the coal-field into two nearly equal portions; on the north-east and west the strata dip distinctly towards the centre of the basin, but at Coleford near Mells on the south, where they abut against the Mendip hills (the calcareous strata of which are there nearly vertical) the stratification of the incumbent coal-measures is much deranged, becoming vertical and frequently thrown backward, and bent into the form of the letter Z. Much undulation of the coal strata also prevails in many points in the interior of the basin, so that in describing the general arrangement of its beds as constituting a single basin, the expression must be construed largely and applied to its configuration when viewed on the great scale, making allowance for much of local irregularity. On the east, west and south, the coal-measures are generally covered up and concealed by the more recent horizontal strata already described, leaving a few tracts only denuded and exposed to the surface: of these the most extensive reaches from Iron Acton across the Avon to Brislington on the south of that river; another very small occurs near Newton on the east of the field, a third of more importance in the vale of Pensford on the south of Dundry hill, a fourth near Clutton, and a fifth at Coleford at the south extremity of the field; but numerous shafts are sunk through the overlying strata, some even from the inferior oolite through lias to the coal-measures, especially throughout the interval which separates the denudations of Pensford and Coleford.*

* The following is a section of the coal-pit which is most interesting in this respect, as passing through the greatest number of formations; it is situated on the brow of the hill about three-quarters of a mile south-east of the village of Paulton, and belongs to Mr. Hill.

	Fs.	Y.	Ft.	In.
Inferior oolite	3	—	—	—
Lias marle and lias, as detailed in the sections of that formation	26	1	1	6
Red marle and sand	22	—	—	—
Calcareo-magnesian conglomerate called millstone	1	—	—	—

All these strata are horizontal; the following, which belong to the coal-measures, are highly inclined.

Shale, variable in thickness.	Fs.	Y.	Ft.	In.
Great vein	—	—	2	4
Shale and grit	6	—	—	—
Little vein	—	—	1	8
Shale and grit	4	—	—	—
Three Coal-veins	—	1	—	—
Shale and grit	7	—	—	—
Middle vein	—	—	1	8
Shale, &c.	12	—	—	—
Sliving vein	—	-	2	4
Shale, &c.	7	—	—	—
Little vein under Sliving	—	—	1	6
Shale, &c.	6	—	—	—
Bull vein	—	—	—	10
Shale, &c.	2	—	—	—
Peaw vein	—	—	—	10

The irregular undulations of the strata in this district, and their concealment through extensive tracts by overlying deposits, present formidable obstacles to the attempt to trace the series of beds which constitute this coal-field; it probably exhibits between fifty and sixty coal-seams, most of them very thin, hardly any exceed one yard, or could be profitably worked, were it not for the highly improved state of machinery. As in the south Welsh-field, shale predominates in the lower, and the Penant grit Rock in the middle part of the series; the shale beds frequently contain beautiful impressions of ferns and muscle shells.

There are several extensive faults, some of which elevate the strata more than one hundred feet.

The outer edge of this basin consists, on every side but a small portion of the east (where the horizontal and more recent depositions conceal its substrata) of more or less elevated ridges (hereafter to be described) of carboniferous limestone and old red sandstone, millstone-grit being occasionally interposed. Round Nailsea, on the west of this coal-field, a depression in the limestone ridges forming its border on that side, includes a smaller attendant coal basin about four miles long and one and a half broad; here ten seams of coal have been discovered.

(*b*) *Millstone Grit.*

A conglomerate and coarse grit of this formation may be traced in the South Welsh coal-field, near Merthyr, and the head of the Neath valley, on the north of the South Welsh coal-field; and near Bridgend and Caerphilly on the south interposed between the coal-measures and subjacent lime. In the forest of Dean it has not yet been described. In the Gloucester and Somerset basin, beds of conglomerate in a hard compact cherty sandstone, often highly ferruginous, occupy a similar position; these may be seen on the west of the coal-field at Brandon hill close to Bristol, and lying upon the limestone on both sides the gorge at Clifton, at the eastern extremity of the field resting upon the limestone in the defile of Wick Rocks; at the northern near Croomhall, and at the southern near Coleford.

(*b*) *Carboniferous Limestone.*

Strata of this rock, closely agreeing in all their characters and contents with those of Derbyshire, circle round all these coal-fields; its average thickness cannot fall short of

500 or 600 feet, and it seems tolerably uniform in this respect.

The course of this rock round the South Welsh basin has been sufficiently defined in treating of the limits of the basin itself; it forms a zone varying from two miles to a furlong in breadth, according as the position of the beds approaches to an horizontal or vertical position; to the west of Milford-haven however these limestone beds seem generally cut off, the limestone rests throughout on old red sandstone; on its southern line it throws off several branches which diverge more or less from the main chain, or sometimes accompany it in nearly parallel lines; thus the peninsula of Pembroke exhibits between the coal-field and the sea no less than three parallel zones of limestone separated by two ridges of old red sandstone, the limestone lying in troughs between them. Caldy Island, near Tenby, exhibits magnificent sections of the middle limestone zone, and the sandstone on which it rests, the strata being almost vertical; the peninsula of Gower (between Carmarthen and Swansea bay) has a central ridge of old sandstone with parallel zones of limestone resting on either side, and the south-east of Glamorganshire exhibits a similar arrangement, with the exception that here the sandstone, which is traversed by the vale of Ely, is often concealed by more recent horizontal deposits of calcareo-magnesian conglomerates, new red sandstone and lias, which formations also rest on the back of the most southerly zone of carboniferous lime along the coast from the mouth of the River Ogmore to the Taafe.

The carboniferous limestone of the forest of Dean forms a zone encircling, and dipping beneath, the coal-basin of that forest as before described; thence it crosses the Wye near Tintern Abbey, and is prolonged on the west as far as Penhow castle, about seven miles distant. Its beds rest on the old red sandstone. It presents the usual fossils and minerals; and in some of its beds ferruginous ores, accompanied by brown spar and pearl spar, are worked.

The carboniferous limestone surrounding what we have termed the Somersetshire and South Gloucester coal-field, presents more complicated details. On the south it appears in the long ridge of the Mendip hills, extending from Mells to the Bristol channel, with a line of bearing from east-south-east to west-north-west. This chain consists of a central axis of old red sandstone, flanked on its opposite declivities by parallel bands of mountain limestone, dipping from it in opposite directions in angles varying from 30° to 70°. This central axis is not however visible throughout its whole course, being occasionally entirely over-arched, and concealed by the calcareous strata; but it appears in four ridges, forming the most elevated points of the chain, and disposed nearly at equal distances through its length. The cavern of Wokey hole,* and the defile of Cheddar cliffs, with its long line of stupendous mural precipices, certainly among the most magnificent objects of this kind in Britain, are the well-known features of this chain.

The western calcareous boundary of this coal-field is not formed by a single continuous ridge, but by a series of successive elevations emerging from the plain of the more recent horizontal strata. Of these, the first in proceeding from the south (where the Mendip chain ends) to the north, is Broadfield down, a calcareous mass about three miles long; its strata dip every way from the centre; it presents the precipitous defiles of Cleve and Brockley Combes; which, though greatly inferior in magnificence to Cheddar, yet derive from their luxuriant woods a contrasted character of beauty which is there absent. Broadfield down approaches within two miles of the Mendips; it is succeeded on the north, after an interval of half a mile only in the narrowest part, by Leigh down; but as the calcareous chains rapidly recede from each other, the interval quickly increases, and in the depression between these chains lies the little basin of the Nailsea coal-field.

Leigh Down extends in a north-east direction to the river Avon below Bristol, and is prolonged on the opposite side of that river (which here flows through that narrow and precipitous gorge, well known by the name of St. Vincent's rocks) by Durdham down to Westbury. The gorge of the river, lined as it is by an almost uninterrupted succession of mural precipices, affords an unrivalled opportunity of studying the various beds of this formation, which are here elevated in an angle of 45° to the north. More than 200 of these beds are enumerated in the 5th volume of the Geological Transactions by Mr. Cumberland.

This calcareous series rests on conformable beds of old red sandstone, which may be seen on both sides the river near Cook's folly, extending on the south under Leigh down and Weston down, a calcareous hill skirting the channel below Portishead point, and apparently thrown forward beyond the general line of Leigh down by an enormous fault. On the north of the river this old red sandstone extends towards Westbury, but is much covered up and concealed by the more recent horizontal deposits of the newer sandstone and magnesian conglomerates, resting on the truncated edges of its inclined strata, as is also the case at Portishead point above-mentioned. From

* This cavern, however, is not situated in the carboniferous limestone, but in the calcareo-magnesian conglomerates, here abutting against the chain.

Westbury the carboniferous limestone mantles round this older sandstone, extending by Blaze castle (where another of those abrupt dells, which form so beautiful a characteristic of this formation, occurs) to King's Weston park. Thus the disposition of the ground immediately on the north of the Avon resembles that of the Mendips; exhibiting a central nucleus of old red sandstone, skirted on either side by the strata of carboniferous limestone, dipping from it to the north-west and to the south-east. Near Blaze castle the strata are almost vertical: on the north of Westbury these calcareous ridges subside, and are concealed by the new sandstone and lias: the carboniferous limestone emerges again, however, after an interval of about three miles, near Almondsbury, and continues to form without further interruption the remainder of the western, and by the inflection of its course the whole of the northern and the north part of the eastern boundary of our coal-field; ranging by Thornbury to Tortworth its most northern point, and thence trending southwards to Wickwar and Chipping Sodbury. On the north-western edge of this calcareous chain, the old red sandstone may be traced underlying its strata; beneath which the yet more ancient beds of transition limestone may be seen in the same vicinity. Near Tortworth two parallel dykes of trap (an amygdaloid with calcareous nests) traverse these latter rocks, and produce some remarkable changes in them. On the east the horizontal strata of lias, &c. which (indeed throughout this tract occasion much embarrassment to the observer, by partially concealing its structure,) approach to close contact with the calcareous ridge, and finally entirely overlie it at Sodbury, overflowing, as it were, into the area of the coal-field; and no extended ridge of carboniferous limestone can be traced through the remainder of the eastern boundary of that area, till we arrive at its south extremity, and the Mendip hills. Occasional and short protrusions of this rock, however, appear in several points along this line; at Wapley, at the romantic defile of Wick rocks, and at Tracey park, and thus continue to indicate its extension beneath the superstrata which conceal it on the surface.

The usual minerals of the limestone tract bounding this coal-basin, are galena, blende, calamine, sulphate of barytes, &c., and its organic remains agree with those of the other carboniferous limestone tracts; but among the former, sulphate of strontian may be mentioned, which, though found in Gloucestershire, is elsewhere of rare occurrence in this rock; and among the latter, the palates of fishes found in St. Vincent's rocks, of which no other example has yet been noticed.

(*d*) OLD RED SANDSTONE OF THE SOUTH-WESTERN COAL-DISTRICT.

It is in the districts connected with the coal-basins, which form the subject of the present chapter, that this formation is exhibited on the largest scale. Hitherto we have only seen it constituting a few not very important beds towards the base of the Penine chain, but shall now survey it as exclusively constituting lofty and extensive mountain groups, and occupying entire counties.

In this quarter, then, the old red sandstone constitutes the common base on which all the coal-fields above described rest, thus connecting them into one whole; and together with those coal-districts forms the external chains which border on the south and east the transition mountains of South Wales. We may trace it from the borders of Shropshire and Staffordshire near the Clee hills, sweeping across Herefordshire of which it occupies nearly the whole, with the exception of a ridge of transition limestone which, rising in arched strata, emerges from beneath the sandstone near Shucknell in the south-east of that county, extending to Longhope near Mitchel Dean in Gloucestershire; this ridge creates an undulation in the sandstone dividing it into two troughs, that on the east lying between this interposed ridge of transition limestone, and that of similar formation which forms on the borders of Worcestershire the Abberley hills, and the western slope of the Malvern group, against which the beds of this sandstone crop out.

This great tract of old red sandstone has its western boundary against the transition chains of South Shropshire and the western confines of Herefordshire, following nearly the course of the rivers Corve in the former, and Arrow in the latter county.

From the south-west of Herefordshire, the old red sandstone is prolonged on either side of the vale of the Usk, through Brecknockshire, forming the lofty mountains called the Beacons of Brecon on the south of that river; and those connected with Trecastle beacon on the north; the beds on the south dip beneath the calcareous border of the great coal-basin, which here follows a line nearly coinciding with the confines of Brecon and Glamorganshire; and those on the north rise against the subjacent transition rocks of the Eppynt hills.

This sandstone tract narrows in its progress through Brecon, being more than twenty miles across from north to south near Abergavenny on the east, and not above two or three near Castle Carreg Cennen on the west. This difference arises partly from the higher inclination of the beds, and partly from the thinning off of the formation; in the latter direction from Carreg Cennen the sandstone continues to form only a narrow zone ranging to the west through Caermarthenshire by Laugharne and Narbeth into Pembrokeshire, but it is not to be traced beyond Haverford West on this line (that is on the north of the coal which in that quarter rests on graywacke,) although it still extensively appears on the south: an instance of unconformable position in the coal, as compared with the outgoings of this rock.

We have thus traced the extension of this sandstone from the great Herefordshire district along the northern border of the South Welsh coal basin: let us next pursue its course in returning eastwards along the southern edge of that basin.

On the south-west of Pembrokeshire it occupies all the peninsula between the termination of coal in the south of Brides Bay, near Littlehaven and Milford haven: there is however a range of trap rocks, associated with greywacke, interposed between the coal and this sandstone. We have before observed that the coal and sandstone were in this quarter unconformable; the former having passed beyond the boundary of the latter, so as to become immediately incumbent on transition rocks.

The peninsula of Pembroke, as it may be called, (between Milford Haven and the sea on the south,) presents (from the undulation of the strata) a double ridge of sandstone hills, on either side of which, and in the intermediate valley, rest the incumbent beds of carboniferous limestone.

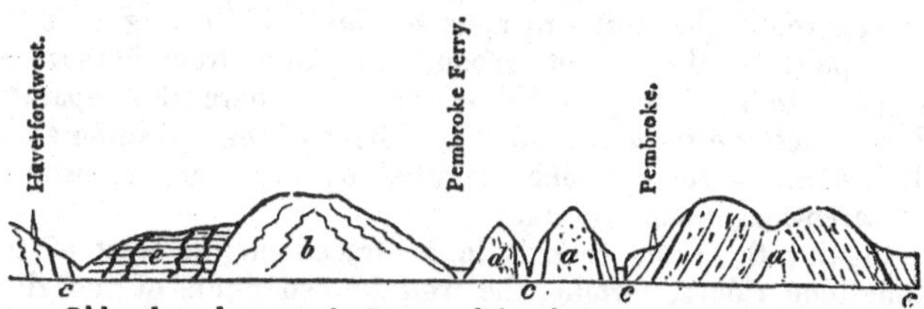

a a Old red sandstone. *b.* Greywacké and trap rocks confusedly mixed. *c c c c.* Limestone. *d.* Old red sandstone & Greywacké. *e.* Coal-formation.

The above section will convey a clearer idea than can be given by description of the phenomena of this district.

The deep indentation of Caermarthen Bay cuts entirely through the south Welsh coal-basin, and of course conceals its exterior chains on the south.

On the east of that bay, they reappear in the peninsula of Gower, presenting a central ridge of old red sandstone in arched strata, rising into considerable hills, flanked on the north and south by the incumbent limestone.

Swansea bay again cuts off these chains for a considerable interval; beyond which on the east they are resumed in the south of Glamorganshire. On this side, however, the limestone alone is at first seen (the subjacent sandstone not rising to the surface) until we arrive at the head of the river Ely near Hensol Park, where the old red sandstone again emerges, forming the outer and lower chain of the mountains into which the coal-basin here swells. This sandstone is probably the fundamental rock throughout the vale of Ely, but it is concealed by overlying deposits of the newer sandstone and lias; nor does it appear in any great quantity till the river Taafe is crossed. Beyond this river it stretches through Monmouthshire, circling round the east end of the south Welsh coal-basin, and (with the exception of a small tract near Uske which exhibits the subjacent transition limestone,) occupying the whole interval between this and the corresponding coal-basin of the forest of Dean; on the north of which it rejoins the great Herefordshire sandstone district which we have before traced, being here interposed between the carboniferous limestone of that field, and the transition limestone chain of Mayhill. The old red sandstone circles round the eastern edge of the forest of Dean, approaching within half a mile of the Severn, but touching it only in one point. Overlying beds of newer red sandstone abut against it in this direction.

We have lastly to trace this rock in the exterior chains of the Somerset and South Gloucester coal-basin. It may be traced round the northern apex of this field, emerging in the escarpment of the hills of carboniferous lime, from Sodbury on the east to Thornbury on the west; and it here also separates that limestone from the calcareous beds of the transition suite. It is here, however, much obscured by overlying deposits of calcareo-magnesian breccia.

Again, the same rock is to be traced on the west of the limestone ridges, forming the well known defile of the river Avon below Clifton, but similarly obscured; and in the prolongation of the exterior ridges of this coal-field on the west, this sandstone may be traced along the coast of the æstuary of the Severn, emerging from the carboniferous limestone from Portishead point to Clevedon.

Lastly, the Mendip hills (the southern boundary of this coal-field) exhibit an axis of this sandstone which emerges from beneath their calcareous strata, not in a continuous line indeed, but so as to form all their most elevated summits: namely, Black down, Nine barrow down, Pen hill and Maseberry beacon.

CHAPTER V.

TRAP ROCKS OCCURRING IN ASSOCIATION WITH THE COAL-MEASURES.

It is our intention to present in this article a combined view of the phœnomena exhibited by this class of rocks, as they occur in the English coal-fields; phœnomena it is well known, of the highest interest and importance to the theoretical geologist. Although the plan we have prescribed excludes all unnecessary discussion of theoretical views, yet the generalization of scattered facts which we propose to give, cannot fail to extend and to render more solid, the true basis of theory: and we shall also find it necessary to depart, in some measure, from the rigour of our own laws, in order to illustrate the bearings of the facts we record.

It is to be observed, as a general fact, in the first place, that the coal-fields afford the first instance in descending the series, in which any of the great formations of England appear to be strikingly connected with rocks of this family; few, or no traces of them being visible in the districts occupied by the strata of more recent origin, with the exception of the beds of amygdaloidal trap said to occur in the new red sandstone near Exeter, and already described under that article; and of the prolongation of a basaltic dyke near Cleaveland, extending from the coal across the oolitic chains.

Although so generally limited in occurrence to the coal-districts, it would be hasty to infer that these trap rocks are of contemporaneous origin with the other members of the coal series. The trap occurring in the dykes which intersect the coal-measures, must obviously be of posterior origin; and the single instance of the Cleveland dyke, is sufficient to prove that, sometimes at least, the difference of age is very considerable. When the trap occurs in overlying masses, of course no certain inference as to age can be drawn; excepting that it is more recent than the rocks which it overlies; and the only case in which a probable inference of contemporaneous origin can be drawn, is when the coal-measures regularly alternate with conformable strata of trap. Even here, also, as we shall soon have occasion to perceive, the appearance of alternation is frequently only partial and delusive; and the inference, therefore, defective, if not erroneous.

In Scotland, where, on the mainland, the same seeming connexion as in England may be traced between the trap rocks and the coal districts, Dr. Mac Culloch has shewn, that in the Western Islands, the very same varieties of trap occupy a position superior to the lias; and if we pass thence to the opposite coast of Ireland, between which countries a perfect geological analogy subsists, we find the trap overlying chalk.*

Whether in all these instances, the trap rocks be the production of a single epoch, overlying indifferently all the anterior formations, or whether they should be considered as instances of the repeated and recurrent production of analagous rocks in successive periods, contemporaneously with the formations near which they occur, is a question which it is foreign to the purpose of the present work to discuss.

Before we enter into a particular detail of the local phœnomena presented by our trap rocks, we shall premise some general remarks: 1st. On their mineralogical character and varieties. 2dly. On the modes of their arrangement and connexion with the rocks among which they occur. And 3rdly, On the changes occasionally produced in these rocks, near the points of contact.

(*a*) *Mineralogical character*. The rocks of this family appear to consist essentially of felspar, combined either with hornblende or augite, or both. Where the hornblende predominates, they are referable to the class of greenstones. Where the augite prevails, they fall into the new class of dolerite, formed by the French geologists for the purpose of receiving these compounds, which were formerly, from neglecting to distinguish the characteristic mineral, confounded with the greenstones. The term augite rock, is similarly employed by Dr. Mac Culloch; and as being already naturalized in our language, will be retained in these outlines.

The structure of these compounds varies from the granitoidal, in which the constituents are distinctly crystallized, to the granular, and lastly the compact, in which every trace of distinct grains vanishes, and the whole assumes the aspect of an homogeneous paste. In the granitoidal varieties, it is easy to recognise the distinct characters of the greenstone and augite rock. But in the more finely grained varieties, this is often nearly impossible; for the form of the crystals being obliterated, and chemical analysis affording no sufficient means of discriminating between hornblende and augite, more especially, as parts of a compound

* Trap rocks also occur in Ireland, associated with the limestone, underlying the coal-formation in all the modes described as prevailing in England. Mr. Weaver has given a most able and interesting description of these.

rock, an obscurity almost hopeless involves formations of this class. Thus, in those rocks of a finely granular texture, to which the term basalt is applied, the precise determination of the constituent principles is a matter of the greatest difficulty. Since the augite rock is often seen to pass by insensible gradations into well characterized basalt, many continental geologists define it as having for constituent principles. felspar, augite, and oxidulous iron, occasionally blended with olivine, hornblende, &c. or even as complete augite: yet a considerable portion of the rocks to which this term has been applied in England, appear to consist principally of hornblende, as their characteristic ingredient, although they often also contain imbedded crystals of augite. Besides the inherent difficulty of the subject, it is to be regretted that little attention has been paid by our English geologists with the exception of Dr. Mac Culloch, to the precise determination of the mineralogical characters of these rocks: we are therefore not able at present to speak without great hesitation concerning them; and shall regard ourselves as fortunate, if these observations may have any effect in directing enquiry to this point.

The most usual characters of the basaltic rocks of England, are, an iron-grey colour approaching to black; a considerable tenacity and hardness, a sharp and sometimes conchoidal fracture, a granular aspect often reflecting light from a number of brilliant spots or striæ, some of which seem to be felspar, others hornblende or augite; very liable to superficial decomposition, in which case, the colour passes from the further oxydation of the contained iron, to a rusty brown, often mingled with spots of green, arising apparently from the grains of hornblende. Often this decomposition penetrates to a considerable depth and distance into the mass of these rocks; in which case numerous spherical masses, interspersed throughout the mouldering mass, occur; having a nucleus of unaltered or less altered basalt, surrounded by concentric coats in successive stages of disintegration. These rocks are fusible at a low degree of heat; and attract the needle strongly in consequence of the low state of oxydation of the iron which they contain. The specific gravity of the basalt of Staffordshire is 2.86; and on the analysis of 100 parts by Dr. Withering, yielded silex 47.5, alumine 32.5, oxyd of iron 20. From the less perfect state of chemical analysis at that time, the soda probably contained in it escaped observation.

Besides the distinctly crystallized varieties of trap, as greenstone and augite rock, and the granular, as basalt, two other species occur; one merely arising from variety of texture, being a porphyroidal trap, formed by crystals of felspar disseminated through a paste, resembling a compact basalt. This may be sometimes seen in Derbyshire, but is not common.

The fourth species, which is of very common occurrence, is amygdaloidal trap, or toadstone: this is formed by a vesicular paste, apparently consisting of a fine-grained basalt of a looser and more earthy texture, the cavities being filled with drusy geodes of calcareous spar or green earth, and more rarely zeolite, chalcedony, &c. All these varieties of trap usually are associated with, and pass into, each other.

(*b*) *Mineral contents.* These are, 1. Crystals of hornblende; these are common. 2. Crystals of augite: these are more rare, but are not uncommon in the basalt of Teesdale, and have been found in the toadstone of Derbyshire. 3. Olivine, found at Teesdale. 4. Green earth, common in the Derbyshire toadstone. 5. Calcareous spar, in the toadstone, passim. 6. Quartz, in Derbyshire toadstone, and in Northumberland. 7. Chalcedony and onyx, ibidem. 8. Jasper, ibidem. 9. Prehnite, in the basalt of Staffordshire. 10. Mesotype, ibidem, and in Derbyshire. 11. Adularia, in Northumberland.

It must be remarked however, that the zeolitic minerals are of much rarer occurrence in the English trap rocks, than in those of most other countries.

Where the traps alternate, as in Derbyshire, with others containing metalliferous veins, they usually intercept those veins; so that with the exception of the oxidulous iron, and a few crystals of iron pyrites, (which are rare) no minerals of this class are found in them.

(*c*) *Organic remains.* No organic remains, either mineral, or vegetable, have yet been found in any rock of this class in England. Mr. Weaver, however, mentions that he has discovered shells of the Terebratula in the greenstone associated with the carboniferous limestone of the centre of Ireland, and exactly corresponding in all its relations, with the trap rocks we are now describing. With every deference to the authority of this writer, which justly deserves to be placed in the highest class, it must still be observed, that a fact hitherto of single occurrence, must necessarily be received with some caution; and that some possible source of error may be still suspected, either as to the nature of the supposed remains, or the rock containing them: the latter especially (when the great changes of character often assumed by other rocks in approaching the trap, is considered) appearing open to ambiguity. Be the fact however as it may, it cannot be considered as pregnant with any very important theoretical inferences, since shells have been found preserved in recent, and decided lavas, at the points where they have flowed into the ocean. And those who contend for the volcanic origin of these rocks, always ascribe them to sub-marine volcanoes, acting while the ocean still covered the surface of our present continents, or at least the districts in which they are found.

(*d*) *Modes of arrangement and connexion with the rocks among which they occur.* The trap rocks which are found within the limits occupied by the coal-measures and subjacent limestone, occur under three distinct modes of position, of which two appear to indicate an origin distinct from that of the strata with which they are geographically associated: while the third seems to countenance the opposite inference of their contemporaneous formation. These modes of position are, I. As overlying masses resting unconformably on the subjacent strata. This appears to be the position of the trap rocks at Clee Hill and at Dudley. II. As dykes irregularly intersecting and traversing the strata; of these, numerous examples are found in Northumberland and Durham, and one remarkable instance will be described in Staffordshire. III. As beds conformably interstratified and regularly alternating with the other strata. The Great Whinsill of Northumberland, and the toadstone strata of Derbyshire, illustrate this case. It does not appear that there are any circumstances in the stratification and arrangement of such beds of trap, which decidedly differ from the appearances presented by the neighbouring strata. It has indeed been asserted that they are subject to much greater and more rapid alterations in thickness; and this certainly appears in some instances at least, to be the case. But this character seems scarcely sufficient to afford foundation for inferring a distinct origin. It has also been said that the metalliferous veins traversing the other strata, are themselves cut off by these, which must therefore be of subsequent formation. But although such is often the fact, yet the exceptions which occur require this argument to be modified, though not perhaps entirely withdrawn; and it must also be remembered, that metalliferous veins are subject to similar contractions, in traversing many other strata; expanding for instance in those of limestone, and shrinking in those of shale which alternate with them.

Between the trap rocks, however, occurring in these three so different positions, no mineralogical or external features of distinction have been shewn to exist; but the same varieties seem to occur indiscriminately in each; so that the indications of their belonging to different formations, afforded by a diversity of position, are balanced by opposite probabilities resulting from identity of character.

We ought not to close this article, without noticing some

important observations of Dr. Mac Culloch, which satisfactorily prove that the inferences derived from the position of trap rocks, cannot always be safely relied on; and that portions of one and the same mass of trap, may be found under all these seemingly different relations. Although the observations refer to a district not falling within the limits of the present work, they are so intimately connected with the subject now under consideration, that their introduction cannot be thought irrelevant. This able geologist observed in the cliffs on the east coast of Trotternish, (Isle of Sky,) several interesting sections of the sandstone strata, traversed by trap, in various manners. Among these he noticed a large mass of trap rising through, and on one side overlying, the sandstone strata. From this mass an horizontal bed was detached running conformably through the midst of the strata; and this ultimately became subdivided into three smaller beds, also conformable and alternating with the sandstone. Dykes also proceeded in several points from the single bed of trap. So that in this instance we have all the three modes of position described in the beginning of this article, assumed by different portions of the very same mass of trap; and these appearances too were exhibited on a large scale. A delineation of them is given in plate 17, fig. 2. Western Islands. A contracted sketch is here subjoined.

It should be remembered that the question of the volcanic origin of these rocks, is not in fact affected by their occasional occurrence as conformable beds; since on the supposition of volcanic agency prevailing contemporaneously with the formation of the strata, the lavas produced would necessarily occur thus interstratified; and no theorist ever refers these rocks to volcanic agency of recent date; but makes it a necessary condition of his hypothesis that they flowed before the continents assumed their present form, and while they were buried beneath the ocean. It is evidently quite consistent with this hypothesis, to admit a successive series of such volcanic agencies, exerted contemporaneously with the epochs of the several geological formations. This observation is not made for the purpose of maintaining a theory; but only to shew that it is necessary to admit with some caution, the arguments brought against it by those, who though of another school, are equally theorists.

(*e*) *Alterations induced on the stratified rocks in the neighbourhood of the trap.* These phenomena are familiar to all geologists, together with the theoretical views they are thought to favour. They consist in the greater degree of induration assumed by many of the neighbouring rocks near the walls of trap dykes; by which loose grits pass into compact quartz rock, and shale into flinty slate; coal being converted under similar circumstances into coak, as if by the volatilization of its bituminous matter. Many geologists dwell on both these effects, as evidently resulting from the proximity of matter in a state of igneous fusion; and consider them as decisive of the volcanic origin of these rocks. Instances of these phenomena are afforded by several of the dykes of Northumberland, Durham, and Staffordshire. The effect attributed to the toadstone beds of Derbyshire in cutting off the metalliferous veins, has been already noticed; but it will be seen that this is not a general fact. Should it appear however, that any such veins are decidedly broken through by them, the proof of their subsequent formation must in every such instance be admitted: and we shall be obliged to ascribe to the uncut veins, a still more recent origin. But this subject cannot be considered as having yet been fully investigated

Particular description and localities of the Trap rocks associated with the Coal-formations, &c.

I. *Beds, overlying masses, and dykes, of trap in Northumberland and Durham.**

The mass of trap which first claims our attention in this quarter, is that named the great Whinsill; this forms a stratified mass, conformably arranged near the middle of the carboniferous limestone formation; and may be traced for many miles about half way up the great western escarpment of the mountain range, connected with Cross fell; particularly near the lead mines on Dufton fell. The thickness of this stratum is, however very irregular; being only six fathoms in some places,

* Mr. Trevelyan has recently laid before the Wernerian Society sections of all those parts of the Northumberland coast which present masses of trap; and has detected the same passage of dykes into apparent beds noticed by Dr. Mac Culloch. Professor Sedgewick also has lately examined the trap of Northumberland, and considers the evidence of its igneous formation as complete.

Monsieur Boué, 'Geologie de l'Ecosse,' while supporting in its most extravagant degree the igneous theory of trap rocks, has yet questioned some of these inferences; one of his arguments is that the flinty slate found near trap dykes cannot be indurated shale, because it is fusible without great difficulty; had he tried the same experiment with ordinary lias shale he would have fonnd the same result.

and twenty or even thirty in others. It consists of basalt, coarse grained in texture, and composed of white felspar and black hornblende: the latter mineral predominating, and giving the rock a dark greenish grey colour.

The basaltic rocks exhibited in the valley of the Tees (which traverses the above mountain group) are believed to be the prolongation of the same stratum, laid open by denudation. These rocks extend from near the source of the Tees to Egglestone.

At Caldron snout, situated on the moors ten miles above Middleton, a basaltic ridge crosses the river, and occasions a succession of cascades for the space of 596 yards, which form a fine contrast with the pool of still water or *wheel*, above the falls. It was here immediately under the basalt that the Rev. J. Harriman discovered some garnets? crystallized in dodecahedrons, and imbedded in a thin stratum of pale red hornstone or chert with particles of calcareous spar.

Near the steep acclivity which terminates Cronkley Fell, another range of basalt interrupts the course of the Tees, and causes the cataract called the High or Mickel force, where the water is precipitated from the height of 56 feet.

The rock which here crosses the river, is apparently an *overlying mass* of coarse-grained grey basalt, the hornblende and the felspar which compose it not being intimately combined. It rests upon the lead-mine sills, and shoots, on the banks of the Tees, into regular columns of considerable magnitude and elevation. A few miles below this cascade, and about three above Middleton, perpendicular basaltic rocks again form the bank of the river. To these, iron chains have been fastened, for supporting Winch bridge.

This remarkable structure, if it can be so called, is a plank two feet in breadth, with low hand-rails, suspended 56 feet above the Tees, which is here 63 feet wide. Some miners contrived it for the purpose of passing from the county of Durham to Holwick in Yorkshire.

In he fragments of basalt which are found scattered over the surface in Teesdale, and in other parts of the district, small grains of yellow olivine, and of greenish black augite, are found imbedded. (G. T. vol. iv. p. 73.)

A bed of basalt is penetrated at the depth of 159 fathoms in the Aldstone Moor mines: this is also referred by the miners to the above stratum of the great Whinsill. But Mr. Winch considers their identity as problematical; and remarks that the miners regard all the beds of basalt which occur in the mountainous district, as ramifications of the great Whinsill.

At Dunstanborough castle, and at Gunwarden castle, near

Barwesford on the north, and at Wratchiff crag near Alnwick, basalt also occurs conformably *interstratified* with other rocks. At Dunstanborough, the cliff consists of columnar basalt, 8 to 10 feet; sandstone, 2 feet; shale or slate-clay, 6 feet; basalt to below the water's edge.

At Gunwarden, strata of dark bluish crystalline limestone, from three to four feet thick, alternate twice with compact basalt. At Wratchiff crag, basalt alternates with limestone and slate clay.

In *overlying masses*, basalt occurs in the general form of a long range, crossing the country in the direction of north-east and south-west, on the north of the lead mines. Close to the edge of this range, the Romans constructed their wall, which is now standing four feet high in many places; this basalt is occasionally columnar. Higher north, other masses are also visible; and still higher, basaltic eminences form a striking feature in the country between Alnwick and Berwick. These eminences have frequently been chosen for the sites of castles, as at Dunstanborough, Bamborough, and Holy Island; some of the small islands near the coast are also composed of this rock. At Bamborough a well was sunk in the Castle hall to the depth of 150 feet, by which it was ascertained that the overlying rock of basalt is 75 feet thick, and rest upon a fine grained red and white sandstone.

Basaltic dykes traverse alike the subjacent formations of limestone and millstone grit, and the coal-measures; in one instance it will be seen even penetrating rocks of far later date; the phenomena accompanying those dykes are thus generally stated by Mr. Winch.

Limestone is often *rendered highly crystalline and unfit for lime*, when in the vicinity of this rock, as is the case with the two lowermost strata at Wratchiff crag, but not with the upper one. *Slate-clay* or shale is turned into a substance like *flinty-slate or porcelain Jasper*, as is the case with the stratum lying immediately beneath the upper bed of basalt at Wratchiff crag; and *coal* is *invariably charred* when in contact with it. The *Sandstone* on which it sometimes reposes, is changed for some depth to a brick-red colour.

Examples of such dykes traversing the carboniferous limestone may be seen in Aldstone moor, Allendale, and in Weardale. One is quoted as being 36 or 37 feet wide. In the crevices beside the dykes, strings of lead ore are frequently observed, but are never known to pass through the dykes.

In the 4th volume of the Geological Transactions, there is a short communication by the Hon. H. G. Bennet on a *Whin* dyke traversing the limestone and other strata in Beadnel bay on the north-east coast of Northumberland; by which it appears that the limestone is so altered in quality in those parts of it which lie in the immediate vicinity of the dyke, that it will not burn into lime of any value, nor within 20 feet of it. A stratum composed of felspar and carbonate of lime, is so altered near the dyke, as to resemble the substance of the dyke itself, which is unlike that of the dykes in the *north-western* part of the county; these are composed principally of hornblende.

The number of *veins*, or *dykes*, *traversing the strata of the Coal-measures*, is very considerable, and there is no uniformity in their direction. The circumstances attending them, are in many cases very extraordinary, and the most considerable basaltic dyke in the immediate neighbourhood of Newcastle is that which passes through Coley hill, about four miles west of the town. A long range of quarries has here been opened upon it, in some places to the depth of 50 feet, and laying bare the entire width of the dyke, which is 24 feet. The dyke in this place appears to be *vertical*. The basalt of which it is composed, lies in detached masses which are coated with yellow ochre. The removal of these, brings to view thin layers of indurated clay with which the fissure is lined, and which, breaking into small quadrangular prisms, are used by the country people for whet-stones; in this substance, clay-ironstone impressed with the figures of ferns, is very abundant.

The upper seam of coal is here found at about 35 feet from the surface, and where in contact with the dyke, is completely charred, forming an ash-grey porous mass, which breaks into small columnar concretions, exactly resembling the coak obtained by baking-coal in close iron cylinders in the process of distilling coal-tar. Calcareous spar and sulphur are disseminated through the pores of this substance.

The basalt itself, when broken, is of a greenish-black colour, and of a coarse-grained fracture. It contains quartz, calcareous spar, and another mineral possessing the following characters. The colour is wax-yellow passing into olive green; the lustre vitreous, resembling that of glassy felspar; the fracture foliated. It resists the action of the blowpipe with borax, but with it melts into a white glass. The latter circumstance, and the foliated fracture, distinguish this substance from olivine, which gives a dark green bead with borax, and presents a fracture more or less conchoidal.

Passing to the east-south-east of the Coley hill dyke in the line of its direction, a vein is found traversing Walker colliery, and crossing the Tyne at Walker. This dyke is well defined; it occasions no alteration in the level of the coal strata, and the depth at which it intersects them is unknown. It has been cut through by horizontal drifts in four places. On each side of it, the *coal is converted into coak*, which on one side, in some places, was found to be 18 feet thick, and on the opposite side upwards of nine feet. A firm, hard, and unbroken vein of basalt, on an average about 13 feet thick, was in immediate contact with the coak on each side; and between these two veins, lay nodules of basalt and sandstone, upwards of nine feet in thickness, imbedded in a cement of blue slate.

At Walbottle Dean, five and a half miles west of Newcastle, below the bridge on the western road, a double vein of basalt (represented by Fig. 2, Pl. 4, of the 4th vol. of G. Trans.) crosses the ravine in a diagonal direction, passing nearly due east and west. It hades or underlies at an angle of 78 degrees, and cuts the coal-strata without altering their dip. On the eastern bank of the ravine, it is laid bare from the level of the brook to the height of about 60 feet. The northern and southern basaltic portions of the vein, the one five and the other six feet in thickness, are there 13 feet apart, and are separated from one another by a confused heap of fragments of sandstone and shale, broken from the coal-strata. With these fragments are found balls of basaltic tufa parting into concentric layers, and of a lightish brown colour; the balls are most abundant on the sides of the rubble near to the basalt. In a neighbouring colliery, both portions of the vein hold their course through the seam of coal, which is *charred* by their influence. This basalt contains nodules of quartz and chalcedony, but not adularia, which is abundant in the basalt of Coley hill.

A string of galena fills a crevice beside a vein of basalt about two miles beyond Durham.

A dyke called the Cockfield dyke, which is 17 feet wide, hades or underlies to the south, and throws up the coal-measures on that side 18 feet. The low main coal contiguous to the basalt, is only nine inches thick, but enlarges to six feet at the distance of 150 feet from it. The *coal is reduced to a cinder*, and the sulphur is sublimed from the pyrites near to the dyke.

A dyke is seen on the banks of the Tees a little below Yarm. It there passes into the newer red sandstone, and continuing its course in the same direction, is well known to traverse the north-eastern part of Yorkshire, near the still more recent formations of lias, and the sandstone of the inferior oolite, in the eastern Moorlands in its way to the German Ocean. This dyke is rendered highly interesting by its great length, and the proof it affords by thus penetrating later rocks, that it must have owed its origin to causes in action at a period long subsequent to the formation of the coal; a proof which yields a strong analogical presumption that the other dykes of the coal-field, are likewise subsequent, and not contemporaneous phœnomena.

These circumstances render the course of this dyke through the more recent formations, a point of much geologicial importance, and Mr. Bakewell has fortunately published a sufficiently detailed account of this. His statement is as follows.

From Berwick on the Tees it may be traced in an easterly direction, near the villages of Stanton, Newby, Nunthorp and Ayton. At Langbath ridge a quarry is worked in it; it passes south of the remarkable hill called Roseberry Toppin, near Stokesly, and from thence by Lansdale to Kildale; it may be seen on the surface nearly all the way in the above track. From Kildale it passes to Denbigh Dale end, and through the village of Egton bridge, and hence over Leace ridge through Gothland, crossing the turnpike road from Whitby to Pickering near the seven mile stone, at a place called Sillow Cross on a high moor. I examined it at this place, where it is quarried for the roads, and is about ten yards wide. From hence it may be traced to Blea Hill near Harwood Dale, in a line towards the sea, near which it is covered with alluvial soil; but there can be no doubt that it extends into the German ocean. It is a dark greyish brown basalt which turns brown on exposure to the atmosphere; it is the principal material for mending the roads in the district called Cleveland.

It is only necessary to add to this account that the dyke enters the lias near Nunthorp and the sand of the inferior oolite near Roseberry Topping.

II. *Trap rocks of Derbyshire.*

These chiefly occur in three strata, conformably alternating in the formation of carboniferous limestone, which, by their interposition, they divide into four separate beds; but other masses of trap, of more limited extent, are occasionally found contained within the limestone beds; especially in that usually called the third limestone (the numbers following a descending series.) An instance of a dyke is also mentioned by Whitehurst. It does not appear to have been ascertained, whether these occasional masses of trap are or are not in any manner connected with the principal strata of that class; though a knowledge of this fact would be of the utmost importance in determining the relations of the whole formation. It should also be observed that the upper toadstone exhibits in Hockley lime quarry, south of Ashover, portions of limestone imbedded in its mass, near the entrance of the quarry; and some other probable instances of the same nature are cited by Mr. Farey.* Before we proceed to examine the course and position of these beds in detail, it will be convenient to give a general view of their mineralogical character: this, as is usual in such formations, is subject to frequent and great changes. The prevailing variety is an amygdaloidal trap; consisting of a compact iron-coloured paste, (probably a fine grained basalt) containing nodules of various sizes, from small granular spots, to that of an hazel nut, or larger, of whitish calcareous spar, and of green earth. Agatine nodules are of more rare occurrence, affording specimens of onyx, chalcedony, jasper, and the quartz crystals, locally termed Derbyshire Diamonds. The varieties of zeolite common in rocks of this family elsewhere, have also occasionally been found in these. The decomposition of the imbedded nodules, frequently occasions the amygdaloid to assume a vesicular and lava like character. The amygdaloid sometimes passes into ordinary basalt, which is as usually characterised by its tendency to decompose into large spherical masses; and by occasionally assuming an irregularly columnar texture (an instance of which may be seen in the deep ravine called Cave dale near Castleton. Greenstone distinctly crystallised is also seen in the same ravine and other places. From the retentive nature of the clay a line of ponds is often found along the bassets of the toadstone; a circumstance which assists in tracing their course. A variety of a finely gritty texture, and yellow colour, which may perhaps be called a trap tufa, occurs in Harborough rocks near Brassington, where it appears to have been quarried as a freestone. On decomposition, the trap passes into a clay of a bluish grey colour.

The prevailing structure is massive; but a laminated structure sometimes though very rarely occurs.

Mr. Whitehurst maintained that the metalliferous veins of the limestone strata were universally cut through by those of toadstone; and produced this as an argument proving their subsequent origin. Mr. Farey admits this to be the general fact, but adduces several exceptions; mentioning no less than nineteen instances in which the toadstone beds carried ore; usually however only in thin strings. Since the period of his publication, the veins of the Seven Rakes Mine near Matlock, which had before been worked in the 2nd or 3rd limestone, have been pursued with success and profit in the intervening

* Professor Buckland observed the nodules of the limestone thus imbedded in the trap of Derbyshire, to be occasionally surrounded by a thin crust of fibrous calcareous spar, exactly resembling that produced in instances of incipient fusion under pressure in Sir James Hall's experiments.

toadstone. Still, if in any authenticated instances, the metallic veins be decidedly cut off by the toadstone, the argument of Mr. W. must, *pro tanto* be held good; and the excepted cases will only prove, that some of the metallic veins of this district are of later formation than others. The average thickness of the first or upper toadstone, is about 60 feet; of the two lower 75 ft. each. This however is subject to very great variations; but the statements of Mr. Whitehurst of the appearances in the mines on Tideswell Moor, whence he inferred the total irregularity or the occasional absence of these beds, are referred by Mr. Farey, to the having mistaken some of the occasional masses before mentioned for the regular beds.

In attempting to give a brief sketch of the course* of these toadstone strata, which range in nearly parallel irregular curves from its north-east to its south-east angle, sweeping round with outlines convex towards the west, through the limestone district of Derbyshire, it must be remembered that that district is described by Mr. Farey as being bounded on three sides (the south, west and north,) by an extensive fault; the effect of which is in these directions, to place the fourth or lowest bed of limestone on the same level with the shale, whose true place, as ascertained on the eastern side, is above the whole limestone series. Hence, while the three toadstone strata form nearly parallel bands along the eastern side, where the beds crop out in regular succession, we must not look for any repetition of them on the western side, where the lowest formation is thus brought abruptly into contact with the shale formation, to the necessary exclusion of the intervening beds.

The only exception to this arrangement, is on the north-western side of the district, where the line of this fault so ranges as to include in two parts of its course small portions of the third limestone, (the next above the lowest) and consequently allows the repetition of the third toadstone.

We shall begin by describing (still following Mr. Farey) the course of the third toadstone, and before we pursue its continuous basset, where it has an eastern or south-eastern dip, we may first notice its occurrence on the north-western border of the district, where it is repeated with an opposite, or north-western dip, underlying the insulated portions of the third limestone just mentioned. Of these portions, one extends from Dove hole, about a mile and a half to the eastward, to Sparrow pit near the celebrated ebbing and flowing rock. The other,

* It may be useful to desire the reader in following this sketch, to fix his attention particularly on the valley of the Wye, and the dales opening into it on the north, and in Bonsall Dale on the South. In these all the 3 strata are finely displayed. See Sections, plate 2.

which adjoins closely to the north-western angle of the limestone tract, includes the villages of Buxton and Fairfield. The toadstone underlying the former, ranges between the points above named, with a small sweep to the south. That underlying the latter, ranges south from Buxton, on the east of Pool's hole, passes on the north of Sherbrook dale, (containing a branch of the river Wye,) and crosses at Mill dale the main branch of the same river, a little below Buxton. Here it has been particularly described by Faujas de St. Fond, (in his English travels,) who mistook it probably from its rapid dip for a dyke; hence it proceeds first south-east, then north-east, and lastly north-west encircling Fairfield Town; its whole course having been about two miles.

Near this there is another insulated portion of the third toadstone at Staden Hill, on the south side of Sherbrook dale; a similar one occurs at Peak Forest Town, about two miles east of that near Dove hole.

We have now to pursue the general basset of this bed, as it crops out westwards beneath the continuous zone of the third limestone on the east; and in this we shall proceed from north to south.

In the north-east then the third toadstone commences on the border of the limestone tract, near Castleton, where it is first seen in Cowler hills, and then crosses the ravine called Cavedale, (which must not be confounded with that in which the great cavern is situated, and which is separated from this by the Cragg Hill on which the castle stands;) here it assumes in one spot a columnar form, and affords basalt passing into greenstone and containing red jasper. Hence it ranges in a sinuous line south-west towards the river Wye, passing by Portaway mine. Knowl, Copt, Dalehead, $\frac{1}{8}$ of a mile north-west of Wheston near Smalldale, to Great Rocks west of Wormhill, and thence by the eastern skirt of Flagdale, on the south-west of that village crossing the river Wye.* A little east of the opening of that dale into the principal valley, where it may be seen both on the east of its upper part (where called Wyedale) and on the west of its middle portion (called Millers dale.) Proceeding down the Wye,† it is also seen in the two next dales branching northwards from Millers dale: viz. in Monks dale, as well as its prolongation, Grass dale, and its eastern branch Thatch dale, and in Tideswell dale and its western

* The course of the toadstone strata in the valley of the Wye and lateral dales will be clearly understood by referring to the section.

† The position of the toadstone crossing Wye Dale will be clearly understood from the section of that dale in the large plate of Sections.

branch Meadow dale; being elevated in the latter instance by a fault. Crossing the Wye above Millers dale, a little east of the remarkable limestone cragg of Chee Tor, it may be traced in the upper end of Sandy dale, which there branches south from that river; thence proceeding to the south-west and south, near the north-west of Blackwell, and east of Chelmerton, west of Cronkstone, Hardlow and Benty Grange, thence running in a nearly straight line south-east by Pike Hall, to Grange Mill in the upper part of Bonsall dale near Cromford; then it skirts along the northern side of that dale, ranging near the top of the hill near Slaley; until, by the dip of the strata to the level of the valley, where it meets a branch of the great fault (already described as nearly surrounding the limestone tract), it is for an interval cut off; but is shortly resumed (beyond an angle of depressed strata included by this fault) in the hills south of Bonsal dale, along which it ranges to the south-west, and skirts the eastern side of the south branch of that dale (called the Gellia dale.) Here are situated Harborough rocks already mentioned as affording a peculiar variety of the trap; and near these, it finally meets the southern edge of the limestone tract.

The second toadstone is the next in order, in ascending the series. The north-eastern extremity of this stratum does not commence exactly on the eastern border of the limestone; which, being there cut by a branch of the great fault before mentioned, has the first or highest limestone brought down on the east side of that fault; whilst the substrata cropping out on its western side, are thus made to abut against the depressed edge of the superior limestone, as is the case with the bed of toadstone we are now tracing: under these circumstances it first presents itself about one-third of a mile south-south-west of Windmill house, a small hamlet on the north-east of Tideswell; proceeding westward, passing south of Tidslow-top-hill, and thence south-west and south, till taking the line of Brookbottom dale, it surrounds and underlies Tideswell town; afterwards skirting the eastern side of Tideswell dale in its course to the river Wye, which it crosses near the mouth of Crossbrook dale, (see section). Before however we trace its progress on the south of that river, we must pause to mention some insulated hills in which it occurs beyond the general line of its basset, forming parts of the ridges dividing the dales that open towards the Wye from the north: one of these is on the south-south-east of Wheston near Tideswell, between Tideswell dale and Monk's dale; the other forms the base of a ridge extending near two miles close to Wormhill, including that village, Bole End hill, and Tavistead, and bearing three caps of the superjacent or second limestone. This ridge is situated between Monk's dale and Flat dale. It should also be noticed, that in Tideswell dale, south of the town at the entrance of a small dale called Brees dale, several alternations of such occasional masses of toadstone (as were formerly mentioned as proved in the neighbouring mines on Tideswell moor) are to be seen in the cliffs composed of the third limestone. In pursuing the regular basset of the second toadstone on the south of the Wye, we find it skirting on the south that portion of the valley of that river distinguished as Rivers dale, and thus ranging westward till it approaches the eastern border of Sandy dale; here it turns southward and takes henceforth a course parallel to, and at an average distance of about a mile east from, the course of the third toadstone, (before described,) keeping to the west of Taddington and of Arborlow, and thence arriving by a south-easterly course near the north of the village of Bonsal, where it appears on the south-west slope of Masson low, one of the hills bounding Matlock dale on the west. In Matlock dale itself this stratum is exposed by denudation on the banks of the river near the foot of the high Tor, being brought down to this level by its dip: from Masson low it ranges south-west and south, passing near Bonsal church, and soon after meets the branch of the great fault before described as running through Bonsal dale, and in consequence of the depression of the strata included by the angular course of that fault, is no more seen.

The first, or highest toadstone, commences on the north-east from the same branch of the fault, as the second about half a mile south of it, and under exactly similar circumstances; it ranges west surrounding Lane-head houses, and thence turns south-east by Litton town proceeding across the north end of Cressbrook dale and by the north and east of Hay dale to the river Wye, on the banks of which it appears about half way up the cliffs of Fin Copt hill (one mile and a quarter north-west of Ashford) which exactly resembles Matlock high Tor in its appearance and the strata it exhibits. This toadstone crosses the river a little below this Tor, and traversing the middle of Dimins dale, ranges south-west to Moneyash, where, like the preceding beds, it trends first south and then south-east running parallel to them, and about three-quarters of a mile east of the second toadstone. This course carries it on the south side of Blakelow Tor (above Bonsal) and thence down the eastern slope of the hills to the valley of the Derwent, which it passes near the north foot of Matlock high Tor, and thence keeps for some distance on the east of that river sweeping from the rise of the strata in an ascending curve through the precipitous escarpment of the Tor, till it attains its middle region;

whence descending on the south, it re-crosses the Derwent, but soon crosses it a third time, and again within two-thirds of a mile lower down finally re-crosses it at the paper-mill near Matlock bath, thence ranging across what is called the Wallet to Bonsal dale, where it abuts against the fault so often described; the depression however of strata included in the angle bounded by that fault is not sufficient entirely to conceal this toadstone, but only throws it a little west of its former line of bearing, so that it ascends through the wood on the south side of the dale, and traversing that angular area in a south-westerly direction, finally abuts against its southern side so nearly at the same point where the third toadstone (as before described) occurs on the opposite side of the fault, the strata being on that side elevated, that without attentive examination it might easily be mistaken for a continuation of the same basset: this point is one-third of a mile west of Middleton.

Such being the course of this stratum along its entire line of basset, it remains only to mention its occurrence in two other situations, where on the east of that regular basset it is thrown up prematurely by a line of fault* ranging round three sides of a parallelogram including the villages of Bastow, Bakewell, Yolgrave and Beeley, on its north-west and south borders; and elevating the included strata, by what is technically termed a trap, up to the east: by this fault (which Mr. Farey calls the Bakewell fault) the first toadstone is thus thrown up just above Bakewell on the course of the river Wye; (here this fault, and the effect it produces, are represented in the accompanying section,) more to the south at Over Haddon where the second limestone is also seen; and still further in the same direction at Yolgrave. II. The last instance we have to mention of the appearance of this toadstone is at some distance from the main limestone area, where the valley of Ashover (three miles north-east from Matlock) exhibits a deep denudation, cutting entirely through the millstone-grit and shale, and thus exposing the subjacent limestone, which also rises by a flexure and saddle, to the same point; here the toadstone is seen as usual underlying the first limestone; and it is believed here to possess a thickness much greater than ordinary.

We have entered into a much more minute detail than we usually allow ourselves of the localities of these strata, since the important theoretical inferences which have been built on them render it desirable that every facility should be given to

* Mr. Watson, however, explains these repetitions of the beds by a mere undulation of the strata unaccompanied by any fault.

future observers who may have opportunity of examining them, and the work whence these notices are principally derived, being rather of a statistical than scientific character, is probably in the hands of but few of the readers of this volume; and those who possess it and are acquainted with the extreme difficulty of reference which the author's arrangement of his materials (in numerous detached lists of hills, vallies, &c. alphabetically disposed) occasions, will doubtlessly find the present attempt to combine and collect into one mass, the scattered information it affords on this subject, a useful aid. Much hesitation has been displayed in admitting the existence of the enormous faults which Mr. Farey believes are proved to exist by the phenomena of this district; with regard however to those which are above alluded to, it is absolutely necessary to concede their existence, if it be allowed that the toadstones are regularly conformable strata; both these positions must stand or fall together: and it may be added, that enormous as these disturbances are, they have in favour of their reality, the analogy of others equally great at the foot of Cross fell, (see Professor Buckland's paper,) and near Ingleton, where the coal-measures are repeated beneath the foot of Ingleborough on the south-west and below the level of the junction of the limestone of that mountain and the greywacké slate, (see the case as represented in the general section from Cumberland to Sussex.)

A section across the middle of Derbyshire from south-east to north-west is given at the end of this volume.

It commences in the denudation of limestone and toadstone in the Ashover valley; and crosses the intervening hill of shale grit and lower coal-measures to the valley of the Derwent at the confluence of the Wye; thence it entirely follows the course of that river upward to Buxton; twice crossing the curved line of the Bakewell fault, and exhibiting its effect in elevating the strata. It then shows the successive regular outcrops of several strata along the north bank of the Wye, and the lateral dales branching from it on that side; lastly, crossing and displaying the great fault, it terminates in Combes Moss on the north-east of Buxton, capped with the coal-measures; thus ending and beginning in the same strata. The general section from Cumberland to Sussex also passes from north to south across the Derbyshire limestone tract near its east border, and thus twice crosses the curved bassets of these strata, and passing Bonsal dale, shews the effect of the fault in that quarter. It is hoped that these sections will render every point in the above descriptions easily intelligible.

We have thus described minutely the course and phenomena of the regular toadstone beds; it has also been stated that occasional masses of the same rock occur in the different limestone strata. Those on Tideswell moor and dale in the third limestone have been already mentioned. The same appearances also occur at Mockshaw and Haredale mine west of Bakewell in the 1st lime.

Mr. Whitehurst mentions dykes of toadstone on Bonsal moor, and says that in the mine called Slack in that quarter one of these formed a cross course, intersecting the metalliferous vein; the description if correct proves that this must have been a true dyke, although Mr. F. observes that the only instance of such a phenomenon which had fallen under his observation, was a small vein of hard toadstone (only six inches wide) partly filling a fissure on the 3rd toadstone stratum. The upper part of the fissure was occupied, not by toadstone, but a brittle and very inflammable coal; which perhaps may have been indurated bitumen, traversed by septa of calcareous spar.

V. GREENSTONE BEDS OF GRIFFE IN THE WARWICKSHIRE COAL-FIELD, AND NEAR THE LICKEY.

A. Pursuing the road to Coventry from Nuneaton, a hill of slight elevation is ascended, about half a mile from the latter town. This consists, as far as the summit, of the common shale of the coal-measures dipping south-west in an angle of between 33 and 40°. At the very summit the greenstone may first be traced on the western side of the road, which in its descent towards the adjoining valley on the south, called Griffe Hollow, is deeply cut in the greenstone; but it is there in so decomposing a state that its characters can scarcely be discerned; towards the bottom, the greenstone appears to be covered by, as at the top of the hill it appeared to rest upon, shale; and its position at this junction is such as seems to indicate, that it forms a conformable bed in the shale, rather than a dyke; below in Griffe hollow itself, which crosses the road at right angles, a second mass of greenstone, very well characterised with distinct but small crystals of felspar, appears a little on the west of the road, and crosses the hollow in a south-east direction (the general bearing of the strata) and proceeding a little way east in the ravine, greenstone rocks again appear on its north border, exactly in the line of bearing which the greenstone first mentioned must have pursued if a regular stratum, underlying the second, and separated from it by interstratified shale. On the whole there seems therefore every reason to believe these masses to be here at least conformably interposed among the other beds. The greatest distance to which the author of these observations traced them, did not exceed one third of a mile; but he was informed they extended nearly a mile further in a south-east direction. On the north-west he could not learn that they had been ever noticed. No dykes were known to exist in the neighbouring collieries, and they were rather free than otherwise from faults. No alteration was noticed in the shale near its contact with the greenstone at the only spot where the junction could be observed; namely the section presented by the deep cutting of the Nuneaton road. The thickness of these beds cannot be accurately observed; but they probably exceed thirty feet each.

B. Professor Buckland noticed some dykes of trap near the broken traces of the coal formation on the north of the Lickey on the opposite side of Warwickshire.

VI. TRAP ROCKS OF STAFFORDSHIRE.

A. *Rowley.*

The principal mass of these occurs in the southern part of the county; overlying the coal-field which surrounds the town of Dudley. It there constitutes the material of a group of hills, beginning on the south of that town, and terminating about half way between Hales Owen and Oldbury, a little beyond the village of Rowley. They appear at first sight to be a continuation of the limestone hills to the north of Dudley, as they proceed nearly in the same direction, and have the same elevation, though their size is not so great. But these hills are composed of very different constituents, and lie in a very different position with respect to the coal-formation of this country. They consist of very pure basalt, which in the neighbourhood of Birmingham is called *Rowley rag*, because the village of Rowley is situated on one of these basalt hills, and this hill appears to the eye to be the highest of the whole range. The names of these hills, beginning at Dudley, and proceeding in the order of their position, are as follow: Corney, Tansley, Bare, Cook's Rough, Ash or Cox's Rough, Turner's, Pearl, Hailstone, Timmins, Rowley, Whitworth. These hills are all covered with soil; but quarries have been opened in several of them, and the basalt of which they are composed is employed for mending the roads. The streets of Birmingham are likewise paved with it.

This basalt has a greyish-black colour. Its fracture is small conchoidal, and nearly even, with here and there a little tendency to the splintery. Its lustre is glimmering, owing to very small black crystals being interspersed. They appear to be prisms. Their lustre would indicate them to be augite; but, as their colour is black, I rather consider them as hornblende. The basalt is opake, brittle, not easily frangible, breaks into fragments with sharp edges, and the paviours in Birmingham complain that they cannot break it into the shape adapted for paving the streets. It is hard enough to cut glass, and to strike fire with a steel. It melts before the blow-pipe; and, when heated in an open fire, becomes magnetic, and loses three per cent. of its weight.

Dr. Withering seems to have missed the lime and soda, which no doubt exist in Rowley rag. But the analysis of minerals was at that time in so imperfect a state, that we have more reason to admire the accuracy which he actually attained, than to be surprised at the mistakes into which he fell. The basalt in these hills has a very distinct columnar structure.

The above account, which gives a clear summary of the geology of these hills, is extracted from a memoir of Dr. Thomson, but it should be in justice added that Mr. Keir, the first investigator of this district, whose observations are inserted in the 1st volume of Shaw's history of Staffordshire, left nothing to succeeding enquirers but the task of compilation. These hills at their border evidently overlie the coal-formation which has been pursued to some distance beneath them. But it is very possible that they may nevertheless be connected with a vast dyke, penetrating the strata beneath their centre; and overlying only at its edges. This was the opinion of Mr. Kier.

It should be observed that the columnar structure, though very frequent, is far from universal in this trap, which very commonly occurs in large spherical masses decomposing on the surface into concentric layers, as described in the general account of the rocks of this formation at the beginning of the article. An amygdaloidal variety containing calcareous spar and zeolite occurs on the south of the town of Dudley. The highest point of the Rowley Hills is stated by Dr. Thomson to be 900 feet above the Thames at Brentford, this height having been deduced from observations on the canal levels.

B. *Trap in the Colliery of Birch Hill near Walsall.*

This colliery, as described by Mr. Aikin, presents many interesting facts connected with our present subject; it is bounded on the north-west by a fault called by the miners the green rock fault, and on the surface above this fault appears a low narrow hillock of greenstone; all the beds rise with an angle rapidly increasing as they approach this fault, advancing in a few yards from 6° to 25°. Among the beds worked in the colliery, is one of greenstone similar to that of the neighbouring hillock. This bed is in appearance conformably interstratified where it is pierced in the workings; but varies remarkably in thickness, being in one shaft 24, and in a second only 12 feet. In a third shaft, which crosses the strata usually above and below it, it is entirely wanting; and it is observable that its thickness increases in approaching the green rock fault. From these appearances Mr. Aikin concludes that the green rock hillock is part of a thick vertical dyke of greenstone; and that this seeming bed is only a wedge-shaped prolongation of that dyke intruded among the regular strata; if so, it affords an example of the case represented in the beginning of this article on Dr. Mac Culloch's authority; but it is fair to state that the absolute junction of the bed and fault have never been laid open. The faults affecting the other bed of the colliery, affect this likewise; which is an important fact, as proving that it existed among the coal-measures (however introduced) before the convulsions producing those faults took place.

The substrata immediately beneath this bed of greenstone appear to be altered by it, differing in their character where it does not exist; thus the sandstone immediately beneath it is indurated, and the shale and coal beneath (though preserved from actual contact by that sandstone,) are yet deprived of their bitumen and much altered in other respects.

The rock in question appears to be a greenstone consisting principally of felspar and amorphous hornblende, and also contains carbonate of lime. In the Annals of Philosophy for Sept. 1818, is a farther account of the ridge of greenstone mentioned in the beginning of this article. This ridge appears to extend a mile in length, varying in breadth from half a mile to 40 yards: at the point of it called Pouck hill a quarry has been opened, which exhibits columnar masses of basalt dipping towards a central point which is traversed by a vertical dyke. Prehnite, mesotype, and sulphate of barytes occur in this quarry.

VII. SECONDARY TRAPS OF SHROPSHIRE.

In the south-east angle of this county, the twin mountains called the Titterstone Clee Hill and the brown Clee Hill rise to an imposing height: they have before been mentioned in this volume on account of the coal-measures of their central regions. Both these mountains have their summits composed of a basalt exactly resembling that of Rowley, and in each this basalt forms a double ridge.

The trap here clearly occurs in overlying masses, forming tops to the mountains, and distinctly reposing on the coal-measures. It thus agrees in position with the Rowley Hills; and both their points are distinctly visible from each other. Mr. Bakewell states that he observed in connexion with the basaltic cap of the Titterstone Clee, a vast fissure or dyke more than 100 yards wide filled with the same basalt which intersected the hill, cutting through the coal-fields. It rises from an unknown depth and appears to have forced a part of the coal to the surface. Where the basalt comes in contact with the coal, it has injured its quality and reduced it to a sooty state.

In the lower grounds, on the east of the Clee hills at Hewlet, an amygdaloid occurs, with calcareous glands; but its position and relations are not stated.

The greenstone which occurs on the limits of the Coalbrookdale Coal-field at Steeraway and Little Wenlock Hills, appears rather to be associated with the transition limestone than with the coal-measures; and therefore will most properly be classed and described among the traps associated with the transition rocks. To many it may indeed appear that the trap rocks have so many marks of a peculiar origin, as to render the subdivision of them, according to the formations into which they have been perhaps intruded, improper. But that subdivision may (however the theoretical question be viewed) be defended on the ground of convenience; and to neglect it would be to pronounce a decided opinion on a much controverted hypothesis; a province which this work entirely disclaims.

On the south of the Clee hills, in Worcestershire, the summits of the Abberley Hills also present trap rocks: but in this instance again, they appear to be associated with transition limestone.

CHAPTER VI.

Comparative view of the distribution of the great Coal-formation in other countries.

Having in the preceding chapter concluded our examination of the English coal-districts, we may usefully close the present book with a rapid survey of the distribution of this important mineral in other countries.

To begin with Scotland, as most nearly connected with our proper field of enquiry, we find in Dumfriesshire (where the great central Penine chain of Northern England joins, almost at right angles, the great southern transition chain of Scotland) many limited coal-fields reposing against, or forming narrow basins in, the vallies of the latter chain; these are associated with, and rest upon as usual, thick beds of the carboniferous limestone. (See Jameson's Mineralogy of Dumfries; and for an account of the limestone quarries, Sowerby's Min. Conchology, vol. 3.)

But the principal coal district of Scotland, occupies the tract forming what may be called the great central valley of Scotland (speaking relatively, for considered in itself its surface is very considerably varied), which lies between the great transition chain on the south, and the still loftier primitive ranges of the highlands on the north. The whole of this wide tract is occupied by the coal-measures, the carboniferous limestone, and the old red sandstone, associated in every possible manner with vast accumulations of every variety of trap. A good general description of this tract will be found in Bouet's 'Geologie de l'Ecosse'; a memoir by Mr. Bald, in the third volume of the Wernerian Transactions, on the Clackmannan coal-field, though referring to one point only, furnishes the most precise information concerning the more detailed structure of the coal-measures; and many particulars, together with a good list of the organic remains found in the carboniferous limestone (which agree with those occurring in England) may be seen in the history of the parishes of Ruthenglen and Kilbride, near Glasgow. Williams's Mineral history (republished by Dr. Miller) furnishes some materials concerning the eastern part of this tract, but such as are rather interesting to the practical miner than the geologist.

In the low district on the east of Sutherland, where the secondary formations again intrude among the primitive highland chains, coal has been discovered at Brora; but from the slight description incidentally given of this tract in the memoir of Mr. Bald (3 vol. Trans. Wern. Soc.), before referred to, it may be conjectured that this does not belong to the principal coal-formation, but to those beds which occasionally occur in more recent formations, being perhaps of the same æra with the coal of the Cleaveland district in Yorkshire.

In the north-east of Ireland we may trace the prolongations of the two great primitive and transition chains which traverse Scotland, and are interrupted on its south-west border by the channel between the two islands; the mica-slate of the former chain appearing to cross* from Cantire to Cushendal in the north-east angle of Antrim, and (after a partial concealment by the overlying of the great basaltic area) to range along the northern counties of Donegal, &c.; while the greywacké of the latter, crossing from Portpatrick to Donaghadee, extends thence towards the centre of the island. The formations of the great central valley of Scotland here likewise intervene, and among them the coal-measures, which may be seen emerging from beneath the overlying basalt at Fairhead on the north-east, and again just beyond the south-west of the basaltic area at Coal Island and Dungannon; but other parts of Ireland present far more important coal districts. Indeed the island may generally be described (with the exception of the north-east basaltic area) as being almost surrounded with a series of primitive and transition groups, including a great central area entirely occupied by the old red sandstone, carboniferous lime, and coal-measures. The structure of these districts very closely corresponds with the tracts of the same formation in England.†

* This connection between the two countries is more fully pointed out and illustrated by a map in the introduction to Dr. Berger's memoir on the north-eastern counties of Ireland, Geol. Trans. vol. 3. drawn up by the present writer.

† The following summary of the Irish coal-fields is extracted from Mr. Griffiths' admirable report on the Leinster coal-district. Mr. Weaver's excellent memoir on the south-east of Ireland (Geo. Trans. vol. iii.) should also be consulted, as affording much important information concerning the first of the coal-districts here mentioned.

'If we except the Leinster district, my knowledge of the coal districts of Ireland is as yet very limited. And, though each in its turn will form the subject of a separate report, I think it right to draw attention to them in this place, by giving such general information as I possess, respecting their situation and circumstances. Coal has been discovered in more or less quantity in seventeen counties * of Ireland; but I believe the island

* The counties are, Antrim near Ballycastle; Donegal north of Mount Charles; Tyrone in the Ulster coal district, and at Drumquin; Fermanagh, north continuation of the Connaught coal district, and at Petigoe; Monaghan near Carrickmacross; Cavan near Belturbet; Leitrim and Ros-

We may now pass to the Continent, and will first mention Sweden because it seems scarcely too bold a generalisation to

contains but four principal coal districts, viz. the Leinster, the Munster, the Connaught, and the Ulster. The two former contain carbonaceous or stone-coal,† and the latter bituminous or blazing coal.

The Leinster coal district is situated in the counties of Kilkenny, Queen's county, and county of Carlow. It also extends a short distance into the county of Tipperary, as far as Killenaule. This is the principal carbonaceous coal district. It is divided into three detached parts, separated from each other by a secondary limestone country, which not only envelopes, but in continuation passes under the whole of the coal district; a fact, which was indisputably though accidentally, proved by the Grand Canal Company, who sunk a pit through eighteen yards of black slate-clay, and flinty slate, into the limestone in search of coal. The Leinster coal district is therefore of subsequent formation to the limestone.

The Munster coal district occupies a considerable portion of the counties of Limerick and Kerry, and a large part of the county of Cork. It is by much the most extensive in Ireland; but as yet there is not sufficient information respecting the number, extent, or thickness of the beds of coal it may contain.

Coal and culm for near a century have been raised in the neighbourhood of Kanturk in the county of Cork. At Dromagh colliery I understand the work has been carried on to a very considerable extent, and its annual supplies of coal and culm have materially contributed to the agricultural improvement of an immense extent of the great maritime and commercial counties of Cork and Limerick, which must otherwise have continued neglected and unreclaimed.

Many circumstances combine to make the examination of this district of peculiar interest and importance,—and as a recent application has been made by the Cork Institution to the Dublin Society, to aid the undertaking, it is probable that this immense district will shortly be minutely explored; from all that has been ascertained it is very clear that the dip of the beds and the quality of coal differ materially from those of the Leinster district. In the Munster district the beds run east and west, and dip to the south, forming an angle of forty-five degrees. In the Dromagh colliery, where all the beds which have been discovered have been successively and in general successfully wrought, four beds incline on each other, and at no greater distance than 200 yards. The first of these beds is a three feet stone-coal, and is the leading bed. All faults, checks, and dislocations, similar to those which are discoverable in this bed, are in general to be encountered in the other three; the names of the four beds are, the *coal-bed*; this lies furthest to the north; the *rock-coal*; so called from its being comparatively of harder quality than the other beds; the *bulk bed*; so called from its contents being found in large masses or bulks; and *Bath's bed*; so called from the name of a celebrated English miner, by whom it had been many years ago discovered and worked; the coal-bed consists of three feet solid coal, and is not sulphureous; the rock-coal, is nearly of the same thickness with the leading bed, but is very sulphureous, and having the soundest roof is the most easily wrought. The other beds are of the culm

common in the Connaught coal district; Westmeath near Athlone; Queen's county, Kilkenny and Carlow, in the Leinster coal district; Tipperary, continuation of the same; and Clare, Limerick, Kerry, and Cork, in the Munster coal district.

† Slaty Glantz coal of Werner.

consider the primitive ranges which, pervading Norway, extend over the greater part of that country into Finland, as a con-

species, but of peculiar strength. Each barrel of culm has been ascertained to burn from nine to ten barrels of lime. The bulk bed forms immense bulks and masses of culm, in which the miners have frequently been unable to retain the ordinary directions of roof and seat.

No work has been undertaken in the Munster coal district to a greater depth than 80 yards. The present work at the Dromagh colliery is at that depth. It is heavily watered, and consequently expensively wrought. The quality of the coal and culm improves as the work descends. Mr. Leader, who is the proprietor of the Dromagh colliery, has kindly communicated to me all the surveys and reports, which, from time to time, have been received from the persons who have inspected and directed the works. The peculiar proximity of the beds, and the extraordinary diversity of their contents in the Munster district make the minute examination of this coal-field of deserved interest and importance to this country. The time I trust is not remote, when the great coal-field on the left bank of the Blackwater will be found to contain mineral treasure altogether inexhaustible, and by the liberality of Parliament, and the judicious and patriotic exertions of the Directors of Inland Navigation, fully adequate to all purposes of domestic convenience, and of national utility.

The Connaught coal district stands next in order, of value, and importance, to the Leinster and Munster, and possibly may be found to deserve the first place, when its subterranean treasures shall be explored. At present nothing is known, except, that the outer edges of several beds of coal have been observed, but they have not been traced to any distance; so that their extent is by no means ascertained. The coal is of the bituminous species. This coal is particularly adapted to the purposes of iron works, foundries, &c. &c.; and the grey pig iron made at the Arigna iron works is among the very best smelted in the empire. Lough Allen, the source of the river Shannon, forms a basin in the midst of this great district, which on this account appears to be formed by nature for industry and commerce; but the navigation has been neglected, and is at present in such a state, that none but boats of trifling burden, can ply on the river, between Lough Allen, and Lanesborough. However, the Directors General of inland navigation in Ireland are at present employed in completing the Royal canal, between Coolnahay and Tarmonbarry on the river Shannon; and it is to be hoped, that they will at the same time complete a navigable communication between Lough Allen and the Shannon; and repair and improve the navigation of that river downward to Lanesborough. This would be a sufficient inducement to the proprietors of coal in the neighbourhood of Lough Allen to open their works to a considerable extent, as they could not fail of having a very large demand for their produce, whenever the navigation between the lake, and the Shannon, shall be completed.

The Ulster coal district is of trifling importance, when compared with the foregoing. It commences near Dungannon, in the county of Tyrone, and extends in a northern direction to Coal island, and in continuation to the neighbourhood of Cookstown. No beds of coal worth working have hitherto been discovered between Coal island and Cookstown, but certainly the coal strata extend there. The principal collieries are at Coal island and Dungannon. The coal of this district is bituminous. I understand that indications of coal have been observed at Drumquin, in the county of Tyrone, and also at Petigoe to the north of Lough Erne. Possibly the coal-formation may extend from the neighbourhood of Cookstown, westward to the north of Lough Erne.

Besides the foregoing principal coal districts, there are others of less continuation of the range of the Scotch highlands, over against which they lie, and with which they closely agree in character. And if this be allowed, the coal which occurs on the south of the primitive tract near Helsingborg at the entrance to the Baltic, will be found on the same geological line* with that of the central Scotch district. There exists coal also, as it appears, on the prolongation of this line in the island of Bornholm; its farther extension might perhaps be found in Russia; but we know too little of that country to speak with precision concerning it, and it will be found more convenient to postpone

sequence. Bituminous coal has been found in the neighbourhood of Belturbet, in the county of Cavan, and at the collieries of Ballycastle, in the county of Antrim; but the Antrim coal district is not very extensive. These collieries have been wrought for a number of years. The coal is of a slaty nature, and greatly resembles both the coal, and the accompanying rocks, which occur in Ayrshire, and probably they belong to the same formation. A very extraordinary discovery was made at these collieries about the year 1770. It is thus described by the Rev. Dr. Hamilton, in his Letters on the north coast of the county of Antrim. "The miners in pushing forward an adit or level toward the bed of coal, at an unexplored part of the Ballycastle cliff, unexpectedly discovered a passage, cut through the rock. This passage was very narrow, owing to incrustations formed on its sides. On being sufficiently widened, some workmen went through it. In minutely examining this subterranean wonder, it was found to be a complete gallery, which had been driven forward, many hundred yards, into the bed of coal; it branched out into 36 chambers, where coal miners had carried forward their works; these chambers were dressed quite square, and in a workman-like manner; pillars were left at proper intervals to support the roof; and in short it was found to be an extensive mine, wrought by a set of people, at least as expert as those of the present generation. Some remains of the tools, and even of the baskets used in the works, were discovered, but in such a decayed state, that on being touched they fell into pieces. Some of the tools appear to have been wood thinly shod with iron.

The great antiquity of this work is evident, from the fact, that there does not exist the most remote tradition of it in the country; but it is more strongly demonstrable from the sides and pillars being found covered with sparry incrustations, which the present workmen do not observe to be deposited in any definite portion of time.

The whole of the coal districts, which, as far as I know, occur in Ireland, have now been mentioned. Trials have, however, been made at Slane on the river Boyne, and also in the neighbourhood of Balbriggan and Rush. These trials were however on the edge of the district, near the junction of the limestone. If the country contain coal, it will more probably be found in the interior than on the edge of the district. From this brief account of the coal districts it appears, that very extensive tracts of coal country exist in Ireland; but none, if we except the Leinster district, have been examined; yet the Munster coal district is in extent greater than any in England, and may probably contain inexhaustible beds of coal.

* We must not be understood to convey the idea that the coal forms continuous strata having a regular basset along these extensive lines; but rather that it constitutes a series of insulated deposits, still preserving a general direction regulated by the great primitive chains.

the few notices we possess, till we have first considered those parts of Europe where better information is to be procured.

As we have noticed in the above paragraph, coal-fields in the north of Europe, apparently connected with the central system of Scotland, so we might naturally look for the resumption of those of the south-western English counties in the opposite regions of France. It is true, indeed, that in the south of Somerset and through Devonshire, no coal has been observed; being probably concealed by the advance of the overlying deposits of new red sandstone which are there in close contact with the transition chains. These transition chains cross, as is well known, from Devon to the peninsula of the Cotentin in Brittany; and we find, as might be expected, a small coal-field reposing against their eastern side at Litry on the south-west of Bayeux; farther south, where the Loire enters between the continuation of these chains, between Angers and Nantes, are more extensive deposits of this formation.

In the centre and south of France there are some limited coal-deposits lying in the vallies of the Loire, the Allier, the Creuse, and the Dordogne, the Aveyron and the Ardeche between ridges proceeding from the primitive central group connected with the Cevennes.

Several particulars concerning some of these districts may be found in the account of the geological speculations of Mr. Rouelle in the first vol. of the Geographie Physique, forming part of the great Encyclopedie Methodique; the Annales des Mines for 1821 contains several particulars of those near St. Etienne, department of the Loire, and a full account of those of the Aveyron.

From the south of France we may proceed to Spain, which could not so conveniently be included in any other part of our survey. Coal is here mentioned as occurring in eight places in Catalonia, in three in Arragon, and one in New Castile, but no particulars are given: a list of these localities may be found in Laborde's view of that country.

We now return to the consideration of the great carboniferous tract of Northern France and the Netherlands.*

It may generally be described as extending westwards from

* Some Continental Geologists, generalising without a sufficient acquaintance with the particular circumstances, have endeavoured to refer some of the English coal-fields to a prolongation of this line; but as the whole intervening breadth of England from Kent to Somerset, is occupied by more recent formations which effectually conceal the true connections of this substrata, and as there is nothing in the circumstances of any part of our coal districts indicating any relation with the above beyond that which always exists between formations of the same age, this view cannot be maintained.

Hardinghen near Boulogne (only a few miles from the coast of the channel) by Valenciennes, and thence up the Scheldt and down the Meuse to Eschweiler beyond Aix la Chapelle; and still further west, many of the coal districts of Northern Germany may with great probability be considered as a prolongation of it.

On the east and north, the great deposits of chalk and the strata above the chalk, skirt and partially (particularly within the limits of France) overlie this tract. On the south it is bounded by the transition ridges (of slate, grauwacke, &c.) which occupy the forest of Ardennes, overhang the magnificent defile of the Rhine from Bingen to Bonn, and thence extend to the Westerwald. This tract does not consist of a single continuous coal-field, but of many insulated and basin-shaped deposits of this formation, encircled by carboniferous limestone and old red sandstone. In many respects it bears, even down to the character of its picturesque scenery, a remarkable analogy to the coal districts (likewise consisting of many insulated basins) in the south-west of England.

We find the most westerly point of this extended chain of coal-fields at Hardinghen, in the great denudation exposing the beds beneath the chalk, which comprises the Boulonnais on the French side of the channel and the Weald of Kent and Sussex on the English; of this we have before given a general description. These coal-mines, and the quarries of the carboniferous limestone associated with them, which appear at Marquise, are situated at the very foot of the escarpment of the environing zone of chalk hills; for the outcrop of all the intermediate formations crosses this part of the denudation to the south, and as it were withdraws to expose the coal; proceeding westwards, the coal is worked at several places within the general limits of the overlying chalk-formation. The environs of Aniche near Douay, and of Monchy le preux near Arras, present deposits of this nature; the mines surrounding Valenciennes are still more extensive.

In the environs of Mons, Charleroy, and Namur, in a tract surrounding Liege, and lastly close to Eschweiler on the east of Aix la Chapelle, other very considerable coal-fields are worked.

A general account of this line of coal-formation, may be found in Omalius d'Halloy 'Geologie du Nord de la France,' Journal des Mines, and in Von Raumer's 'Geognostich Versuche'; many interesting details are also given in Villefosse sur la richesse minerale, (tom. 2. p. 432 & seq.) and illustrated in the magnificent atlas of that work by sections exhibiting the contortions, &c. of the beds, and the mode of working them, from Pl. 25 to 27.*

Proceeding still further along the northern border of the same transition chain, against which all these deposits of coal repose, we find the more recent formations (probably of the tertiary class) intruding upon it, and concealing the coal till we cross the Rhine near Bonn. On the right bank of that river these again recede to the north, and in this direction we again find an extensive coal-field proceeding along the small river Ruhr a little above its junction with the Rhine: on the south the beds of this coal-field describe the segment of a circle, cropping out against alternations of limestone, shale, and what is called grauwacké (our old red standstone probably), which separate them from the regular transition slate: on the north they are bounded by the overlying and more recent deposits. An account and plans of this district may be found in Villefosse, tom. 2. page. 424, & plate 24.

A little on the south of the same district of transition rocks, whose northern border we have been hitherto pursuing, limited coal-fields occur in the country between the Moselle and Rhine; 1. beween Sarrebruck and Sarre Louis on the river Sarre, and (secondly) near Waldmohr on the banks of the Glane, extending to its confluence with the Nahe: the beds of the former coal-field are described as ranging south-west and north-east, and dipping north-west; they are covered with red sandstone, and also surrounded by the same formation on the points to which they rise, appearing therefore to repose upon it; but these appearances are very vaguely described; "sur ces grès rouges semblent quelquefois s'appuyer les couches de houille;" they are probably deceptive. A careful examination of the country between this coal-field and the northern transition chain, is necessary to ascertain its true relations. Villefosse, tom. 2. p. 447. & Pl. 27. may be consulted on this district.

Pursuing the line of Northern Germany, a tract containing

* The series of beds between the coal-measures and the transition chains on the south, exactly corresponds with those occurring in England. The following list is extracted from Omalius d'Halloy, with the English equivalents annexed.

Coal strata	the same.
Shale and grit	Millstone grit and shale.
Limestone alternating with slate-clay	Carboniferous limestone.
Red and variegated grits alternating with slate-clay	Old red sandstone.
Yellow granular quartz	Old red sandstone.
Grit and breccia	Old red sandstone.
Clay-slate	Transition slate.

The organic remains and metalliferous deposits of the limestone underlying the coal, exactly coincide with those of England.

coal appears to range many miles on the south-west and south of Hanover between Osnabruck and Hildesheim, but we cannot refer to any description of it; it may probably form the prolongation of the northern line of coal-fields which we lately traced as far as the Ruhr.

On the north-east and south-east of the Hartz mountains, near Ballenstadt and Neustadt, coal-measures repose on the transition rocks of that group; they are covered by porphyry associated with the rothe todte liegende, to which succeed the cupriferous marle-slate, &c. corresponding with the newer red sandstone and magnesian limestone of England. Coal-measures occur beneath the porphyries of the environs of Halle under similar circumstances. An account of these districts may be found in Freisleben's "Kupfer schiefer gebirges."

In Saxony, coal-measures are found in many places skirting the northern base of the Erzegebirge mountains. There are extensive mines at Schonfeld near Zwickau, and at Planenschen grund near Dresden. At Schonfeld the coal alternates with porphyry, above which occur the following beds. 1. A conglomerate of porphyry and gneiss. 2. Bituminous shale with vegetable impressions. 3. Red sand. At Zwickau the beds (beginning with the lowest) 1. Wacke. 2. Basalt. 3. Nine or ten coal-beds alternating with white grit and shale containing vegetable impressions. 4. Sandstone. 5. Red grit. The coal-field of Planenschen grund is more extensive than either of the former; here a range of sienite, extending on the right bank of the Weisseritz, forms the fundamental rock on which a secondary porphyry reposes; then succeed four beds of coal alternating with grit and shale, and inclined at an angle of 65°.

There is no bitumen in the coal of Schonfeld; although it abounds at Planenschen grund: the same fossil vegetation is presented in all these mines. (See Sternberg 'Flora zur Vorwelt.')

We are not able to refer to any particular account of the prolongation of the coal-measures on the northern side of the great primitive chain which traverses central Europe, proceeding westwards from the Erzegebirge under the names of the Riesengebirge and Carpathian mountains. Count Sternberg, however, is of opinion that such a continuation may be traced, extending through Poland into Moldavia and Wallachia.

On the south of the Erzegebirge and Riesengebirge the basin of Bohemia, extending along the course of the Beraun and Iser, and including the adjoining parts of Upper Silesia, between Landshut and Silberberg, presents a very extensive coal-district.

This may generally be described as lying between the great primitive chain above-mentioned on the north, and the great district of primitive slate which occupies the larger part of Bohemia south of the Beraun and upper Elbe; but on its northern boundary, near the point where the Elbe breaks out across the Erzegebirge, the range of the Mittlewalde, consisting of summits of the flœtz-trap formation resting on wood-coal or lignite (of much more recent origin), is interposed between the true coal and the primitive chains: trap summits are likewise scattered over this edge of the coal-field.

On the south-west, the coal of Bohemia appears to form a number of small detached basins scattered along the line of the Beraun, and reposing on what are called transition rocks, but which may possibly answer to our own carboniferous limestone and old red sandstone. On the north-west towards Silesia, the coal-district becomes more continuous and extensive; it is here covered by red sandstone associated with porphyry as in the neighbourhood of Hartz and Saxony: we have before assigned our reasons for believing these to be connected with the newer red sandstone of England. (See the section on the foreign localities of that formation.)

This coal-district of Bohemia and Upper Silesia appears to be bounded on all sides by primitive and transition rocks; by those we have already mentioned on the north and south, by the Fichtelgebirge and Bohemer Wald on the west, and on the east by the chains which extend from Glatz round the head of the Adler to join the primitive slate district of Southern Bohemia.

More than forty beds of coal are supposed to be worked in the Bohemian district.

The geological Map and Sections of Bohemia by Riepb (Vienna 1819), and the 'Flora Zur Vorwelt' by Count Sternberg, afford much information concerning the Bohemian part of this coal district; and the Map and description of Von Raumur add the particulars of its extension into Upper Silesia.

The coal-strata and grit of this district appear to agree closely with those of England, and the vegetable impressions preserved in them to be generally of the same genera and species.

Coal is said to occur on the north of Constantinople.

In Russia there are said to be indications of coal in several provinces, but the notices are too vague and uncertain to be relied upon.

In Asia coal occurs extensively, and has long been worked in China.

America is known to afford very extensive coal-fields on the west of the Alleghany chain towards the plains of the Missisippi.

In the Australasian Archipelago, coal occurs very plentifully associated with sandstone in Van Dieman's Land.

END OF PART I.

Phillips, Printer, London.

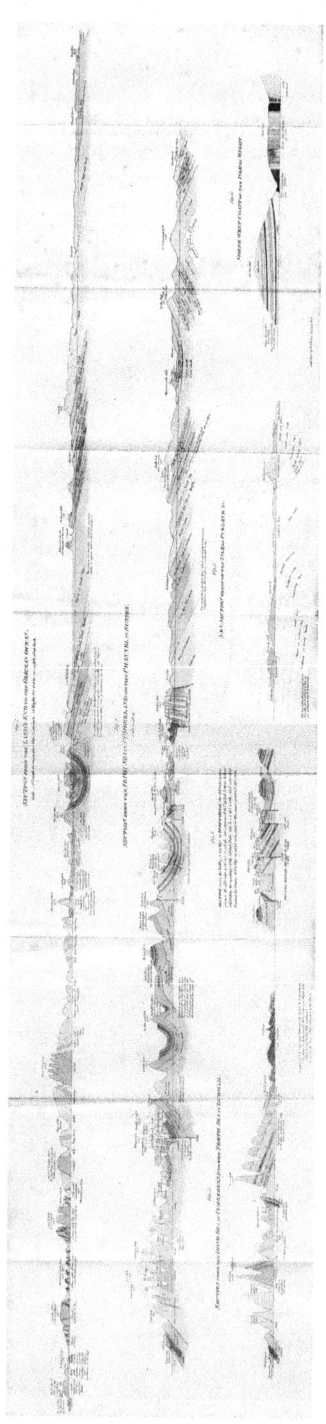

The material originally positioned here is too large for reproduction in this reissue. A PDF can be downloaded from the web address given on page iv of this book, by clicking on 'Resources Available'.

DESCRIPTION OF THE ANNEXED PLATE,

chiefly of

Geological and Mineralogical Instruments and Apparatus,

Taken from the Catalogue of the Manufacturers,

R. & G. KNIGHT, Foster Lane, London.

1 & 2.—Geological Hammers of the form recommended by Dr. MacCulloch, as the best adapted for detaching specimens from the obtusely angular surfaces of rocks. These vary in size from 2 to 5 lb. 5s to 10s each.

3.—Resembles a Miner's Pick, and is useful for detaching a fossil or other object by removing the surrounding matrix. In order to render the hammer more portable, the handle is made detachable from the head without difficulty.

4.—Is a hammer for breaking minerals; one face is square and the other sharp or chissel-shaped, and usually weighs from 12 to 20 ounces.

5 & 6.- Trimming Hammers of forms best adapted for reducing specimens of minerals and rocks to suitable forms and dimensions for the cabinet.

7.—A small square-headed Hammer, with a steel chissel-shaped handle in length 8 inches, is particularly adapted for the pocket, and may be had with a leather case. It also forms a part of the Portable Mineralogical Chest for travellers. 4s to 7s.

8.—A Mineralogical Anvil, an instrument particularly well adapted for facilitating the reduction of large or ill-formed specimens to a more desirable shape for the cabinet. It is of iron, and weighs about 9 lb. Price 5s.

9.—An improved Clinometer for determining with greater facility the direction, inclination and dimensions of strata. 15s to 20s.

10.—The Goniometer improved by Mr. Pepys Is perhaps the best pocket instrument of the kind in use. It consists of two pair of forceps detached from the scale, which greatly facilitate their application. 15s.

11.—A Steel Mortar upon an improved principle, for the reduction of gems and other hard substances for analysis. 14s.

12.—Iron Mortars with iron or steel Pestels, from a quarter of a pint to 1 gallon. 4s to 40s each.

13.—Clarke's Condensing Blow-pipe, for reducing refractory substances by means of the mixed gasses. £3 13s 6d to £4 4s.

14.—An Hydraulic Blow-pipe, which enables the operator to keep up the blast without the con tant application of the lungs. £1 1s to £2 2s.

15.—An improved Blow-pipe of the ordinary form, with a variety of jets. 8s.

16.—Pepys' Improved Gas Holder. The great facility with which all the various experiments on Oxygen and Hydrogen Gas are performed by means of this apparatus, renders it a most valuable acquisition to the Laboratory and Lecture-room: the subsequent addition of the long-necked funnel makes it also an excellent Hydraulic Blow-pipe. £2 12s 6d to £3 13s 6d.

17.—Knight's Universal Table Furnace Is composed of strong sheet iron and lined with ure lute: its capacity 6 inches in diameter and 12 inches deep. Is adapted for performing the various operations of smelting metals in the crucible, subliming, distillation either by the naked fire or sand bath, digestion, evaporation, decomposition of water, the oxidation of metals, the assay of gold and silver by quartation and cupellation, and all other operations upon a smaller scale than can be effected by the means of charcoal. £5 15s 6d to £6 6s.

18.—Knight's Improved Black's Portable Furnace. This furnace is adapted for performing all the before mentioned operations, and being of much larger dimensions, and altogether more substantial, is calculated for exciting an intense heat, and therefore better adapted for the reduction of metallic ores, &c. £6 6s to £7 7s.

19.—Aikin's Portable Blast Furnace, for raising a sudden and intense heat by dividing a strong blast of air from a double bellows and driving it forcibly through a number of small holes in the bottom of the furnace. £1 12s to £3.
20.—Evaporating Basins of Wedgewood's ware glazed inside. 6d to 5s each.
21.—Crucibles of Iron, Silver, and Platina. 10s to £5 each.
22.—Evaporating Basins of Silver and Platina. 10s to £5 each.
23 —Hessian Crucibles in nests, various. 6d to 2s.
24.—Skittle-shaped Crucibles. 3d to 2s.
25.—Calcining Ports or Crucibles which divide in the middle, for sublimation, &c. 9d to 2s.
26.—Muffles for assaying and roasting Metallic Ores, &c. 1s to 5s.
27.—Enameling Pans. 3d to 3s.
28.—Two Neck'd Glass Vessels for Hydrogen Gas, with stopper and bent tube ground in. 7s to 10s.
29.—Glass and Porcelain Funnels.
30.—Test Glasses of various sizes. 1s to 2s.
31. The same, with a wide bottom for precipitating.
32.—A small neck'd Bottle with ground stopper made thin for the purpose of taking specific gravities. 3s.
33.—Pepys' Bottle with spiral tube, for taking the aerial contents of calcareous earth. 5s to 7s.
34.—Bent long-neck'd Funnel for filling glass retorts without soiling their necks. 3s to 5s.
35.—Davy's Glass Apparatus for the analysis of soils. £1 11s 6d to £2 2s.
36.—Bell Glass with cork and sliding wire cup for shewing the combustion of substances in oxygen. 7s to 10s.
37.—Mercurial Pneumatic trough, with a graduated air jar and glass bottle.
38.—A Digesting Flask with capital and conical stopper. 5s to 10s.
39.—A Chemical Argand's Lamp. 10s 6d to 15s.
40.—Howard's Portable Rain Guage, with graduated Glass Measure. £1 1s to £1 5s.
41.—A Glass Stopper'd Retort. 3s to 10s.
42.—Glass Receiver for ditto.
43.—A Glass Filtering Funnel. 1s to 3s.
44.—Glast Precipitating Jar. 1s 6d to 4s.
45.—Glass Spirit Lamps. 5s to 7s 6d.
46.—Glass Evaporating Dishes. 6d to 3s.
47.—Guyton's Lamp Apparatus, the stand consisting of a brass round pillar with foot of the same, and three rings sliding thereon of different apertures, with an Argand's Fountain Lamp. £2 2s to £2 12s 6d.
48.—Welter's Glass Tube of Safety.
49.—Pepys' Apparatus for drying Precipitates at the uniform temperature of boiling water. 7s 6d to 12s.
50.— A Glass Eudiometer containing a cubic inch divided into 100 parts. 5s.
51.— Hope's Eudiometer for the analysis of Atmospheric air, divided in ike manner. 8s.
52.—Davy's Pocket ditto for the same purpose. 10s 6d.
53.—Pepys' ditto ditto. 10s 6d.
54.—An Eudiometer for shewing the formation of Water by the combustion of Oxygen and Hydrogen gas by the Electric spark. 10s 6d to 14s.
55.—Davy's improved ditto attached to a stand with spiral spring. £1 16s.
56.—A small Glass Funnel with long neck for exhibiting the combustion of Phosphorus under water.
57.—A Magnetic Needle and Centre. 3s to 5s.

Note.—In addition to the above, a variety of Pocket and Portable Cases of Instruments, with appropriate tests, for the analysis of Minerals by the Blow-pipe, as described by Professor Berzelius in his treatise recently published on that subject, and since translated by Mr. Children.. From £1 1s to £10 10s.

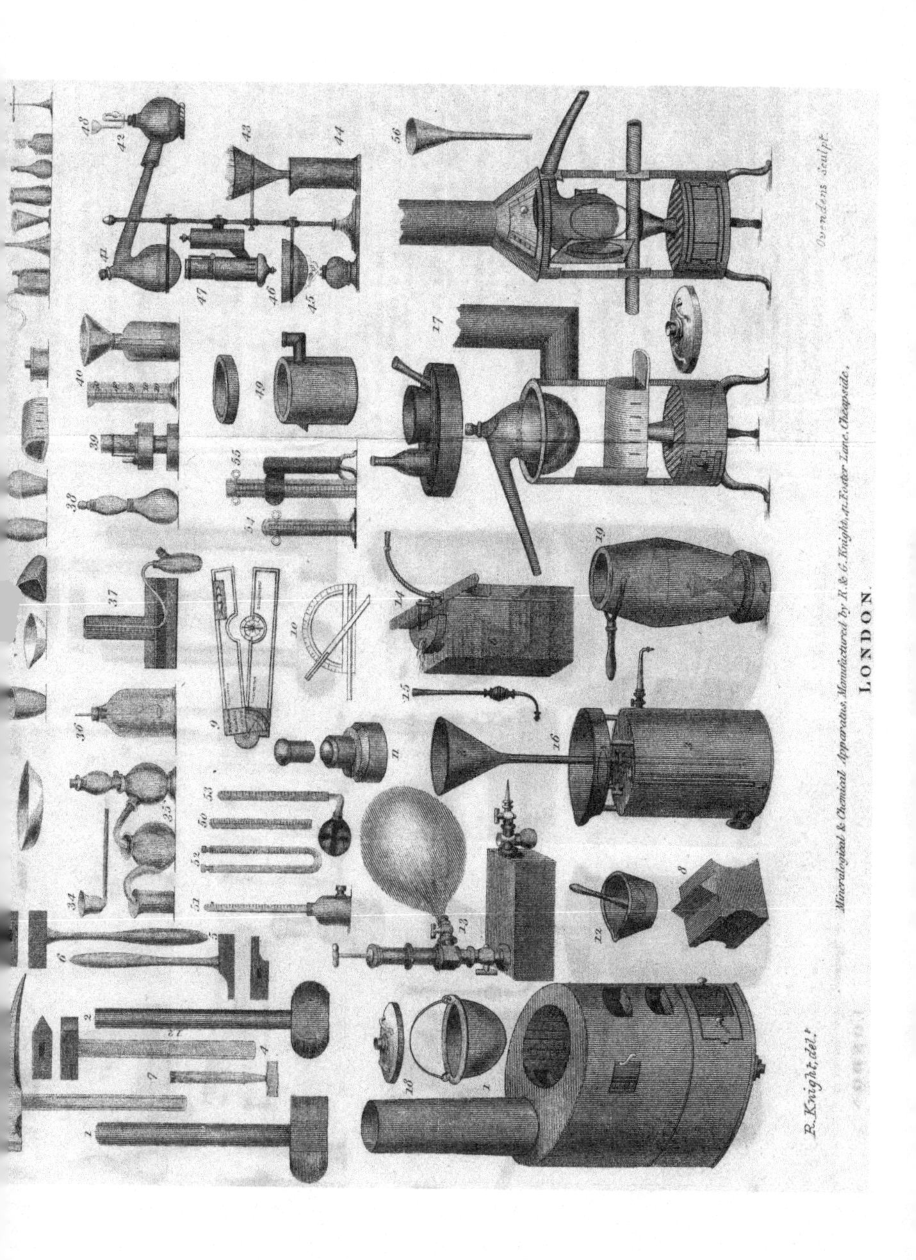

The material originally positioned here is too large for reproduction in this reissue. A PDF can be downloaded from the web address given on page iv of this book, by clicking on 'Resources Available'.

Printed in the United States
By Bookmasters